Topics in
Current Physics

18

Topics in Current Physics

Founded by Helmut K. V. Lotsch

1 **Beam-Foil Spectroscopy**
Editor: S. Bashkin

2 **Modern Three-Hadron Physics**
Editor: A. W. Thomas

3 **Dynamics of Solids and Liquids by Neutron Scattering**
Editors: S. W. Lovesey and T. Springer

4 **Electron Spectroscopy for Surface Analysis**
Editor: H. Ibach

5 **Structure and Collisions of Ions and Atoms**
Editor: I. A. Sellin

6 **Neutron Diffraction**
Editor: H. Dachs

7 **Monte Carlo Methods** in Statistical Physics
Editor: K. Binder

8 **Ocean Acoustics**
Editor: J. A. DeSanto

9 **Inverse Source Problems** in Optics
Editor: H. P. Baltes

10 **Synchrotron Radiation**
Techniques and Applications
Editor: C. Kunz

11 **Raman Spectroscopy** of Gases and Liquids
Editor: A. Weber

12 **Positrons in Solids**
Editor: P. Hautojärvi

13 **Computer Processing of Electron Microscope Images**
Editor: P. W. Hawkes

14 **Excitons** Editor: K. Cho

15 Physics of **Superionic Conductors**
Editor: M. B. Salamon

16 **Aerosol Microphysics I**
Particle Interactions
Editor: W. H. Marlow

17 **Solitons**
Editors: R. Bullough, P. Caudrey

18 **Magnetic Electron Lenses**
Editor: P. W. Hawkes

19 Theory of **Chemisorption**
Editor: J. R. Smith

20 **Inverse Scattering Problems** in Optics
Editor: H. P. Baltes

21 **Coherent Nonlinear Optics**
Recent Advances
Editors: M. S. Feld and V. S. Letokhov

22 **Electromagnetic Theory of Gratings**
Editor: R. Petit

23 **Structural Phase Transitions I**
Editors: K. A. Müller and H. Thomas

24 **Amorphous Solids**
Low-Temperature Properties
Editor: W. A. Phillips

25 **Mössbauer Spectroscopy II**
The Exotic Side of the Method
Editor: U. Gonser

26 **Crystal Cohesion and Conformational Energies** Editor: R. M. Metzger

27 **Dissipative Systems in Quantum Optics**
Resonance Fluorescence, Optical Bistability, Superfluorescence
Editor: R. Bonifacio

28 **The Stratospheric Aerosol Layer**
Editor: R. C. Whitten

29 **Aerosol Microphysics II**
Chemical Physics of Microparticles
Editor: W. H. Marlow

30 **Real-Space Renormalization**
Editors: T. W. Burkhardt and J. M. J. van Leeuwen

31 **Hyperfine Interactions of Radioactive Nuclei**
Editor: J. Christiansen

32 **Superconductivity in Ternary Compounds I**
Structural, Electronic, and Lattice Properties
Editors: Ø. Fischer and M. B. Maple

33 **Molecular Collision Dynamics**
Editor: J. M. Bowman

34 **Superconductivity in Ternary Compounds II**
Editors: M. B. Maple and Ø. Fischer

Magnetic Electron Lenses

Edited by P.W. Hawkes

With Contributions by
P. W. Hawkes   E. Kasper   F. Lenz
T. Mulvey   W. D. Riecke

With 240 Figures

Springer-Verlag Berlin Heidelberg New York 1982

Dr. Peter W. Hawkes

Laboratoire d'Optique Electronique du C.N.R.S., B.P. 4347,
F-31055 Toulouse Cedex, France

ISBN-13:978-3-642-81518-8 e-ISBN-13:978-3-642-81516-4
DOI: 10.1007/978-3-642-81516-4

Library of Congress Cataloging in Publication Data. Main entry under title: Magnetic electron lenses. (Topics in current physics ; v. 18) Includes bibliographical references and index. 1. Magnetic lenses. I. Hawkes, P. W. II. Series. QC793.5.E622P76 535'.33 82-833 AACR2

Softcover reprint of the hardcover 1st edition 1982

2153/3130-543210

Preface

No single volume has been entirely devoted to the properties of magnetic lenses, so far as I am aware, although of course all the numerous textbooks on electron optics devote space to them. The absence of such a volume, bringing together information about the theory and practical design of these lenses, is surprising, for their introduction some fifty years ago has created an entirely new family of commercial instruments, ranging from the now traditional transmission electron microscope, through the reflection and transmission scanning microscopes, to columns for micromachining and microlithography, not to mention the host of experimental devices not available commercially.

It therefore seemed useful to prepare an account of the various aspects of magnetic lens studies. These divide naturally into the five chapters of this book: the theoretical background, in which the optical behaviour is described and formulae given for the various aberration coefficients; numerical methods for calculating the field distribution and trajectory tracing; extensive discussion of the paraxial optical properties and aberration coefficients of practical lenses, illustrated with curves from which numerical information can be obtained; a complementary account of the practical, engineering aspects of lens design, including permanent magnet lenses and the various types of superconducting lenses; and finally, an up-to-date survey of several kinds of highly unconventional magnetic lens, which may well change the appearance of future electron optical instruments very considerably after they cease to be unconventional. The authors have all been intimately concerned with the subjects of their chapters for many years. Thus much material not readily available elsewhere is presented here, and recent developments that have not yet found their way into review articles or textbooks are put in context. It was clear from the outset that the chapter on practical lens design would be somewhat longer than the others but not that it would occupy quite so large a part of the book. Retrospectively, however, its length does seem justified by the range of topics to be covered and by the lack of any comparable account elsewhere. Dr. Riecke and I agreed that if this opportunity to record such details of practical design as are still available or retrievable were not taken, much valuable material might well be lost for ever.

The text is complemented by two appendices. The first contains an anthology of lens-design curves that have been widely used in the past. It was felt that, although Professor Lenz gives a very full range of such curves, all newly prepared and largely recalculated for this book, readers might like to have available earlier work as well. The curves reproduced here are, of course, only a small selection from those available, but they are felt to be reasonably representative. The second appendix contains a bibliography of papers dealing with the subject of magnetic lens properties, interpreted in the following somewhat restrictive sense: papers are listed if they include some information (however slight) about the paraxial properties and aberration coefficients of magnetic lenses and, in particular, about the relation between lens geometry and excitation and these optical properties. The subjects of Chap.2 and, to a lesser extent, Chap.4 are therefore largely excluded. In the case of Chap.2, old precomputer papers are for the most part mainly of historic interest nowadays, and furthermore, most of the information about numerical methods is to be found in works that are of much wider application than electron optics. Professor Kasper's own bibliography provides an excellent starting point for further exploration of the literature. The case of Chap.4 is slightly different. Quite apart from the fact that the list was threatening to become too long to be useful, the practical details of lens design have changed appreciably over the years, and although a (correct) formula for an aberration coefficient remains unaffected by the passage of time, the design of lenses and their accessories is in a continual state of evolution. Finally, this bibliography has enabled us to draw attention to many early papers that would not otherwise have been mentioned, since limitations of space did not permit inclusion of an account of the history of the magnetic lens.

Toulouse, May 1982 *Peter W. Hawkes*

Note. Full bibliographical details of the series of international and European conferences on electron microscopy are not normally given in the individual lists of references but are to be found at the end of Appendix B (p.449).

Contents

List of Contributors

Hawkes, Peter W.

Laboratoire d'Optique Electronique du C.N.R.S., B.P. 4347,
F-31055 Toulouse Cedex, France

Kasper, Erwin

Institut für angewandte Physik der Universität, Auf der Morgenstelle,
D-7400 Tübingen, Fed. Rep. of Germany

Lenz, Friedrich

Lehrstuhl für theoretische Elektronenphysik, Auf der Morgenstelle,
D-7400 Tübingen, Fed. Rep. of Germany

Mulvey, Thomas

Department of Physics, University of Aston, Gosta Green,
GB-Birmingham, B4 7ET, England

Riecke, W. Dieter

Fraunhofer Institut für Informations-und Datenverarbeitung,
Sebastian-Kneipp-Straße 12-14, D-7500 Karlsruhe, Fed. Rep. of Germany

1. Magnetic Lens Theory

P. W. Hawkes

With 10 Figures

The first attempts to build an electron microscope were a direct consequence of the realization that a rotationally symmetric magnetic or electrostatic field (or a combination of both) exerts a focusing effect on electrons. If electrons are incident on the field in the vicinity of the axis of rotational symmetry and not too steeply inclined to it, then a uniformly magnified image of the electron distribution in one plane will be formed in some other plane downstream. For various reasons, both technological and optical, magnetic lenses are almost exclusively used in electron microscopes and related instruments such as microanalysers, except of course in the accelerating structure. In this chapter, we give a reasonably full and self-contained account of the theory of the lens action for rotationally symmetric magnetic fields. After briefly recapitulating the laws governing electron motion in magnetic fields, we explore in detail the implications of the paraxial equations of motion. We then consider departures of various kinds from the perfect image formation predicted by these linear, homogeneous, second-order differential equations; this leads us to study the various geometrical and chromatic aberrations of magnetic lenses. Finally, we indicate briefly how the paraxial properties and aberrations of lens combinations can be calculated.

1.1 Derivation of the Equations of Motion of Electrons in Rotationally Symmetric Magnetic Fields

We take as point of departure the expression for the Lagrangian L, from which the equations of motion of electrons in static magnetic fields may be derived with the aid of the variational formula,

$$\delta \int L \, dt = 0 \quad , \tag{1.1}$$

in which the end-points remain fixed during the variation. The function L is given by

$$L = m_0 c^2 [1 - (1 - v^2/c^2)^{\frac{1}{2}}] + e\varphi - e\underline{A} \cdot \underline{v} \quad ; \tag{1.2}$$

in which -e, m_0, and $\underline{v}$ are the charge, rest mass, and velocity of the electron respectively; c is the velocity of light; φ is the electrostatic potential corresponding to the electron velocity $\underline{v}$; and $\underline{A}$ is the magnetic vector potential, $\underline{B} = \text{curl}\ \underline{A}$. The origin of φ is chosen so that $\varphi = 0$ corresponds to zero velocity.

Since the time variable is of no interest in the present context, we shall replace (1.1) by a variational formula involving the space coordinate z, which coincides with the axis of rotational symmetry, the optic axis of our system. First, however, we derive a number of useful relations that lead us to the principle of least action and hence to a form of Fermat's principle.

From (1.1) we may deduce immediately that

$$\frac{d}{dt}\left(\frac{\partial L}{\partial \underline{v}}\right) = \frac{\partial L}{\partial \underline{r}} \quad . \tag{1.3}$$

Furthermore,

$$\begin{aligned} \frac{dL}{dt} &= \frac{\partial L}{\partial t} + \frac{\partial L}{\partial \underline{r}} \cdot \underline{v} + \frac{\partial L}{\partial \underline{v}} \cdot \dot{\underline{v}} \\ &= \frac{\partial L}{\partial t} + \frac{d}{dt}\left(\frac{\partial L}{\partial \underline{v}}\right) \cdot \underline{v} + \frac{\partial L}{\partial \underline{v}} \cdot \dot{\underline{v}} \end{aligned} \tag{1.4}$$

or

$$\frac{d}{dt}\left(v \cdot \frac{\partial L}{\partial \underline{v}} - L\right) = -\frac{\partial L}{\partial t} \tag{1.5}$$

so that for static fields

$$\underline{v} \cdot \frac{\partial L}{\partial \underline{v}} - L = E \quad , \tag{1.6}$$

where E is a constant, the total energy,

$$E = m_0 c^2 \left(\frac{1}{(1 - v^2/c^2)^{\frac{1}{2}}} - 1\right) - e\varphi \quad . \tag{1.7}$$

From (1.1) and (1.6) we obtain the principle of least action,

$$\delta \int \underline{v} \cdot \frac{\partial L}{\partial \underline{v}}\, dt = 0 \tag{1.8}$$

or

$$\delta \int \frac{\partial L}{\partial \underline{v}} \cdot d\underline{r} = 0 \quad , \tag{1.9}$$

in which we have written $\underline{v} = d\underline{r}/dt$. The canonical momentum is given by $\partial L/\partial \underline{v}$ but it is convenient to scale this with respect to $[2em_0\varphi(1 + \varepsilon\varphi)]^{\frac{1}{2}}$, where

$$\varepsilon = e/2m_0c^2 \approx 1\ \text{MV}^{-1} \quad . \tag{1.10}$$

We therefore set

$$\begin{aligned} \underline{P} &= \frac{1}{[2em_0\varphi(1 + \varepsilon\varphi)]^{\frac{1}{2}}} \frac{\partial L}{\partial \underline{v}} \\ &= \left[\frac{m_0}{2e\varphi(1 + \varepsilon\varphi)}\right]^{\frac{1}{2}} \frac{\underline{v}}{(1 - v^2/c^2)^{\frac{1}{2}}} - \left[\frac{e}{2m_0\varphi(1 + \varepsilon\varphi)}\right]^{\frac{1}{2}} \underline{A} \quad . \end{aligned}$$

Writing

$$\eta = (e/2m_0)^{\frac{1}{2}} \tag{1.11}$$

$$U = \varphi(1 + \varepsilon\varphi) \tag{1.12}$$

and noting that φ is the same as the accelerating voltage, Φ, we have

$$\underline{p} = \frac{\underline{v}}{2\eta U^{\frac{1}{2}}(1 - v^2/c^2)^{\frac{1}{2}}} - \frac{\eta\underline{A}}{U^{\frac{1}{2}}} \quad . \tag{1.13}$$

Writing $d\underline{r} = \underline{s}\, ds$, where ds is an element of arc length, (1.9) becomes

$$\delta \int \underline{p}\cdot\underline{s}\, ds = 0 \quad , \tag{1.14}$$

which is exactly analogous to Fermat's principle. Substituting for $\underline{v}$ in (1.13) from (1.7), we find

$$\underline{p}\cdot\underline{s} = 1 - \eta\underline{A}\cdot\underline{s}/U^{\frac{1}{2}} \quad . \tag{1.15}$$

Finally, we replace the element of arc length ds in (1.14) by (ds/dz)dz, giving

$$\delta \int m\, dz = 0 \tag{1.16}$$

with

$$m = (1 + X'^2 + Y'^2)^{\frac{1}{2}} - \frac{\eta}{U^{\frac{1}{2}}}(A_X X' + A_Y Y' + A_z) \quad , \tag{1.17}$$

in which X, Y, and z form a system of cartesian coordinates, the z axis coinciding with the axis of rotational symmetry, the optic axis, as mentioned above.

The paraxial equations of motion are obtained by expanding m as a power series in the off-axis coordinates, X and Y and their derivatives, and retaining only the lower-order terms. For this we need the power-series expansions of A_X, A_Y, and A_z. For rotationally symmetric fields we have

$$\begin{aligned} A_X &= -\frac{1}{2} Y\left[B(z) - \frac{1}{8} B''(z)(X^2 + Y^2) + \ldots\right] \\ A_Y &= \frac{1}{2} X\left[B(z) - \frac{1}{8} B''(z)(X^2 + Y^2) + \ldots\right] \end{aligned} \tag{1.18}$$

with $A_z = 0$ and hence

$$m = m^{(0)} + m^{(2)} + m^{(4)} + \ldots \quad , \tag{1.19}$$

where

$$m^{(0)} = 1 \tag{1.20a}$$

$$m^{(2)} = \frac{1}{2}(X'^2 + Y'^2) - \frac{\eta B(z)}{2U^{\frac{1}{2}}}(XY' - X'Y) \tag{1.20b}$$

$$m^{(4)} = -\frac{1}{8}(X'^2 + Y'^2)^2 + \frac{\eta B''(z)}{16U^{\frac{1}{2}}}(XY' - X'Y)(X^2 + Y^2) \quad . \tag{1.20c}$$

Setting $m \approx m^{(2)}$ in (1.16), we obtain the following Euler equations:

$$\frac{d}{dz}\frac{\partial m^{(2)}}{\partial X'} = \frac{\partial m^{(2)}}{\partial X}$$

$$\frac{d}{dz}\frac{\partial m^{(2)}}{\partial Y'} = \frac{\partial m^{(2)}}{\partial Y} \quad . \tag{1.21}$$

These yield coupled second-order differential equations when we substitute for $m^{(2)}$. In order to obtain separated equations, we introduce a new cartesian coordinate system rotated about the optic axis with respect to the fixed axes X, Y through an angle that varies with z. Writing

$$X = r\cos\psi \qquad Y = r\sin\psi$$

$$x = r\cos\chi \qquad y = r\sin\chi$$

$$\psi = \chi + \theta \quad , \tag{1.22}$$

so that

$$X^2 + Y^2 = x^2 + y^2$$

$$X'^2 + Y'^2 = x'^2 + y'^2 + 2\theta'(xy' - x'y) + \theta'^2(x^2 + y^2)$$

$$XY' - X'Y = xy' - x'y + \theta'(x^2 + y^2) \quad ,$$

we finally obtain

$$m^{(2)} = \frac{1}{2}\left(\theta'^2 - \frac{\eta B\theta'}{U^{\frac{1}{2}}}\right)(x^2 + y^2) + \frac{1}{2}(x'^2 + y'^2)$$

$$+ \left(\theta' - \frac{\eta B}{2U^{\frac{1}{2}}}\right)(xy' - x'y) \quad . \tag{1.23}$$

This form of $m^{(2)}$ will yield uncoupled Euler equations only if the coefficient of the term involving $(xy' - x'y)$ vanishes. For this, we must select $\theta(z)$ so that

$$\frac{d\theta(z)}{dz} = \frac{\eta B(z)}{2U^{\frac{1}{2}}} \tag{1.24}$$

or

$$\theta(z) = \frac{\eta}{2U^{\frac{1}{2}}}\int B(z)dz \quad , \tag{1.25}$$

whereupon

$$m^{(2)} = -\frac{\eta^2 B^2}{8U}(x^2 + y^2) + \frac{1}{2}(x'^2 + y'^2) \quad . \tag{1.26}$$

The paraxial equations of motion,

$$\frac{d}{dz}\frac{\partial m^{(2)}}{\partial x'} = \frac{\partial m^{(2)}}{\partial x}$$

$$\frac{d}{dz}\frac{\partial m^{(2)}}{\partial y'} = \frac{\partial m^{(2)}}{\partial y} \quad , \tag{1.27}$$

are thus

$$x'' + \frac{\eta^2 B^2}{4U} x = 0$$

$$y'' + \frac{\eta^2 B^2}{4U} y = 0 \quad . \tag{1.28}$$

Before going on to analyse the consequences of (1.28) in detail, we draw attention to a differential relation, the perturbed form of which is very important when we go beyond the paraxial approximation. Suppose that we vary $\int m dz$ without keeping the end-points fixed. Suppose, furthermore, that this variation is so chosen that the paraxial equations are satisfied both before and after the variation. For an arbitrary variation we should have

$$\delta \int_{z_1}^{z_2} m\, dz = \int \left(\frac{\partial m}{\partial x}\delta x + \frac{\partial m}{\partial y}\delta y + \frac{\partial m}{\partial x'}\delta x' + \frac{\partial m}{\partial y'}\delta y'\right)dz$$

$$= \left[\frac{\partial m}{\partial x'}\delta x + \frac{\partial m}{\partial y'}\delta y\right]_{z_1}^{z_2} + \int_{z_1}^{z_2}\left(\frac{\partial m}{\partial x} - \frac{d}{dz}\frac{\partial m}{\partial x'}\right)\delta x\, dz$$

$$+ \int_{z_1}^{z_2}\left(\frac{\partial m}{\partial y} - \frac{d}{dz}\frac{\partial m}{\partial y'}\right)\delta y\, dz \quad ,$$

so that when the paraxial equations remain satisfied despite the variation

$$\delta \int_{z_1}^{z_2} m\, dz = \left[\frac{\partial m}{\partial x'}\delta x + \frac{\partial m}{\partial y'}\delta y\right]_{z_1}^{z_2} \quad . \tag{1.29}$$

Writing $p = \partial m/\partial x'$, $q = \partial m/\partial y'$, and denoting $\int_{z_1}^{z_2} mdz$ by V_{12}, (1.29) may be written

$$\delta V_{12} = p_2\delta x_2 + q_2\delta y_2 - (p_1\delta x_1 + q_1\delta y_1) \quad . \tag{1.30}$$

The function V_{12} must therefore be a function of the position coordinates of the end-points only.

Throughout this chapter we use the (rotating) cartesian coordinates (x,y,z). The formalism can be rendered distinctly more compact by introducing complex coordinates, X + iY for the fixed coordinates and x + iy for the rotating set. We have not adopted these complex coordinates for, although they are very suitable in a formal account of the subject or for a lecture course, their use makes it more difficult for the reader to consult the text in search of information on some specific question. Nevertheless, the decision whether or not to use complex coordinates is to a large extent arbitrary, and they are extensively employed in numerous texts [e.g. 1.10,15,99-101,107,125,139]. They also were used by GLASER in his early papers but not in his treatises [1.21-22].

All textbooks of electron optics devote space to the topics discussed in the foregoing pages; the present account is similar in spirit to those of GLASER [1.21-22] and STURROCK [1.125]. A good idea of the throes of development through which the subject passed can be gained from the early text of BRÜCHE and SCHERZER [1.11] and the collection edited by BUSCH and BRÜCHE [1.15]; the textbooks of MYERS [1.87], von BORRIES [1.9] and ZWORYKIN et al. [1.139] contain very extensive lists of references to those early years.

1.2 Paraxial Properties

1.2.1 General Remarks

All the paraxial properties of magnetic lenses are derived from (1.28), which we repeat here for convenience:

$$x'' + \frac{\eta^2 B^2}{4U} x = 0$$

$$y'' + \frac{\eta^2 B^2}{4U} y = 0 \quad . \tag{1.31}$$

These differential equations are linear, second order, and homogeneous, and their general solutions may therefore be written in the form

$$x(z) = x_1 s(z) + x_2 t(z)$$

$$y(z) = y_1 s(z) + y_2 t(z) \quad , \tag{1.32}$$

where s(z) and t(z) are a pair of linearly independent solutions of both paraxial equations. The solutions s(z) and t(z) may be identified with various pairs of rays that prove to be particularly convenient and which are characterized by various sets of boundary conditions, as we shall see. For each choice the constants x_1, x_2, y_1, and y_2 will be identified with some convenient set of coordinates.

Most of the lenses used in electron microscopes and related devices are used to form images not of real objects but of intermediate images, either of the gun crossover in the case of condenser lenses and probe-forming lenses or of the specimen in the case of intermediate and projector lenses. The image is then the "object" for the next lens or, in the case of the final projector of a transmission microscope, the final image formed in field-free space at the fluorescent screen, the photographic plate, or the image intensifier. The only exceptions are objective lenses, in which the specimen is very likely to lie within the lens field, and probe-forming lenses, particularly in scanning transmission instruments, in which the probe (the image of the crossover) may be formed and scanned across the specimen within the field of the lens. For objectives therefore and, mutatis mutandis, probe-forming lenses, the part of the lens field before the specimen belongs to the condenser system, and only the part downstream from the object participates in the image-forming process. It is hardly surprising therefore that a slightly different treatment is needed to describe the properties of objective lenses from that appropriate for condensers, intermediates, and projectors. In the case of the latter, we define quantities that characterize the entire field of the lens, whereas for objectives these quantities describe only part of it. For objective lenses, we speak of "real" quantities – real focal lengths, real aberration coefficients – and for the remainder, of "asymptotic" values because we are interested in the relation between the asymptote to a ray incident on the lens field from object space and the asymptote to the same ray as it emerges into image space. (The terms object and image space are both local and are defined with respect to the lens in question; image space for one lens will be object space for its neighbour downstream.)

1.2.2 Real Cardinal Elements

We now consider image formation when the object is a real physical specimen, which may be immersed in the field of the lens itself. Denoting the object plane by $z = z_o$, we choose the rays $s(z)$ and $t(z)$ so that x_1, y_1 and x_2, y_2 coincide with coordinates of position and slope in the object plane. This is achieved by writing

$$\begin{aligned} s(z) &\to g(z) \quad ; \quad g(z_o) = 1 \quad , \quad g'(z_o) = 0 \\ t(z) &\to h(z) \quad ; \quad h(z_o) = 0 \quad , \quad h'(z_o) = 1 \quad , \end{aligned} \tag{1.33}$$

so that (1.32) become

$$\begin{aligned} x(z) &= x_o g(z) + x_o' h(z) \\ y(z) &= y_o g(z) + y_o' h(z) \quad . \end{aligned} \tag{1.34}$$

An alternative pair of solutions widely used in the past characterizes the ray pair s and t, not by their position and slope in a single plane, the object plane, but by their positions in two planes, the object plane and some aperture plane $z = z_a$. It is then usual to write

$$\begin{aligned} s(z) &\to r_\gamma(z) \quad ; \quad r_\gamma(z_o) = 1 \quad , \quad r_\gamma(z_a) = 0 \\ t(z) &\to r_\alpha(z) \quad ; \quad r_\alpha(z_o) = 0 \quad , \quad r_\alpha(z_a) = 1 \quad . \end{aligned}$$

These rays will not be used here; they are convenient when the aberrations are to be expressed in terms of x_o, y_o, x_a, and y_a, as we see briefly in Sect.1.3. A hybrid pair has also been used, but these are not entirely satisfactory for aberration calculation; here we write

$$s(z) \to r_\gamma(z) \quad ; \quad r_\gamma(z_o) = 1 \quad , \quad r_\gamma(z_a) = 0$$

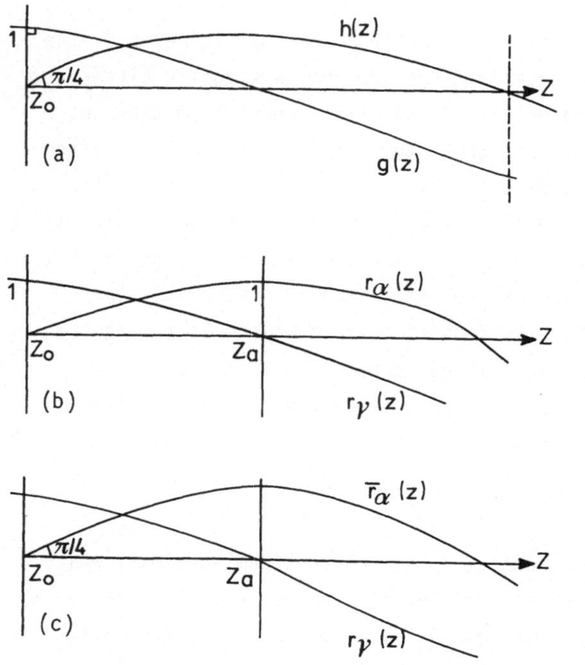

Fig.1.1a-c. Pairs of linearly independent solutions of the paraxial equations of motion. (a) g(z) and h(z): boundary conditions at $z=z_0$; (b) $r_\alpha(z)$ and $r_\gamma(z)$: boundary conditions at $z=z_0$ and $z=z_a$; (c) $\bar{r}_\alpha(z)$ and $r_\gamma(z)$: hybrid boundary condition:

as before but

$$t(z) \rightarrow \bar{r}_\alpha(z) \quad ; \quad \bar{r}_\alpha(z_0) = 0 \quad , \quad \bar{r}'_\alpha(z_0) = 1 \quad ,$$

so that $\bar{r}_\alpha$ and h are identical. Clearly $\bar{r}_\alpha(z) \propto r_\alpha(z)$ (Fig.1.1).

Returning to (1.34), we see that if h(z) vanishes in some other plane $z = z_i$, then

$$\begin{aligned} x(z_i) &= g(z_i)x_0 \\ y(z_i) &= g(z_i)y_0 \quad . \end{aligned} \tag{1.35}$$

A stigmatic image is therefore formed with magnification M:

$$M = g(z_i) \quad . \tag{1.36}$$

Let us now consider the family of rays for which $x'(z_0)$ vanishes,

$$x(z) = x_0^{(k)}g(z)$$

$$y(z) = y_0^{(k)}g(z) \quad ,$$

for a range of values of the label k. If g(z) vanishes in some plane $z = \bar{z}_{Fi}$, then all rays parallel to the axis in z_0 will intersect the axis at $z = \bar{z}_{Fi}$. Furthermore, any family of rays that are parallel (but not necessary parallel to the axis) in the plane $z = z_0$ will intersect one another in $z = \bar{z}_{Fi}$. This can readily be seen as follows. We represent the family of parallel rays by the equations

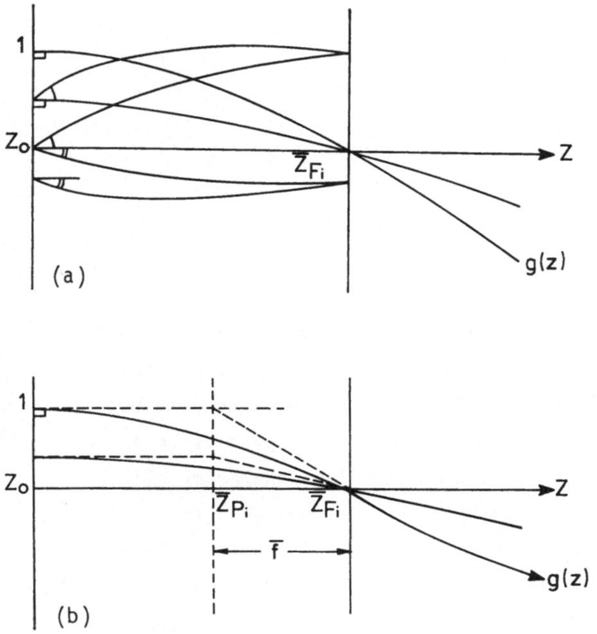

Fig.1.2. (a) Families of rays with the same slope in $z=z_0$ converge to the same point in $z=\bar{z}_{Fi}$. (b) Local focal plane, principal plane, and focal length; these cardinal elements are not used in practice since they vary with object position

$$x(z) = x_0^{(k)} g(z) + x_0' h(z)$$

$$y(z) = y_0^{(k)} g(z) + y_0' h(z) \quad , \tag{1.37}$$

where k is an index labelling the different values of x_0 while x_0' remains constant. In the plane $z = \bar{z}_{Fi}$, we have $g(\bar{z}_{Fi}) = 0$ and so (Fig.1.2a)

$$\left.\begin{aligned} x(\bar{z}_{Fi}) &= x_0' h(\bar{z}_{Fi}) \\ y(\bar{z}_{Fi}) &= y_0' h(\bar{z}_{Fi}) \end{aligned}\right\} \quad \text{for all } k \quad . \tag{1.38}$$

The family of rays $x(z) = x_0^{(k)} g(z)$, $y(z) = y_0^{(k)} g(z)$ intersect the axis at $z = \bar{z}_{Fi}$ with slope

$$x'(\bar{z}_{Fi}) = x_0^{(k)} g'(\bar{z}_{Fi}), \; y'(\bar{z}_{Fi}) = y_0^{(k)} g'(\bar{z}_{Fi}) \quad .$$

The tangent to the kth ray at $z = \bar{z}_{Fi}$ therefore intersects the tangent to the same ray (Fig.1.2b) at $z = z_0$ in the plane $z = \bar{z}_{Pi}$, where $\bar{z}_{Pi} - \bar{z}_{Fi} = 1/g'(\bar{z}_{Fi})$. We write

$$\bar{z}_{Fi} - \bar{z}_{Pi} = \bar{f} = -1/g'(\bar{z}_{Fi}) \quad . \tag{1.39}$$

From (1.31) and (1.34), we see that

$$\frac{d}{dz}(gh' - g'h) = 0 \quad ,$$

so that

$$gh' - g'h = \text{constant} = 1 \quad .$$

Hence

$$h(\bar{z}_{Fi}) = \bar{f} \quad , \tag{1.40}$$

and (1.38) may be written as

$$x(\bar{z}_{Fi}) = \bar{f}x'_o$$

$$y(\bar{z}_{Fi}) = \bar{f}y'_o \quad . \tag{1.41}$$

Although the plane $z = \bar{z}_{Fi}$ and the distance $\bar{f}$ appear to exhibit many of the characteristics associated with a focal plane and a focal length, they are not suitable for characterizing these quantities in a general way since they vary with object position when the object is immersed within the field of the objective lens. In practice, the problem is avoided by tabulating the quantities characteristic of high-magnification operation of the lens, since this is the normal situation. The alternative is to introduce the notion of osculating cardinal elements, to which we return below. The properties of high-magnification objective lenses (or high-demagnification probe-forming lenses) are defined in terms of the rays $G(z)$ and $\bar{G}(z)$ (which we shall meet again in connection with asymptotic properties):

$$\lim_{z\to-\infty} G(z) = 1$$

$$\lim_{z\to\infty} \bar{G}(z) = 1 \quad . \tag{1.42}$$

These rays intersect the optic axis at $z = z_{Fi}$ and $z = z_{Fo}$, respectively (Fig. 1.3):

$$G(z_{Fi}) = 0$$

$$\bar{G}(z_{Fo}) = 0 \quad , \tag{1.43}$$

and only if the lens field is symmetric about the mid-plane of the lens are z_{Fi} and z_{Fo} equidistant from this mid-plane: if $B(z) \neq B(-z)$ (setting the origin of z at the mid-plane), then $z_{Fo} \neq -z_{Fi}$. The plane $z = z_{Fo}$ is known as the real object focal plane, and for high magnification the specimen must be placed in the vicinity of this plane. The corresponding object focal length is then given by

$$f_o = 1/\bar{G}'(z_{Fo}) \quad . \tag{1.44a}$$

The plane $z = z_{Fi}$ is known as the real image focal plane. The real image focal length is then defined in terms of $G(z)$:

$$f_i = -1/G'(z_{Fi}) \quad . \tag{1.44b}$$

With these definitions, the real object and image focal lengths are equal only if $B(z)$ is symmetrical about the mid-plane.

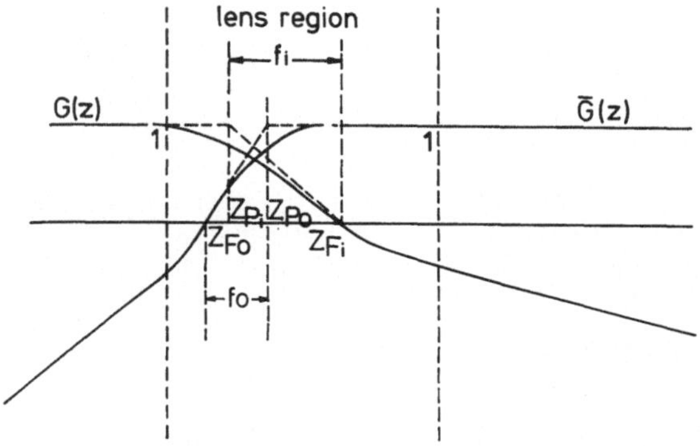

Fig.1.3. The pair of linearly independent solutions of the paraxial equations of motion that are used to define the real (and asymptotic) cardinal elements, $G(z)$ and $\bar{G}(z)$. The real image focus is at $z=z_{Fi}$, the real object focus at $z=z_{Fo}$; the object and image focal lengths are equal only if the lens field is symmetric

We now consider a family of rays parallel in object space but not parallel to the optic axis, which we may describe thus:

$$x(z) = c^{(k)}G(z) + \alpha\bar{G}(z) \quad ,$$

in which the $c^{(k)}$ are a set of constants and α is proportional to the common gradient in object space. In the plane $z = z_{Fi}$, we see that

$$x(z_{Fi}) = \alpha\bar{G}(z_{Fi})$$

so that all the rays belonging to the family intersect in this plane. The electron diffraction pattern will be formed in this plane if the electron source is distant from the objective (as seen from the object space of the latter), since the diffraction pattern plane is conjugate to the source. If, however, the effective source is situated a finite distance from the lens, the plane of the diffraction pattern will not coincide with $z = z_{Fi}$. This point is important in connection with condenser - objective lenses. If the specimen is immersed in the lens field but the condenser system is such that all the electrons are incident on the specimen essentially normal to it, the diffraction pattern will be formed in the plane $z = \bar{z}_{Fi}$.

If the specimen is immersed within the lens field and the magnification is not high, a knowledge of these real cardinal elements is insufficient. The properties of the lens change with object position, as we can readily understand since a different segment of the field distribution acts on the electrons when the specimen is shifted. There is a category of field distributions for which cardinal elements can be defined that do not depend on object position. Such fields are said to be Newtonian since we require at the outset that an equation having the form of Newton's lens equations should be satisfied. It so happens that one of the most widely used model fields, Glaser's bell-shaped field (discussed fully in Chap.3), is Newtonian. A thorough study of Newtonian fields requires the notion of osculating cardinal elements which are invariant under a small change of object position. We now examine these very briefly, referring to the work of GLASER and BERGMANN [1.21-23] for a detailed account or to STURROCK [1.125] for a concise discussion.

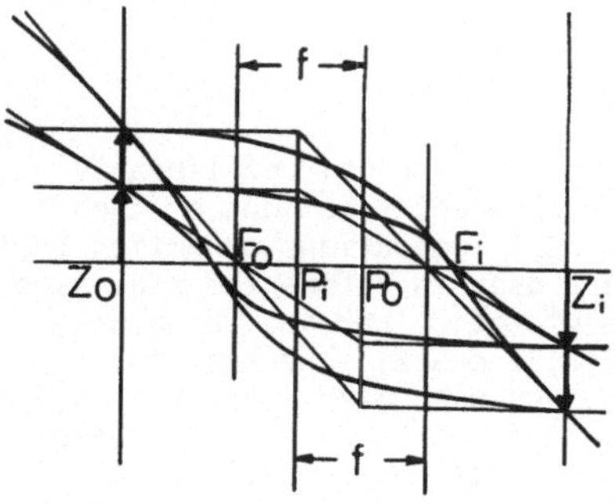

Fig.1.4. Osculating cardinal elements. The rays shown intersect the object plane (z_o) or the image plane (z_i) at right angles. The osculating foci are F_o and F_i, the principal planes are P_o and P_i. The focal length is the distance between F_o and P_o or between F_i and P_i (these distances are equal)

Clearly, we are always at liberty to define cardinal elements for which the lens equations in the form

$$\frac{f}{z_o - z_{Fo}} = -\frac{z_i - z_{Fi}}{f} = M \tag{1.45}$$

are satisfied and which coincide with the asymptotic cardinal elements defined below when object and image lie in field-free space. Otherwise, these elements will be functions of object position.

If osculating cardinal elements are defined as in Fig.1.4, then the usual Newtonian lens equation is found to be satisfied for a range of object positions around that for which the osculating elements are defined: In other words, the values of these osculating cardinal elements may be expected to vary reasonably slowly with object position. We note that the object focus, F_o on Fig.1.4, is the point at which tangents in the object plane $z = z_o$ to the ray that intersects the image plane parallel to the optic axis themselves intersect the optic axis. We note too that with these definitions, object and image focal lengths are equal.

Suppose now that the magnification is high so that the image is distant from the lens. The osculating focus and the real object focus – the point at which a ray emerging parallel to the axis, such as $\bar{G}(z)$, strikes the axis – now coincide. We thus recognise that the high-magnification real object focal length and object focus coincide with the corresponding high-magnification osculating elements or, for Newtonian fields, with the (stationary) osculating quantities.

1.2.3 Asymptotic Cardinal Elements[1]

Only when we are concerned with a real specimen immersed in the magnetic field are the real cardinal elements of interest. The latter are therefore needed in conventional transmission electron microscopy only for the objective lens and in SEM and STEM for the probe-forming lens, as we have already mentioned. For the other lenses, the "object" is an intermediate image in the system, and we need to know the relation between the asymptotes to rays incident on the lens and the

(Footnote 1 see next page)

asymptotes to rays emerging from it. We therefore consider the pair of solutions of the paraxial equations $G(z)$ and $\bar{G}(z)$ that satisfy the following boundary conditions (1.42):

$$\lim_{z\to-\infty} G(z) = 1$$

$$\lim_{z\to\infty} \bar{G}(z) = 1 \quad ,$$

and we write

$$\lim_{z\to\infty} G(z) = G'(z - z_{Fi})$$

$$\lim_{z\to-\infty} \bar{G}(z) = \bar{G}'(z - z_{Fo}) \quad . \tag{1.46}$$

These rays are illustrated in Fig.1.3. Clearly, all rays with incident asymptotes parallel to the axis emerge with asymptotes that intersect the axis at $z = z_{Fi}$ and likewise, mutatis mutandis, for $z = z_{Fo}$. These are therefore the asymptotic image and object foci. We note that the family of rays whose incident asymptotes all have the same slope likewise emerge with asymptotes that intersect in the plane $z = z_{Fi}$. To see this we consider the general ray

$$x(z) = aG(z) + b\bar{G}(z) \tag{1.47}$$

which has incident asymptote

$$x(z) = a + b\bar{G}'(z - z_{Fo}) \tag{1.48a}$$

and emergent asymptote

$$x(z) = aG'(z - z_{Fi}) + b \quad . \tag{1.48b}$$

The latter intersects the plane $z = z_{Fi}$ at b for all values of a. For common incident slope, however, b is fixed so that the family with common incident slope (b fixed, a variable) all intersect $z = z_i$ at b.

The planes $z = z_{Po}$ and $z = z_{Pi}$ are the object and image principal planes, and we see that (Fig.1.5)

---

1 In any general discussion of electron lens properties, where the real and asymptotic properties are discussed together, it is obviously necessary to employ a notation that distinguishes between the two. Unfortunately, no single convention is in widespread use, though the suffix p (for projector) is often used for the asymptotic quantities. In the present chapter, real and asymptotic cardinal elements and aberration coefficients are discussed separately and therefore no special notation has been introduced. In the chapter by Lenz, however, a notation indicating the distinction is essential and is defined there. It is perhaps worth stressing that despite some loss in mnemonic convenience, the introduction of a new letter (replacing f for focal length, etc.) can have distinct advantages, leaving subscripts and superscripts free for other purposes. In the chapter by Mulvey, the suffix p is employed on the few occasions where it is necessary for understanding work from Mulvey's laboratory that is published elsewhere.

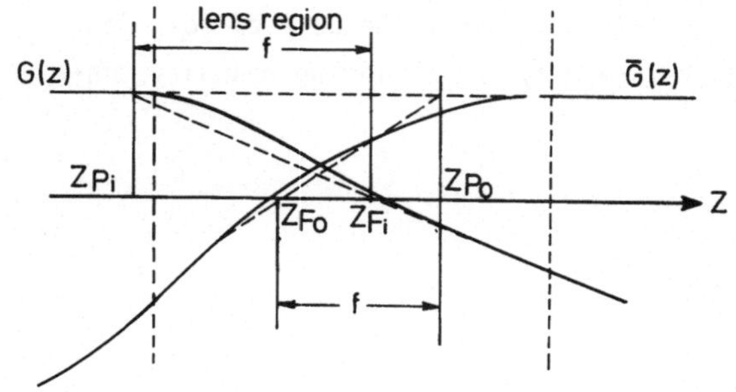

Fig.1.5. Definition of the asymptotic cardinal elements. Notice that they are defined in terms of asymptotes to rays, not the rays themselves

$$z_{Fi} - z_{Pi} = f$$

$$z_{Fo} - z_{Po} = -f \quad , \tag{1.49}$$

where f is the asymptotic focal length.

We now investigate the relation between object and image. The general ray

$$x = aG(z) + b\bar{G}(z) \tag{1.50}$$

tends to the incident asymptote

$$x_1 = a + b\bar{G}'(z - z_{Fo}) \tag{1.51}$$

and to the emergent asymptote

$$x_2 = aG'(z - z_{Fi}) + b \quad . \tag{1.52}$$

If $x_1(z) = 0$ at $z = z_o$, so that

$$z_o = z_{Fo} - a/b\bar{G}'$$

$$= z_{Fo} - af/b \quad , \tag{1.53}$$

then the emergent asymptote $x_2(z)$ intersects the axis at $z = z_i$, where

$$z_i = z_{Fi} - b/aG'$$

$$= z_{Fi} + fb/a \quad . \tag{1.54}$$

Returning to the family of rays incident parallel to the axis, $x(z) = aG(z)$, with emergent asymptotes $x(z) = aG'(z - z_{Fi})$ we see that the emergent and incident asymptotes intersect in the plane $z = z_{Pi}$, where

$$G'(z - z_{Fi}) = 1 \tag{1.55}$$

$$z_{Pi} - z_{Fi} = 1/G' \quad . \tag{1.56}$$

We denote this constant by -f (1.49),

$$1/G' = -f \quad . \tag{1.57}$$

Likewise, the family that emerges parallel to the axis, $x(z) = b\,\bar{G}(z)$, corresponds to the incident asymptotes $x(z) = b\,\bar{G}'(z - z_{Fo})$; these intersect at $z = z_{Po}$,

$$z_{Po} - z_{Fo} = 1/\bar{G}' \quad . \tag{1.58}$$

From the invariance of $G(z)\bar{G}'(z) - G'(z)\bar{G}(z)$ we know that

$$\bar{G}' = -G' \tag{1.59}$$

and so

$$1/\bar{G}' = f \quad . \tag{1.60}$$

Equations (1.57,60) have been anticipated in (1.53,54). We note that

$$(z_o - z_{Fo})(z_i - z_{Fi}) = -f^2 \quad , \tag{1.61}$$

which is Newton's lens equation. If $z = z_o$ and $z = z_i$ are conjugate in the sense that points in $z = z_o$ are stigmatically imaged in $z = z_i$, we have to show that the family of rays with incident asymptotes concurrent at a point in $z = z_o$ have emergent asymptotes concurrent in $z = z_i$.

The family of incident rays with asymptotes passing through x_o in $z = z_o$ are of the form

$$x_1 = a'G(z) + b'\bar{G}(z) \quad , \tag{1.62}$$

where

$$\begin{aligned} x_o &= a' + b'\bar{G}'(z_o - z_{Fo}) \\ &= a' - b'a/b \end{aligned} \tag{1.63}$$

or

$$a' = x_o + b'a/b \quad . \tag{1.64}$$

The corresponding emergent asymptotes intersect $z = z_i$ at

$$\begin{aligned} x_2 &= a'G'(z_i - z_{Fi}) + b' \\ &= -(x_o + b'a/b)(b/a) + b' \\ &= -bx_o/a \quad . \end{aligned} \tag{1.65}$$

A stigmatic image of each point in $z = z_o$ is therefore formed in $z = z_i$ if the planes are related by (1.61). The magnification M is given by

$$M = -b/a \quad . \tag{1.66}$$

All these results can be summarized very compactly in terms of the transfer matrix, which relates the position and slope of the asymptote to an arbitrary emergent ray in some plane $z = z_2$ to the corresponding quantities for the incident ray in $z = z_1$. We have:

incident

$$\begin{aligned} x_1 &= a + b(z_1 - z_{Fo})/f \\ x_1' &= b/f \quad , \end{aligned} \tag{1.67a}$$

emergent

$$x_2 = -a(z_2 - z_{Fi})/f + b$$

$$x_2' = -a/f \quad , \tag{1.67b}$$

or

$$\begin{pmatrix} x_1 \\ x_1' \end{pmatrix} = \begin{pmatrix} 1 & \frac{z_1 - z_{Fo}}{f} \\ 0 & 1/f \end{pmatrix} \begin{pmatrix} a \\ b \end{pmatrix}$$

$$\begin{pmatrix} x_2 \\ x_2' \end{pmatrix} = \begin{pmatrix} -\frac{z_2 - z_{Fi}}{f} & 1 \\ -1/f & 0 \end{pmatrix} \begin{pmatrix} a \\ b \end{pmatrix}$$

or

$$\begin{pmatrix} x_2 \\ x_2' \end{pmatrix} = \begin{pmatrix} -\frac{z_2 - z_{Fi}}{f} & 1 \\ -1/f & 0 \end{pmatrix} \begin{pmatrix} 1 & -(z_1 - z_{Fo}) \\ 0 & f \end{pmatrix} \begin{pmatrix} x_1 \\ x_1' \end{pmatrix}$$

$$= \begin{pmatrix} -\frac{z_2 - z_{Fi}}{f} & \frac{(z_2 - z_{Fi})(z_1 - z_{Fo})}{f} + f \\ -\frac{1}{f} & \frac{z_1 - z_{Fo}}{f} \end{pmatrix} \begin{pmatrix} x_1 \\ x_1' \end{pmatrix} \quad . \tag{1.68}$$

Thus if $z_2 = z_i$ and $z_1 = z_o$, we have

$$\begin{pmatrix} x_i \\ x_i' \end{pmatrix} = \begin{pmatrix} -\frac{z_i - z_{Fi}}{f} & 0 \\ -\frac{1}{f} & \frac{z_o - z_{Fo}}{f} \end{pmatrix} \begin{pmatrix} x_o \\ x_o' \end{pmatrix} \quad , \tag{1.69}$$

and we see that

$$z_i - z_{Fi} = -fM$$

$$z_o - z_{Fo} = f/M \tag{1.70}$$

[as in (1.53) and (1.54), using (1.66)].

Other pairs of planes z_1 and z_2 are of interest. If we set $z_1 = z_2$ and take this common plane as the origin of z coordinates, we find

$$\begin{pmatrix} x_2 \\ x_2' \end{pmatrix} = \begin{pmatrix} z_{Fi}/f & z_{Fi}\, z_{Fo}/f + f \\ -1/f & -z_{Fo}/f \end{pmatrix} \begin{pmatrix} x_1 \\ x_1' \end{pmatrix} \quad . \tag{1.71}$$

If $z_1 = z_{Po}$ and $z_2 = z_{Pi}$, we have

$$\begin{pmatrix} x_2 \\ x_2' \end{pmatrix} = \begin{pmatrix} 1 & 0 \\ -1/f & 1 \end{pmatrix} \begin{pmatrix} x_1 \\ x_1' \end{pmatrix} , \qquad (1.72)$$

and as expected, we see that the principal planes are conjugate with unit magnification.

These matrix forms are particularly useful when we wish to calculate the overall optical properties of a chain of lenses – the combined effect of accelerator and condensers in a high-voltage instrument, for example, or of the intermediate-projector combination downstream from the objective. (In the first of these examples, the matrices corresponding to accelerating structures are required; these are slightly different from those of purely magnetic lenses.) We shall see that matrices can be devised that allow us to calculate the aberration coefficients of groups of lenses and hence to eliminate or minimize some of the aberrations. We note that if matrix (1.71) is used, relating incident and emergent ray height and slope in a single plane, we need the transfer matrix for a lens-free drift space of length D:

$$\begin{pmatrix} x_2 \\ x_2' \end{pmatrix} = \begin{pmatrix} 1 & D \\ 0 & 1 \end{pmatrix} \begin{pmatrix} x_1 \\ x_1' \end{pmatrix} . \qquad (1.73)$$

In order to obtain the positions of the foci and the focal length of a doublet, it is simplest to set z_1 at the object focus (z_{Fo}) of the first lens and z_2 at the image focus ($\bar{z}_{Fi}$) of the second. Then, at the image focus of the first lens (z_{Fi}) we have

$$\begin{pmatrix} x_{Fi} \\ x_{Fi}' \end{pmatrix} = \begin{pmatrix} 0 & f \\ -1/f & 0 \end{pmatrix} \begin{pmatrix} x_1 \\ x_1' \end{pmatrix} ;$$

at the object focus ($\bar{z}_{Fo}$) of the second lens

$$\begin{pmatrix} x_{\bar{F}o} \\ x_{\bar{F}o}' \end{pmatrix} = \begin{pmatrix} 1 & D \\ 0 & 1 \end{pmatrix} \begin{pmatrix} x_{Fi} \\ x_{Fi}' \end{pmatrix} = \begin{pmatrix} -D/f & f \\ -1/f & 0 \end{pmatrix} \begin{pmatrix} x_1 \\ x_1' \end{pmatrix} ,$$

and so at the image focus of the second lens

$$\begin{pmatrix} x_{\bar{F}i} \\ x_{\bar{F}i}' \end{pmatrix} = \begin{pmatrix} 0 & \bar{f} \\ -1/\bar{f} & 0 \end{pmatrix} \begin{pmatrix} x_{\bar{F}o} \\ x_{\bar{F}o}' \end{pmatrix} = \begin{pmatrix} -\bar{f}/f & 0 \\ D/f\bar{f} & -f/\bar{f} \end{pmatrix} \begin{pmatrix} x_1 \\ x_1' \end{pmatrix} .$$

Denoting the object and image foci of the doublet by ζ_{Fo} and ζ_{Fi}, respectively, and the focal length by φ, we have (1.68):

$$-(\bar{z}_{Fi} - \zeta_{Fi})/\varphi = -\bar{f}/f$$

$$(\bar{z}_{Fo} - \zeta_{Fo})/\varphi = -f/\bar{f}$$

$$-1/\varphi = D/f\bar{f}$$

or

$$\varphi = -f\bar{f}/D$$

$$\zeta_{Fi} = \bar{z}_{Fi} + \bar{f}^2/D$$

$$\zeta_{Fo} = z_{Fo} - f^2/D \tag{1.74}$$

with

$$D = \bar{z}_{Fo} - z_{Fi} \quad . \tag{1.75}$$

1.3 Methods of Calculating Aberration Coefficients

We consider two types of aberrations. Both can conveniently be analysed by studying the effect on the paraxial solutions $x(z)$ and $y(z)$ of perturbing $m^{(2)}$ to $m^{(2)} + m^{(P)}$. There are essentially two methods of attacking this problem: one, due to GLASER and perfected by STURROCK, is known as the eikonal or characteristic function method and has its roots in the work of HAMILTON; the other, introduced into electron optics by SCHERZER, is commonly referred to as the "trajectory method". The results are of course the same, and a comparable amount of effort is needed to obtain them. Nevertheless, the method of characteristic functions, although not so straightforward as the trajectory method at first sight, has the advantage that interrelations among the aberration coefficients emerge naturally and do not need to be proved, as they do in the trajectory method.

1.3.1 Characteristic Functions

If the function $m^{(2)}$ characterizing the paraxial properties of the system is altered to $m^{(2)} + m^{(P)}$, the point characteristic function changes from $V_{12}^{(2)}$ to $V_{12}^{(2)} + v_{12}^{(P)}$. The increment $v^{(P)}$ contains two terms, one arising from the integral $\int m^{(P)}dz$ taken along the paraxial ray and the other from the change from the paraxial ray to the perturbed ray. The latter, given by (1.30), is equal to $p_2x_2^{(P)} + q_2y_2^{(P)} - (p_1x_1^{(P)} + q_1y_1^{(P)})$. Denoting the first contribution by V_{12} (with no superscript, since we shall not use V_{12} in any other context), we see that

$$\delta v_{12}^{(P)} = \delta p_2 \cdot x_2^{(P)} + \delta q_2 \cdot y_2^{(P)} + p_2 \cdot \delta x_2^{(P)} + q_2 \cdot \delta y_2^{(P)}$$
$$- (\delta p_1 \cdot x_1^{(P)} + \delta q_1 \cdot y_1^{(P)} + p_1 \cdot \delta x_1^{(P)} + q_1 \cdot \delta y_1^{(P)}) + \delta V_{12} \quad . \tag{1.76}$$

The perturbed form of (1.30), given in the text above, provides us with another expression for $\delta v_{12}^{(p)}$, so that finally we obtain

$$\begin{aligned}\delta V_{12} = {} & p_2^{(P)}\cdot\delta x_2 + q_2^{(P)}\cdot\delta y_2 \\ & - (x_2^{(P)}\cdot\delta p_2 + y_2^{(P)}\cdot\delta q_2) \\ & - (p_1^{(P)}\cdot\delta x_1 + q_1^{(P)}\cdot\delta y_1) \\ & + x_1^{(P)}\cdot\delta p_1 + y_1^{(P)}\cdot\delta q_1 \quad .\end{aligned} \tag{1.77}$$

This very important relation is known as the first-order perturbation relation. It enables us to calculate the change in the point of arrival of an electron when $m^{(2)}$ is altered to $m^{(2)} + m^{(p)}$. Suppose, for example, that we consider an electron trajectory specified by the point of departure and momentum in some plane (the object plane) $z = z_1$. Then after altering $m^{(2)}$, we keep x_1, y_1, p_1, and q_1 constant so that $\delta x_1 = \delta y_1 = \delta p_1 = \delta q_1 = 0$. Hence

$$\delta V_{12} = p_2^{(P)}\cdot\delta x_2 + q_2^{(P)}\cdot\delta y_2 - x_2^{(P)}\cdot\delta p_2 - y_2^{(P)}\cdot\delta q_2 \quad . \tag{1.78}$$

In order to calculate the geometrical aberrations, we have only to set $m^{(p)}=m^{(4)}$ and $V_{12} = \int m^{(4)}dz$; for the paraxial chromatic aberrations, $m^{(p)} = (\partial m^{(2)}\partial U)\Delta U$ or $m^{(p)} = \partial m^{(2)}/\partial\Phi)\Delta\Phi$, where $\Delta\Phi$ or $\Delta U = (1 + 2\varepsilon\Phi)\Delta\Phi$ is the variation in question. As we shall see below, the variation in lens current can likewise be incorporated.

Like the cardinal elements, the aberration coefficients may be either real or asymptotic, depending on the role of the lens: real for objectives and probe-forming lenses, asymptotic for the remainder. In both cases, the aberration coefficients can be calculated straightforwardly with the aid of the appropriate characteristic function, but this function is not quite the same in the two situations.

For real aberrations specified by position and momentum in the object plane, we have (1.77)

$$\delta V_{12} = p_2^{(P)}\cdot\delta x_2 + q_2^{(P)}\cdot\delta y_2 - x_2^{(P)}\cdot\delta p_2 - y_2^{(P)}\cdot\delta q_2 \quad . \tag{1.79}$$

This gives

$$\frac{\partial V_{12}}{\partial x_o} = p_2^{(P)}\,\frac{\partial x_2}{\partial x_o} + q_2^{(P)}\,\frac{\partial y_2}{\partial x_o} - x_2^{(P)}\,\frac{\partial p_2}{\partial x_o} - y_2^{(P)}\,\frac{\partial q_2}{\partial x_o} \tag{1.80}$$

with similar expressions for $\partial V_{12}/\partial y_o$, $\partial V_{12}/\partial x_o'$ and $\partial V_{12}/\partial y_o'$, so that with

$$\begin{aligned} x(z) &= x_o g(z) + x_o' h(z) \\ y(z) &= y_o g(z) + y_o' h(z) \end{aligned} \tag{1.81}$$

(1.34), we find

$$\frac{\partial V_{o2}}{\partial x_o} = p_2^{(P)} g(z_2) - x_2^{(P)} g'(z_2)$$

$$\frac{\partial V_{o2}}{\partial x_o'} = p_2^{(P)} h(z_2) - x_2^{(P)} h'(z_2) \quad , \tag{1.82}$$

in which we have set $z_1 = z_o$ and we have used the result that $p = x'$, $q = y'$. Solving for $x_2^{(p)}$, $y_2^{(p)}$, $p_2^{(p)}$, and $q_2^{(p)}$ and recalling that $gh' - g'h = 1$, we finally obtain

$$x_2^{(P)} = h(z_2)\frac{\partial V_{o2}}{\partial x_o} - g(z_2)\frac{\partial V_{o2}}{\partial x_o'}$$

$$y_2^{(P)} = h(z_2)\frac{\partial V_{o2}}{\partial y_o} - g(z_2)\frac{\partial V_{o2}}{\partial y_o'} \tag{1.83}$$

and

$$p_2^{(P)} = h'(z_2)\frac{\partial V_{o2}}{\partial x_o} - g'(z_2)\frac{\partial V_{o2}}{\partial x_o'}$$

$$q_2^{(P)} = h'(z_2)\frac{\partial V_{o2}}{\partial y_o} - g'(z_2)\frac{\partial V_{o2}}{\partial y_o'} \quad . \tag{1.84}$$

In the image plane, $z_2 = z_i$, where $h(z_i) = 0$ and $g(z_i) = M$,

$$x^{(P)}(z_i) = -M\frac{\partial V_{oi}}{\partial x_o'}$$

$$y^{(P)}(z_i) = -M\frac{\partial V_{oi}}{\partial y_o'} \tag{1.85}$$

and

$$p^{(P)}(z_i) = \frac{1}{M}\frac{\partial V_{oi}}{\partial x_o} - g_i'\frac{\partial V_{oi}}{\partial x_o'}$$

$$q^{(P)}(z_i) = \frac{1}{M}\frac{\partial V_{oi}}{\partial y_o} - g_i'\frac{\partial V_{oi}}{\partial y_o'} \quad . \tag{1.86}$$

The only modification necessary when the asymptotic aberrations are required concerns the definition of the characteristic function V_{o2}, which we now write in the explicit form $V(z_o, z_2)$. This function now contains three contributions:

$$V(z_o, z_2) = V(z_o, -\infty) + V(-\infty, \infty) + V(\infty, z_2) \quad . \tag{1.87}$$

In $V(z_o, -\infty)$ we substitute the incident asymptotes to the general ray; in $V(\infty, z_2)$ we substitute the emergent asymptotes; and in $V(-\infty,\infty)$ we substitute the general solutions $x(z) = x_oG(z) + x_o'H(z)$, $y(z) = y_oG(z) + y_o'H(z)$, where

$$\lim_{z\to-\infty} G(z) = 1$$

$$\lim_{z\to-\infty} H(z) = z - z_o \quad . \tag{1.88}$$

Furthermore, the function V required for $V(z_o, -\infty)$ and $V(\infty, z_2)$ is the free-space expression. Reasoning analogous to that given above for the real aberrations leads straightforwardly to

$$x^{(P)}(z_2) = H(z_2)\frac{\partial V}{\partial x_o} - G(z_2)\frac{\partial V}{\partial x_o'} - \frac{1}{2}H(z_2)x_o'(x_o'^2 + y_o'^2) \tag{1.89}$$

with a similar expression for $y^{(p)}(z_2)$, in which V denotes $V(-\infty,\infty)$. $H(z_2)$ and $G(z_2)$ denote the heights of the emergent asymptotes to H(z) and G(z), respectively, in the plane $z = z_2$. In the image plane, where H(z) vanishes and G(z) is the magnification M, we have

$$x_i^{(P)} = -M\frac{\partial V}{\partial x_o'}$$

$$y_i^{(P)} = -M\frac{\partial V}{\partial y_o'} \quad . \tag{1.90}$$

In the image plane, therefore, real and asymptotic aberration coefficients differ only in that integrals from object to image plane (V_{oi}) are replaced by integrals over the entire optic axis [$V(-\infty,\infty) = V$] and that the paraxial solutions g(z) and h(z) give place to G(z) and H(z).

The method of analysing aberrations by means of characteristic functions was introduced into electron optics by GLASER in the 1930s; references to his early work may be found in his textbook [1.21] or in the improved condensed text in the *Handbuch der Physik* [1.22]. The method was reexamined and put on a more formal basis by STURROCK [1.122], whose contributions are conveniently collected in book form in [1.125]. Many other papers concerned with aberrations appeared during the 1930s; [1.20,25,89,98,134] form a representative selection of those not discussed elsewhere in this chapter. KANAYA's early work may also be listed in this context [1.46].

Unlike GLASER, STURROCK also considered briefly the question of calculating aberrations when the object is virtual or "asymptotic". The difference between the two situations was first analysed thoroughly by LENZ [1.68,69], who adopted the trajectory method, discussed in the next section. A full list of formulae for these aberrations, obtained by both methods, has been given by the present writer [1.35,37], who has also examined quadrupole lenses. There, asymptotic aberrations are particularly important, since quadrupoles are rarely used singly and asymptotic astigmatic objects and images may occur [1.35,39].

Many authors have attempted to construct very general theories for calculating the aberrations of higher order and for lower symmetries than that of round lenses. From the numerous older papers on this topic, we draw attention to [1.26,56,124, 132]. More recently, ROSE has investigated these questions further, paying particular attention to systems with straight optic axes. From his work has emerged a complicated theory which is, however, less unmanageable than that of many of his predecessors. In particular, it can be applied to the calculation of misalignment errors and of fifth-order aberrations of round lenses (see Sect.1.4.1 for references to the latter). His work is described in [1.99-101,105], which includes a number of papers concerned exclusively with non-rotationally symmetric systems.

1.3.2 The Trajectory Method

If we wish to derive the third-order or primary geometrical aberrations, we substitute $m = m^{(2)} + m^{(4)}$ into the original variational relation, $\delta \int m dz = 0$. The corresponding Euler equations are now

$$\frac{d}{dz}\left(\frac{\partial m^{(2)}}{\partial x'}\right) - \frac{\partial m^{(2)}}{\partial x} = -\frac{d}{dz}\left(\frac{\partial m^{(4)}}{\partial x'}\right) + \frac{\partial m^{(4)}}{\partial x} \tag{1.91}$$

with a similar equation for y. If we seek solutions of the form

$$\begin{aligned} \bar{x}(z) &= x(z) + x^{(3)}(z) \\ \bar{y}(z) &= y(z) + y^{(3)}(z) \quad , \end{aligned} \tag{1.92}$$

where x(z), y(z) are the paraxial solutions, we find that $x^{(3)}$ and $y^{(3)}$ satisfy inhomogeneous second-order differential equations, the homogeneous parts of which are identical with the paraxial equations:

$$\begin{aligned} \frac{d^2x^{(3)}}{dz^2} + \frac{\eta^2 B^2}{4U} x^{(3)} &= R_x \\ \frac{d^2y^{(3)}}{dz^2} + \frac{\eta^2 B^2}{4U} y^{(3)} &= R_y \quad , \end{aligned} \tag{1.93}$$

in which R_x and R_y contain numerous terms of third degree in $\bar{x}$, $\bar{y}$, and their derivatives. In order to obtain the third-order aberrations, we replace $\bar{x}$ and $\bar{y}$ in R_x and R_y by the paraxial solutions, x(z) and y(z); R_x and R_y are then known functions and (1.93) can be solved by variation of parameters. We write

$$\begin{aligned} x^{(3)} &= A_x(z)s(z) + B_x(z)t(z) \\ y^{(3)} &= A_y(z)s(z) + B_y(z)t(z) \quad , \end{aligned} \tag{1.94}$$

where A_x, A_y, B_x, and B_y are to be determined and s, t are paraxial solutions (1.32).

Substituting (1.94) into the differential equations (1.93) and requiring that

$$A_x's + B_x't = 0$$
$$A_y's + B_y't = 0 \quad , \tag{1.95}$$

we obtain

$$A_x's' + B_x't' = R_x$$
$$A_y's' + B_y't' = R_y \quad , \tag{1.96}$$

and hence

$$A_x'(st' - s't) = -R_x t$$
$$B_x'(st' - s't) = R_x s \tag{1.97}$$

with similar equations for A_y' and B_y'.

For objective lenses the aberrations may be expressed in terms of position and slope at the object or in terms of position in the object and aperture planes. For the first of these choices, we write s = g and t = h (1.34) so that

$$x^{(3)}(z) = h(z)\int_{z_o}^{z} gR_x\,dz - g(z)\int_{z_o}^{z} hR_x\,dz$$
$$y^{(3)}(z) = h(z)\int_{z_o}^{z} gR_y\,dz - g(z)\int_{z_o}^{z} hR_y\,dz \quad , \tag{1.98}$$

and in the image plane, where h = 0 and g = M,

$$x_i^{(3)} = -M\int_{z_o}^{z_i} hR_x\,dz$$
$$y_i^{(3)} = -M\int_{z_o}^{z_i} hR_y\,dz \quad . \tag{1.99}$$

For the other choice we write $s = r_\gamma$, $t = r_\alpha$ and

$$x = x_o\,r_\gamma(z) + x_a r_\alpha(z) \quad . \tag{1.100}$$

The aberrations must vanish in the object and aperture planes, giving

$$x^{(3)}(z) = r_\alpha(z)\int_{z_a}^{z} r_\gamma R_x\,dz - r_\gamma\int_{z_o}^{z} r_\alpha R_x\,dz$$
$$x_i^{(3)} = -M\int_{z_o}^{z_i} r_\alpha R_x\,dz \tag{1.101}$$

with similar expressions for y.

We noted in Sect.1.2.2 that the hybrid ray pair $r_\gamma(z)$ and $\bar{r}_\alpha(z)$ may not be satisfactory for aberration calculations (particularly in connection with a type of lens not considered here, the electron optical analogue of cylindrical glass lenses). This can be seen when we attempt to apply the method of variation of parameters. If we write

$$x^{(3)}(z) = \bar{r}_\alpha(z) \int_{z_a}^{z} r_\gamma R_x \, dz - r_\gamma(z) \int_{z_o}^{z} r_\alpha R_x \, dz \quad , \tag{1.102}$$

we find that $x^{(3)}(z_o) = x^{(3)}(z_a) = 0$, which is satisfactory so far as r_γ is concerned since it is characterized by $r_\gamma(z_o)$ and $r_\gamma(z_a)$.

How then are we to picture the aberrations? With g and h we answer the following question (Fig.1.6a): given a paraxial ray specified by its position and slope in the object plane, how is its point of arrival in some other plane affected by aberrations? With r_α and r_γ we answer the same question, but is is now the positions in object and aperture plane that are specified (Fig.1.6b). For r_γ and $\bar{r}_\alpha$, however, we appear to have a specified position in two planes, object and aperture, in addition to slope in the object plane, which overdetermines the system. However, no problem arises if we consider only the image plane, even in this hybrid situation.

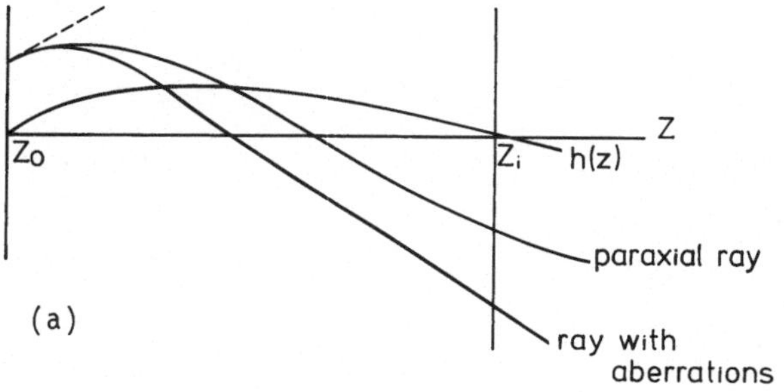

(a)

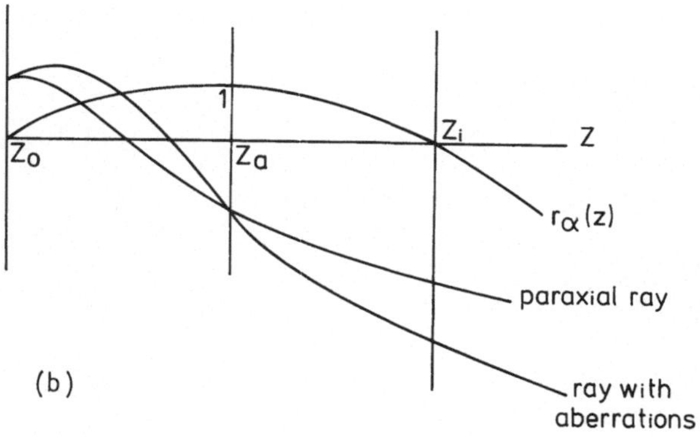

(b)

Fig.1.6a,b. Aberrations and boundary conditions. (a) The paraxial ray and the aberrated ray leave the object plane at the same position and slope: aberrations of both position and gradient vanish at $z=z_o$. (b) The aberrations of position now vanish in $z=z_o$ and in $z=z_a$ but the aberrations of gradient vanish in neither of these planes

Discussion of the aberrations in terms of object and aperture plane coordinates is to be found in, for example, [1.21-22,122,125]. GLASER, in particular, gave a detailed account of the dependence of each aberration on aperture position, showing which coefficients can be made to vanish by suitable choice of aperture position and the conditions in which these suitable choices coincide for two or more aberrations.

In order to calculate the asymptotic aberrations, we use the rays G(z) and H(z), which are, we recall, characterized by their asymptotes in object space:

$$\lim_{z\to-\infty} G(z) = 1$$

$$\lim_{z\to-\infty} H(z) = z - z_o \quad , \tag{1.103}$$

where z_o is the object plane as seen from the object space of the lens in question. The general solutions of the third-order equations are now of the form

$$x(z) = x_oG(z) + x_o'H(z) + H(z)\int_{-\infty}^{z} GR_x\,dz - G(z)\int_{-\infty}^{z} HR_x\,dz$$

$$y(z) = y_oG(z) + y_o'H(z) + H(z)\int_{-\infty}^{z} GR_y\,dz - G(z)\int_{-\infty}^{z} HR_y\,dz \quad . \tag{1.104}$$

The boundary conditions require that the incident asymptotes to the aberrated rays are the same as those to the corresponding paraxial rays (Fig.1.7a).

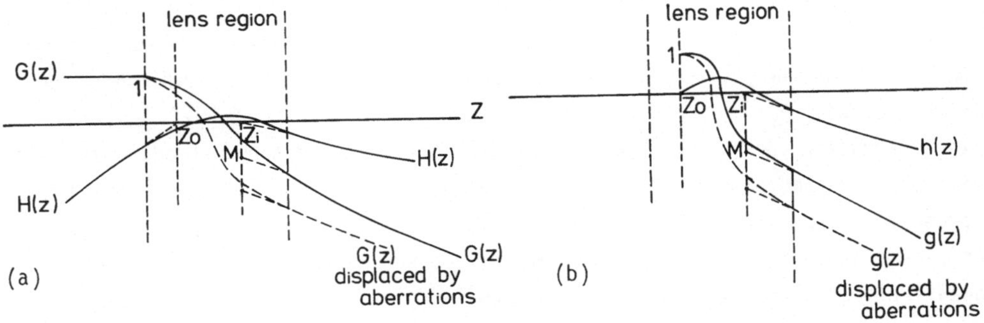

Fig.1.7. (a) Definition of asymptotic aberrations. The aberration is now the distance between the points at which the emergent asymptotes to a ray with and without aberrations intersect the (asymptotic) image plane. The ray with and without aberrations has the same incident asymptote. The figure illustrates the case of G(z), but the construction is similar for any ray; (b) asymptotic aberrations with a real object. On the image side, the construction is the same as in (a), but the ray with and without aberrations is now such that position and slope are the same in the real object plane, $z=z_o$. The figure shows the case of g(z), but again, the construction is applicable to any ray

We now calculate the point at which the emergent asymptote to the ray intersects the (asymptotic) image plane, $z = z_i$. The emergent asymptote is given by

$$x(z) \to x_o[M + G_i'(z - z_i)] + \frac{x_o'}{M}(z - z_i)$$

$$+ \frac{z - z_i}{M}\int_{-\infty}^{\infty} GR_x\,dz - [M + G_i'(z - z_i)]\int_{-\infty}^{\infty} HR_x\,dz \tag{1.105}$$

with a similar expression for $y(z)$, where G_i' is the slope of the emergent asymptote to $G(z)$. In the image plane we therefore have

$$\frac{x_i - Mx_o}{M} = -\int_{-\infty}^{\infty} HR_x \, dz$$

$$\frac{y_i - My_o}{M} = -\int_{-\infty}^{\infty} HR_y \, dz \quad . \tag{1.106}$$

For a very strong objective lens, an intermediate situation could arise in which the first image was formed within the magnetic field of the objective; we would ther require the aberrations for a real object but an asymptotic image. It is readily seen that these are obtained with the aid of the g and h rays and formulae similar to (1.99):

$$\frac{x_i - Mx_o}{M} = -\int_{z_o}^{\infty} hR_x \, dz$$

$$\frac{y_i - My_o}{M} = -\int_{z_o}^{\infty} hR_y \, dz \quad , \tag{1.107}$$

in which M is the height of the emergent asymptote to $g(z)$ in the asymptotic image plane defined as the plane in which the emergent asymptote to $h(z)$ intersects the axis (Fig.1.7b).

The aberration coefficients were calculated by the trajectory method, using variation of parameters, in the 1930s by SCHERZER [1.107,108]. Often found easier to grasp than the method of characteristic functions, the trajectory method has been widely used, but the two methods are of course equivalent, as GRATSIATOS showed explicitly [1.25]. The first thorough study of asymptotic aberrations, made by LENZ in 1956-1957 [1.68-69], used the trajectory method, as we mentioned in Sect.1.3.1.

1.4 Aberration Coefficients

1.4.1 Real (Objective) Aberration Coefficients

a) *Geometrical Aberrations*

We begin with the expression for $m^{(4)}$ (1.20c) and introduce the rotating coordinates (x,y) with the aid of (1.22). We find that

$$m^{(4)} = -\frac{1}{4}L(x^2+y^2)^2 - \frac{1}{2}K(x^2+y^2)(x'^2+y'^2) - \frac{1}{4}N(x'^2+y'^2)^2$$
$$- P(x^2+y^2)(xy'-x'y) - Q(x'^2+y'^2)(xy'-x'y) - K(xy'-x'y)^2 \quad , \tag{1.108}$$

in which

$$K = \frac{\eta^2 B^2}{8U} \tag{1.109}$$

$$L = \frac{\eta^4 B^4}{32U^2} - \frac{\eta^2 BB''}{8U} \quad , \quad N = \frac{1}{2} \tag{1.110}$$

$$P = \frac{\eta^3 B^2}{16U^{3/2}} - \frac{\eta B''}{16U^{1/2}} \quad , \quad Q = \frac{\eta B}{4U^{1/2}} \quad . \tag{1.111}$$

The notation has been chosen to agree with that of GLASER [1.22], except that in the magnetic case to which the present discussion is restricted, it is more convenient to incorporate the term $\sqrt{U}$, which here includes the relativistic correction, into L and N since there is no need to distinguish between accelerating voltage and lens potential distribution. The factors involving U have likewise been absorbed into K, P, and Q. This slight change has rendered the functions K and M used by GLASER equal and, we have therefore used K everywhere to avoid confusion between the function $M = \eta^2 B^2/8U$ and the magnification.

We now form the integral

$$V_{oi} = V(z_o, z_i) = \int_{z_o}^{z_i} m^{(4)} \, dz \tag{1.112}$$

and obtain the aberration coefficients by evaluating $\partial V_{oi}/\partial x_o'$ and $\partial V_{oi}/\partial y_o'$, since

$$\frac{x_i^{(3)}}{M} = \frac{x_i - Mx_o}{M} = -\frac{\partial V_{oi}}{\partial x_o'}$$

$$\frac{y_i^{(3)}}{M} = \frac{y_i - My_o}{M} = -\frac{\partial V_{oi}}{\partial y_o'} \tag{1.113a}$$

We find that

$$\frac{x_i^{(3)}}{M} = (x_o' \quad x_o \quad -y_o)A\underline{r}$$

$$\frac{y_i^{(3)}}{M} = (y_o' \quad y_o \quad x_o)A\underline{r} \quad , \tag{1.113b}$$

in which the aberration matrix A is given by

$$A = \begin{pmatrix} B & 2F & D & 2f \\ F & 2C & E & c \\ f & c & e & 0 \end{pmatrix} \tag{1.114}$$

and the column vector $\underline{r}$ by

$$\underline{r} = \begin{pmatrix} x_o'^2 + y_o'^2 \\ x_o x_o' + y_o y_o' \\ x_o^2 + y_o^2 \\ x_o y_o' - x_o' y_o \end{pmatrix} . \qquad (1.115)$$

The matrix elements B, C, D, E, F, c, e, f are the real aberration coefficients; they are given by the following integrals, all of which run from z_o to z_i.

Spherical aberration:

$$B = \int (Lh^4 + 2Kh^2h'^2 + Nh'^4)dz \quad . \qquad (1.116)$$

Isotropic astigmation and field curvature:

$$C = \int (Lg^2h^2 + 2Kgg'hh' + Ng'^2h'^2 - K)dz \qquad (1.117)$$

$$D = \int [Lg^2h^2 + K(g^2h'^2 + g'^2h^2) + Ng'^2h'^2 + 2K]dz \quad . \qquad (1.118)$$

Isotropic distortion:

$$E = \int [Lg^3h + Kgg'(gh)' + Ng'^3h']dz \quad . \qquad (1.119)$$

Isotropic coma:

$$F = \int [Lgh^3 + K(gh)'hh' + Ng'h'^3]dz \quad . \qquad (1.120)$$

Anisotropic astigmatism

$$c = 2 \int (Pgh + Qg'h')dz \quad . \qquad (1.121)$$

Anisotropic distortion:

$$e = \int (Pg^2 + Qg'^2)dz \quad . \qquad (1.122)$$

Anisotropic coma:

$$f = \int (Ph^2 + Qh'^2)dz \quad . \qquad (1.123)$$

This classification of the aberrations is based on the relative importance of the electron position and slope in the aberration term. We now consider each in turn.

Spherical aberration. The aberration that depends only on ray gradient at the specimen,

$$x^{(3)} = MBx_o'(x_o'^2 + y_o'^2)$$
$$y^{(3)} = MBy_o'(x_o'^2 + y_o'^2) \quad , \qquad (1.124)$$

is the spherical aberration and is always denoted by C_s. From now on we write

$$B = C_s \quad . \tag{1.125}$$

Clearly, this aberration has the same effect on the image of an object point, whereever the latter may be situated. In particular, it does not vanish when the object point lies on the axis, unlike all the other aberrations. If a cone of electrons, $0 \leq x_o', y_o' \leq \theta$, leaves any point in the object plane, the image of that point will be a disc of radius r,

$$r = MC_s\theta^3 \quad , \tag{1.126}$$

in the image plane. There is, however, a plane close to the image plane in which the disc is smaller, as we can see by writing

$$\begin{aligned} x(z_i - \zeta) &= x_o'h(z_i - \zeta) + g(z_i - \zeta)C_s x_o'(x_o'^2 + y_o'^2) \\ &\approx - x_o'h_i'\zeta + MC_s x_o'(x_o'^2 + y_o'^2) \\ &= -x_o'\zeta/M + MC_s x_o'(x_o'^2 + y_o'^2) \quad . \end{aligned} \tag{1.127}$$

For a given M, there is a value of ζ for which the ray most distant from the axis is closer than $MC_s\theta^3$. This extreme ray,

$$x_m(z_i - \zeta) = -\theta\zeta/M + MC_s\theta^3 \quad , \tag{1.128}$$

and the general ray of (1.127) are equidistant from the axis in the plane for which

$$\frac{\zeta}{M}(x_o' - \theta) = MC_s(x_o'^3 - \theta^3) \quad , \tag{1.129}$$

where we have set $y_o' = 0$, since we are not concerned with the actual point of intersection of the rays but only with the plane in which they are equidistant from the axis. For this plane therefore

$$\zeta = M^2C_s(x_o'^2 + x_o'\theta + \theta^2) \quad , \tag{1.130}$$

and the off-axis distance is

$$x(z_i - \zeta) = -MC_s x_o'\theta(x_o' + \theta) \quad , \tag{1.131}$$

which is least when

$$2x_o' + \theta = 0 \quad . \tag{1.132}$$

For this value of x_o' we have

$$\zeta = \frac{3}{4}M^2C_s\theta^2 \quad , \tag{1.133}$$

and the rays approach the axis at a distance

$$x(z_i - \zeta) = \frac{1}{4}MC_s\theta^3 \quad . \tag{1.134}$$

The radius of the "circle" or "disc of least confusion" is thus one quarter that of the aberration disc in the paraxial image plane. This disc is formed in the plane three-quarters the way from the paraxial image plane to the marginal focus, the point at which the outermost rays of the cone leaving the object point intersect the axis.

The fact that the effect of spherical aberration does not vary with the position of the object point renders this aberration of extreme importance in the electron microscope and related instruments. Of the geometrical aberrations, this aberration limits the resolution of conventional transmission electron microscopes and the probe size in scanning instruments of both types. In order to estimate the size of the probe formed by a lens with spherical aberration coefficient C_s, we proceed as follows.

In the case of a high-magnification objective, the radius of the aberration disc is given by $MC_s\theta^3$, where C_s is the high-magnification value of the spherical aberration coefficient. A ray from the point $x = x_o$ will therefore intersect the axis at $z = z_i$ if

$$Mx_o + MC_s\theta^3 = 0$$

or

$$x_o = -C_s\theta^3 \qquad (1.135)$$

(Fig.1.8). The radius of the probe formed at the focus of the same lens (operated in the reverse direction if the field distribution is not geometrically symmetric about its mid-plane) is thus given by

$$r = C_s\theta^3 \quad , \qquad (1.136)$$

where C_s is, we repeat, the high-magnification value of the spherical aberration coefficient, the value normally tabulated.

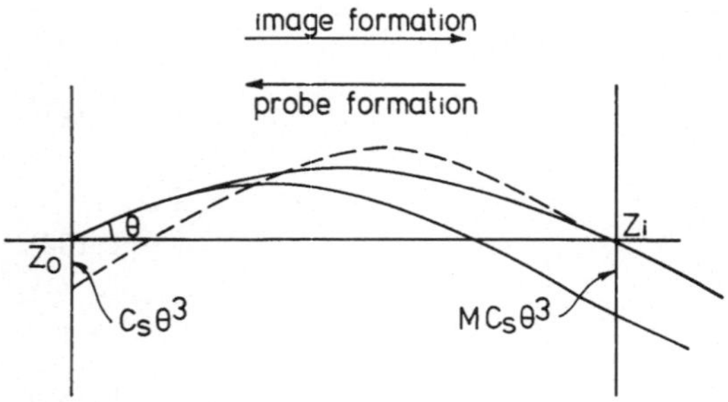

Fig.1.8. Spherical aberration at an image (left-to-right) and a probe (right-to-left)

The coefficient C_s can be written in many different ways by partial integration, using the paraxial equation satisfied by h(z) to eliminate second derivatives of h. The most useful expressions are as follows [the first of which is identical with (1.116)].

$$C_s = \int \left[\left(\frac{\eta^4 B^4}{32U^2} - \frac{\eta^2 BB''}{8U}\right)h^4 + \frac{\eta^2 B^2}{4U} h^2 h'^2 + \frac{1}{2} h'^4\right]dz$$

$$= \int \left[\left(\frac{\eta^4 B^4}{32U^2} - \frac{\eta^2 BB''}{8U}\right)h^4 + \frac{5\eta^2 B^2}{8U} h^2 h'^2\right]dz$$

$$= \int \left(\frac{\eta^4 B^4}{12U^2} + \frac{5}{48}\frac{\eta^2 B'^2}{U} - \frac{\eta^2 BB''}{48U}\right)h^4 \, dz$$

$$= \int \left[\left(\frac{3\eta^4 B^4}{32U^2} + \frac{\eta^2 B'^2}{8U}\right)h^4 - \frac{\eta^2 B^2}{8U} h^2 h'^2\right]dz$$

$$= \int \left[\left(\frac{\eta^4 B^4}{16U^2} + \frac{\eta^2 B'^2}{8U}\right)h^4 + \frac{\eta^2 BB'}{4U} h^3 h' + \frac{\eta^2 B^2}{4U} h^2 h'^2\right]dz \quad . \qquad (1.137)$$

The final expression has had an immense influence on electron optics since it was first derived, in a more general form that included electrostatic lenses, by SCHERZER in 1936 [1.108]. Its importance resides in the fact that a simple rearrangement of the integrand yields

$$C_s = \int \left[\frac{\eta^4 B^4}{16U^2} h^4 + \frac{\eta^2 h^2}{8U} (hB' + h'B)^2 + \frac{\eta^2 B^2}{8U} h^2 h'^2\right]dz \quad , \qquad (1.138)$$

and it is immediately clear that C_s can never change sign. The spherical aberration coefficient is therefore always positive and cannot be reduced below some minimum value given the practical constraints that must be imposed. This result is commonly known as *Scherzer's Theorem*.

What is this minimum value? It is governed by certain practical limitations, for example on the field distribution. The absolute minimum was obtained by TRETNER [1.128-130], whose work has been essentially confirmed by MOSES [1.82]. A good discussion of this point, in the context of measurements on actual lenses, is to be found in the lengthy review by MULVEY and WALLINGTON [1.84].

A very general form of the spherical aberration integral (and indeed of each of the aberration integrals) can be derived, from which any of the above forms may be obtained immediately. The technique required was first introduced into electron optics by SEMAN but passed largely unnoticed owing to the extreme inaccessibility of the recondite journal in which the work was published [1.111]; even in the standard Russian treatise on electron optics by KEL'MAN and YAVOR [1.62] not all the relevant papers are listed! The general technique was applied to the quadrupole lens characteristic function by HAWKES [1.34] and to individual aberration coefficients in [1.36]. Details of the procedure are to be found in these papers — here we simply give the result:

$$C_s = \frac{1}{32} \int \left\{ (16 + p)h'^4 + \frac{\eta^2 B^2}{U} \left(8 - \frac{3}{4}p + 3q\right) h^2 h'^2 \right.$$

$$+ \frac{\eta^2 BB'}{U} (2q + 4r) h^3 h' + \left[\frac{\eta^4 B^4}{U^2} \left(1 - \frac{1}{4}q\right) \right.$$

$$\left. \left. + \frac{\eta^2 BB''}{U} (r - 4) + \frac{\eta^2 B'^2}{U} r \right] h^4 \right\} dz \quad . \qquad (1.139)$$

The coefficients p, q, and r may be chosen arbitrarily to satisfy some prerequisite: for example that no derivatives of h should appear, or to obtain the Scherzer expression. The latter is given by $p = -16$, $q = -4$, $r = 4$.

Ever since Scherzer showed that, subject to various conditions, round lenses never have vanishingly small spherical aberration, ways of correcting this aberration have been sought. The various possibilities were outlined in [1.110] and most have been explored, though with varying degrees of thoroughness. If and when satisfactory correction is achieved, the fifth-order spherical aberration will become of interest. The mathematics involved in calculating this is extremely laborious, and it has been correspondingly little studied. Various formulae, some approximate, others exact, are to be found in [1.3,33,76,131]. The method was first given by STURROCK [1.120] and subsequently formulated differently by ROSE [1.100]; see footnote in [Ref.1.105, p.152].

The spherical aberration coefficient has been measured and calculated almost exclusively for high magnification. An approximate formula for the variation of C_s with magnification was derived by PETRIE [1.88] and the exact dependence, for Glaser's bell-shaped field, was later pointed out by HAWKES [1.37a], who also investigated the dependence for a non-Newtonian field, the Grivet-Lenz distribution [1.37b]. An early paper by BECKER and WALLRAFF is devoted to this coefficient [1.4a].

Coma. The coma depends linearly on object position coordinates and quadratically on the gradients. It may therefore be expected to be the next most important geometrical aberration for objective lenses after spherical aberration. It is characterized by two coefficients, the isotropic coma coefficient F and the anisotropic coefficient f. If we consider the effect of these alone on the image of an object point (x_o, y_o), we have

$$x^{(3)} = [F(2 + \cos 2\psi) + f \sin 2\psi] x_o r^2$$

$$+ [F \sin 2\psi - f(2 + \cos 2\psi)] y_o r^2$$

$$y^{(3)} = [F(2 - \cos 2\psi) - f \sin 2\psi] y_o r^2$$

$$+ [F \sin 2\psi + f(2 - \cos 2\psi)] x_o r^2 \quad , \qquad (1.140)$$

in which we have written

$$x_o' = r \cos\psi \quad , \quad y_o' = r \sin\psi \quad . \qquad (1.141)$$

It is not difficult to show that

$$(x^{(3)} - 2\,Fr^2x_o + 2fr^2y_o)^2 + (y^{(3)} - 2\,Fr^2y_o - 2fr^2x_o)^2$$

$$= (Fr^2x_o - fr^2y_o)^2 + (Fr^2y_o + fr^2x_o)^2$$

or

$$(x^{(3)} - \bar{x})^2 + (y^{(3)} - \bar{y})^2 = r^4(F^2 + f^2)(x_o^2 + y_o^2) \quad , \qquad (1.142)$$

where

$$\bar{x} = (2Fx_o - 2fy_o)r^2$$

$$\bar{y} = (2Fy_o + 2fx_o)r^2 \quad . \qquad (1.143)$$

For given r the aberration figure corresponding to a given object point is a circle of radius ρ

$$\rho = r^2(x_o^2 + y_o^2)^{1/2}(F^2 + f^2)^{1/2} \qquad (1.144)$$

centred on $(\bar{x}, \bar{y})$. For a range of values of r, that is, for cones of rays leaving (x_o, y_o) at various semiangles, a series of circles are described, their centres lying on the line determined by (1.143). The distance of the centre from the Gaussian image point is given by $(\bar{x}^2 + \bar{y}^2)^{1/2}$, which is equal to 2ρ, and the envelope of the circles is thus a pair of straight lines inclined to one another at 60° (Fig.1.9). From (1.143) we see that the line of centres is inclined with respect to the line joining the axis to the object point unless f vanishes. The possibility of observing the aberration pattern due to coma alone has been discussed by LENZ [1.67].

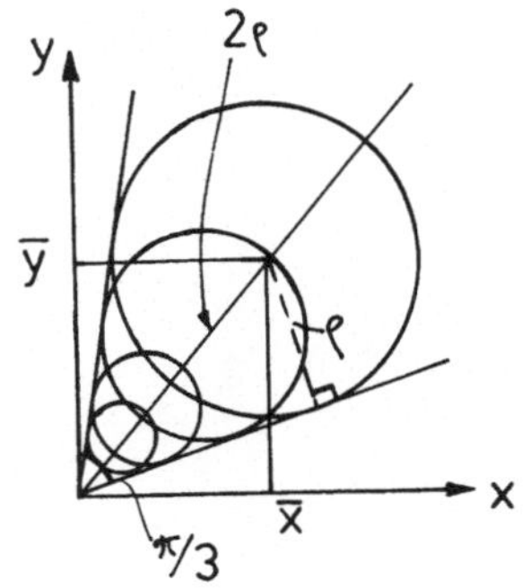

Fig.1.9. The pattern produced by coma in the absence of other aberrations for a family of conical pencils around an (off-axis) image point. The radius of a typical "circle" is ρ and the centre is 2ρ from the image point

The coefficients F and f can be rewritten in various forms. Rather than list the numerous possible versions of F, we derive a general formula analogous to (1.139), which also illustrates the method of obtaining such general formulae. We set out from expression (1.120) for F, noting that the terms of which it is composed conform to a simple pattern: (a) terms in B^4, g and h^3; (b) terms in B^2, g, and h^3, amongst which (d/dz) operates twice; (c) terms in g and h^3, amongst which (d/dz) operates four times. We therefore explore the set of terms generated by (d/dz), B^2, g, and h^3 and by $(d/dz)^3$, g, and h^3, using the paraxial equations to eliminate g" and h". We find that (d/dz), B^2, g, and h^3 generate

$$BB'gh^3, \quad B^2g'h^3, \quad \text{and} \quad B^2gh^2h'$$

and that $(d/dz)^3$, g and h^3 generate the following additional terms:

$$g'hh'^2 \quad \text{and} \quad gh'^3 \quad .$$

We now establish five identities involving these terms:

$$\frac{d}{dz}(BB'gh^3) = B'^2gh^3 + BB''gh^3 + BB'g'h^3 + 3BB'gh^2h'$$

$$\frac{d}{dz}(B^2g'h^3) = 2BB'g'h^3 - \frac{\eta^2B^4}{4U}gh^3 + 3B^2g'h^2h'$$

$$\frac{d}{dz}(B^2gh^2h') = 2BB'gh^2h' + B^2g'h^2h' + 2B^2ghh'^2 - \frac{\eta^2B^4}{4U}gh^3$$

$$\frac{d}{dz}(g'hh'^2) = -\frac{\eta^2B^2}{4U}ghh'^2 + g'h'^3 - \frac{\eta^2B^2}{2U}g'h^2h'$$

$$\frac{d}{dz}(gh'^3) = g'h'^3 - \frac{3\eta^2B^2}{4U}ghh'^2 \tag{1.145}$$

and add them to the integrand of F in the form [d/dz( ) - r.h.s. = 0], first multiplying each by an arbitrary constant multiplier. We obtain

$$\begin{aligned} F = \int_{z_0}^{z_i} \Bigg[& gh^3\left(\frac{\eta^4B^4}{32U^2} - \frac{\eta^2BB''}{8U} + p\,\frac{\eta^2B'^2}{U} + p\,\frac{\eta^2BB''}{U}\right. \\ & \left. - q\,\frac{\eta^4B^4}{4U^2} - t\,\frac{\eta^4B^4}{4U^2}\right) \\ & + \frac{\eta^2BB'}{U}\,g'h^3(p + 2t) \\ & + \frac{\eta^2BB'}{U}\,gh^2h'(3p + 2q) \\ & + \frac{\eta^2B^2}{U}\,g'h^2h'\left(\frac{1}{8} + q - \frac{1}{2}r + 3t\right) \\ & + \frac{\eta^2B^2}{U}\,ghh'^2\left(\frac{1}{8} + 2q - \frac{1}{4}r - \frac{3}{4}s\right) \\ & + g'h'^3\left(\frac{1}{2} + r + s\right)\Bigg]dz - s[gh'^3]_{z_0}^{z_i} \quad , \end{aligned} \tag{1.146}$$

and the integrated term may be written

$$s[gh'^3]_{z_o}^{z_i} = s(1/M^2 - 1) \quad . \tag{1.147}$$

If we wish to obtain an expression containing no derivatives of g and h, we choose

$$p = \frac{5}{48} , \quad q = -\frac{5}{32} , \quad r = -\frac{3}{8} , \quad s = -\frac{1}{8} , \quad t = -\frac{5}{96} \quad .$$

The formula for the anisotropic coma coefficient f is so simple that any partial integrations can be performed directly without difficulty. Eliminating h'^2, for example, we obtain

$$f = \frac{1}{16} \int \left(\frac{\eta B''}{U^{\frac{1}{2}}} + \frac{2\eta^3 B^3}{U^{3/2}}\right) h^2 \, dz \quad . \tag{1.148}$$

Particular attention has been paid to coma and to the design of coma-free lenses by ROSE [1.102-104]. The minimum spherical aberration of a coma-free lens has been derived by MOSES [1.81,82]. For a general account of MOSES' work on aberration minimization procedures, see [1.82].

Astigmatism and Field Curvature. These aberrations are rarely of interest in electron optics, partly because the dependence on angle at the object plane is now only linear and also because imperfections in lens roundness and homogeneity create a paraxial astigmatism, corrected by the stigmator, which is likely to dominate over any third-order effect. It is easy to show that a cone of rays of semi-angle θ at the object plane collapses to two line images in the vicinity of the paraxial image plane; these lines are separated by the astigmatic distance. A full discussion of the effect of these aberrations is given by GLASER [1.21], who also considers the case of curved object and image surfaces, of little interest in electron microscopy but very important in deflection systems.

The coefficients C, D, and c can be transformed by partial integration into various simpler forms. We note that the difference D-C proves to be independent of the rays g and h, since

$$\begin{aligned} D - C &= \int [K(gh' - g'h)^2 + 3K] dz \\ &= 4 \int K \, dz \\ &= \frac{\eta^2}{2U} \int B^2 dz \quad . \end{aligned} \tag{1.149}$$

This quantity is known as the Petzval coefficient, as in classical optics.

The elimination or minimization of third-order astigmatism has been considered by KAS'YANKOV et al. [1.18,28,55,57,126,127], and much earlier by BECKER and WALLRAFF [1.4c]. The latter also studied field curvature [1.4b], as has SEMAN [1.112a]. The Petzval coefficient is also the suject of a number of papers [1.15a,24a,58,58a,113].

Distortions. Just as the spherical aberration is the most important aberration at the objective, where the ray gradients x_o' and y_o' are comparatively large and the ray coordinates x_o and y_o small in a microscope system, the distortions are the most important in projectors where the converse is the case. It is therefore the asymptotic formulae for E and e that are commonly required, and we return to these in Sect.1.4.2. We see from (1.119,120,122,123) that the formulae for E and F, e and f are identical in form except that g and h change places. The general formula (1.146) can therefore be used, except that integrated terms that vanish for F persist in E. These are as follows: to obtain the general formula for E from (1.146), we must interchange g and h and replace

$$- s[gh'^3]_{z_o}^{z_i}$$

by

$$-rg_i'^2 - t\left[\frac{\eta^2B^2}{U}g^2\right]_{z_o}^{z_i} \quad .$$

The anisotropic coefficient e is given by

$$e = \frac{1}{16}\int\left(\frac{\eta B''}{U^{1/2}} + \frac{2\eta^3B^3}{U^{3/2}}\right)g^2\,dz + \left[\frac{\eta B}{4U^{1/2}}gg' - \frac{\eta B'}{8U^{1/2}}g^2\right]_{z_o}^{z_i} \quad . \tag{1.150}$$

As their name indicates, distortions do not blur the image but destroy the simple linear relation between object and image coordinates:

$$\begin{aligned} x_i &= M[x_o + Ex_o(x_o^2 + y_o^2) - ey_o(x_o^2 + y_o^2)] \\ y_i &= M[y_o + Ey_o(x_o^2 + y_o^2) + ex_o(x_o^2 + y_o^2)] \quad . \end{aligned} \tag{1.151}$$

The isotropic coefficient E produces the barrel and cushion distortion familiar in light optics; the anisotropic term e creates a distortion with a noticeable twist, which Sturrock evocatively names "pocket-handkerchief distortion" (Fig.1.10).

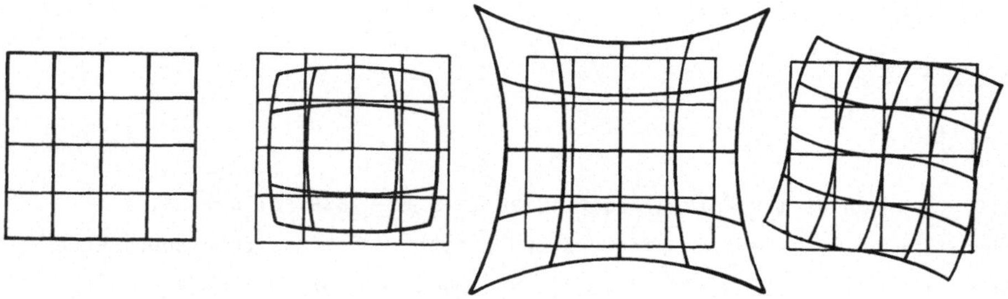

Fig.1.10. The effect of distortion on a pattern of squares. (a) Original pattern, (b) negative and (c) positive isotropic distortion, (d) anisotropic or "pocket-handkerchief" distortion

Distortions are studied in particular in the following papers [1.1,17,19,43, 63,64,71,74,75,83,90,137].

(b) *Chromatic Aberrations*

The ubiquitous presence of the relativistic accelerating voltage U suggests that electron lens properties will be very sensitive to fluctuations in electron beam energy. This is indeed the case, and for many years the chromatic aberration of the objective lens caused by the energy and hence wavelength spread of the incident beam was the major factor limiting electron microscope resolution. This spread was, furthermore, exacerbated by energy losses in the specimen. Nowadays, the accelerating voltage and lens currents are so highly stabilized that the resulting chromatic aberration is only perceptible at very high resolution, but energy loss in the specimen still has an effect unless the latter is sufficiently thin.

The chromatic aberration coefficients are calculated with the aid of (1.85), but instead of perturbing $m^{(2)}$ to $m^{(2)} + m^{(4)}$, we write $m^{(2)} \rightarrow m^{(2)} + m^{(c)}$, where

$$m^{(c)} = \frac{\partial m^{(2)}}{\partial U}\,\Delta U + \frac{\partial m^{(2)}}{\partial B_0}\,\Delta B_0 \quad , \tag{1.152}$$

in which B_0 is the maximum field strength, $B(z) = B_0 b(z)$.
From (1.20b) we have

$$m^{(2)} = \frac{1}{2}(X'^2 + Y'^2) - \frac{\eta B}{2U^{1/2}}(XY' - X'Y) \tag{1.153}$$

so that

$$\frac{\partial m^{(2)}}{\partial U} = \frac{\eta B}{4U^{3/2}}(XY' - X'Y)$$

$$\frac{\partial m^{(2)}}{\partial B_0} = -\frac{\eta b}{2U^{1/2}}(XY' - X'Y) \quad . \tag{1.154}$$

Introducing the rotating coordinates (x,y) we find that

$$\begin{aligned} m^{(c)} &= \frac{\eta B}{4U^{3/2}}\left[(xy' - x'y) + \frac{\eta B}{2U^{1/2}}(x^2 + y^2)\right]\Delta U \\ &\quad - \frac{\eta B}{2U^{1/2}}\left[(xy' - x'y) + \frac{\eta B}{2U^{1/2}}(x^2 + y^2)\right]\frac{\Delta B_0}{B_0} \quad , \end{aligned} \tag{1.155}$$

and with $x = x_0 g + x_0' h$, $y = y_0 g + y_0' h$, we obtain

$$\frac{\partial V^{(c)}}{\partial x_0'} = \int\left[\frac{\eta B}{4U^{1/2}}(-y_0) + \frac{\eta^2 B^2}{4U}(x_0 gh + x_0' h^2)\right]\frac{\Delta U}{U}\,dz$$

$$- \int \left[\frac{\eta B}{2U^{1/2}}(-y_o) + \frac{\eta^2 B^2}{2U}(x_o gh + x_o' h^2)\right] \frac{\Delta B_o}{B_o}\, dz$$

$$\frac{\partial V^{(c)}}{\partial y_o'} = \int \left[\frac{\eta B}{4U^{1/2}} x_o + \frac{\eta^2 B^2}{4U}(y_o gh + y_o'\, h^2)\right] \frac{\Delta U}{U}\, dz$$

$$- \int \left[\frac{\eta B}{2U^{1/2}} x_o + \frac{\eta^2 B^2}{2U}(y_o gh + y_o' h^2)\right] \frac{\Delta B_o}{B_o}\, dz \quad . \tag{1.156}$$

Hence

$$x^{(c)}/M = \frac{\Delta U}{U}\left(- x_o \int \frac{\eta^2 B^2}{4U}\, gh\, dz - x_o' \int \frac{\eta^2 B^2}{4U}\, h^2 dz + y_o \int \frac{\eta B}{4U^{1/2}}\, dz\right)$$

$$- \frac{\Delta B_o}{B_o}\left(- x_o \int \frac{\eta^2 B^2}{2U}\, gh\, dz - x_o' \int \frac{\eta^2 B^2}{2U}\, h^2 dz + y_o \int \frac{\eta B}{2U^{1/2}}\, dz\right)$$

or

$$x^{(c)}/M = x_o\left(\frac{\eta^2}{4U}\right) \int B^2 gh\, dz \left(\frac{2\Delta B_o}{B_o} - \frac{\Delta U}{U}\right)$$

$$+ x_o' \left(\frac{\eta^2}{4U}\right) \int B^2 h^2\, dz \left(\frac{2\Delta B_o}{B_o} - \frac{\Delta U}{U}\right)$$

$$- y_o \left(\frac{\eta}{4U^{1/2}}\right) \int B\, dz \left(\frac{2\Delta B_o}{B_o} - \frac{\Delta U}{U}\right) \quad , \tag{1.157a}$$

and

$$y^{(c)}/M = y_o\left(\frac{\eta^2}{4U}\right) \int B^2 gh\, dz \left(\frac{2\Delta B_o}{B_o} - \frac{\Delta U}{U}\right)$$

$$+ y_o'\left(\frac{\eta^2}{4U}\right) \int B^2 h^2\, dz \left(\frac{2\Delta B_o}{B_o} - \frac{\Delta U}{U}\right)$$

$$+ x_o\left(\frac{\eta}{4U^{1/2}}\right) \int B\, dz \left(\frac{2\Delta B_o}{B_o} - \frac{\Delta U}{U}\right) \quad . \tag{1.157b}$$

We write

$$x^{(c)} = - M(C_c x_o' + C_D x_o - C_\theta y_o)\left(\frac{\Delta U}{U} - 2\,\frac{\Delta B_o}{B_o}\right)$$

$$y^{(c)} = - M(C_c y_o' + C_D y_o + C_\theta x_o)\left(\frac{\Delta U}{U} - 2\,\frac{\Delta B_o}{B_o}\right) \quad , \tag{1.158}$$

and so

$$C_c = \frac{\eta^2}{4U} \int B^2 h^2\, dz$$

$$C_D = \frac{\eta^2}{4U} \int B^2 gh\, dz$$

$$C_\theta = \frac{\eta}{4U^{1/2}} \int B \, dz \quad . \tag{1.159}$$

The signs in (1.158) have been chosen so that the coefficients are not preceded by minus signs. Since ΔU and ΔB_0 are fluctuations rather than steady drifts, the sign is of little importance.

Of these, C_c is the chromatic aberration coefficient of greatest importance for objective lenses since it does not vanish for points on the axis. The chromatic aberration of magnification C_D and the anisotropic chromatic aberration C_θ are of particular interest in intermediate and projector lenses. Clearly, C_c never vanishes since the integrand is positive definite; C_θ is equal to half the image rotation. Partial integration gives the following alternative expressions:

$$C_c = - \int hh'' \, dz = \int h'^2 \, dz$$

$$C_D = - \int hg'' \, dz = \int g'h' \, dz \quad . \tag{1.160}$$

1.4.2 Asymptotic (Projector) Aberration Coefficients

a) *Geometrical Aberrations*

Asymptotic aberrations are considerably easier than their real counterparts to classify and handle since they are characterized, for all object positions, by a limited number of coefficients. As we shall see, each coefficient can be written as a polynomial of at most fourth degree in the reciprocal magnification $(1/M)$.

The first thorough study of asymptotic aberrations was made by LENZ [1.68,69] although earlier workers had not been unaware of the distinction (for example [1.112] and [Ref.1.139, p.563]). This work was continued [1.37,38] and extended to quadrupole lenses [1.35,37,39] by the present author, who also established the polynomial representation for the individual coefficients [1.37]; this latter stage was inspired by the work of VERSTER [1.133].

These polynomial expression can be established in various ways. Before investigating them in detail, however, we point out that the general form of the aberration coefficients leads us to expect them. Each aberration coefficient is given by a formula similar to those of (1.83) except that G and H (1.88) replace g and h. $G(z)$ is independent of object position and $H(z)$ is defined by

$$\lim_{z \to -\infty} H(z) = z - z_o \quad . \tag{1.161}$$

H may be written as a linear combination of G and $\bar{G}$ (1.42):

$$H(z) = -(z_o - z_{Fo})G(z) + \bar{G}(z)/\bar{G}'_o \quad , \tag{1.162}$$

where

$$\lim_{z \to -\infty} \bar{G}'(z) = \bar{G}'_o \quad . \tag{1.163}$$

Furthermore, $z_o - z_{Fo} = f/M$ (1.70) so that integrals of the form $\int f(z)G^jH^k\,dz$, $j + k \leq 4$, can be transformed into sums of integrals each of the form $(1/M)^k \int f(z)G^j\bar{G}^k\,dz$, $j + k \leq 4$. The rays G and $\bar{G}$ are both independent of object position and hence of magnification so that each coefficient can be written as a polynomial in 1/M.

Although we could derive these polynomials by suitably modifying the formulae for the real aberrations, it is more instructive to use rays such as G and $\bar{G}$ that are independent of object position form the outset. Since

$$\lim_{z\to\infty} G(z) = -(z - z_{Fi})/f$$

$$\lim_{z\to-\infty} \bar{G}(z) = (z - z_{Fo})/f \quad , \tag{1.164}$$

we consider the aberrations between the object and image asymptotic foci. If the object plane coincides with the object focus, we have $H(z) = f\bar{G}(z)$.

In the image focal plane,

$$x^{(3)}(z_{Fi}) = f\frac{\partial V_F}{\partial x_o} - \frac{1}{2} fUx_o'(x_o'^2 + y_o'^2)$$

$$y^{(3)}(z_{Fi}) = f\frac{\partial V_F}{\partial y_o} - \frac{1}{2} fUy_o'(x_o'^2 + y_o'^2)$$

$$x'^{(3)}(z_{Fi}) = \frac{1}{f}\frac{\partial V_F}{\partial x_o'} - \frac{1}{2}\frac{x_o}{f^3}(x_o^2 + y_o^2)$$

$$y'^{(3)}(z_{Fi}) = \frac{1}{f}\frac{\partial V_F}{\partial y_o'} - \frac{1}{2}\frac{y_o}{f^3}(x_o^2 + y_o^2) \quad , \tag{1.165}$$

in which V_F denotes $\int_{-\infty}^{\infty} m^{(4)}\,dz$ when x and y have been replaced by

$$x(z) = x_oG(z) + fx_o'\bar{G}(z)$$

$$y(z) = y_oG(z) + fy_o'\bar{G}(z) \quad . \tag{1.166}$$

The aberrations of slope, $x'^{(3)}$ and $y'^{(3)}$, are obtained from (1.84). It is necessary to distinguish carefully between $(x'^{(3)}, y'^{(3)})$ and $(p^{(3)}, q^{(3)})$; see [1.37] for a full discussion of these questions.

We write

$$V_F = \begin{pmatrix} r \\ s \\ u \end{pmatrix} \begin{pmatrix} 2000 & 1100 & 1010 & 1001 \\ 0 & 0200 & 0110 & 0101 \\ 0 & 0 & 0020 & 0011 \end{pmatrix} \begin{pmatrix} r \\ s \\ u \\ v \end{pmatrix} \quad , \tag{1.167}$$

in which

$$r = (x_o^2 + y_o^2)/f^2$$

$$s = x_o'^2 + y_o'^2$$

$$u = (x_o x_o' + y_o y_o')/f$$

$$v = (x_o y_o' - x_o' y_o)/f \quad . \tag{1.168}$$

The matrix elements (pqrs) are obtained by substituting (1.166) for x(z) and y(z) into the general formula (1.109); the resulting forms are given in [Ref.1.38, p.217]. These can be simplified considerably by partial integration, and it is these compact expressions that we now list.

$$(2000) = f^4 \int S_1 G^4 \, dz$$

$$(0200) = f^4 \int S_1 \bar{G}^4 \, dz$$

$$(0020) = 4f^4 \int (S_1 G^2 \bar{G}^2 - 2S_2/3f^2) \, dz$$

$$(1100) = 2f^4 \int (S_1 G^2 \bar{G}^2 + 4S_2/3f^2) \, dz$$

$$(1010) = 4f^4 \int S_1 G^3 \bar{G} \, dz + f/8$$

$$(0110) = 4f^4 \int S_1 G \bar{G}^3 \, dz + f/8$$

$$(1001) = f^3 \int S_3 G^2 \, dz$$

$$(0101) = f^3 \int S_3 \bar{G}^2 \, dz$$

$$(0011) = 2f^3 \int S_3 G \bar{G} \, dz \quad , \tag{1.169}$$

where

$$S_1 = \frac{1}{192} \left(\frac{\eta^2 BB''}{U} - 4 \frac{\eta^4 B^4}{U^2} - 5 \frac{\eta^2 B'^2}{U} \right)$$

$$S_2 = -\frac{1}{16} \frac{\eta^2 B^2}{U}$$

$$S_3 = -\frac{1}{16} \left(\frac{\eta B''}{U^{1/2}} + 2 \frac{\eta^3 B^3}{U^{3/2}} \right) \quad . \tag{1.170}$$

After evaluating $x^{(3)}(z_{Fi})$ and $y^{(3)}(z_{Fi})$ and the corresponding aberrations of gradient with the aid of (1.89), we transfer to an arbitrary pair of conjugate planes by writing

$$x(z_o) = x(z_{Fo}) + f x_o'/M$$

$$y(z_o) = y(z_{Fo}) + f y_o'/M$$

$$x(z_i) = x(z_{Fi}) - f x_i' M$$

$$y(z_i) = y(z_{Fi}) - f y_i' M \quad . \tag{1.171}$$

This leads to the following expressions for the aberrations:

$$\frac{x_i - Mx_o}{M} = (x_o' \quad x_o \quad -y_o)A\underline{r}$$

$$\frac{y_i - My_o}{M} = (y_o' \quad y_o \quad x_o)A\underline{r} \quad , \tag{1.172}$$

where as before, (1.114-115),

$$\underline{r} = \begin{pmatrix} x_o'^2 + y_o'^2 \\ x_o x_o' + y_o y_o' \\ x_o^2 + y_o^2 \\ x_o y_o' - x_o' y_o \end{pmatrix} \tag{1.173}$$

and[2]

$$A = \begin{pmatrix} B & 2F & D & 2\varphi \\ F & 2C & E & c \\ \varphi & c & e & 0 \end{pmatrix} \quad . \tag{1.174}$$

The asymptotic aberration coefficients B, C, D, E, F, c, e, and φ are now given by the following formulae.

Spherical aberration

$$B = -4(2000)m^4 + [4(1010) - f/2]m^3 - 4[(1100) + (0020)]m^2 + [4(0110) - f/2]m - 4(0200) \quad . \tag{1.175}$$

Isotropic astigmatism and field curvature

$$f^2C = -4(2000)m^2 + [2(1010) - f/2]m - (0020) \tag{1.176a}$$

$$f^2D = -4(2000)m^2 + [2(1010) - f/2]m - 2(1100) \quad . \tag{1.176b}$$

Distortion

$$f^3E = 4(2000)m + f/2 - (1010) \quad . \tag{1.177}$$

Isotropic coma

$$fF = 4(2000)m^3 - [3(1010) - f/2]m^2 + 2[(1100) + (0020)]m - (0110) \quad . \tag{1.178}$$

---

2 In order to avoid confusion between f (anistropic coma) and f (focal length), we denote the former by φ.

Anisotropic astigmatism

$$f^2c = 2(1001)m - (0011) \quad . \tag{1.179}$$

Anisotropic distortion

$$f^3e = -(1001) \quad . \tag{1.180}$$

Anisotropic coma

$$f\varphi = -(1001)m^2 + (0011)m - (0101) \quad . \tag{1.181}$$

We have written

$$m = 1/M \quad . \tag{1.182}$$

Similar expressions may be obtained for the aberrations of ray gradient:

$$x'(z_i) = -\frac{x_o}{f} + \frac{x'_o}{M} + \frac{1}{f}\left(x'_o \quad x_o \quad \left(\frac{y_o}{f} - \frac{y'_o}{M}\right)\right)\bar{A}\underline{r}$$

$$y'(z_i) = -\frac{y_o}{f} + \frac{y'_o}{M} + \frac{1}{f}\left(y'_o \quad y_o \quad \left(\frac{x_o}{f} - \frac{x'_o}{M}\right)\right)\bar{A}\underline{r} \quad , \tag{1.183}$$

where

$$\bar{A} = \begin{pmatrix} \bar{B} & 2\bar{F}' & \bar{D} & 2\bar{\varphi} \\ \bar{F} & 2\bar{C} & \bar{E} & \bar{c} \\ \bar{\varphi}' & \bar{c}' & \bar{e} & 0 \end{pmatrix} \tag{1.184}$$

and

$$\bar{B} = -[(1010) - f/2]m^3 + 2[(0020) + (1100)]m^2$$
$$- 3(0110)m + 4(0200)$$
$$f^2\bar{C} = -[(1010) - f/2]m + (0020)$$
$$f^2\bar{D} = -[(1010) - f/2]m + 2(1100)$$
$$f^3\bar{E} = (1010) - f/2$$
$$f\bar{F} = [(1010) - f/2]m^2 - 2(0020)m + (0110)$$
$$f\bar{F}' = [(1010) - f/2]m^2 - [(0020) + 2(1100)]m + (0110)$$
$$f^2\bar{c} = (0011) \qquad f^2\bar{e} = (1001)$$
$$f\bar{\varphi} = -(0011)m + 2(0101) \qquad f\bar{c}' = -2(1001)m + (0011)$$
$$\bar{\varphi}' = (1001)m^2 - (0011)m + (0101) \quad . \tag{1.185}$$

Many general results can be derived with the aid of these formulae, which simplify further if the lens in question is symmetric, since for this case $B(z) = B(-z)$ and $G(z) = \bar{G}(-z)$. We then have

$$(2000) = (0200)$$
$$(1010) = (0110)$$
$$(1001) = (0101) \tag{1.186}$$

and the spherical aberration, for example, is characterized by only three independent coefficients. We return to these general expressions in Sect.1.4.3, where we obtain formulae for the aberration coefficients of a series of lenses.

Approximate thin-lens expressions have been established by DER-SHVARTS and MAKAROVA [1.17a][3] for the various geometrical aberration coefficients, which are accurate to within 10-15% if the lens is used to provide demagnification ($|M| \leq 1$). These coefficients correspond to the aberrations expressed in terms of object and aperture coefficients and hence vary with the position of the aperture. We reproduce these formulae for the case in which the aperture is situated in the centre of the gap, after which we indicate how values for other aperture positions can be obtained [Ref.1.21, Eq.115.4]. DER-SHVARTS and MAKAROVA expressed their main results in terms of aperture and image plane parameters, as is usual in the case of probe-forming and other demagnifying systems. This convention is followed here and the reader is referred to the original paper for full definitions. The suffix i attached to the coefficients is intended as a reminder of this.

Spherical aberration

$$B_i = V^3(1+|M|)p_B(x)\left\{\frac{1+|M|}{V}q_B(x)+1\right\} \quad ,$$

where

$$p_B(x) = \frac{2.46}{0.47+x} - 0.28$$

$$q_B(x) = 0.26x - 0.25 \quad .$$

Astigmatism

$$C_i = -\left(\frac{1+|M|}{V}\right)^2 p_C(x) - \frac{1+|M|}{V} q_C(x) \quad ,$$

where

$$p_C(x) = 1.65 + 0.39\sin[(x-1.96)\pi/3.12]$$

$$q_C(x) = 0.107 + \sin[(x-1)\pi/6]$$

$$c_i = \left(\frac{1+|M|}{V^3}\right)^{\frac{1}{2}}(1-|M|)\left[p_c(x) + \frac{1+|M|}{V} q_c(x)\right] \quad ,$$

where

$$p_c(x) = 0.246 + 0.071x$$

$$q_c(x) = 0.1 + 0.114(x - 0.48) \quad .$$

3 This is one of several papers devoted to this topic by these authors; full details are to be found in Appendix B. See also [1.41a].

Field curvature

$$D_i = 2.08 \frac{1 + |M|}{v} \quad .$$

Distortion

$$E_i = -\frac{1 - M^2}{v^4}\left[p_E(x) + \frac{1 + |M|}{v} q_E(x)\right] \quad ,$$

where

$$p_E(x) = 0.06 + 0.04(x - 1)^{1.7}$$

$$q_E(x) = 0.066 + 0.034 \sin[(x - 2.12)\pi/2.96]$$

$$e_i = -\left(\frac{1 + |M|}{v^5}\right)^{\frac{1}{2}}\left[p_e(x) + \left(\frac{1 + |M|}{v}\right)^{\frac{1}{2}} q_e(x)\right] \quad ,$$

where

$$p_e(x) = 0.34 + 0.071(x - 1)$$

$$q_e(x) = 0.36 + 0.25 \sin[(x - 2.6)\pi/4.4] \quad .$$

Coma

$$F_i = -(1 - M^2)p_F(x) \quad ,$$

where

$$p_F = \frac{0.16}{x - 0.2} + 0.4$$

$$\varphi_i = -(1 + |M|)^{3/2} v^{1/2}\left[p_\varphi(x) + \frac{1 + |M|}{v} q_\varphi(x)\right] \quad ,$$

where

$$p_\varphi = 0.72 - 0.0625(x + 0.6)$$

$$q_\varphi = 0.1 + 0.114(x - 0.48) \quad .$$

The lens geometry is characterized by x,

$$x = D/S \quad ,$$

which may take values between 0.5 and 5. $|M|$ denotes the modulus of the magnification, $0 \leq |M| \leq 1$. Finally, v denotes the distance between the image plane and the centre plane of the lens, divided by the gap S, $2 \leq v \leq 10$.

(b) *Chromatic Aberrations*

The asymptotic chromatic aberration coefficients may be written down directly by replacing the integrals from object to image in (1.159) by integrals from $-\infty$ to ∞ and replacing g(z) and h(z) by G(z) and H(z), respectively. In order to establish polynomial forms, we have only to substitute for H(z) with the aid of (1.162). We obtain

$$C_c = \frac{\eta^2}{4U} \int B^2H^2 \, dz$$

$$C_D = \frac{\eta^2}{4U} \int B^2GH \, dz$$

$$C_\theta = \frac{\eta}{4U^{1/2}} \int B \, dz \quad , \tag{1.187}$$

or

$$\begin{pmatrix} C_c \\ C_D \\ C_\theta \end{pmatrix} = \begin{pmatrix} f^2C_2 & -2f^2C_1 & f^2C_0 \\ 0 & -fC_2 & fC_1 \\ 0 & 0 & C_\theta \end{pmatrix} \begin{pmatrix} m^2 \\ m \\ 1 \end{pmatrix} \quad , \tag{1.188}$$

where

$$C_2 = \frac{\eta^2}{4U} \int B^2G^2 \, dz$$

$$C_1 = \frac{\eta^2}{4U} \int B^2G\bar{G} \, dz$$

$$C_0 = \frac{\eta^2}{4U} \int B^2\bar{G}^2 \, dz \quad . \tag{1.189}$$

1.4.3 Aberration Matrices and the Aberrations of Lens Combinations

We have seen in Sect.1.4.2 that the asymptotic aberration coefficients can be written as polynomials in reciprocal magnification m of at most fourth degree. It is of considerable interest to know how the coefficients of the terms in these polynomials for a lens combination are related to those of the individual lenses. These relations are most easily obtained by introducing the notion of aberration matrices, the third-order counterparts of the paraxial transfer matrices of Sect. 1.2.3.

The calculation can be performed in terms of the real (as opposed to complex) coordinates x_o, y_o, x_o', y_o' that we have used hitherto, but owing to the dimensions of the matrices involved, it is considerably more convenient to introduced complex coordinates at this point. Setting

$$w = x + iy$$

$$t = x' + iy' \quad , \tag{1.190}$$

we write the relation between coordinate position and slope in a pair of conjugate planes, z_m and z_o, as follows:

$$\underline{w}_m = \underline{M}\underline{w}_o \quad , \tag{1.191}$$

in which $\underline{w}_m$ and $\underline{w}_o$ are column vectors and $\underline{M}$ is a 10 × 10 matrix containing paraxial properties and aberration coefficients. (We use the suffix m rather than i to indicate that the magnification is M, since more than one image plane is in question.) The column vector $\underline{w}_o$ is given by

$$\underline{w}_o = \begin{pmatrix} w_o \\ t_o \\ w_o(x_o^2 + y_o^2) \\ t_o(x_o^2 + y_o^2) \\ w_o(x_o x_o' + y_o y_o') \\ t_o(x_o x_o' + y_o y_o') \\ w_o(x_o'^2 + y_o'^2) \\ t_o(x_o'^2 + y_o'^2) \\ w_o(x_o y_o' - x_o' y_o) \\ t_o(x_o y_o' - x_o' y_o) \end{pmatrix} \tag{1.192}$$

and the matrix $\underline{M}$ divides naturally into four block matrices:

$$\underline{M} = \begin{pmatrix} \underline{M}_1 & \underline{M}_2 \\ \underline{M}_3 & \underline{M}_4 \end{pmatrix} . \tag{1.193}$$

$\underline{M}_1$ is the 2 × 2 paraxial matrix

$$\underline{M}_1 = \begin{pmatrix} M & 0 \\ k & m \end{pmatrix} , \tag{1.194}$$

in which we have introduced the convergence k:

$$k = -1/f \quad , \tag{1.195}$$

and as usual, $m = 1/M$. The matrix $\underline{M}_2$ has two rows and eight columns, the first row containing the various position aberration coefficients, the second, the coefficients defining the aberrations of gradient. We write

$$\underline{M}_2 = \begin{pmatrix} Mm_{11} & Mm_{12} & Mm_{13} & Mm_{14} & Mm_{15} & Mm_{16} & Mm_{17} & Mm_{18} \\ m_{21} & m_{22} & m_{23} & m_{24} & m_{25} & m_{26} & m_{27} & m_{28} \end{pmatrix} \tag{1.196}$$

and

$$m_{11} = -k^3(E + ie) \quad , \qquad m_{15} = -k(F + i\varphi)$$

$$m_{12} = k^2 D \quad , \qquad m_{16} = B$$

$$m_{13} = k^2(2C + ic) \quad , \qquad m_{17} = k^2c$$
$$m_{14} = -2kF \quad , \qquad m_{18} = -2k\varphi \quad . \tag{1.197}$$

The block matrix $\underline{M}_3$ is of course null. Finally, $\underline{M}_4$, obtained from $\underline{M}_1$, contains all the information required for adding the aberrations of lenses in tandem. We find that

$$\underline{M}_4 = \begin{pmatrix} M^3 & 0 & 0 & 0 & 0 & 0 & 0 & 0 \\ kM^2 & M & 0 & 0 & 0 & 0 & 0 & 0 \\ kM^2 & 0 & M & 0 & 0 & 0 & 0 & 0 \\ k^2M & k & k & m & 0 & 0 & 0 & 0 \\ k^2M & 0 & 2k & 0 & m & 0 & 0 & 0 \\ k^3 & k^2m & 2k^2m & 2km^2 & km^2 & m^3 & 0 & 0 \\ 0 & 0 & 0 & 0 & 0 & 0 & M & 0 \\ 0 & 0 & 0 & 0 & 0 & 0 & k & m \end{pmatrix} . \tag{1.198}$$

Suppose now that we have two lenses characterized by matrices $\underline{M}$ and $\underline{M}'$. The elements of $\underline{M}'$ are denoted by the same symbols as those of $\underline{M}$, primes being added to those of $\underline{M}'$. Each "lens" may of course be a combination of several lenses, $\underline{M}$ or $\underline{M}'$ characterizing the combination. If we denote the total magnification by P,

$$P = MM' \quad , \tag{1.199}$$

we may write

$$\underline{w}_p = \underline{M}'\underline{w}_m = \underline{M}'\underline{M}\underline{w}_o = \underline{P}\underline{w}_o \quad . \tag{1.200}$$

The matrix $\underline{P}$ has the same structure as $\underline{M}$ or $\underline{M}'$:

$$\underline{P} = \begin{pmatrix} \underline{P}_1 & \underline{P}_2 \\ \underline{P}_3 & \underline{P}_4 \end{pmatrix} \quad , \tag{1.201}$$

where $\underline{P}_1$ is the paraxial transfer matrix for the doublet and $\underline{P}_2$ contains the aberrations. $\underline{P}_3$ is null and $\underline{P}_4$ is identical with $\underline{M}_4$ with the changes $m \rightarrow p$, $M \rightarrow P$, $k \rightarrow k_p$. The doublet convergence k_p can be read directly from the paraxial matrix $\underline{P}_1$:

$$\underline{P}_1 = \begin{pmatrix} P & 0 \\ k_p & p \end{pmatrix} = \begin{pmatrix} MM' & 0 \\ Mk' + km' & 1/MM' \end{pmatrix} \tag{1.202}$$

and so

$$\begin{aligned} k_p &= Mk' + km' \\ &= D_p kk' \quad , \end{aligned} \tag{1.203}$$

where D_p is the distance between the image focus of $\underline{M}$ and the object focus of $\underline{M}'$:

$$D_p = z'_{Fo} - z_{Fi} \quad . \tag{1.204}$$

The foci of the doublet are located at $z^{(p)}_{Fo}$, $z^{(p)}_{Fi}$, where, see (1.74,75),

$$z^{(p)}_{Fi} = z'_{Fi} + 1/D_p k'^2$$

$$z^{(p)}_{Fo} = z_{Fo} - 1/D_p k^2 \quad . \tag{1.205}$$

With a notation similar to that of (1.196) for the elements of $\underline{P}_2$,

$$\underline{P}_2 = \begin{pmatrix} Pp_{11} & Pp_{12} & Pp_{13} & Pp_{14} & Pp_{15} & Pp_{16} & Pp_{17} & Pp_{18} \\ p_{21} & p_{22} & p_{23} & p_{24} & p_{25} & p_{26} & p_{27} & p_{28} \end{pmatrix} \quad , \tag{1.206}$$

it is easy to see that

$$p_{11} = m_{11} + M^2 m'_{11} + Mk(m'_{12} + m'_{13}) + k^2(m'_{14} + m'_{15}) + k^3 mm'_{16}$$

$$p_{12} = m_{12} + m'_{12} + kmm'_{14} + k^2m^2m'_{16}$$

$$p_{13} = m_{13} + m'_{13} + km(m'_{14} + 2m'_{15}) + 2k^2m^2m'_{16}$$

$$p_{14} = m_{14} + m^2m'_{14} + 2km^3m'_{16}$$

$$p_{15} = m_{15} + m^2m'_{15} + km^3m'_{16}$$

$$p_{16} = m_{16} + m^4m'_{16}$$

$$p_{17} = m_{17} + m'_{17} + kmm'_{18}$$

$$p_{18} = m_{18} + m^2m'_{18} \quad . \tag{1.207}$$

Each coefficient has a characteristic dependence on magnification, as we have seen (1.175-181). We now introduce a more compact notation:

$$\kappa = -4(2000) \quad , \quad \nu = -4(0200) \quad , \quad \rho = (1001)$$

$$\lambda = -2(1100) \quad , \quad \xi = (0110) \quad , \quad \sigma = (0101)$$

$$\mu = (1010) \quad , \quad \pi = -2(0020) \quad , \quad \tau = -(0011) \quad , \tag{1.208}$$

so that

$$m_{11} = k^3(\kappa m + \mu + 1/2k + i\rho)$$

$$m_{12} = k^2[\kappa m^2 + (2\mu + 1/2k)m + \lambda]$$

$$m_{13} = k^2[2\kappa m^2 + 2(2\mu + 1/2k)m + \pi + i(2\rho m + \tau)]$$

$$m_{14} = 2k[\kappa m^3 + (3\mu + 1/2k)m^2 + (\lambda + \pi)m + \xi]$$

$$m_{15} = k[\kappa m^3 + (3\mu + 1/2k)m^2 + (\lambda + \pi)m + \xi + i(\rho m^2 + \tau m + \sigma)]$$

$$m_{16} = \kappa m^4 + (4\mu + 1/2k)m^3 + 2(\lambda + \pi)m^2 + (4\xi + 1/2k)m + \nu$$

$$m_{17} = k^2(2\rho m + \tau)$$

$$m_{18} = 2k(\rho m^2 + \tau m + \sigma) \qquad (1.209)$$

with similar expressions for $p_{11}, p_{12}, \ldots, p_{18}$ except that $k \rightarrow k_p$, $m \rightarrow p$, and the suffix p is added to each of the coefficients $\kappa, \lambda, \ldots, \sigma$. After some manipulation we find that $\kappa_p, \lambda_p, \ldots, \sigma_p$ are related to the coefficients of the individual components of the doublet by the following formulae:

$$\kappa_p = \kappa' + 4\mu'\bar{k} + 2(\lambda' + \pi')\bar{k}^2 + 4\xi'\bar{k}^3 + (\kappa + \nu')\bar{k}^4 + \bar{k}(1 + \bar{k}^2)/2k'$$

$$\lambda_p = \lambda'\bar{k}'^2 + \lambda\bar{k}^2 + 2\xi'\bar{k}\bar{k}'^2 + 2\mu\bar{k}^2\bar{k}' + (\kappa + \nu')\bar{k}^2\bar{k}'^2 + \bar{k}\bar{k}'/2k_p$$

$$\mu_p = \mu'\bar{k}' + (\lambda' + \pi')\bar{k}\bar{k}' + 3\xi'\bar{k}^2\bar{k}' + \mu\bar{k}^3 + (\kappa + \nu')\bar{k}^3\bar{k}' + \bar{k}^2/2k_p$$

$$\nu_p = \nu + 4\xi\bar{k}' + 2(\lambda + \pi)\bar{k}'^2 + 4\mu\bar{k}'^3 + (\kappa + \nu')\bar{k}'^4 + \bar{k}'(1 + \bar{k}'^2)/2k$$

$$\xi_p = \xi\bar{k} + (\lambda + \pi)\bar{k}\bar{k}' + \xi'\bar{k}'^3 + 3\mu\bar{k}\bar{k}'^2 + (\kappa + \nu')\bar{k}\bar{k}'^3 + \bar{k}'^2/2k_p$$

$$\pi_p = \pi'\bar{k}'^2 + \pi\bar{k}^2 + 4\xi'\bar{k}\bar{k}'^2 + 4\mu\bar{k}^2\bar{k}' + 2(\kappa + \nu')\bar{k}^2\bar{k}'^2 + \bar{k}\bar{k}'/k_p$$

$$\rho_p = (\rho\bar{k} + \bar{\sigma}\bar{k}')\bar{k}^2 + \tau'\bar{k}\bar{k}' + \rho'\bar{k}'$$

$$\tau_p = 2(\rho\bar{k} + \sigma'\bar{k}')\bar{k}\bar{k}' + \tau'\bar{k}'^2 + \tau\bar{k}^2$$

$$\sigma_p = (\rho\bar{k} + \sigma'\bar{k}')\bar{k}'^2 + \tau\bar{k}\bar{k}' + \sigma\bar{k} \quad , \qquad (1.210)$$

where

$$\bar{k} = k/k_p \quad , \quad \bar{k}' = k'/k_p \quad .$$

The foregoing results were obtained by HAWKES [1.38,40]. They have been applied to practical lens systems by MacLACHLAN [1.72,73], some of the results being reported in [1.41].

1.5 Parasitic Aberrations

In practice, electron lenses are very liable to suffer not only from the geometrical and chromatic aberrations discussed above but also from a variety of parasitic aberrations caused by imperfections in construction or alignment, magnetic inhomogeneities, and any other external influences that cause departures from perfect rotational symmetry about the optic axis. It is very difficult to establish a general theory of such effects, and in practice, general conclusions about the types of aberration that each type of imperfection is most likely to generate are often more useful than formulae. Thus we will merely comment briefly on the effect of such parasitic aberrations and give a representative selection of references dealing with them. Further comment is also to be found in the chapter by Riecke in the present volume.

The types of departure from rotational symmetry are legion: tilts and other asymmetries in the pole pieces, imperfections in the magnetic circuit, shifts and tilts of the entire lens relative to some optic axis defined by other components in the system, ellipticity of the bore – this by no means exhausts the possibilities. Fortunately, all these affect the field symmetry in ways that can be described by means of comparatively few coefficients so that the number of essentially different kinds of parasitic aberration is much smaller. Provided that the cause of the aberrations is not too serious, their effect can be cancelled or reduced to a harmless level by a suitable alignment procedure [1.96,97] and a stigmator, which may be quite complex if a high degree of correction is required. This can readily be understood once we realise that one of the purposes of alignment procedures is to minimize the harmful effect of lateral misalignments, which introduce a slight curvature into the optic axis. Moreover, if we examine the angular dependence of the field asymmetry, we see that the dominant defect will normally consist of a weak quadrupole component, which can be directly cancelled by means of a suitably orientated weak quadrupole: this is the simplest type of stigmator.

For further discussion of these questions in general, we refer to [1.2,5-8, 12-14,16,24,27,30-32,42,45,47-49,51,59,70,77-79,94-97,100,106,114,115,121,123, 136,138]. Stigmators are specifically dealt with in [1.4,29,44,50,52-54,56a, 60,61,65,66,80,91-93,109,116-120,135]. Some recent attempts to calculate or measure the relation between pole piece displacement or imperfect roundness of pole piece or field is to be found in [1.45a,85-86]; of the earlier papers, those by STURROCK [1.123], ARCHARD [1.2], GLASER and SCHISKE [1.24], MEYER [1.77,78], and HAHN [1.32] are particularly relevant; for a survey see [1.97]. (A number of these references are concerned with electrostatic lenses, but are nevertheless germane.)

References

1.1 A. Alshwaikh, T. Mulvey: In *Developments in Electron Microscopy and Analysis, 1977*, ed. by D.L. Misell (Institute of Physics, Bristol 1977) pp.25-28

1.2 G.D. Archard: J. Sci. Instrum. *30*, 352-358 (1953)

1.3 G.D. Archard: Brit. J. Appl. Phys. *11*, 521-522 (1960)

1.4 E. Bärnighausen: Optik *24*, 10-17 (1966/1967)

1.4a H. Becker, A. Wallraff: Arch. Elektrotech. *32*, 664-675 (1938)

1.4b H. Becker, A. Wallraff: Arch. Elektrotech. *33*, 491-505 (1939)

1.4c H. Becker, A. Wallraff: Arch. Elektrotech. *34*, 43-48 (1940)

1.5 F. Bertein: C.R. Acad. Sci. Paris *224*, 106-107, 560-562, 737-739 (1947); *225*, 801-803, 863-865 (1947); J. Phys. Radium *9*, 104-112 (1948)
1.6 F. Bertein: Ann. Radioélectr. *2*, 379-408 (1947); *3*, 49-62 (1948)
1.7 F. Bertein, E. Regenstreif: C.R. Acad. Sci. Paris *228*, 1854-1856 (1949)
1.8 F. Bertein, H. Bruck, P. Grivet: Ann. Radioélectr. *2*, 249-252 (1947)
1.9 B. von Borries: *Die Übermikroskopie: Einführung, Untersuchung ihrer Grenzen und Abriss ihrer Ergebnisse* (Editio Cantor, Aulendorf/Württ. 1949)
1.10 L. de Broglie: *Optique électronique et corpusculaire* (Hermann, Paris 1950)
1.11 E. Brüche, O. Scherzer: *Geometrische Elektronenoptik* (Springer, Berlin 1934); continued by E. Brüche, W. Henneberg: Ergeb. Exakten Naturwiss. *15*, 365-421 (1936)
1.12 H. Bruck: C.R. Acad. Sci. Paris *224*, 1628-1629, 1818-1820 (1947)
1.13 H. Bruck, P. Grivet: C.R. Acad. Sci. Paris *224*, 1768-1769 (1947); Rev. Opt. *29*, 164-170 (1950)
1.14 H. Bruck, R. Remillon, L. Romani: C.R. Acad. Sci. Paris *226*, 650-652 (1948)
1.15 H. Busch, E. Brüche (eds.): *Beiträge zur Elektronenoptik* (Barth, Leipzig 1937)
1.15a Chiang Man-Ying: Acta Phys. Sinica *12*, 439-446 (1956)
1.16 M. Cotte: C.R. Acad. Sci. Paris *228*, 377-378 (1949) and *Comptes Rendus du 1er Congrès International de Microscopie Electronique, Paris, 1950* (Edns Revue d'Optique, Paris 1953) pp.155-157
1.17 M.L. De, D.K. Saha: Indian J. Phys. *28*, 263-268 (1954)
1.17a G.V. Der-Shvarts, I.S. Makarova: Izv. Akad. Nauk SSSR (Ser. Fiz.) *36*, 1304-1311 (1972) [English transl.: Bull. Acad. Sci. USSR (Phys. Ser.) *36*, 1164-1174 (1972)]; Radiotekh. Elektron. *18*, 2374-2378 (1973) [English transl.: Radio Eng. Electron. Phys. (USSR) *18*, 1722-1725 (1973)]
1.18 K.P. Dutova, P.P. Kas'yankov: Izv. Akad. Nauk SSSR (Ser. Fiz.) *27*, 1127-1130 (1963) [English transl.: Bull. Acad. Sci. USSR(Phys. Ser.) *27*, 1108-1111 1963)]
1.19 H.H. Elkamali, T. Mulvey: In *Developments in Electron Microscopy and Analysis 1977*, ed. by D.L. Misell (Institute of Physics, Bristol 1977) pp.33-34
1.20 P. Funk: Monatsh. Math. Phys. *43*, 305-316 (1936); *45*, 314-319 (1937)
1.21 W. Glaser: *Grundlagen der Elektronenoptik* (Springer, Vienna 1952)
1.22 W. Glaser: "Elektronen- und Ionenoptik", in *Korpuskularoptik*, Handbuch der Physik, Vol. XXXIII (Springer, Berlin, Göttingen, Heidelberg 1956) pp.123-395
1.23 W. Glaser, O. Bergmann: Z. Angew. Math. Phys. *1*, 363-379 (1950); *2*, 159-188 (1951)
1.24 W. Glaser, P. Schiske: Z. Angew. Phys. *5*, 329-339 (1953)
1.24a L.S. Goddard: Proc. Cambridge Philos. Soc. *42*, 127-131 (1946)
1.25 J. Gratsiatos: Z. Phys. *102*, 641-651 (1936)
1.26 G.A. Grinberg (Grünberg): C.R. (Dokl.) Acad. Sci. URSS *37*, 172-178, 261-268 (1942); *38*, 78-81 (1943); Zh. Tekh. Fiz. *13*, 361-388 (this is essentially a Russian translation of the articles in English in C.R. Acad. Sci. URSS); this material is reproduced with only minor changes in *Izbrannye Voprosy Matematicheskoi Teorii Elektricheskikhk i Magnitnykh Yavlenii* (Izd. Akad. Nauk SSSR, Moscow 1948) Part 5, pp.507-535
1.27 P. Grivet, F. Bertein, E. Regenstreif: *Proc. Conf. Electron Microscopy, Delft, 1949*, ed. by A.L. Houwink, J.B. Le Poole, W.A. Le Rütte (Hoogland printer, Delft 1950) pp.86-88
1.28 G.G. Gurbanov, P.P. Kas'yankov: Izv. Akad. Nauk SSSR (Ser. Fiz) *30*, 735-738 (1966) [English transl.: Bull Acad Sci USSR (Phys. Ser.) *30*, 762-765 (1966)]
1.29 E. Gütter: *Proc. 5th Int. Cong. Electron Microscopy, Philadelphia, 1962*, ed. by S.S. Breese (Academic, New York, London 1962) communication D-4
1.30 E. Hahn: Jenaer Jahrb. 63-75 (1954:I)
1.31 E. Hahn: Jenaer Jahrb. 86-114 (1959:I)
1.32 E. Hahn: Jenaer Jahrb. 145-172 (1966)
1.33 P.W. Hawkes: Phil. Trans. Roy. Soc. London Ser. A *257*, 523-552 (1965)
1.34 P.W. Hawkes: Optik *24*, 252-262 and 275-282 (1966/1967)
1.35 P.W. Hawkes: Optik *25*, 315-331 (1967)

1.36 P.W. Hawkes: J. Microscopie *6*, 917-932 (1967)
1.37 P.W. Hawkes: Optik *27*, 287-304 (1968)
1.37a P.W. Hawkes: J. Phys. D *1*, 131-133 (1968)
1.37b P.W. Hawkes: J. Microscopie *9*, 435-454 (1970; *Proc. 7th Int. Cong. Electron Microscopy, Grenoble, 1970*, ed. by P. Favard (SFME, Paris 1970) Vol.2, pp.17-18
1.38 P.W. Hawkes: Optik *31*, 213-219 (1970)
1.39 P.W. Hawkes: Optik *31*, 302-314 (1970); *32*, 50-60 (1970)
1.40 P.W. Hawkes: Optik *31*, 592-599 (1970)
1.41 P.W. Hawkes: "Computer-aided design of electron lens combinations" in *Image Processing and Computer-aided Design in Electron Optics*, ed. by P.W. Hawkes (Academic, London, New York 1973) pp.230-248
1.41a P.W. Hawkes: Optik *56*, 293-320 (1980)
1.42 J. Hillier: J. Appl. Phys. *17*, 307-309 (1946)
1.43 J. Hillier: J. Appl. Phys. *17*, 411-419 (1946)
1.44 J. Hillier, E.G. Ramberg: J. Appl. Phys. *18*, 48-71 (1947)
1.45 Y. Inoue: Shimadzu Hyoron *5*, 196-200 (1948); *C.R. 1er Congrès Int. Microscopie électronique, Paris 1950* (Edns Revue d'Optique, Paris 1953) pp.199-200
1.45a J. Janse: Optik *33*, 270-281 (1971)
1.46 K. Kanaya: Researches of the Electrotechnical Laboratory, no. 495, 37 pp. (1949); Bull. Electrotech. Lab. *15*, 86-91, 91-94, 193-198, 199-202 (1951); *16*, 25-30, 135-142, 184-191 (1952)
1.47 K. Kanaya: J. Electron Microsc. *1*, 7-12 (1953); Researches of the Electrotechnical Laboratory no. 548, 70 pp. (1955); Bull. Electrotech. Lab. *22*, 615-622 (1958); *26*, 161-172 (1962)
1.48 K. Kanaya, A. Ishikawa: Bull. Electrotech. Lab. *22*, 641-646; J. Electron Microsc. *7*, 13-15 (1959)
1.49 K. Kanaya, A. Kato: Bull. Electrotech. Lab. *15*, 827-834 (1951)
1.50 K. Kanaya, H. Kawakatsu: J. Electron Microsc. *8*, 1-3 (1960); *9*, 71-73 (1960); *10*, 218-221 (1961); Bull. Electrotech. Lab. *24*, 721-732 (1960); *25*, 641-656 (1961); *26*, 241-250 (1962)
1.51 K. Kanaya, A. Kato, I. Yamaji: Bull. Electrotech. Lab. *33*, 281-298 (1958)
1.52 K. Kanaya, H. Kawakatsu, A. Kato: Bull. Electrotech. Lab. *23*, 431-440, 801-816 (1959)
1.53 K. Kanaya, H. Kawakatsu, K. Tanaka: Bull. Electrotech. Lab. *25*, 481-494 (1961)
1.54 K. Kanaya, H. Kawakatsu, S. Matsui: Bull. Electrotech. Lab. *26*, 881-896 (1962)
1.55 P.P. Kas'yankov: Zh. Tekh. Fiz. *20*, 1426-1434 (1950); *22*, 80-83 (1952)
1.56 P.P. Kas'yankov: *Teoriya Elektromagnitnykh Sistem s Krivolineinoi Os'yu* (Leningrad University Press, Leningrad 1956)
1.56a P.P. Kas'yankov: Izv. Akad. Nauk SSSR (Ser. Fiz.) *23*, 711-715 (1959) [English transl.: Bull. Acad. Sci. USSR (Phys. Ser.) *23*, 706-710 (1959)]
1.57 P.P. Kas'yankov: "[Methods of calculation for axially symmetric electron optical systems]" in *Chislennye Metody Rascheta Elektronno-opticheskikh Sistem*, ed. by G.I. Marchuk (Nauka, Novosibirsk 1967) pp.4-10
1.58 P.P. Kas'yankov, N.P. Rynkevich, N.S. Cheremisina: "[Systems with corrected Petzval curvature]" in *Metody Rascheta Elektronno-opticheskikh Sistem*, ed. by G.I. Marchuk (Novosibirsk 1970) pp.62-67
1.58a P.P. Kas'yankov, N.S. Cheremisina, N.P. Rynkevich: Opt.-Mekh. Prom. *37*, No. 11, 67 (1970) [English transl.: Sov. J. Opt. Technol. *37*, 757-758 (1970)]
1.59 S. Katagiri: J. Electron Microsc. *8*, 13-16 (1960); *9*, 119-120 (1960)
1.60 S. Katagiri, B. Tadano: *Proc. 5th Int. Cong. Electron Microscopy, Philadelphia, 1962*, ed. by S.S. Breese (Academic, New York, London 1962) communication KK-3
1.61 H. Kawakatsu, K. Kanaya: Bull. Electrotech. Lab. *25*, 801-814 (1961)
1.62 V.M. Kel'man, S.Ya. Yavor: *Elektronnaya Optika* (Izd. Akad. Nauk SSSR, Moscow 1959; 3rd. ed. Izd. Nauka, Leningrad 1968)
1.63 D. Kynaston, T. Mulvey: *Proc. 5th Int. Cong. Electron Microscopy, Philadelphia, 1962*, ed. by S.S. Breese (Academic, New York, London 1962) communication D-2; Brit. J. Appl. Phys. *14*, 199-206 (1963)

1.64 E. Lambrakis, F.Z. Marai, T. Mulvey: In *Developments in Electron Microscopy and Analysis, 1977*, ed. by D.L. Misell (Institute of Physics, Bristol 1977) pp.35-38
1.65 S. Leisegang: Optik *10*, 5-14 (1953); *11*, 49-60 (1954)
1.66 S. Leisegang: "Elektronenmikroskope", in *Korpuscularoptik*, Handbuch der Physik, Vol. XXXIII (Springer, Berlin, Göttingen, Heidelberg 1956) pp.396-545
1.67 F. Lenz: *Proc. 3rd Int. Conf. Electron Microscopy, London, 1954*, ed. by R. Ross (Royal Microscopical Society, London 1956) pp.86-88
1.68 F. Lenz: *Proc. Stockholm Conference on Electron Microscopy, 1956*, ed. by F.J. Sjöstrand, J. Rhodin (Almqvist and Wiksell, Stockholm 1957) pp.48-51
1.69 F. Lenz: Optik *14*, 74-82 (1957)
1.70 F. Lenz, M. Hahn: Optik *10*, 15-27 (1953)
1.71 G. Liebmann: Proc. Phys. Soc. B *65*, 94-108 (1952)
1.72 M.E.C. Maclachlan: In *Electron Microscopy and Analysis*, ed. by W.C. Nixon (Institute of Physics, London 1971) pp.98-99
1.73 M.E.C. Maclachlan, P.W. Hawkes: *Proc. 7th Int. Cong. Electron Microscopy, Grenoble, 1970*, ed. by P. Favard (SFME, Paris 1970) Vol.2, pp.23-24
1.74 F.Z. Marai, T. Mulvey: In *Developments in Electron Microscopy and Analysis*, ed. by J.A. Venables (Academic, London, New York 1976) pp.43-44
1.75 F.Z. Marai, T. Mulvey: Ultramicroscopy *2*, 187-192 (1977)
1.76 W.E. Meyer: Optik *13*, 86-91 (1956)
1.77 W.E. Meyer: Optik *18*, 69-91 (1961)
1.78 W.E. Meyer: Optik *18*, 101-114 (1961)
1.79 N. Morito: Hitachi Hyoron *37*, 817-822 (1955)
1.80 N. Morito, K. Koizumi: Denshikenbikyo [Electron Microscopy] *2*, 84-86 (1952)
1.81 R.W. Moses: *Proc. 5th Eur. Cong. Electron Microscopy, Manchester, 1972* (Institute of Physics, London 1972) pp.86-87
1.82 R.W. Moses: "Lens optimization by direct application of the calculus of variations", in *Image Processing and Computer-aided Design in Electron Optics*, ed. by P.W. Hawkes (Academic, London, New York 1973) pp.250-272
1.83 T. Mulvey, L. Jacob: Nature *163*, 525-526 (1949)
1.84 T. Mulvey, M.J. Wallington: Repts. Prog. Phys. *36*, 347-421 (1973)
1.85 E. Munro: "Computer-aided Design Methods in Electron Optics"; Dissertation, Cambridge, 1971
1.86 R. Murillo: "Contribution à l'étude des lentilles magnétiques utilisées en microscopie électronique à très haute tension"; Thèse, Toulouse (1978)
1.87 L.M. Myers: *Electron Optics, theoretical and practical* (Chapman and Hall, London 1939)
1.88 D.P.R. Petrie: *Proc. 5th Int. Cong. Electron Microscopy, Philadelphia, 1962*, ed. by S.S. Breese (Academic, New York, London 1962) communication KK-2
1.89 E.G. Ramberg: J. Opt. Soc. Am. *29*, 79-83 (1939)
1.90 O. Rang: Optik *4*, 251-257 (1948)
1.91 O. Rang: Phys. Bl. *5*, 78-80 (1949)
1.92 O. Rang: Optik *5*, 518-530 (1949)
1.93 A. Recknagel, G. Haufe: Wiss. Z. Tech. Hochsch. Dresden *2*, 1-10 (1952/1953)
1.94 E. Regenstreif: Ann. Radioélectr. *6*, 244-267, 299-317 (1951); C.R. Acad. Sci. Paris *232*, 1918-1920 (1951); *233*, 854-856 (1951)
1.95 W.D. Riecke: *Proc. 3rd. Reg. Conf. Electron Microscopy, Prague, 1964*, ed. by M. Titlbach (Czechoslovak Academy of Sciences, Prague 1964) Vol.A, pp.7-8
1.96 W.D. Riecke: Optik *24*, 397-426 (1966/1967); *36*, 66-84, 288-308, 375-398 (1972)
1.97 W.D. Riecke: "Instrument operation for microscopy and microdiffraction", in *Electron Microscopy in Materials Science*, ed. by U. Valdrè, E. Ruedl (Commission of the European Communities, Luxemburg 1976) pp.19-111
1.98 W. Rogowski: Arch. Elektrotech. *31*, 555-593 (1937)
1.99 H. Rose: Optik *24*, 36-59, 108-121 (1966/1967)
1.100 H. Rose: Optik *27*, 466-474, 497-514 (1968)
1.101 H. Rose: Optik *28*, 462-474 (1968/1969)

1.102 H. Rose: Optik *33*, 1-24 (1971)
1.103 H. Rose: Optik *34*, 285-311 (1971)
1.104 H. Rose, R.W. Moses: Optik *37*, 316-336 (1973)
1.105 H. Rose, U. Petri: Optik *33*, 151-165 (1971)
1.106 Y. Sakaki, S. Maruse: J. Electron Microsc. *2*, 8-9 (1954)
1.107 O. Scherzer: Z. Physik *80*, 193-202 (1933)
1.108 O. Scherzer: Z. Physik *101*, 593-603 (1936)
1.109 O. Scherzer: Phys. Bl. *2*, 110 (1946
1.110 O. Scherzer: Optik *2*, 114-132 (1947)
1.111 O.I. Seman: Trudy Inst. Fiz. Astron. Akad. Nauk Est. SSR No.2, 3-29, 30-49 (1955); Uchenye Zapiski Rostov.-na-Donu Gos. Univ. (Ser. Fiz.) *68*, No.8, 77-90 (1958)
1.112 O.I. Seman: Uchenye Zapiski Rostov.-na-Donu Gos. Univ. (Ser. Fiz.) *68*, No.8, 63-75 (1958)
1.112a O.I. Seman: Opt. Spektrosk. *7*, 113-115 (1959) [English transl.: Opt. Spectrosc. *7*, 68-69 (1959)]
1.113 O.I. Seman: Radiotekh. Elektron. *13*, 907-912 (1968) [English transl.: Radio Eng. Electron. Phys. (USSR) *13*, 784-788 (1968)]
1.114 S. Shirai: Denshikenbikyo [Electron Microscopy] *8*, 93-98 (1959)
1.115 Si-Men Gi-ie, Xi Zhung-xo: Acta Electron. Sinica *3*, No.9, 24-35 (1964)
1.116 P.A. Stoyanov: Zh. Tekh. Fiz. *25*, 2537-2541 (1955)
1.117 P.A. Stojanow: *Verhandl. 4. Int. Kong. Elektronenmikroskopie, Berlin 1958*, ed. by W. Bargmann, G. Möllenstedt, H. Niehrs, D. Peters, E. Ruska, C. Wolpers (Springer, Berlin, Heidelberg, New York 1960) Vol.1, pp.61-66
1.118 P.A. Stoyanov: Opt.-Mekh. Prom. *25*, No.4, 40-49 (1958)
1.119 P.A. Stoyanov: Izv. Akad. Nauk SSSR (Ser. Fiz) *23*, 467-472 (1959) [English transl.: Bull. Acad. Sci. USSR (Phys. Ser.) *23*, 449-455 (1959)]
1.120 P.A. Stoyanov, E.A. Shulyak, S.F. Zelev, A.M. Klimovitskii, R.A. Grishin: Izv. Akad. Nauk. SSSR (Ser. Fiz) *32*, 1118-1119 (1968) [English transl.: Bull. Acad. Sci. USSR (Phys. Ser.) *32*, 1041-1042 (1968)]
1.121 P.A. Sturrock: *Proc. Conf. Electron Microscopy, Delft, 1949*, ed. by A.L. Houwink, J.B. Le Poole, W.A. Le Rütte (Hoogland, printer, Delft 1950) pp.89-93
1.122 P.A. Sturrock: Proc. Roy. Soc. London A *210*, 269-289 (1951)
1.123 P.A. Sturrock: Phil. Trans. Roy. Soc. London Ser. A *243*, 387-429 (1951)
1.124 P.A. Sturrock: Phil. Trans. Roy. Soc. London Ser. A *245*, 155-187 (1952)
1.125 P.A. Sturrock: *Static and Dynamic Electron Optics* (Cambridge University Press, Cambridge 1955)
1.126 I.N. Taganov, P.P. Kas'yankov: Opt.-Mekh. Prom. *31*, No.11, 14-16 (1964); *32*, No.12, 21-23 (1965); Radiotekh. Elektron. *11*, 1329-1330 (1966) [English transl.: Radio Eng. Electron. Phys. (USSR) *11*, 1160-1162 (1966)]
1.127 I.N. Taganov, P.P. Kas'yankov: [Use of the direct method for calculating electron optical systems with corrected third-order aberrations] in *Chislennye Metody Rascheta Elektronno-opticheskykh Sistem*, ed. by G.I. Marchuk (Nauka, Novosibirsk 1967) pp.11-22
1.128 W. Tretner: Optik *7*, 242 (1950); *11*, 312-326 (1954); *12*, 293 (1955)
1.129 W. Tretner: Optik *13*, 516-519 (1956)
1.130 W. Tretner: Optik *16*, 155-184 (1959)
1.131 U My-Chzhen': Sci. Sin. *6*, 833-846 (1953)
1.132 Yu.V. Vandakurov: Zh. Tekh. Fiz. *26*, 2578-2594 (1956) [English transl.: Sov. Phys. Tech. Phys. *1*, 2491-2507 (1956)]; Zh. Tekh. Fiz. *27*, 1850-1862 (1957) [English transl.: Sov. Phys. Tech. Phys. *2*, 1719-1733 (1957)]
1.133 J.L. Verster: Philips Res. Repts. *18*, 465-605 (1963)
1.134 H. Voit: Z. Instrumentenkde. *59*, 71-82 (1939)
1.135 M. Watanabe, T. Someya: Optik *20*, 99-108 (1963)
1.136 R.A. Watkins: "Axial astigmatism of the electron microscope objective", Dissertation, Ohio State University (1953); Dissertation Abstr. *20*, 704-705 (1959)

1.137 L. Wegmann: Helv. Phys. Acta *26*, 448-449 (1953); Optik *11*, 153-170 (1954)
1.138 T. Yanaka, K. Shirota: *Proc. 7th Int. Cong. Electron Microscopy, Grenoble, 1970*, ed. by P. Favard (SFME, Paris 1970) Vol.2, pp.59-60
1.139 V.K. Zworykin, G.A. Morton, E.G. Ramberg, J. Hillier, A.W. Vance: *Electron Optics and the Electron Microscope* (Wiley, New York; Chapman and Hall, London 1945)

2. Magnetic Field Calculation and the Determination of Electron Trajectories

E. Kasper

With 23 Figures

In this chapter we shall consider the use of a digital computer in the design of magnetic lenses. Problems which can be solved satisfactorily without the use of a computer are treated in the preceding and the subsequent chapters.

Generally, the computer-aided design of an electron optical system proceeds in the following three steps. The first is the calculation of the electric or magnetic field from given boundary conditions. The next is the calculation of representative electron trajectories from given initial or boundary conditions. The last is the determination of imaging properties and aberration quantities from the calculated trajectories.

The first step, the field calculation, is by far the most complicated, and the greater part of this chapter will therefore deal with this subject. Among the numerous methods of field calculation, the most familiar, the finite - difference method and the finite - element method, are treated in some detail. Some analytical methods are also briefly described in Sect.2.5, the most general being that based upon the solution of an integral equation.

We shall deal almost exclusively with rotationally symmetric systems. In Sect. 2.6 we shall treat some systems of lower symmetry in so far as they are of interest in electron microscopy. Section 2.7 gives a survey of the most important methods for the calculation of electron trajectories and aberration coefficients. In this context two types of equations of motion are considered, the paraxial equation and the Lorentz equation.

In the last decade the amount of scientific work concerned with the subjects discussed in this chapter has increased enormously. Thus the list of references given in this context can never be complete. Far more references can be found in the works cited. Survey articles treating the electron optical aspects of many of the following themes are given by HAWKES [2.23] and WEBER [2.85]. Many details of current problems in magnetic field calculation are given in [2.7a,60]. Electron optical design problems are treated in a series of papers appearing in [2.24].

2.1 Basic Concepts

In the present chapter we shall mostly confine our considerations to the numerical calculation of magnetostatic fields in devices with rotationally symmetric field boundaries. It is not necessary that these boundaries should completely coincide with the yoke surfaces nor that the boundary values should also be rotationally symmetric. These assumptions essentially include the treatment of round lenses and of toroidal deflection systems.

2.1.1 General Field Equations

In the special case of magnetostatic fields, Maxwell's equations reduce to

$$\operatorname{div} \underline{B} = 0 \quad , \tag{2.1}$$

$$\operatorname{curl} \underline{H} = \underline{j} \quad , \tag{2.2}$$

where $\underline{B}(\underline{r})$ is the magnetic flux density, $\underline{H}(\underline{r})$ the magnetic field strength, and $\underline{j}(\underline{r})$ the electric current density fulfilling the condition

$$\operatorname{div} \underline{j} = 0 \quad . \tag{2.3}$$

Furthermore, $\underline{B}$ and $\underline{H}$ are related by a material equation

$$\underline{H} = \underline{H}(\underline{B}) \tag{2.4}$$

which will be very complicated in the most general case. Here we shall only consider isotropic media, (2.4) then simplifying to

$$\underline{H} = \nu(B)\, \underline{B} \quad , \tag{2.5}$$

where $\mu(B) = 1/\nu(B)$ is the magnetic permeability of the material and $B = |\underline{B}|$. In unsaturated materials μ and ν are practically constant.

The well-known boundary conditions to be fulfilled by the solutions of (2.1) and (2.2) are that on the yoke surfaces the normal component of $\underline{B}$ and the tangential component of $\underline{H}$ are to be continuous, and furthermore, the vectors $\underline{B}$ and $\underline{H}$ are to be zero on an infinite surface enclosing the whole field system. Due to these boundary conditions, a unique solution of (2.1) and (2.2) is defined.

It is customary to introduce a magnetic vector potential $\underline{A}(\underline{r})$ by

$$\underline{B} = \operatorname{curl} \underline{A} \quad , \tag{2.6}$$

$$\operatorname{div} \underline{A} = 0 \quad , \tag{2.7}$$

$$\operatorname{curl}(\nu \operatorname{curl} \underline{A}) = \underline{j} \quad . \tag{2.8}$$

In unsaturated materials, (2.8) reduces to the vector Poisson equation,

$$\nabla^2 \underline{A} = -\mu \underline{j} \quad . \tag{2.9}$$

The boundary conditions to be imposed on the vector potential are described later.

Very often one is only interested in the magnetic field in vacuo outside the sources of the field. Since $\underline{H}$ is then irrotational, it is convenient to introduce a magnetic scalar potential $\psi(\underline{r})$, the vector $\underline{H}$ being related to this by

$$\underline{H} = -\mathrm{grad}\psi \quad . \tag{2.10}$$

Recalling that $\nu = 1/\mu_0$ = const. and introducing (2.5) into (2.1), one obtains Laplace's equation,

$$\nabla^2\psi = 0 \quad . \tag{2.11}$$

The boundary conditions to be imposed on this potential are comparatively simple in the case of yoke materials with very high permeability μ. If the yoke contains a sufficiently large gap for saturation effects not to occur, the field vectors will be practically normal to the yoke surface on its vacuum side. This means that the scalar potential varies extremely slowly on such a surface. It is then a good approximation to regard sufficiently small coherent parts of the yoke surface as equipotentials. The whole region of solution must be chosen in such a manner that it is finite and that its surface does not enclose the sources of the field. As will be shown in examples, reasonable boundary values can be defined over the whole surface. The corresponding Dirichlet problem then has a unique solution.

Empirical rules for the onset of saturation effects as a function of the lens geometry and the excitation are given in Sects.2 and 3.

With respect to the finite-element method, it is of importance that the field calculation can be formulated as the solution of a variational problem. As is well known in classical electrodynamics, for any region G in three-dimensional space, the functional

$$F = \int_G [U(|\mathrm{curl}\ \underline{A}|) - \underline{j}\cdot\underline{A}\ ]dV \tag{2.12}$$

is to be minimized, the constraints being (2.7) and $\delta\underline{A} = 0$ on the surface of G. Physically, the functional is the stored magnetic energy. The field vectors of interest are again defined by (2.6) and (2.5), the material function $\nu(B)$ now being given by

$$\frac{1}{\mu} = \nu(B) = \frac{1}{B}\frac{dU(B)}{dB} \quad .$$

Equation (2.8) or (2.2) is now obtained as the Euler-Lagrange equation of the variational principle.

For unsaturated materials, (2.12) simplifies to

$$F = \int_G \left[\frac{1}{2\mu}(\mathrm{curl}\ \underline{A})^2 - \underline{j}\cdot\underline{A}\right]dV = \mathrm{min.} \tag{2.13}$$

Equation (2.11) can also be derived from a variational principle,

$$F = \frac{\mu_0}{2}\int(\mathrm{grad}\psi)^2 dV = \mathrm{min.} \tag{2.14}$$

2.1.2 Fourier Series Expansions of the Scalar Potential

We shall now consider systems with a straight electron optic axis. It is convenient to introduce cylindrical coordinates z, r, φ with respect to this axis. Laplace's equation (2.11) is now given by

$$\frac{\partial^2 \psi}{\partial z^2} + \frac{\partial^2 \psi}{\partial r^2} + \frac{1}{r} \cdot \frac{\partial \psi}{\partial r} + \frac{1}{r^2} \cdot \frac{\partial^2 \psi}{\partial \varphi^2} = 0 \quad . \tag{2.15}$$

In order to reduce the number of variables, it is usual to expand $\psi(z,r,\varphi)$ into a Fourier series with respect to the azimuth φ [Ref.2.17,p.99]:

$$\psi(z,r,\varphi) = \sum_{m=-\infty}^{\infty} r^{|m|} \psi_m(z,r) e^{im\varphi} \quad . \tag{2.16}$$

The fact that the potential ψ is real implies that

$$\psi_{-m}(z,r) = \psi_m^*(z,r) \quad , \quad m = 0,1,2,\ldots\infty \quad , \tag{2.17}$$

where the asterisk denotes complex conjugation. Introducing (2.16) into (2.15) and invoking the linear independence of different Fourier components, one obtains a set of differential equations; for any integer m we have

$$\frac{\partial^2 \psi_m}{\partial z^2} + \frac{\partial^2 \psi_m}{\partial r^2} + \frac{2|m| + 1}{r} \cdot \frac{\partial \psi_m}{\partial r} = 0 \quad . \tag{2.18}$$

The real and imaginary parts of ψ_m separately obey this differential equation. Thus it is no loss of generality to study only real solutions.

The simplification obtained by this series expansion is considerable, since the solution of a three-dimensional problem is transformed into that of an uncoupled set of two-dimensional differential equations. Due to the rapid decrease as $r \to 0$ caused by the factors $r^{|m|}$, (2.16) can always be truncated after a few essential terms of low order, since usually only a narrow region in the vicinity of the axis is of physical interest.

In order to solve (2.18) one has to know the boundary conditions for the component ψ_m. These are found as Fourier transforms of the boundary values of $\psi(z,r,\varphi)$. The latter must be known on a closed rotationally symmetric surface S, defined by a parametric representation $z = z_s(s)$, $r = r_s(s)$ of its contour in any axial section. Then the required boundary values for ψ_m are to be calculated according to

$$\psi_m(z_s,r_s) = \frac{1}{2\pi r_s^{|m|}} \int_0^{2\pi} \psi(z_s,r_s,\varphi) e^{-im\varphi} d\varphi \quad . \tag{2.19}$$

The axis, $r = 0$, is an open part of the boundary, where for reasons of regularity the solutions have to fulfil the Neumann condition $\partial\psi_m/\partial r = 0$. The problem, defined by the whole set of boundary conditions, then has a unique solution.

For reasons of simplicity the integer constant $\alpha = 2|m| + 1$ is introduced. In the following sections we shall omit the subscript m wherever it is not essential. We thus have obtained differential equations of the type

$$\frac{\partial^2\psi(z,r)}{\partial z^2} + \frac{\partial^2\psi(z,r)}{\partial r^2} + \frac{\alpha}{r} \cdot \frac{\partial\psi(z,r)}{\partial r} = 0 \quad . \tag{2.20}$$

The most important special case, $\alpha = 1$, is Laplace's equation for rotationally symmetric potentials. Poisson's equation corresponding to (2.20), defined by

$$\frac{\partial^2 P(z,r)}{\partial z^2} + \frac{\partial^2 P(z,r)}{\partial r^2} + \frac{\alpha}{r} \cdot \frac{\partial P(z,r)}{\partial r} = -g(z,r) \quad , \tag{2.21}$$

is also of importance as will be obvious later. The function $g(z,r)$ is the source density of the field.

2.1.3 Paraxial Expansions

We only consider those solutions of (2.20) which are regular on the axis ($r = 0$), since these are the only ones of physical interest. It is possible to seek a solution of (2.20) by means of an expansion into a power series [Ref.2.17, p.99]

$$\psi(z,r) = \sum_{n=0}^{\infty} \frac{(-1)^n}{(2n)!} \phi_n(z) r^{2n} \quad , \tag{2.22}$$

the functions $\phi_n(z)$ being unknown at the moment. This expansion already fulfils all regularity requirements on the axis. Substitution into (2.20) results in recurrence relations for the functions $\phi_n(z)$ and their derivatives,

$$\left(1 + \frac{\alpha}{2n + 1}\right)\phi_{n+1}(z) = \phi_n''(z) \quad . \tag{2.23}$$

The function $\phi_0(z) = \psi(z,0)$ can be chosen arbitrarily provided that it is regular in z and $\phi_0(\pm\infty)$ remains finite. By means of (2.22) all higher coefficients can be expressed as derivatives of $\phi_0(z)$. The first few terms of (2.22) are then explicitly given by

$$\psi(z,r) = \phi_0(z) - \frac{r^2}{2!(1 + \alpha)} \phi_0''(z) + \frac{r^4}{4!(1 + \alpha)(1 + \alpha/3)} \phi_0^{(4)}(z) - \frac{r^6}{6!(1 + \alpha)(1 + \alpha/3)(1 + \alpha/5)} \phi_0^{(6)}(z) + O(r^8) \quad . \tag{2.24}$$

In the case of rotationally symmetric fields, $\alpha = 1$, (2.24) specializes to

$$\psi(z,r) = \phi_0(z) - \frac{r^2}{4} \phi_0''(z) + \frac{r^4}{64} \phi_0^{(4)}(z) - \frac{r^6}{2304} \phi_0^{(6)}(z) \quad \dots \quad . \tag{2.25}$$

Then $\phi_0(z)$ is the potential on the axis. A similar paraxial expansion can be derived for the solutions of (2.21). This will not be given here, since in magnetic lenses the sources of the field are far distant from the axis.

Unfortunately, (2.22) cannot be used in practice to calculate the potential in the whole plane from the function $\phi_0(z)$. There are different reasons for this. Extrapolation from the axial values to values of ψ for large r is numerically unstable; this means that rounding errors, initially very small, may increase rapidly, thus making the solution invalid. Another reason is that it will be practically impossible to find reasonable electrode or pole-piece shapes for the experimental realization of the field if $\phi_0(z)$ is chosen arbitrarily. Therefore (2.24) can only be used to derive analytical relations between paraxial quantities.

2.1.4 Circular Vector Potentials and Related Fields

The application of differential equation (2.8) or (2.9) to round magnetic lenses requires their explicit representation in cylindrical coordinates. Owing to the rotational symmetry of the field, it is necessary to assume the current density $\underline{j}$ to be circular and it is convenient to assume the vector potential $\underline{A}$ to be circular too,

$$\underline{j}(\underline{r}) = j(z,r)\underline{u}_\varphi \quad , \tag{2.26}$$

$$\underline{A}(\underline{r}) = A(z,r)\underline{u}_\varphi \quad , \tag{2.27}$$

$\underline{u}_\varphi$ being the unit vector in the azimuthal direction. Then (2.3) and (2.7) are automatically satisfied. It is convenient to introduce two new functions

$$P(z,r) = 2\,A(z,r)/r \quad , \tag{2.28}$$

$$F(z,r) = 2\pi r \cdot A(z,r) \quad . \tag{2.29}$$

For reasons of regularity P(z,r) must remain finite, thus A(z,r) must vanish linearly and F(z,r) as r^2 for $r \rightarrow 0$. The components of $\underline{B} = \text{curl}\,\underline{A}$ in cylindrical coordinates are given by

$$B_z = \frac{\partial A}{\partial r} + \frac{A}{r} = P + \frac{r}{2} \cdot \frac{\partial P}{\partial r} = \frac{1}{2\pi r} \cdot \frac{\partial F}{\partial r} \quad , \tag{2.30}$$

$$B_r = -\frac{\partial A}{\partial z} = -\frac{r}{2} \cdot \frac{\partial P}{\partial z} = -\frac{1}{2\pi r} \cdot \frac{\partial F}{\partial z} \quad . \tag{2.31}$$

Integration of (2.30) with respect to the radial coordinate yields

$$F(z,r) = \int_0^r B_z(z,r') \cdot 2\pi r'dr' = \int_C B_z\,da \quad .$$

This shows that F(z,r) is the magnetic flux through a coaxial circular disc C of radius r in the plane z = const. Hence the lines F(z,r) = const. are identical with the lines of magnetic flux. The physical meaning of the quantity P is obtained by evaluation of (2.30) for r = 0, resulting in

$$B_z(z,0) = P(z,0) \quad . \tag{2.32}$$

Hence P(z,r) is related to A(z,r) in such a way that on the axis, P coincides with the flux density B, a result which is of special importance in electron optics.

The partial differential equation to be satisfied by A(z,r) is obtained by introduction of (2.30) and (2.31) into (2.8) and evaluation of all vector differentiations in cylindrical coordinates, resulting in

$$\frac{\partial}{\partial z}\left(\nu\,\frac{\partial A}{\partial z}\right) + \frac{\partial}{\partial r}\left(\nu\,\frac{\partial A}{\partial r} + \nu\,\frac{A}{r}\right) = -j \quad , \tag{2.33}$$

where ν is a function of z and r in saturated materials. Since $\nu = \mu^{-1} = \text{const.}$, in all unsaturated materials including the vacuum, (2.33) then simplifies to

$$\frac{\partial^2 A}{\partial z^2} + \frac{\partial^2 A}{\partial r^2} + \frac{1}{r}\cdot\frac{\partial A}{\partial r} - \frac{A}{r^2} = -\mu j \quad .$$

Introducing (2.30) and (2.31) into this equation one obtains more convenient partial differential equations:

$$\frac{\partial^2 P}{\partial z^2} + \frac{\partial^2 P}{\partial r^2} + \frac{3}{r}\cdot\frac{\partial P}{\partial r} = -\frac{2\mu j}{r} \quad , \tag{2.34}$$

$$\frac{\partial^2 F}{\partial z^2} + \frac{\partial^2 F}{\partial r^2} - \frac{1}{r}\cdot\frac{\partial F}{\partial r} = -2\pi\mu r j \quad . \tag{2.35}$$

These are special cases of (2.21) with $\alpha = 3$ and $\alpha = -1$, respectively; the function g(z,r) is identified with the right-hand sides of (2.34) and (2.35) in turn.

The boundary conditions to be imposed on the functions A(z,r), P(z,r), and F(z,r) are as follows. At infinity all these functions must asymptotically vanish. Furthermore, $A(z,0) = F(z,0) = 0$ and $\partial P/\partial r = 0$ for $r = 0$ must be satisfied. At interfaces like yoke surfaces A(z,r), P(z,r) and F(z,r) must remain continuous. Consequently, the normal component of the flux density $\underline{B}$ is also continuous. The continuity of the tangential component of $\underline{H} = \nu\underline{B}$ requires that $\nu\partial F/\partial n$ must be continuous, $\partial F/\partial n$ being the derivative with respect to the local surface normal (generally $\nu\partial P/\partial n$ and $\nu\partial A/\partial n$ are then discontinuous). The boundary-value problem specified in this manner has a unique solution.

In the general case the complete solution of this problem may become very complicated. One therefore often assumes infinite permeability of the yoke material. The field calculation can then be entirely confined to the vacuum part of the lens and the interface conditions are transformed into the familiar boundary conditions, now simplified to

$$\psi = \text{const} \quad , \quad \frac{\partial F}{\partial n} = 0 \quad , \tag{2.36}$$

where ψ is the scalar potential introduced in Sect.2.1.1 and $\partial/\partial n$ denotes differentiation with respect to the surface normal.

Another interesting case arises with superconducting yoke parts. Due to the Meissner-Ochsenfeld effect, the magnetic field is completely expelled from the interior of the superconductor. With respect to the vacuum field, this effect is simulated by the boundary conditions

$$\frac{\partial\psi}{\partial n} = 0 \quad , \quad F = \text{const} \quad , \tag{2.37}$$

which express that the normal component of $\underline{H}$ vanishes. At the surface the tangential component discontinuously decreases from a finite value in vacuo to zero in the superconductor. This jump is accompanied by circular screening currents which do not need to be calculated.

The role of the functions A(z,r), P(z,r), F(z,r) is now clear: A(z,r) is used for the formulation of the variational principle, P(z,r) conveniently represents the field in the paraxial region, while F(z,r) is necessary for computing flux lines and to formulate the boundary and interface conditions.

2.2 Methods of Discretization

A complete analytical solution of an elliptic partial differential equation for given boundary conditions is only possible in very few cases. In a large number of realistic cases one is forced to choose one of the approximately valid numerical solution techniques based upon discretization of the field. Generally, this means that the continuous partial differential equation is replaced by a system of ordinary linear equations of finite rank. This system can be solved by use of standard techniques. After that, any interesting field quantity can be calculated to a certain accuracy by means of interpolation.

There are several different methods for performing this discretization. The most important ones are the finite-difference method and the finite-element method. Common to both methods is the fact that the differential equation is replaced by a linear relation between the potential values in adjacent nodes of a finite-mesh grid. Only the ways of obtaining these linear relations are different. In the following sections we shall describe the basic ideas of these methods and possible ways of generalizing them.

In each case we shall obtain a system of coupled linear equations. Here, we are only concerned with its formulation. The question of the best way of solving it is quite a different matter, which will be discussed in Sect.2.3. The formulation of the discretized equations will be treated in more detail than their solution, since this formulation raises more technical problems.

2.2.1 The Finite-Difference Method

The finite-difference method (FDM), usually combined with iterative solution techniques, is the most frequently used method of discretization. As early as 1918, it was used by LIEBMANN [2.46] and is thus often called "Liebmann's method". Its mathematical theory is readily available in the mathematical literature (for instance [2.1,13,83,80,62]).

For conciseness, we shall confine the following considerations to two-dimensional Dirichlet problems based upon a general elliptic differential equation

$$A(u,v)P_{uu} + B(u,v)P_{vv} + a(u,v)P_u + b(u,v)P_v = C(u,v)P(u,v) + G(u,v) \quad , \qquad (2.38)$$

which will be obtained by transformation of Poisson's equation to general orthogonal curvilinear coordinates. In (2.38) subscripts denote differentiation with respect to the corresponding variable.

The basic idea of the FDM is to cover the entire region of solution by a discrete network of finite mesh-width h (Fig.2.1). Since the choice of the orthogonal coordinates u, v is still arbitrary, it is no loss of generality to assume that the network is square-shaped in u and v. The corresponding network in straight coordinates will then generally be curvilinear. We have to distinguish between regular internal points (A), irregular internal points (C,C'), regular axial points B, irregular axial points (D,D'), and boundary points E,E' as is indicated in Fig.2.1.

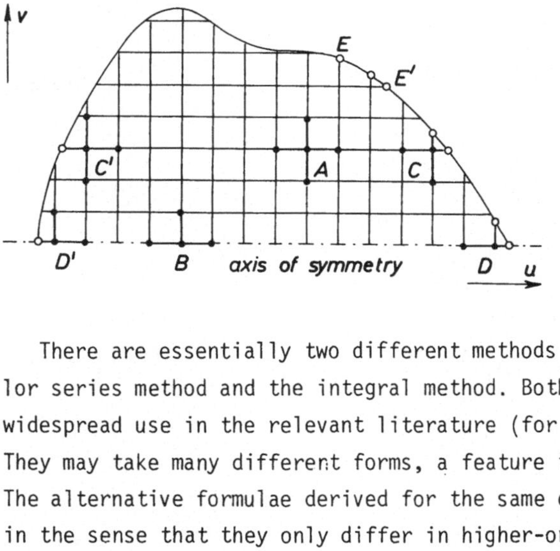

Fig.2.1. Example of a square-mesh grid with different types of points: regular internal point A; regular axial point B; irregular internal points C, C'; irregular axial points D, D'; boundary points E, E'

There are essentially two different methods of deriving mesh formulae, the Taylor series method and the integral method. Both are quite familiar and in very widespread use in the relevant literature (for instance see references cited above). They may take many different forms, a feature that may be confusing at first sight. The alternative formulae derived for the same configuration are, however, equivalent in the sense that they only differ in higher-order terms of the discretization errors.

2.2.1a The Taylor Series Method

For every regular internal mesh point we consider the four closest neighbours, their numbering being indicated in Fig.2.2a. We then expand the function P(u,v) into a Taylor series with respect to the coordinates u,v:

$$P(u,v) = \sum_{n=0}^{\infty} \sum_{j+k=n} \frac{1}{j!k!} \frac{\partial^n P}{\partial u^j \partial v^k} (u - u_0)^j (v - v_0)^k \quad , \qquad (2.39)$$

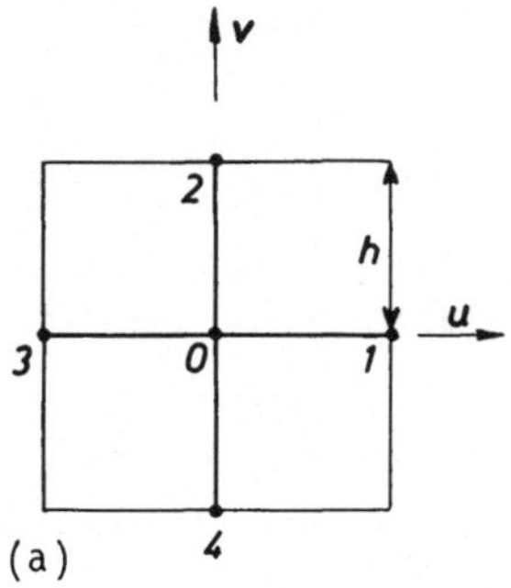

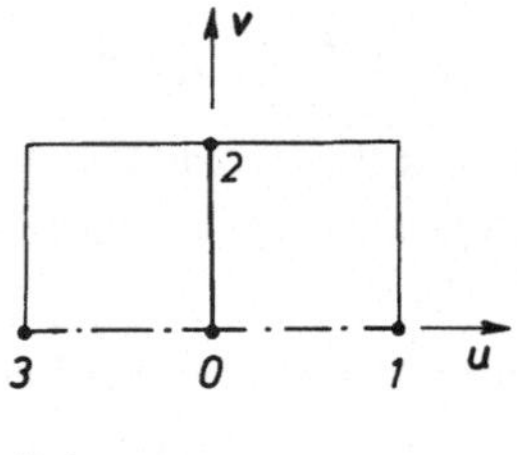

Fig.2.2a,b. Numbering of the mesh points in five- and four-point formulae: (a) for an internal mesh point 0 having four neighbours, (b) for an axial mesh point 0 having three neighbours

the derivatives to be taken at the central point. Introducing the coordinates $u_{1,3} = u_o \pm h$, $v_{1,3} = v_o$ of the points P_1 and P_3 respectively, we find

$$P_{1,3} = P_0 \pm hP_u + \frac{h^2}{2} P_{uu} \pm \frac{h^3}{6} P_{uuu} + O(h^4) \quad . \tag{2.40a}$$

There is no danger of confusion with respect of the subscripts. Subscript letters denote differentiations, while subscript numbers refer to the corresponding point. By addition and subtraction of the two equations (2.40a) we find that

$$P_u = \frac{P_1 - P_3}{2h} + O(h^2) \quad , \quad P_{uu} = \frac{P_1 + P_3 - 2P_0}{h^2} + O(h^2) \quad . \tag{2.40b}$$

In a similar way

$$P_v = \frac{P_2 - P_4}{2h} + O(h^2) \quad , \quad P_{vv} = \frac{P_2 + P_4 - 2P_0}{h^2} + O(h^2)$$

is obtained. Substituting these expressions into (2.38), we find a finite-difference expression. This can be solved for the value P_0, resulting in the formula

$$P_0 = \beta_0 + \beta_1 P_1 + \beta_2 P_2 + \beta_3 P_3 + \beta_4 P_4 + O(h^4) \quad , \tag{2.41}$$

the coefficients being

$$\left.\begin{aligned} \beta_{1,3} &= \frac{A \pm ha/2}{N} \quad , \quad \beta_{2,4} = \frac{B \pm hb/2}{N} \quad , \\ \beta_0 &= -h^2 G/N \quad , \quad N = 2(A + B) + h^2 C \quad . \end{aligned}\right\} \tag{2.42}$$

The special case of most interest in the calculation of magnetic fields is the solution of (2.21). The coordinates u,v are then cylindrical polar coordinates z,r. Identifying the coefficients in (2.38) with those in (2.21), one obtains the mesh formula

$$P_0 = \frac{1}{4}(P_1 + P_2 + P_3 + P_4 + h^2 g_0) + \frac{h\alpha}{8r_0}(P_2 - P_4) + O(h^4) \quad , \tag{2.43}$$

r_0 being the radial coordinate of the central point.

Obviously (2.43) does not hold for mesh points located on the axis of symmetry. In this case only the potential values of three adjacent mesh points are available (Fig.2.2b). In addition to these one has to use the special symmetry of the differential equation. Here we shall only consider (2.21).

In the paraxial region, an expansion like (2.22) must be valid. From this the relation

$$\lim_{r\to 0} \frac{1}{r} P_r(z,r) = P_{rr}(z,0)$$

can be derived. Using this and (2.40b) for $P_{zz} = P_{uu}$ and introducing them into (2.21), one finds

$$P_0 = \frac{1}{2(\alpha + 2)}[P_1 + P_3 + h^2 g_0 + 2(\alpha + 1)P_2] + O(h^4) \quad . \tag{2.44}$$

Equations (2.43,44) include as the special case of most interest, $\alpha = 1$, $g_0 = 0$, the formulae

$$P_0 = \frac{1}{4}(P_1 + P_2 + P_3 + P_4) + \frac{h}{8r_0}(P_2 - P_4) \quad , \quad (r_0 > 0) \quad , \tag{2.45}$$

$$P_0 = \frac{1}{6}(P_1 + 4P_2 + P_3) \quad , \quad (r_0 = 0) \quad , \tag{2.46}$$

which represent the finite-difference approximation for rotationally symmetric Laplace fields.

It is possible to generalize the method described above in such a way that asymmetric mesh formulae, valid for grids with different mesh-widths in different directions can be derived [2.74]. We shall not discuss this matter here, since these formulae are generally not equal but mostly equivalent to those derived in the next section. Detailed investigations on the Taylor series methods are given in [2.12].

2.2.1b The Integral Method

This method can only be applied to self-adjoint elliptic differential equations

$$\frac{\partial}{\partial u}[U(u,v)P_u] + \frac{\partial}{\partial v}[V(u,v)P_v] = W(u,v)P(u,v) + Q(u,v) \quad . \tag{2.47}$$

Every mesh point is now associated with a finite region R having a closed contour C. For reasons of simplicity these regions are chosen as rectangles, as is shown in Fig.2.3a,b. Of course one may choose other shapes if they are more favourable.

We now apply Gauss's integral theorem

$$\iint_R [A_u(u,v) + B_v(u,v)]du dv = \oint (-B\, du + A\, dv) \quad ,$$

valid for all continuously differentiable functions A(u,v), B(u,v). Choosing now $A = U \cdot P_u$, $B = V \cdot P_v$ and making use of (2.47) we obtain

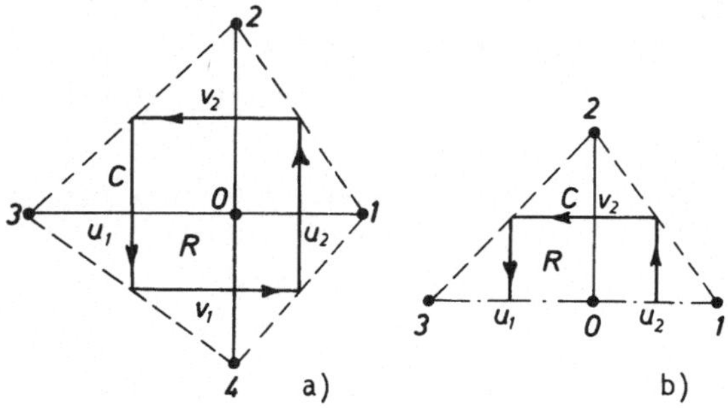

Fig.2.3a,b. The rectangular region R and its directed contour C used for the derivation of mesh formulae (a) for an internal mesh point 0 having four neighbours, (b) for an axial mesh point 0 having three neighbours

$$\iint_R (W \cdot P + Q)du\,dv = \oint (-VP_v du + UP_u dv) \quad . \tag{2.48}$$

This is exactly true; for the practical evaluation, however, simplifying assumptions are necessary. Each rectangle R is assumed to be so small that as a first approximation the function P(u,v) may be regarded as constant and replaced by P_0. Then the left-hand side of (2.48) is approximated by $\iint(W P_0 + Q)$ du dv. On the right-hand side of (2.48) the derivatives of P are regarded as constant on every straight part of the contour, for instance $P_u = (P_1 - P_0)/h_1$ on the right-hand part of the contour in Fig.2.3a, h_1 being the distance between the points numbered 1 and 0. Evaluation of the right-hand side of (2.48) then results in the very general formula

$$\frac{P_1 - P_0}{h_1}\int_{v_1}^{v_2} U(u_2,v)dv + \frac{P_2 - P_0}{h_2}\int_{u_1}^{u_2} V(u,v_2)du$$

$$+ \frac{P_3 - P_0}{h_3}\int_{v_1}^{v_2} U(u_1,v)dv + \frac{P_4 - P_0}{h_4}\int_{u_1}^{u_2} V(u,v_1)du$$

$$= \int_{u_1}^{u_2}\int_{v_1}^{v_2} [P_0W(u,v) + Q(u,v)]dudv \quad , \tag{2.49}$$

valid for all internal mesh points including the irregular ones, but not for axial mesh points. For the latter, one has to apply (2.48) to regions of the type shown in Fig.2.3b.

This theory is also applicable to the solutions of (2.21), since this differential equation is equivalent to

$$\frac{\partial}{\partial z}(r^\alpha P_z) + \frac{\partial}{\partial r}(r^\alpha P_r) = -r^\alpha g(z,r) \quad .$$

For reasons of simplicity the function g(z,r) is assumed to be slowly varying and may be replaced by the proper central value g_0 in every rectangle R. For internal mesh points the resulting formula is then

$$P_0[\eta_0(h_1^{-1} + h_3^{-1}) + \eta_2 h_2^{-1} + \eta_4 h_4^{-1}]$$

$$= \eta_0[P_1h_1^{-1} + P_3h_3^{-1} + \tfrac{1}{2}\, g_0(h_1 + h_3)] + P_2\eta_2h_2^{-1} + P_4\eta_4h_4^{-1} \tag{2.50}$$

with

$$\left.\begin{aligned} &\eta_2 = \left(1 + \frac{h_2}{2r_0}\right)^{\alpha} , \quad \eta_4 = \left(1 - \frac{h_4}{2r_0}\right)^{\alpha} , \\ &\eta_0 = \left[\left(1 + \frac{h_2}{2r_0}\right)^{\alpha+1} - \left(1 - \frac{h_4}{2r_0}\right)^{\alpha+1}\right] \frac{2r_0}{(\alpha + 1)(h_1 + h_3)} , \quad (\alpha \neq -1) , \\ &\eta_0 = \frac{2r_0}{h_1 + h_3} \ln \left[\frac{2r_0 + h_2}{2r_0 - h_4}\right] , \quad (\alpha = -1) . \end{aligned}\right\} \tag{2.51}$$

For axial mesh points the resulting formula is given by

$$P_0[h_1^{-1} + h_3^{-1} + (\alpha + 1)(h_1 + h_3)h_2^{-2}]$$

$$= P_1h_1^{-1} + P_3h_3^{-1} + (\alpha + 1)(h_1 + h_3)h_2^{-2}P_2 + \tfrac{1}{2}\, g_0(h_1 + h_3) \quad . \tag{2.52}$$

By use of these formulae, the discretization can be performed for all the types of mesh points shown in Fig.2.1. In the case of asymmetric configurations ($h_1 \neq h_3$ or $h_2 \neq h_4$) the resultant discretization error is of third order in the mesh-widths, while in the symmetric case it decreases to the fourth order. If, however, the boundary is far distant from the region of electron trajectories, the deterioration of the accuracy in the irregular mesh points is of little, if any, importance.

Equations (2.50) and (2.52), in their present form, have been derived by KASPER [2.34]. As the special case $h_1 = h_3$, $h_2 = h_4$, $g_0 = 0$, they include formulae published by JANSE [2.32], who used them for the numerical calculation of electron optical devices with small perturbations of the rotational symmetry (Sect.2.6.1). In the more special case of equal distances $h_i = h$, (2.52) reduces to (2.44), but (2.43) is only obtained from (2.50) if $\alpha = 0$ or $\alpha = 1$. For $\alpha \geqq 2$ more complicated formulae are obtained. But these now have the advantage that all coefficients β_i are strictly positive, while the coefficient of P_4 in (2.43) becomes negative for $2r_0 < h\alpha$. This is of importance with respect to the application of the successive overrelaxation method (Sect.2.3.2).

Yet other formulae may be derived if the linear relation between the partial derivatives, which is given by the differential equation, is first introduced into the Taylor series (2.39), and after that, this series is applied to the adjacent mesh points [2.85]. As has been pointed out, the process of discretization is not unique. According to KASPER and LENZ [2.34b], there is a three-dimensional degree of freedom in the choice of the coefficients $\beta_1,\ldots,\beta_4$ occurring in the discrete form of (2.20).

2.2.1c Application to a Simple Magnetic Lens

An extremely simplified application of the FDM to a magnetic lens is demonstrated in Fig.2.4. Here the ycke contours are chosen in such a manner that no irregular mesh points occur. The boundary values in the bores are those of the corresponding pole pieces, while in the gap they are simply defined by linear interpolation between the yoke potentials. Having defined reasonable starting values, one can easily solve the linear system of equations based upon (2.45) or (2.46) by use of iteration methods.

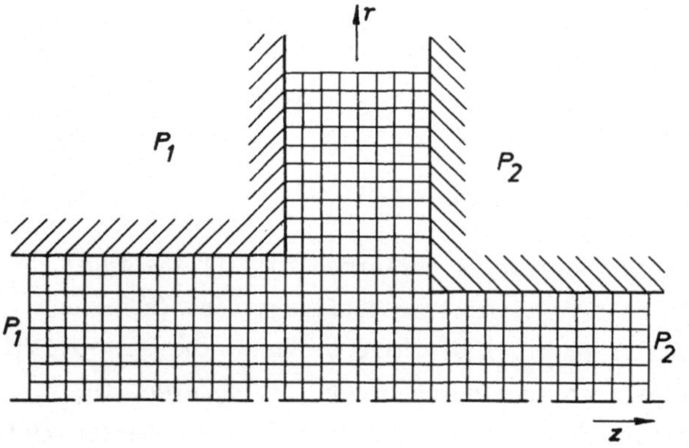

Fig.2.4. Regular mesh for the calculation of a magnetic lens with pole pieces having the scalar potentials P_1 and P_2

Symmetric lenses of this kind were measured by LIEBMANN [2.47] and LIEBMANN and GRAD [2.48], who used an analogue technique based upon a square-shaped resistance network. These and still more complicated lens types with saturation effects had already been calculated in 1950 by LENZ [2.42]. In 1955 a magnetic double-lens was calculated by LAUDET [2.41], who also solved (2.35). Today, geometric configurations like that shown in Fig.2.4 can be successfully treated in student courses. Possible refinements of the FDM are treated in the next section.

2.2.2 Improvement of the Finite-Difference Method

By proper use of the formulae given in Sect.2.2.1 it is possible to perform the discretization for any Dirichlet problem in two dimensions. In order to obtain sufficient accuracy, however, it may be necessary to use a very fine mesh-grid. In practice, the numerical calculation then requires large storage capacity and computing time. In order to keep both reasonably small, one has, depending on the boundary conditions, to make use of some refinements.

2.2.2a Different Mesh Grids

A first refinement is the use of grids with locally different mesh-widths depending on a rough guess of the magnitudes of the higher derivatives neglected in the finite-difference equations. This is demonstrated in Fig.2.5a. It is essential that the internal boundaries of both grids overlap in such a way that the boundary points of the coarse grid are also internal points of the fine one. In those boundary points of the fine grid that are not simultaneously internal points of the coarse grid, the potential values are to be calculated by means of interpolation.

The most convenient way of doing this is to apply the well-known Lagrange interpolation formula to the central interval between the corresponding four equidistant mesh points. This is demonstrated in Fig.2.5b. In the first radial interval close to the axis of symmetry, a three-point formula based upon a paraxial series $P(z,r) = A(z) + B(z)r^2 + C(z)r^4$ is adequate (Fig.2.5b). This formula is given by

$$P(z,r) = P_0\left(1 - \frac{5}{4}v^2 + \frac{v^4}{4}\right) + P_1\,\frac{v^2}{3}(4 - v^2) + P_2\,\frac{v^2(v^2 - 1)}{12}\ , \qquad (2.53)$$

where $v = r/H$ and $P_j = P(z,j\cdot H)$ for $j = 0,1,2$ are the corresponding potential values in the coarse grid. In regions close to external boundaries, asymmetric four-point-interpolation formulae have to be used.

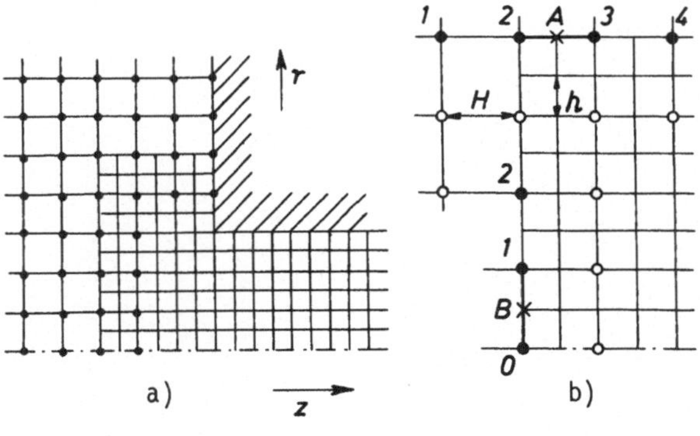

Fig.2.5a,b. Coupling between square-mesh grids of different mesh-widths H and h: (a) overlap between both grids, (b) four-point interpolation with respect to the point A and three-point interpolation with respect to the point B close to the axis

This kind of interpolation is rather flexible in its applicability, since it does not depend on the special structure of the differential equation to be solved. Its accuracy is sufficient for $H/h \leq 5$. Of course it is possible not to use the Lagrange interpolation but to apply special formulae directly derived from the partial differential equation to be solved. We shall, however, not discuss these because they have only very limited applicability.

2.2.2b Nine-Point Formulae

A second refinement is the use of discretization formulae based upon more than the minimum of four neighbours of any internal mesh point. It is then possible to decrease the inherent discretization error. Unfortunately, depending on the structur of the partial differential equation, these formulae soon become very complicated, as the number of mesh points included increases. In practice nine-point formulae are an upper limit. With respect to Laplace's equation in two cartesian coordinates, such a formula is quite common in the mathematical literature (for instance [2.1,83]). For Laplace's equation for rotationally symmetric fields in cylindrical coordinates, a formula is also known [2.12]; this is equivalent to the formula given below if $\alpha = 1$. For the more general case of (2.21) a nine-point formula wit a discretization error of sixth order in the mesh-width was derived by KASPER [2.34]. The numbering of the mesh points is demonstrated in Fig.2.6a for an internal mesh point and in Fig.2.6b for an axial mesh point. For every regular internal mesh point this formula is given by

$$\begin{aligned} P_0 &= C_+(r_0)[2(P_1 + P_5) + P_2 + 4P_3 + P_4] \\ &+ C_-(r_0)[2(P_1 + P_5) + P_6 + 4P_7 + P_8] \\ &+ C_o(r_0)[P_3 + P_7 - P_1 - P_5] \\ &+ D_+(r_0)[8g_o + g_1 + g_3 + g_5 + g_7] + D_-(r_0)(g_3 - g_7) + O(h^6) \quad , \end{aligned} \qquad (2.54$$

the radial coordinate r_0 referring to the central point (0). Using the abbreviatio $\eta = h\alpha/r_0$, the coefficient functions are given by

$$C_o(r_0) = \frac{\eta^2(\alpha - 2)}{80(1 + \eta^2/24)\alpha} \quad ,$$

$$C_\pm(r_0) = \frac{1}{20}\left(1 \pm \frac{\eta}{2} + \frac{\eta^2}{24} \pm \frac{\eta^3}{12\alpha^2}\right)\Big/\left(1 + \frac{\eta^2}{24}\right) \quad ,$$

$$D_+(r_0) = \frac{h^2}{12}\left[\frac{3}{10} - C_o(r_0)\right] \quad ,$$

$$D_-(r_0) = \frac{h^2\eta}{24}\left[\frac{3}{10} - C_o(r_0) \cdot \frac{\alpha + 2}{\alpha}\right] \quad .$$

For every regular axial mesh point the corresponding six-point formula is given by

$$\begin{aligned} P_0 &= [2(2 + \alpha - \beta)]^{-1}\Big[(1 - \beta)(P_1 + P_5) + \beta(P_2 + P_4) \\ &+ 2(1 + \alpha - \beta)P_3 + \frac{h^2}{12}(g_1 + g_5 + 10g_o) \\ &+ \frac{h^2(1 + \alpha)}{2(3 + \alpha)}(g_3 - g_o)\Big] + O(h^6) \quad , \end{aligned} \qquad (2.55$$

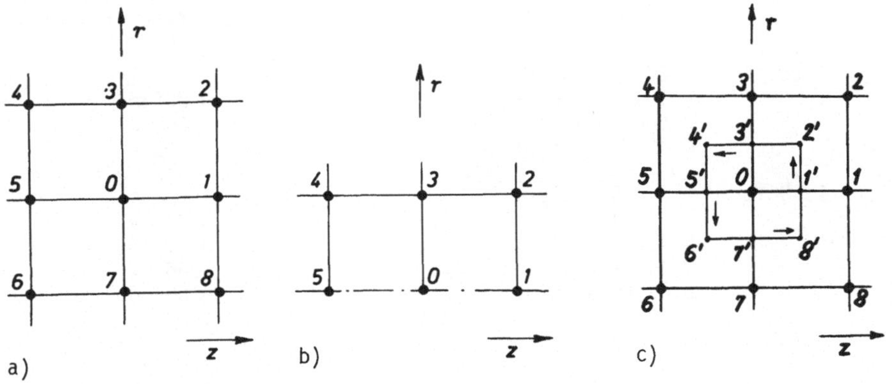

Fig.2.6a-c. Numbering of the mesh points in nine- and six-point formulae: (a) for an internal mesh point 0 having eight neighbours, (b) for an axial mesh point 0 having five neighbours, (c) for an internal mesh point having eight neighbours located on the outer square; the inner square with the points 1',...,8' represents the integration contour

the parameter β being

$$\beta = \frac{(1+\alpha)(6+\alpha)}{6(3+\alpha)} \quad .$$

The method for deriving these formulae is based upon a Taylor series expansion of the potential function P(z,r). By forming appropriate linear combinations of the series expansions at the mesh points in question and by use of the paraxial expansion in some high-order terms, all partial derivatives including the fifth order can be eliminated.

At a first glance, these formulae may appear to be very complicated. They are, however, advantageous with respect to the computation of rotationally symmetric Laplace- and Poisson-type fields and of multipole fields. One of the first successful applications, to the computation of the electric field in electron guns, has been reported by KERN [2.35]. LENZ has tested the accuracy of different discretizations by comparing them with an analytical potential. The function

$$\psi(z,r) = (h/r) \cdot J_1(r/h)\exp(-z/h)$$

is an exact solution of (2.20) for $\alpha = 3$. At each of the nodes with $z = 0$ and $r/h < 6$, the value of ψ was calculated from different discretization formulae, the potentials at the neighbouring nodes being taken from the analytic function. The results are shown in Table 2.1. The radial interpolation is given by $\psi(z,h) = 0.75\,\psi(z,0) + 0.25\,\psi(z,2h)$. Obviously the nine-point formulae has the highest accuracy.

Table 2.1. Accuracies of different discretizations in the finite-difference method

r/h \ z/h	-1	0	+1	Discretization error
0	1.35914	0.50000 Analytic	0.18394	-
		0.50635 Integral		0.00635
		0.50635 Taylor		0.00635
		0.49980 9-point		-0.00020
1	1.19618	0.44005 Analytic	0.16189	-
		0.45554 Integral		0.01549
		0.45724 Taylor		0.01719
		0.44709 rad. interpol.		0.00704
		0.43996 9-point		-0.00009
2	0.78385	0.28836 Analytic	0.10608	-
		0.30043 Integral		0.01207
		0.29943 Taylor		0.01107
		0.28829 9-point		-0.00007
3	0.30722	0.11302 Analytic	0.04158	-
		0.11822 Integral		0.00520
		0.11705 Taylor		0.00403
		0.11296 9-point		-0.00006
4	-0.04488	-0.01651 Analytic	-0.00607	-
		-0.01680 Integral		-0.00029
		-0.01760 Taylor		-0.00109
		-0.01655 9-point		-0.00004
5	-0.17809	-0.06552 Analytic	-0.02410	-
		-0.06804 Integral		-0.00252
		-0.06842 Taylor		-0.00290
		-0.06554 9-point		-0.00002
6	-0.12535	-0.04611	-0.01696	

2.2.2c Fields in Regions with Variable Material Coefficients

In these cases the finite-element method (FEM) is usually applied, especially when the material coefficients are discontinuous at the interface between two homogeneous materials since FDM become very complicated. This is, however, only true if the square-shaped mesh grid does not fit the interface. Since the yoke contours of magnetic lenses are often of a simple rectangular structure – at least in the vicinity of the gap – the mesh grid can in practice be chosen to fit the interface. Even the discontinuity of the reluctance is then no obstacle for the FDM. We wish to derive a reasonably simple nine-point formula for the partial differential equation

$$\frac{\partial}{\partial z}\left[C(z,r)F_z\right] + \frac{\partial}{\partial r}\left[C(z,r)F_r\right] = 0$$

which is a special case of (2.47) with $u = z$, $v = r$, $U = V = C$, $W = Q = 0$, $P = F$. The integral method is applied to the inner square-shaped contour shown in Fig. 2.6c. The values of C at the points 1'-8' will be given and denoted by C_1'-C_8'. The normal derivatives of F at these points are approximated by finite differences just as in Sect.2.2.1b, for instance $F_z = (F_1 - F_0)/h$ at point 1' and $F_z = (F_1 + F_2 - F_3 - F_0)/2h$, $F_r = (F_2 + F_3 - F_0 - F_1)/2h$ at point 2', the other derivatives being treated analogously. The contour integral is then evaluated by application of Simpson's rule to the four sides of the square. The resulting FDM formula is

$$F_0 = \frac{4(C_1'F_1 + C_3'F_3 + C_5'F_5 + C_7'F_7) + C_2'F_2 + C_4'F_4 + C_6'F_6 + C_8'F_8}{4(C_1' + C_3' + C_5' + C_7') + C_2' + C_4' + C_6' + C_8'} \quad .$$

In magnetic lens applications, the potential F(z,r) is preferably chosen to be the magnetic flux potential, see (2.29). The function C(z,r) is then $C(z,r) = \nu(z,r)/r$. By Taylor series expansions, valid on each side of a discontinuity of $\nu(z,r)$, it can be shown that the discretization error is of fourth order in h, provided that the ν value on the interface itself is chosen to be the *mean value* of those for the two materials; this means, when an interface goes through the points 5, 5', 0, 1', 1 in Fig.2.6c, for example, the values $\nu_1' = (\nu_2' + \nu_8')/2$, $\nu_5' = (\nu_4' + \nu_6')/2$ should be chosen. The advantage of this method is that the FDM can also be easily applied to saturated magnetic lenses with sufficiently simple contours. In the vacuum region, $\nu = 1/\mu_0$ = const. is valid, while in the iron part, $\nu = \nu(B^2)$ is a complicated function of $B^2 = |\underline{B}|^2$. This function is to be evaluated with respect to the corners 2', 4', 6', 8' of the configuration (Fig.2.6c), while at the points 1', 3', 5', 7' the mean values are sufficient. For the evaluation of B^2, the expressions concerning F in (2.30,31) can be used. For the point 2' in Fig.2.6c, for example, the resulting finite-difference expression is given by

$$B_{2'}^2 = \frac{1}{8\pi^2 r_{2'}^2 h^2}[(F_2 - F_0)^2 + (F_3 - F_1)^2] \quad , \quad \nu_{2'} = \nu(B_{2'}^2) \quad .$$

In the vicinity of the electron optic axis, the discretization of (2.55) is less accurate than that given by (2.54a,b). This is, however, no disadvantage, since in the bores (2.54a,b) can be used with $\nu = 1/\mu_0$ = const., and $\alpha = 3$. The corresponding potential P(z,r) can be easily related to F(z,r) by $F = \pi r^2 P$, see (2.28,29)

The resulting nonlinear equations for the potentials at the nodes of the mesh grid can be solved iteratively by successive overrelaxation. The only new aspect is that after each iteration the ν values have to be calculated from the last set of potential values.

This procedure is comparatively simple, at least simpler and probably more accurate than MUNRO's method (Sect.2.2.3) but, of course, not so flexible. Our present investigations are still incomplete.

2.2.2d Irregular Mesh Grids

In the vicinity of a boundary not fitting to the square-mesh grid, the corresponding mesh formulae become too complicated to be derived explicitly. If the boundary is far distant from the region of electron trajectories, one will still use the fiv point formulae. If there are parts of the boundary where more accurate approximatic are necessary, the following procedures can be applied.

According to a proposal of DENEGRI et al. [2.10], one can regard the N + 1 lowes derivatives $\partial^n P/\partial u^j \partial v^k$ in (2.39) as unknowns; the factors $(u_i-u_0)^j(v_i-v_0)^k/(j!k!)$, calculated from the given coordinates (u_i,v_i) of N points surrounding the mesh point (u_0,v_0), then form an $N \times N$ matrix if $j + k > 0$. This matrix can be inverted numerically, and thus the required partial derivatives (with $n > 0$) may be expresse by the differences $P_j - P_0$ for $j = 1 \ldots N$, where $P_1 \ldots P_N$ are known, while P_0 is unknown. Introducing these expressions into (2.38), one obtains a linear finite-difference equation for the unknown P_0.

DENEGRI et al. gave examples that seem to demonstrate a higher accuracy than can be obtained with the conventional finite-element method. The main difficulty with their method is that the matrix will often have a rank less than N if the geometric configuration of the mesh points is not suitably chosen. This occurs mainly if some of the locations are regular.

If the differential equation is comparatively simple, for instance (2.20), the following procedure is advantageous. At least some analytical solutions $\phi_\mu(z,r)(\mu = 1,\ldots,N)$ are known that are valid in a finite region G. We now form a linear combination

$$\psi(z,r) = \sum_{\mu=1}^{N} c_\mu \phi_\mu(z,r) \quad , \tag{2.56}$$

which represents a more general solution with initially unknown coefficients. These may be determined by fitting ψ to the known values ψ_j $(j = 1 \ldots N)$ at N discrete points. In order to perform this we first calculate the matrix elements

$$A_{j\nu} = \phi_\nu(z_j,r_j) \ , \quad (j = 1,\ldots,N \quad , \quad \nu = 1,\ldots,N) \quad ,$$

the location of these points to be chosen in such a way that the matrix A is invertible, $B = \{B_{\mu\ell}\}$ being the inverse matrix. Since

$$\psi_j = \sum_{\nu=1}^{N} A_{j\nu} c_\nu \quad , \quad (j = 1,\ldots,N)$$

is valid, the inverse relation is given by

$$c_\mu = \sum_{\ell=1}^{N} B_{\mu\ell}\psi_\ell \ , \quad (\mu = 1,\ldots,N) \quad .$$

Introducing this into (2.56), we find the solution

$$\psi(z,r) = \sum_{\ell=1}^{N} \sum_{\mu=1}^{N} B_{\mu\ell}\psi_\ell\phi_\mu(z,r) \quad .$$

This solution is now available for the calculation of the potential at any point located in the region G, for instance at a mesh point (z_o,r_0). Altogether, we obtain a linear relation

$$\psi(z_o,r_0) = \psi_o = \sum_{\ell=1}^{N} b_\ell\psi_\ell \quad , \tag{2.57}$$

the coefficients being

$$b_\ell = \sum_{\mu=1}^{N} B_{\mu\ell}\phi_\mu(z_o,r_0) \quad . \tag{2.58}$$

The advantage of this method compared to that of DENEGRI et al. is a decrease in the rank of the linear system to be inverted, the order of approximation being equal. Thus the danger of singularity of the matrix A has decreased.

A practical application is demonstrated in Fig.2.7. Equation (2.57) is used as an asymmetric ten-point formula (N = 9) for the calculation of the potential at the point 0.

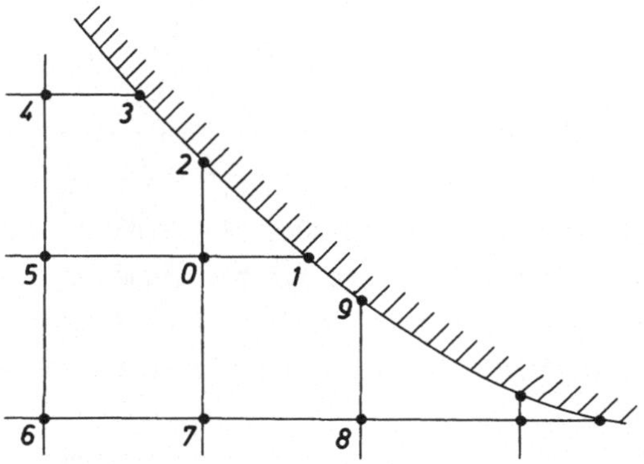

Fig.2.7. Numbering of the mesh points in an asymmetric ten-point formula

For the case of rotationally symmetric solutions of Laplace's equation [(2.20) with α = 1], a suitable set of basic functions ϕ_μ is given by

$$\phi_1 = 1 \ , \quad \phi_2 = z \ , \quad \phi_3 = r_0 \ln\frac{r}{r_0} \quad , \quad \phi_4 = z\phi_3 \quad , \tag{2.59}$$

$$\phi_5 = z^2 - \frac{1}{2}(r^2 - r_0^2) + r_0\phi_3 \quad ,$$

$$\phi_6 = z^3 - \frac{3z}{2}(r^2 - r_0^2) + 3r_0\phi_4 \quad ,$$

$$\phi_7 = \left[z^2 - \frac{1}{2}(r^2 + r_0^2)\right]\phi_3 + \frac{r_0}{2}(r^2 - r_0^2) \quad ,$$

$$\phi_8 = z^4 - 3z^2r^2 + \frac{3}{8}r^4 + 6r_0\phi_7 + 3r_0^2\phi_5 - \frac{3}{2}r_0^3\phi_3 - \frac{3}{8}r_0^4 \ ,$$

$$\phi_9 = \left[z^3 - \frac{3z}{2}(r^2 + r_0^2)\right]\phi_3 + \frac{3r_0z}{2}(r^2 - r_0^2) \ . \qquad (2.60)$$

These functions are linear combinations of simple homogeneous polynomial solutions of Laplace's equation and the well-known cylindrical solution $\phi \sim \ln r$. The combinations are chosen in such a way that all functions are linearly independent and that all, except ϕ_1, have a zero of unique order at the origin (z_o, r_0), $z_o = 0$ being chosen for reasons of conciseness. If this point is identified with the mesh point (z_o, r_0) in (2.57), (2.58) simplifies to $b_k = B_{1k}$, $(k = 1,\dots,N)$.

By use of the procedure described above one obtains a very accurate discretization even in irregular mesh points close to the boundary, the discretization error being decreased to the fifth order in the mesh-width. If the grid has a regular rectangular shape, the function ϕ_9 can be omitted and the procedure then results in a nine-point formula of fifth order. This is not applicable to mesh points located on the axis or on the next row above it, but there, one does not need it in practice

Special formulae valid for the mesh points adjacent to sharp edges are given by LENZ [2.44]. These are derived for straight edges, that means for solutions of Laplace's equation in a plane perpendicular to the edge. Therefore it is not clear whether they give better results than (2.54). Detailed investigations on the vicinity of field singularities have also been made by DURAND [2.12]. We shall not discuss this matter here.

2.2.3 The Finite-Element Method

The FDM is convenient for the solution of Dirichlet problems. It can also be applied to solve boundary-value problems involving Neumann conditions if the field is confined to regions with continuous material parameters (for instance [Ref.2.83, p.181]

Its application to cases of interface conditions like those for the magnetic field at surfaces of yokes with finite permeability becomes very complicated, though this is possible in principle [2.10]. For the computation of such fields the finite-element method (FEM) is more favourable.

This method is familiar in mechanical and electrical engineering (for instance [2.88,89,8] and has been mathematically investigated in great detail by NORRIE and de VRIES [2.58] and AZIZ [2.3]. In electron optics, the FEM was first introduced by MUNRO, who used it for the computation of the magnetic field in round lenses [2.53-55]. Applications to magnets in particle accelerators are described by COLONIAS [2.8], who gives detailed instruction for the use of complete and generally accessible computer programs.

The FEM is based upon two elementary ideas. The first is to use triangular-mesh grids instead of rectangular ones, since arbitrary triangles are the most elementary kinds of elements. For example Fig.2.8 demonstrates the discretization in an axial section through a round magnetic lens treated by KERN.

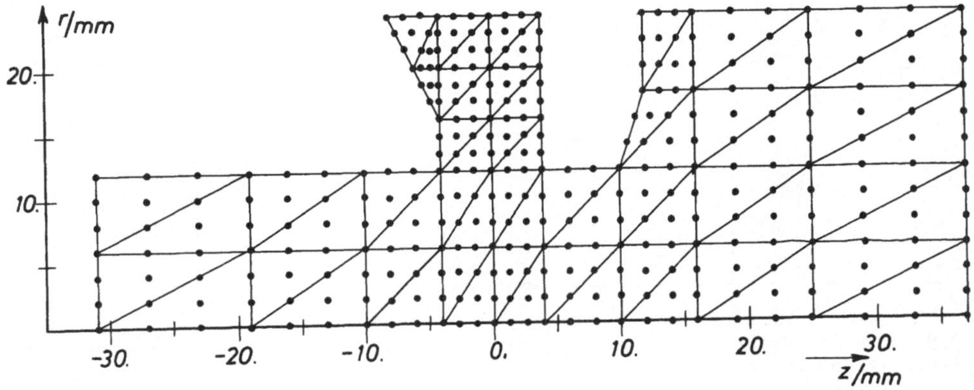

Fig.2.8. A triangular-mesh grid chosen for the application of the finite-element method to a magnetic lens. In the third-order computation, the triangles were used as elements and the potential was computed at the points marked by dots. In the first-order computation all these points were used as vertices of smaller triangles not shown in the figure

The second basic idea is to derive the discretization formulae from a variational principle. First, one has to know the variational relation equivalent to the differential equation to be solved. Then, one has to introduce several simplifications in order to derive a reasonably simple expression for the value of the corresponding functional in an arbitrary triangle. Finally, one obtains a set of discretization formulae from the condition that the functional must be an extremum with respect to the values of the potential in all internal mesh points.

GALERKIN's method can also be used to derive finite-element equations [2.58]. This is essentially equivalent to starting from a variational principle. In the following section we shall use the more traditional variational route. For brevity we shall mostly confine our considerations to computations of circular vector potentials, such as those performed by MUNRO.

2.2.3a Application to Circular Vector Potentials

The basic variational principle is given by (2.13). Expressing this in cylindrical coordinates by means of (2.26,27,30,31), we obtain

$$F = 2\pi \int\int_Q \left\{\frac{1}{2\mu}\left[\left(\frac{\partial A}{\partial z}\right)^2 + \left(\frac{\partial A}{\partial r} + \frac{A}{r}\right)^2\right] - jA\right\} r\, dr\, dz \quad , \tag{2.61}$$

Q being the region of solution in the axial section through the lens. In order to evaluate (2.61), simplifying assumptions on the functions $A(z,r)$, $j(z,r)$ and $\mu(z,r)$ are necessary. It is customary to assume all these functions to be polynomials in each triangle, the polynomial approximation being different for different triangles. At the vertices the functions are to be continuous.

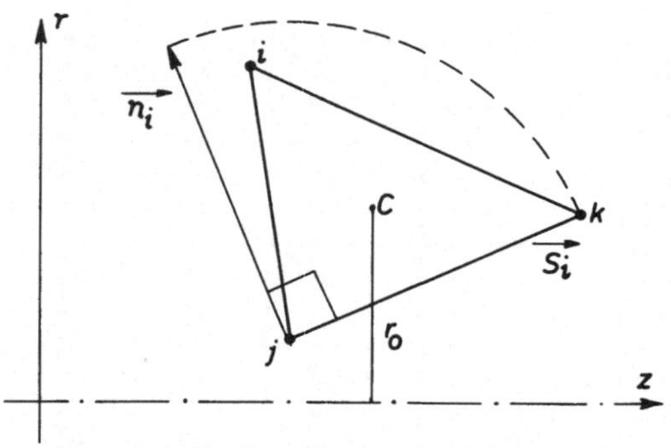

Fig.2.9. Notation referring to an arbitrary triangular finite element. For reasons of simplicity only one oriented side vector $\underline{s}_i$ and the corresponding normal vector $\underline{n}_i$ of equal length are shown. Point C is the centroid of the triangle

In the lowest order, used by MUNRO, the function A(z,r) is assumed to vary linearly in each triangle, while j(z,r) and μ(z,r) are regarded as piecewise constant and are replaced by their values at the corresponding centroid (z_o,r_0). With the further simplification $\iint_\Delta z^m r^m\,dz\,dr = z_o^m r_0^m\,a$, a being the area of triangle Δ, MUNRO derived the following expression for the contribution ΔF of an arbitrary triangle to the integral (2.61):

$$\Delta F = \frac{\pi r_0}{4\mu a}\left[\left(\sum_{i=1}^{3} b_i A_i\right)^2 + \left(\sum_{i=1}^{3} d_i A_i\right)^2\right] - \frac{2\pi}{3} r_0 a j \sum_{i=1}^{3} A_i \quad . \tag{2.62}$$

Here A_i is the value of A(z,r) at the vertex i, $b_i = r_j - r_k$, and $d_i = z_k - z_j + 2a/r_0$ (Fig.2.9).

By differentiating (2.62) with respect to A_i, a linear relation

$$\frac{\partial}{\partial A_i}\Delta F = \sum_k F_{ik}A_k - G_i \tag{2.63}$$

is obtained, the coefficients being

$$F_{ik} = \frac{\pi r_0}{2\mu a}(b_i b_k + d_i d_k) \quad , \quad G_i = \frac{2\pi}{3} r_0 a j \quad . \tag{2.64}$$

The matrix F_{ik} is symmetric. Expressions of this kind hold for each vertex of each triangle in the grid.

In order to generate finite-element equations one has to consider all those triangles having one common vertex in an arbitrary node, as shown in Fig.2.10, the notation being explained there. The essential condition is now that the contribution of this whole set of triangles to the functional F must be minimized with respect to the central value A_o, the marginal values $A_1,\ldots,A_N$ being kept fixed. This means that

$$\frac{\partial}{\partial A_o}\left(\sum_{j=1}^{N} \Delta F_j\right) = 0 \tag{2.65}$$

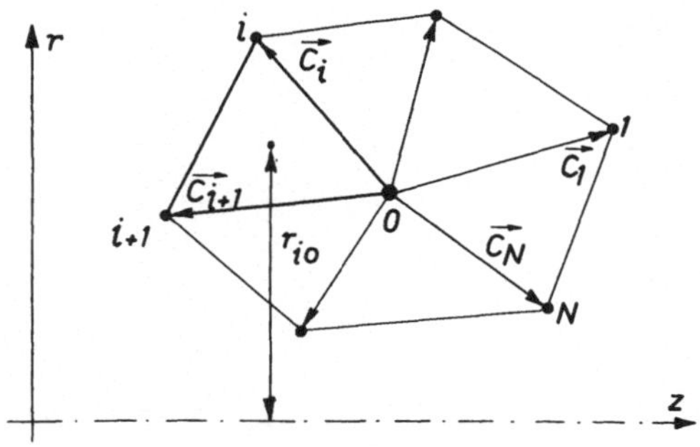

Fig.2.10. The system of triangular finite elements involved in the computation at the arbitrary internal mesh point 0. For reasons of simplicity, only the notation referring to one element i (with $1 \leq i \leq N = 6$) is shown

must be valid. Introducing (2.63) and (2.64) into (2.65) and taking care to change the index notation properly, one obtains a linear relation

$$\sum_{m=0}^{N} p_m A_m = q \quad , \tag{2.66}$$

where p_m is the sum of appropriate matrix elements F_{0j} and q is the sum of appropriate values of G_j.

In this way the value of A(z,r) at each internal node is uniquely related to those at the adjacent nodes. At the axis and the outer boundary the values A = 0 must be used. Interface conditions are simply fulfilled by using the appropriate value of $\mu(z,r)$ at the centroid of each triangle. In summary, a unique system of linear equations is generated which can be solved by use of standard techniques. As a result, one obtains the values of A(z,r) at the nodes of a discrete triangular grid.

2.2.3b Application to Saturated Magnetic Lenses

The application of the FEM to saturated magnetic lenses has also been performed by MUNRO [2.54,55]. Equation (2.12) has now to be evaluated in cylindrical coordinates: $F = 2\pi \iint_Q [U(B) - j\,A] r\,dr\,dz = \min$, where $B = |\underline{B}|$ is given by

$$B = \left[\left(\frac{\partial A}{\partial z}\right)^2 + \left(\frac{\partial A}{\partial r} + \frac{A}{r}\right)^2\right]^{1/2} \quad .$$

The formulation of the finite-element equations is analoguous to that in the previous example. The only important difference to (2.66) is that these new equations become essentially nonlinear. These will not be investigated here, the reader is referred to MUNRO.

In order to calculate the flux density $\underline{B}(\underline{r})$ at any point of the field, one has to perform numerical differentiation and interpolation (Sect.2.4). The results for the lines of magnetic flux and the axial flux density, obtained by MUNRO, are shown in Figs.2.11,12. Figure 2.11 refers to an unsaturated lens and Fig.2.12 to a saturated lens. The main difference between these cases is the appearance of a second broad and flat peak of the axial flux density in a saturated lens.

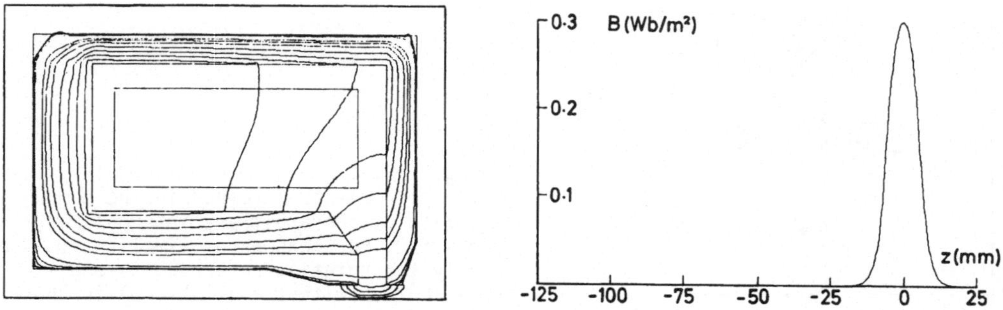

a) b)

Fig.2.11a,b. Results of the application of the finite-element method to an unsaturated magnetic lens: (a) distribution of the magnetic flux, (b) distribution of the axial flux density ([2.54], courtesy of Academic Press)

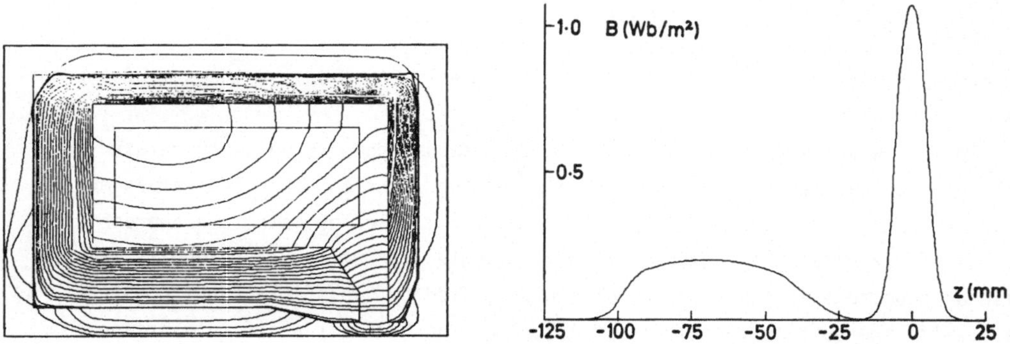

a) b)

Fig.2.12a,b. Results of the application of the finite-element method to a saturated magnetic lens: (a) distribution of the magnetic flux, (b) distribution of the axial flux density, ([2.54], courtesy of Academic Press)

2.2.3c The Computation of Scalar Potential Fields

Here we shall confine our attention to rotationally symmetric potentials $\psi(z,r)$ and only briefly report the results.

The appropriate variational principle, obtained from (2.14), is given by

$$F = \pi\mu_0 \iint\limits_Q \left[\left(\frac{\partial\psi}{\partial z}\right)^2 + \left(\frac{\partial\psi}{\partial r}\right)^2\right] r\, dr\, dz = \min \quad .$$

The results of the finite-element approximation in linear (lowest) order are, according to MUNRO [2.54,55],

$$\Delta F = \frac{\pi\mu_0 r_0}{4a}\left[\left(\sum_{i=1}^{3} b_i\psi_i\right)^2 + \left(\sum_{i=1}^{3} d_i'\psi_i\right)^2 \quad , \tag{2.67}$$

$$\frac{\partial}{\partial\psi_i}\Delta F = \sum_k F_{ik}'\psi_k \quad ,$$

$$F'_{ik} = F'_{ki} = \frac{\pi\mu_o r_0}{2a} (b_i b_k + d'_i d'_k) \quad , \tag{2.68}$$

where $b_i = r_j - r_k$ (unchanged) and $d'_i = z_k - z_j$. The notation is slightly different from that used by MUNRO.

It is now easy to give the explicit expression for the linear relations between the potentials in adjacent nodes. This expression has not been published by MUNRO. The quantities b_i, d'_i appearing in (2.67) and (2.68) are the components of a vector $\underline{n}_i$ orthogonal to the corresponding oriented side vectors $\underline{s}_i$ of the triangle, the lengths being equal, $|\underline{n}_i| = |\underline{s}_i|$. This is demonstrated in Fig.2.9. By use of $b_i b_k + d'_i d'_k = \underline{n}_i \cdot \underline{n}_k = \underline{s}_i \cdot \underline{s}_k$, the linear relation between the potentials can be expressed by side vectors, the result being

$$\psi_0 \cdot \sum_{i=1}^{N} \frac{r_{io}}{a_i} (\underline{c}_{i+1} - \underline{c}_i)^2$$

$$= \sum_{i=1}^{N} \frac{r_{io}}{a_i} [\underline{c}_i^2 \psi_{i+1} + \underline{c}_{i+1}^2 \psi_i - \underline{c}_i \cdot \underline{c}_{i+1}(\psi_i + \psi_{i+1})] \quad . \tag{2.69}$$

Here the vectors $\underline{c}_1,\dots,\underline{c}_N$ are the side vectors directed from the central vertex 0 to the adjacent vertices, as shown in Fig.2.10. The notation is cyclic, so that $\underline{c}_{N+1} \equiv \underline{c}_1$, $\psi_{i+N} \equiv \psi_i$. The quantity a_i is the area of the ith triangle and r_{io} the radial coordinate of its centroid.

2.2.3d Comparison of the Finite-Element Method with the Finite-Difference Method

With respect to differentiation and interpolation it is favourable even in applications of the FEM to use square-shaped or at least rectangular grids as far as possible. In some regions of the field more general quadrangles may be necessary. The necessary triangular grid is now obtained by dissecting each quadrangle along one of its diagonals. There are thus many ways of generating triangular grids based upon the same quadrangular pattern.

Figure 2.13 shows some configurations of triangles based upon a square-shaped grid; these are only a few typical examples. The evaluation of (2.69) for the different configurations now yields the following results. For the configurations A or B or those obtained by a symmetry operation $z \to z_0 - z$ with respect to the corresponding centroid, the off-axis Liebmann formula (2.45) is obtained, the notation being changed appropriately. One might expect that configuration C, due to its higher symmetry, will give a better approximation, but this is not true; the result is surprisingly

$$\psi_0 = (\psi_1 + \psi_3 + \psi_5 + \psi_7)/4 + h(\psi_3 - \psi_7)/(6r_0) \quad ,$$

which is clearly wrong. For a node 0 located on the axis ($r_0 = 0$) only six neighbours are available. Correspondingly, only the upper half of configurations like

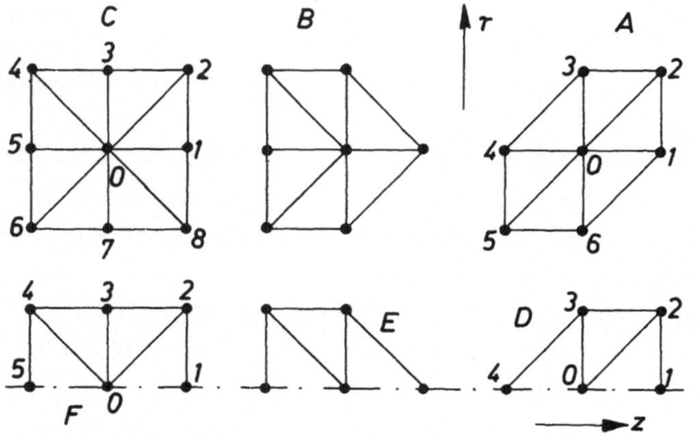

Fig.2.13. Different configurations of triangular finite elements obtained by dissection of square-shaped elements. Only the configurations A, B, and F yield correct results

A, B, or C can be used in (2.69). Thus there remain two essential configurations D and F. But now configuration D gives the wrong result $\psi_0 = (\psi_1 + 3\psi_3 + \psi_4)/5$, while F leads to (2.46), the notation being changed properly. The wrong result mentioned above cannot be avoided by averaging the results calculated from D and E. In every case it can be shown that the potentials at the mesh points in diagonal directions relative to the central point do not contribute to the value of ψ_0, since the corresponding factors in the linear combination vanish.

A reasonably simple evaluation of (2.69) is still possible for triangular mesh grids obtained by dissecting a rectangular pattern with variable mesh-widths in both directions. In this case even configurations with the same topological structure as configuration A in Fig.2.13 yield an inherent discretization error of second order in the mesh-widths. Again the accuracy cannot be improved by symmetrization since the factors of the potentials in diagonal directions with respect to the central point are zero. The second-order part of the discretization error decreases as r_0^{-1}, the remaining part being that of the corresponding asymmetric Liebmann formula. This shows that the terms arising from the rotationally symmetric structure of the field are inaccurate, while for plane Laplace fields, FDM and FEM, both in their lowest order, are equivalent.

From this point of view the FEM seems to be uneconomical, as it uses more neighbours of every mesh point and yet does not achieve a better accuracy than the FDM, as long as the basic mesh grid is rectangular. The best accuracy of the FEM is obtained with a regular grid of equilateral triangles, that is, with a regular hexagonal configuration. Then in regions far distant from the axis ($r_0 \gg h$) the discretization error is of sixth order in the mesh width h, while in paraxial regions it is only slightly increased. For the axial mesh points of a regular hexagonal structure, however, (2.69) yields the wrong result, $\psi_0 = [\psi_1 + 3(\psi_2 + \psi_3) + \psi_4]/8$, which must not be used, while the FDM gives the correct formula,
$\psi_0 = [\psi_1 + 8(\psi_2 + \psi_3) + \psi_4]/18$.

This also gives the answer to the question of what is the best triangulation of grids with skew quadrilateral elements. One should use the configuration that has the smallest differences between the angles since this is closer to the ideal equilateral structure. This choice is better than averaging over all possible configurations. Whenever possible, one should avoid obtuse or very small acute angles. In every case the axial mesh points require special formulae.

The reduced accuracy of the finite-element method just in the most important paraxial region is a consequence of the linear approximation of the potential. For the vector potential this is still reasonable. For the scalar potential this assumption is worst precisely on the axis of rotational symmetry. A much higher accuracy can be achieved by finite-element approximations of higher order. Not only are the original vertices then used but also additional vertices on the circumference and in the interior of each triangle. The potentials at these points are additional variables. The resulting finite-element equations of higher rank are too complicated to be given here. The reader is referred to SILVESTER [2.75], and SILVESTER and KONRAD [2.76], who first derived these equations. For an equal total number of mesh points the FEM of higher order may be more accurate than that of first order. This has been tested by SILVESTER and KONRAD [2.76]. The application of the higher-order approximations in the FEM may, however, be problematical, as will be shown below.

As an illustration for the present chapter, KERN has kindly performed the following investigations on the FEM. For the mesh grid shown in Fig.2.8 a program written by SILVESTER and KONRAD [2.77] was used for computing the scalar potential in the first- and third-order approximation, the yoke potentials being zero and unity, respectively. The axial potential values were differentiated by means of the cubic spline technique. The results are shown in Fig.2.14. The first-order approximation (curve A) gives a slightly larger maximum than the third-order one. The same lens was also computed using a program written by the author; this program is based upon an analytical method explained in Sect.2.5.2. The use of 120 rings and 4 apertures gave a very accurate result. The values of the maximum field strength are

$$H_{max} \cdot m/I = \begin{cases} 51.74 \text{ using the first-order FEM,} \\ 51.11 \text{ using the third-order FEM,} \\ 50.84 \text{ using the analytical method} \end{cases} .$$

The maximum error is about 2%.

Smooth curves like those shown in Fig.2.14 are only obtained if the order of the FEM approximation does not change within the paraxial region of solution. The use of different orders in different domains causes unreasonable oscillations of the field strength in the vicinity of the borders between these domains. The second order gave useless results even if it was used in the whole region. Further-

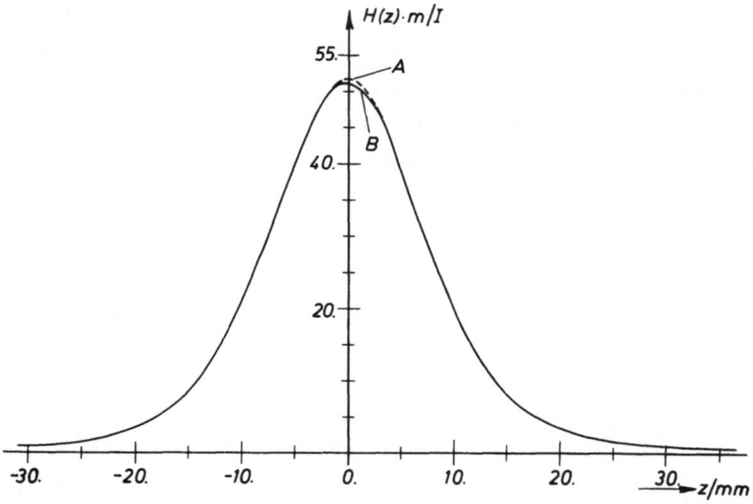

Fig.2.14. Results of the application of the finite-element method to the configuration shown in Fig.2.8. Curves A and B show the axial field strength H(z) of the magnetic lens in a first- and third-order approximation, respectively. The difference between the two curves is very small and can be shown only in the vicinity of the maximum

more, in interpolation procedures and computations of radial derivatives, it was noticed that in all cases the off-axis values of the potential were not accurate enough. Detailed investigations are still necessary to study these effects, but it is obvious that care has to be taken in electron optical applications of the FEM. A direct comparison of the FEM with the FDM has been made by LENZ. The method is similar to that used for calculating the data in Table 2.1. The test function chosen here,

$$\psi(z,r) = J_0(r/2h) \cdot \exp(z/2h) \quad ,$$

is an analytic solution of (2.20) for $\alpha = 1$. The results are shown in Table 2.2.

If one wants to avoid the complicated FEM of higher order, one can combine the FEM of first order with the FDM. In order to do this one has to use a square-shaped grid as much as possible. In the remaining part a triangular grid fitting the interfaces and the boundary is to be generated. Now in the square-shaped part one has to apply the corresponding finite-difference equations, for instance (2.54, 55) with $\alpha = 3$ for the function P(z,r) defined by (2.28). In the triangular part the FEM of first order is to be used, for example that explained in Sect.2.2.3a. This seems to be an advantageous combination of both methods.

Table 2.2. Accuracies of the FDM and of the FEM, the mesh-widths being equal

r/h= \ z/h=	-1	0	+1	Discretization error
0	1.64872	1.00000 Analytic	0.60653	
		1.00152 FDM		0.00152
		1.01413 FEM (conf. D)		0.01413
		0.99999 9-point		-0.00001
1	1.54728	0.93847 Analytic	0.56921	
		0.94107 FDM		0.00260
		0.93129 FEM (conf. C)		-0.00718
		0.94130 rad. interpol.		0.00283
		0.93848 9-point		0.00001
2	1.26160	0.76520 Analytic	0.46412	
		0.76734 FDM		0.00214
		0.75845 FEM (C)		-0.00675
		0.76519 9-point		-0.00001
3	0.84386	0.51183 Analytic	0.31044	
		0.51329 FDM		0.00146
		0.50577 FEM (C)		-0.00606
		0.51183 9-point		0.00000
4	0.36913	0.22389 Analytic	0.13580	
		0.22459 FDM		0.00070
		0.21875 FEM (C)		0.00514
		0.22390 9-point		0.00001
5	-0.07977	-0.04838	-0.02935	
		-0.04842 FDM		-0.00004
		-0.05245 FEM (C)		-0.00407
		-0.04837 9-point		0.00001
6	-0.42875	-0.26005	-0.15773	

Obviously the use of wrong FE configurations (Fig.2.13) causes large discretization errors. The correct FE configurations, being equivalent to five-point FD configurations, are far less accurate than nine-point FD configurations.

2.3 Solution of the Discretized Equations

2.3.1 The General Structure of the Discretized Equations

Altogether the discretized equations derived in Sect.2.2 usually form a linear system of equations for the potential at every internal mesh point. If high accuracy is required, this system will be very large.

For its solution, a unique ordering of the mesh points is necessary. This means that in a scan over the whole region of solution, all internal mesh points including those at the axis of symmetry, all mesh points at interfaces, and all those at the boundary with given Neumann conditions must be encountered exactly once, while boundary points with Dirichlet conditions must be excluded from this scan. There are many equivalent possibilities of ordering, for instance along the rows and columns of a grid.

In the following presentation we do not require any special kind of ordering, and for conciseness we shall use only one subscript for the numbering of the mesh points; this runs from 1 to the total number N of mesh points. The whole set of potential values then forms a linear array $\underline{P} = \{P_1 \ldots P_N\}^T$ which satisfies an ordinary system of linear equations of rank N.

With respect to iterative solution techniques, the form

$$P_i = \sum_{k=1}^{N} C_{ik} P_k + Q_i \quad , \quad (i = 1, \ldots, N) \tag{2.70}$$

with $C_{ii} = 0$ for $i = 1, \ldots, N$ is favourable. Many discretized equations, for instance (2.45,46,54,55,69), already have this form. The $N \times N$ matrix C is built up from the coefficients appearing in the mesh formulae; it is a sparse matrix, which means each row contains only a small number of nonzero elements. The linear array $\underline{Q} = \{Q_1 \ldots Q_N\}^T$ contains the source terms in the mesh formulae and terms arising from the boundary conditions.

There are numerous techniques for solving systems of linear equations. Among these we shall only briefly discuss the successive overrelaxation method as a very convenient one.

2.3.2 The Successive Overrelaxation Method

The successive overrelaxation method (SOR) is an iterative method for the solution of systems like (2.70). It is quite extensively discussed in the mathematical literature (for instance [2.1,13,80,83,85]).

The iteration procedure is defined by

$$S_i^{(n+1)} = \sum_{k=1}^{i-1} C_{ik} P_k^{(n+1)} + \sum_{k=i+1}^{N} C_{ik} P_k^{(n)} + Q_i \quad , \tag{2.71}$$

$$P_i^{(n+1)} = P_i^{(n)} + \omega (S_i^{(n+1)} - P_i^{(n)}) \quad . \tag{2.72}$$

Here superscripts in parentheses denote the number of the corresponding iteration, while subscripts refer to the corresponding variables or matrix elements.

The auxiliary quantity $S_i^{(n+1)}$, introduced in (2.71), is the well-known Gauss-Seidel value of the variable P_i (in the (n+1)th iteration as the index shows). The factor ω appearing in (2.72) is called the relaxation parameter. Usually, $1 < \omega < 2$ is valid. The corresponding amplification of the difference between the new value and the preceding one, expressed by the term in ω in (2.72), is known as successive overrelaxation. This procedure causes a considerable acceleration of the convergence, provided that ω is chosen appropriately.

In order to start the iteration process, one has to define initial values $P_i^{(0)}$, $(i = 1,\dots,N)$. These may be chosen arbitrarily and in practice a rough guess is always sufficient. During the iteration process the maximum error defined by

$$\Delta^{(n)} = \max_i |S_i^{(n+1)} - P_i^{(n)}| \quad , \qquad (i = 1,\dots,N)$$

and, for reasons explained later, the sum

$$D^{(n)} = \sum_{i=1}^{N} |S_i^{(n+1)} - P_i^{(n)}|$$

are calculated. At the end of each iteration over the whole array of variables these quantities are examined. The iteration process can be stopped if $\Delta^{(n)}$ has decreased below a given error limit ε. The convergence is then assumed to be sufficient.

In the most general case the theory of convergence is extremely complicated. The assumptions, described in the mathematical literature, are only necessary in order to perform the corresponding proofs. In practice, however, good convergence can be achieved in general cases which are too complicated to satisfy the conditions required for an exact proof. Here we shall not treat the standard theory of SOR but only illustrate some of its results.

The basic starting point for this theory is the series expansion of the variables $P_i^{(n)}$ with respect to the eigenvectors of the matrix C, defined by

$$C\underline{e}_\mu = \lambda_\mu \underline{e}_\mu \quad , \qquad (\mu = 1,\dots,N) \quad . \tag{2.73}$$

This series expansion has the general form

$$\underline{P}^{(n)} = \underline{P}^{(\infty)} + \sum_{\mu=1}^{\infty} A_\mu \alpha_\mu^{2n} T_\mu \underline{e}_\mu \quad . \tag{2.74}$$

Here the quantities A_μ are the amplitudes of the eigenmodes, depending on the initial conditions; the α_μ are the damping factors for the different eigenmodes. Convergence is only possible for $|\alpha_\mu| < 1$. The expressions T_μ are $N \times N$ matrices depending on the eigenmodes and on the ordering of (2.70) but not on the iteration number n. Equation (2.74) describes a linear superposition of the final solution and of exponentially damped oscillations caused by the wrong initial values.

In order to evaluate (2.74) further assumptions on the matrix C are necessary. The first is that the eigenvalues defined by (2.73) are real and satisfy $|\lambda_\mu| < 1$, the largest being $\lambda_1 > 0$. The second assumption is that the matrix C has property (A). This property, introduced by YOUNG [2.86] in 1954, cannot be discussed here, and the reader is referred to the corresponding literature. The result of these assumptions is an algebraic relation,

$$\alpha_\mu^2 - \omega\lambda_\mu\alpha_\mu + \omega = 1 \quad . \tag{2.75}$$

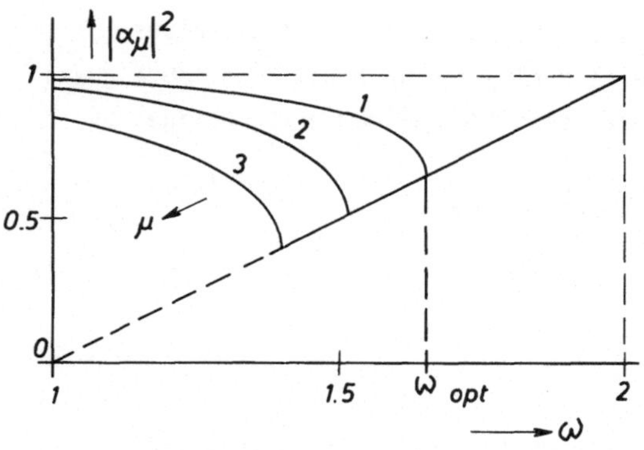

Fig.2.15. The damping factors $|\alpha_\mu|^2$ for $\mu = 1,2,3$ as a function of the relaxation parameter ω. The minimum of $|\alpha_1|^2$ yields the optimum value of ω

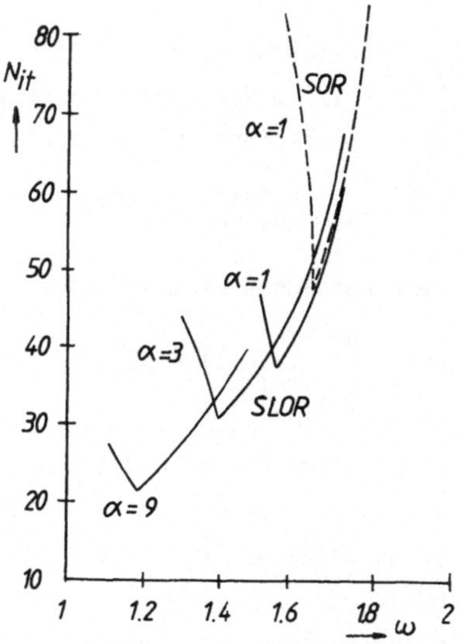

Fig.2.16. Number N_{it} of iterations required versus overrelaxation coefficient ω for SOR and SLOR, the initial conditions and the iteration error limit being equal. The example chosen is a polynomial solution of (2.20), α being the coefficient $2|m| + 1$; field size: $0 < z/h < 20$, $r/h < 10$

From this equation it can be shown that $|\alpha_\mu| < 1$ is only possible for $0 < \omega < 2$. Asymptotically, the eigenmode with the largest value of $|\alpha_\mu|$ is dominant. From the solution of (2.75) it can be seen that this is the mode with the maximum value of $|\lambda|$, thus the first one. Minimizing $|\alpha_1|$ results in

$$\omega_{opt} = 2 \cdot (1 + \sqrt{1 - \lambda_1^2})^{-1} \quad , \tag{2.76}$$

$$|\alpha|^2_{opt} = \omega_{opt} - 1 \quad .$$

Under these conditions the rate of convergence is maximum. The dependence of the damping factors $|\alpha_\mu|^2$ on the choice of ω is qualitatively shown in Fig.2.15 and that of the necessary number of iterations in Fig.2.16. This figure clearly demonstrates that it is essential to choose a value of ω very close to ω_{opt} in order to keep the necessary computing time short.

In order to calculate ω_{opt} one has to solve (2.73), which is even more complicated than solving Poisson's equation. Thus only rough estimates of the true value of ω_{opt} are possible. This has been a major drawback of the SOR for a long time. There have been many attempts to derive reasonable estimates for ω_{opt}. For practical purposes this problem has been solved with sufficient accuracy by CARRÉ [2.7]. The corresponding theory cannot be treated here; we shall only briefly discuss its results.

The approximate value of ω_{opt} is not calculated from a closed theory but successively estimated from the actual behaviour of the iteration process. This is to be started with a value $\omega \approx 1.4$ which must certainly be below ω_{opt}. After a fixed number N' of iterations, say $N' \approx 12$, the rate $|\alpha|^2$ of convergence is estimated from the last three available values of $D^{(n)}$ (if possible, by means of Aitken's acceleration formula). Now a rough estimate for ω_{opt} is given by

$$\omega_0 = 2\{1 + [1 - (|\alpha|^2 + \omega - 1)^2/(|\alpha|^2\omega^2)]^{1/2}\}^{-1} \quad .$$

This value may be too high. In order to avoid exceeding the true value of ω_{opt} the smaller value $\omega = \omega_0 - (2 - \omega_0)/4$ is used for the next N' iterations. This is to be repeated until the difference $|\omega - \omega_0|$ has become sufficiently small to be neglected. In practice ω_{opt} can always be found with sufficient accuracy. The number of iterations actually used to reach the error limit ε is only about 30% larger than the minimum number. This procedure is therefore acceptable.

In practice, the assumptions made in deriving (2.76) are rarely fulfilled. Nevertheless, the application of SOR presents little if any difficulty. For instance, the matrices obtained from nine-point formulae do not have property (A). Also, in (2.54) some of the coefficients may become negative. In spite of this, SOR can be applied satisfactorily [2.34]. If the grid contains irregular mesh points or couplings, as described in Sect.2.2.2a, the conditions for real eigenvalues [2.85] are violated, but in spite of this Carré's method is useful, after some minor corrections have been made [2.35]. The only major difficulty arises in the application of relaxation methods to (2.43). If the coefficient of P_4 becomes negative, local underrelaxation is necessary [2.71].

In summary, SOR is a very convenient method for the solution of large systems of linear equations. Its main advantages are its comparatively easy programming and its very low storage requirements. SOR is the method that requires the minimum of necessary memory capacity. Only the array $\underline{P} = \{P_1 \ldots P_N\}^T$ and some minor sets of coefficients are to be stored. The determination of the optimum relaxation parameter does not make difficulties, as has been shown above.

2.3.3 Survey of Other Methods

The most important direct method for solving systems of linear equations is the common Gauss algorithm of elimination. This is very convenient for smaller matrices, but for large matrices special techniques for saving memory capacities have to be applied [2.14,56,66]. These cannot be discussed here.

For enhancement of the convergence in iteration procedures, numerous special techniques have been developed. The most familiar are the successive line overrelaxation (SLOR) or block iteration method and the alternating direction implicit methods (ADI) (e.g., [2.1,83]). The application of SLOR has been favourably reported by SCHULER [2.72]. More recent developments are the cyclic reduction method [2.4,73] and STONE's method [2.81], which has been favourably reported by HERITAGE [2.26].

KASPER [2.34a] has tested the SLOR by applying it to (2.43,44) and to the linear system of equations resulting from these. Due to the symmetry with respect to the indices 1 and 3, the blocks were chosen to be the radial rows of the field. The algorithm is then as follows. In each row a linear subsystem of tridiagonal structure results if the potentials in the neighbouring rows are regarded as known for the moment. This tridiagonal system can be easily solved by means of the Gauss algorithm. Thereafter, the values obtained are modified by overrelaxation, as usual, after which the same algorithm proceeds to the next radial row; in this way the whole field is scanned. The entire process is continued until it has converged sufficiently. By comparison with SOR using an analytic test function, the following results are obtained.

A: SLOR is more stable. It converges even in those cases where SOR does not converge, especially if the coefficient β_4 becomes negative.

B: SLOR converges slightly faster then SOR, and is less sensitive to small deviations of ω from ω_{opt}.

C: Using the same error limit ε, the node potentials obtained with SLOR differ far less from the corresponding analytical values than those obtained with SOR.

The results are shown in Fig.2.16. They clearly demonstrate the points A and B. Result C is obvious from the fact that with $\varepsilon = 10^{-7}$ the final error with SOR is larger then ε, whereas with SLOR it is about 10^{-9}. This confirms SCHULER's observations. It is possible to find a form of the SLOR algorithm that is finally as simple as SOR. Its derivation cannot be given here. In summary it turns out that SLOR is clearly superior to SOR.

By use of these methods the computing time may be appreciably decreased, but, with the exception of SLOR, their implementation is more complicated than that of SOR. They will thus only become favourable if a program is designed in a very general manner for a large group of users and runs very frequently. If the saving of computing time is not the most important factor, SOR is more favourable.

Traditionally, SOR is used in combination with the FDM while the Gauss algorithm is combined with the FEM, but this is not necessary. In fact there are programs that solve finite-element equations by application of SOR [Ref.2.8, p.15]. From this point of view the combination of FEM and FDM in one program is quite possible and reasonable.

2.4 Differentiation and Interpolation in Mesh Grids

By use of the finite-difference method or of the finite-element method, one obtains the values of the potential on a discrete mesh grid. This is only the first step of the field calculation. For the computation of electron trajectories one needs the value of the field strength at an arbitrary point which is usually not a mesh point.

There are many ways of calculating the required components of the field strength. Most of them have the serious disadvantage of not being continuous along the lines separating adjacent meshes. Here we only discuss those methods that are continuous everywhere. The fulfillment of this requirement is essential for the application of predictor-corrector methods in the calculation of trajectories.

2.4.1 The Calculation of the Axial Flux Density

If the rotationally symmetric scalar potential $\psi(z,r)$ has been computed on the mesh grid, the values at the axial nodes can be used for the calculation of the axial flux density $B(z)$. This can be performed by applying the cubic spline technique [2.18,67] to the discrete values $\phi_i = \psi(z_i,0)$. This ensures the continuity of the functions

$$\phi(z) = \psi(z,0) \quad , \quad B(z) = -\phi'(z) \quad , \quad B'(z) = -\phi''(z) \quad .$$

In the case of equidistant mesh points along the axis, the following simpler procedure is advantageous. At the mesh points, the differentiation is performed numerically by

$$\phi_i' = \frac{1}{12h}\,[8(\phi_{i+1} - \phi_{i-1}) - \phi_{i+2} + \phi_{i-2}] + O(h^4) \quad , \tag{2.77}$$

or more accurately by

$$\phi_i' = \frac{1}{60h}\,[45(\phi_{i+1} - \phi_{i-1}) - 9(\phi_{i+2} - \phi_{i-2}) + \phi_{i+3} - \phi_{i-3}] + O(h^6) \quad . \tag{2.78}$$

(In the case of marginal points special asymmetric formulae are necessary.) The well-known Hermite interpolation is now applied, which means that in the interval $z_i \leqq z < z_{i+1}$ the cubic parabola exactly defined by the marginal values ϕ_i, ϕ_i' at $z = z_i$ and ϕ_{i+1}, ϕ_{i+1}' at $z = z_{i+1}$ is used for interpolation and differentiation. The first derivative is still continuous at the mesh points while the discontinuities of the second derivative can be neglected. The interpolation polynomial defined above is given by

$$\phi(z) = \phi_i \cdot (1 - 3u^2 + 2u^3) + \phi_{i+1} \cdot (3u^2 - 2u^3)$$
$$+ h\phi_i' \cdot (u^3 - 2u^2 + u) + h\phi_{i+1}' \cdot (u^3 - u^2)$$

with $h = z_{i+1} - z_i$ and $u = (z - z_i)/h$.

If the potential $P(z,r)$ defined by (2.28) has been calculated on the mesh grid, one numerical differentiation can be saved since (2.32) immediately yields

$$B(z) = P(z,0) \quad , \quad B'(z) = P'(z,0) \quad .$$

The techniques described above can then again be applied; they now work more accurately.

If the azimuthal component $A(z,r)$ of the vector potential has been computed on the mesh grid, the values of $B(z)$ are to be computed from the values of $A(z,r)$ in the second and third row. For a regular grid this will be explained in Sect.2.4.2, but if the grid is irregular, this differentiation becomes very complicated and will not be discussed here. From this point of view the use of the potential P(z,r) defined by (2.28) is distinctly preferable.

2.4.2 Interpolation in Square-Shaped Grids

Here we consider a two-dimensional function P(u,v) given at the mesh points of a square-shaped grid. The coordinates u,v and the grid will usually be the same as those introduced in Sect.2.2.1. We now want to calculate P(u,v) at an arbitrary point Q with coordinates (u,v). When computing equipotential lines and electron trajectories, the partial derivatives of the potential are also needed. There is a variety of different interpolation methods, which cannot all be described here. Criteria for the choice of the appropriate method are a wide range of applicability, a simple procedure in order to save computing time, and maximum smoothness in all directions. The last criterion turns out to be the most important with respect to the applicability of predictor-corrector methods for the trajectory calculation. In this context, this means that along the mesh lines, as many partial derivatives as possible must be continuous, since discontinuities may cause the algorithm for the adjustment of the step-width during a trajectory calculation to break down.

Most of the older methods, like those based upon nine-point formulae of WEBER [2.85] and upon sixteen-point Lagrange interpolation of LENZ [2.44a], turned out to be unfavourable in these respects. The method of KERN [2.35] and HAUKE [2.22] is more suitable but only applicable to rotationally symmetrical fields and is very complicated. The best method seems to be the bivariate Hermite interpolation, which will be outlined here.

This interpolation is built up from a sequence of one-dimensional Hermite interpolations in the u and v directions. This requires a knowledge of the potential P and its derivatives $\partial P/\partial u$, $\partial P/\partial v$, and $\partial^2 P/\partial u \partial v$ at the corners of the corresponding mesh cell (Fig.2.17). The partial differentiation at the nodes of the mesh grid can be performed by appropriate application of (2.77) or (2.78) or by use of the cubic spline technique with respect to the corresponding one-dimensional sequence of potential values. It is advantageous to store the results in two-dimensional arrays, as has been done for the potential.

The sixteen values corresponding to the configuration of Fig.2.17, $P_j=P(u_j,v_j)$, $U_j=P_u(u_j,v_j)$, $V_j=P_v(u_j,v_j)$, $W_j=P_{uv}(u_j,v_j)$, for j = 1,...,4, uniquely define the coefficients of a polynomial

$$P_{uv} = \sum_{j=0}^{3} \sum_{k=0}^{3} A_{jk}(u - u_1)^j \cdot (v - v_1)^k \quad .$$

The practical computation proceeds in the following way. First, the cell containing the point Q is determined. Then, two one-dimensional Hermite interpolations

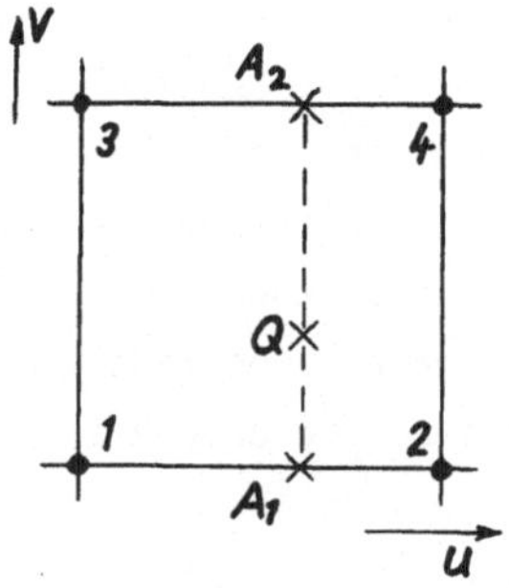

Fig.2.17. Configuration for the two-dimensional Hermite interpolation in square-shaped grids; Q is an arbitrary point in the field

and appropriate differentiations along the mesh lines in the u direction are performed, yielding the values of P, P_u, P_v, P_{uv} at the points A_1, A_2. Now, one interpolation and differentiation are made along the line $\overline{A_1A_2}$, and from these the final values of P, P_u, P_v are obtained. From the properties of the one-dimensional Hermite interpolation it can be concluded that all these functions and, in practice, the second-order derivatives also remain continuous if the point Q is located on a mesh line.

The potential itself is of little importance in magnetic field calculations. It thus appears to be preferable to omit the potential array and to use the values of P_u,P_v and all second-order derivatives P_{uu}, P_{uv}, P_{vv} at the nodes. The values of P_u and P_v at the point Q are then obtained from the Hermite interpolation itself, not by differentiation, and are thus more accurate and smooth. In the case of a partial differential equation of simple structure, e.g. (2.20), the storage of the entire arrays for P_{uu}, P_{uv}, P_{vv} can be avoided. For configurations like that shown in Fig.2.6a with u = z, v = r, the values of the second-order derivatives at a mesh point 0 are given by

$$P_{zz} = \frac{1}{12h}\left[4(Z_1 - Z_5) + Z_2 - Z_4 + Z_8 - Z_6\right] + \frac{h^2\alpha}{24r_0}(Z_2 - Z_4 + Z_6 - Z_8) \quad ,$$

$$K \cdot P_{zr} = \frac{1}{12h}\left[4(R_1 - R_5) + R_2 - R_4 + R_8 - R_6\right] + \frac{h^2\alpha}{24r_0}(R_2 - R_4 + R_6 - R_8) \quad ,$$

$$P_{rr} = -P_{zz} - \frac{\alpha}{r_0}R_0 \quad , \quad K = 1 + \frac{\alpha}{6}\left(\frac{h}{r_0}\right)^2 \quad , \quad r_0 \neq 0 \quad , \tag{2.79a}$$

where $Z_\mu = P_z(z_\mu, r_\mu)$, $R_\mu = P_r(z_\mu, r_\mu)$, $\mu = 0,\ldots,8$ are introduced for abbreviation. For a mesh point 0 located on the z axis, as shown in Fig.2.6b, the corresponding set of formulae is given by

$$P_{zz} = \frac{1}{6h}\left[(1 + \alpha)(Z_2 - Z_4) + (2 - \alpha)(Z_1 - Z_5)\right] \quad ,$$

$$P_{rr} = -(1 + \alpha)^{-1}P_{zz} \quad , \quad P_{zr} = 0 \quad . \tag{2.79b}$$

These formulae can be derived by use of Taylor series expansions about the point 0 and by application of the partial differential equations resulting from partial differentiations of (2.20); they are very accurate.

In the practical application to the bivariate Hermite interpolation, the evaluation of (2.79) refers to that configuration of sixteen points in which the point 0 successively coincides with the four corners of the mesh cell shown in Fig.2.17. Since frequent interpolations in the same cell usually have to be made, the computing time can be kept short if the second-order derivatives are only evaluated once immediately after entering the cell and then stored as long as the point Q remains in the cell. This method of interpolation seems to be the optimum.

2.5 Analytical Field Calculation

The rotationally symmetrical potential $\psi(z,r)$ in a round magnetic lens can be expanded as a series of given analytical solutions of Laplace's equation with coefficients to be determined from the boundary conditions. The method described below is in some respects related to that of Sect.2.2.2d, but here some new points of view will appear. We shall formulate this method for electric fields and then extend it to the magnetic analogue.

2.5.1 Superposition of Aperture Fields

Reasonable analytical solutions, which can be used to describe the field in the far off-axis region of electrodes with plane surfaces, are combinations of the fields of thin circular coaxial apertures. The potential of one such aperture element, located in the plane z = 0, is given by

$$V(z,r) = P + v\left[-\frac{Ru}{2}(E^l + E^r) + \frac{R}{\pi}(E^l - E^r)(1 + u \cdot \arctan u)\right] , \qquad (2.80)$$

where P is the potential of the electrode, R is the radius of the aperture, and E^l and E^r are the asymptotic electric field strengths on the left- and right-hand sides of the aperture plane, respectively. The variables u and v are oblate spheroidal coordinates defined by

$$r = R\sqrt{(u^2+1)(1-v^2)} , \quad z = Ruv , \qquad (2.81)$$
$$-\infty < u < \infty , \quad 0 \leqq v \leqq 1 .$$

A general linear combination of N such elementary fields is given by

$$V^A(z,r) = \sum_{i=1}^{N} V_i(z,r,\alpha_i) , \qquad (2.82)$$

where $\alpha_i = \{z_i, R_i, P_i, E_i^l, E_i^r\}$ is the whole set of parameters characterizing the ith element of this series. Here $z = z_i$ is the plane of the ith electrode, R_i the radius of its aperture, P_i its potential; E_i^l, E_i^r are the corresponding asymptotic field strengths. The parameters P_i, E_i^l, E_i^r have no physical meaning, they only serve to define functions like that in (2.80). The physical quantities are the real electrode potentials $\phi_1,\ldots,\phi_N$, the real homogeneous electric field strengths between the electrodes,

$$F_i = -\frac{\phi_{i+1} - \phi_i}{z_{i+1} - z_i} \quad , \quad z_{i+1} > z_i \quad , \quad i = 1,\ldots,N-1$$

(valid in the far off-axis region), and the asymptotic field strengths F^l, F^r. There are only N + 2 linearly independent parameters, whereas the sets $\alpha_1,\ldots,\alpha_N$ contain 3N parameters. Thus (2.82) contains 2N-2 redundant parameters, which can be chosen arbitrarily.

Superpositions of fields like those described above have been investigated by REGENSTREIF [2.65], LENZ [2.43], DOMMASCHK [2.11] and HOCH et al. [2.28] for example. These authors used different sets of the free parameters. The most convenient choice is that of HOCH et al. It is given in the following table.

Table 2.3. Choice of the electric parameters in a system of apertures

Aperture	P_i	E_i^l	E_i^r
1	ϕ_1	F^l	F_1
i	0	0	$F_i - F_{i-1}$
N	0	0	$F^r - F_{N-1}$

The formal parameters of the first element are identical with the corresponding physical parameters. The following elements are characterized by as many zeros as possible. The whole field is thus uniquely defined.

In the far off-axis region this solution fits very well to the real field if the planes $z = z_i$ are chosen as the surfaces of the electrodes and the parameters ϕ_i as the corresponding electrode potentials. In the vicinity of the bores, however, the equipotentials differ considerably from the surfaces of the electrodes.

2.5.2 Additional Superposition of the Fields of Charged Rings

In order to fit the analytical field to the boundary conditions in the vicinity of the bores, an additional superposition of the fields of charged rings may be used. The potential of one charged ring located coaxially in the plane z = 0 and having the radius a and the charge q is given by the well-known formula

$$V(z,r) = qK(k)/(2\pi^2\varepsilon_0 s) \quad , \tag{2.83}$$

using the abbreviations

$$s = \sqrt{z^2 + (r+a)^2} \quad , \quad k = 2\sqrt{ar}/s \quad . \tag{2.84}$$

The function K(k) is the complete elliptic integral of the first kind. The superposition of such fields can be performed in different ways. Here we shall briefly describe the method of HOCH et al. [2.28].

In this method each electrode is represented by two apertures and a finite sequence of charged rings, all located inside the electrode. The corresponding potentials of the apertures are calculated by appropriate linear extrapolations. In order to fit the surface of each electrode to the equipotential of the corresponding potential value, a sequence of control points on all essential parts of the surfaces is introduced, the total number being M. In the axial section through the system each point (z_i^o, r_i^o) representing the trace of a ring is associated with a control point (z_i, r_i) at the corresponding surface, as demonstrated in Fig.2.18.

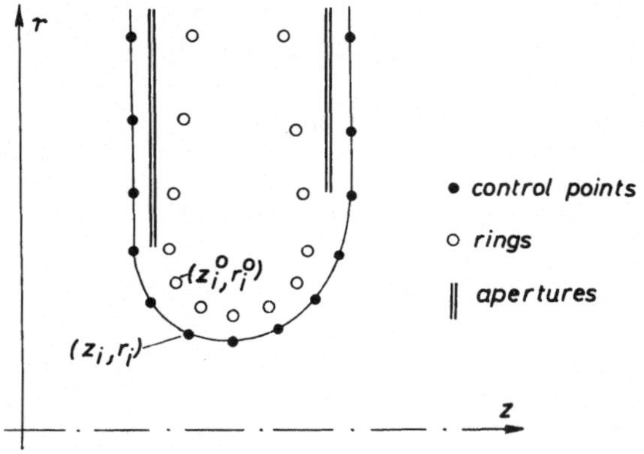

Fig.2.18. Example of a geometrical configuration used in an analytical field calculation

Let $V_k^o(z,r,z_k^o,r_k^o)$ be the potential produced at an arbitrary point (z,r) by a ring having unit charge and the coordinates (z_k^o, r_k^o). Then by solution of the linear system

$$\sum_{k=1}^{M} V_k^o(z_i,r_i,z_k^o,r_k^o)q_k = U_i - V^A(z_i,r_i) \quad (i = 1...M) \tag{2.85}$$

for the charges q_k, we can ensure that the potential of the whole system of apertures and rings is adjusted in such a manner that each control point (z_i,r_i) has the given electrode potential U_i. Now the potential at every point of interest can be calculated from

$$V(z,r) = V^A(z,r) + \sum_{k=1}^{M} q_k V_k^o(z,r,z_k^o,r_k^o) \quad .$$

The derivatives of this potential with respect to z and r can also be calculated without any difficulty since the basic potentials, defined by (2.80,81,83,84), can be differentiated analytically [2.28].

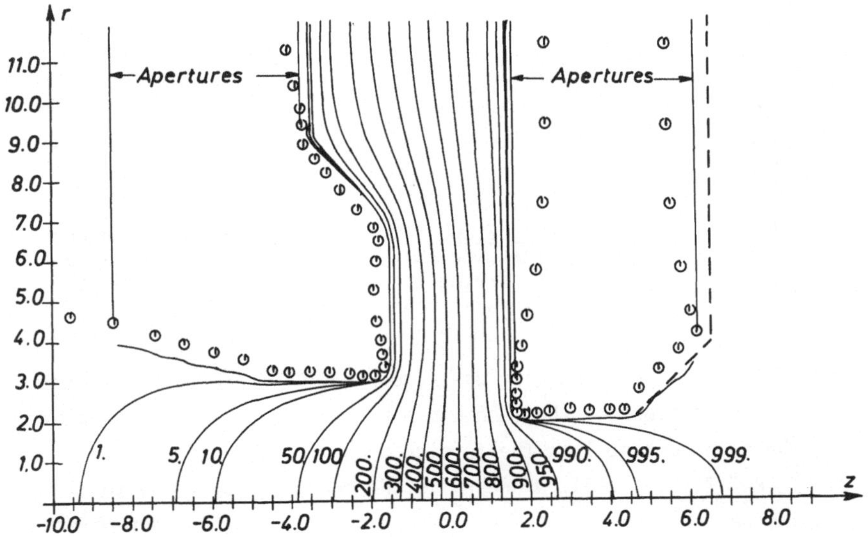

Fig.2.19. Analytical computation of equipotentials in a magnetic lens. The figure shows only a part of the 70 rings used in the computation. In almost field-free space, the potential may shift very slowly with an extremely slow gradient. This effect is quite unimportant, but due to it, the equipotentials cannot be reasonably drawn beyond the given limits

Superposition of the fields of charged rings had already been used before the work of HOCH et al., for instance [2.9,63]. Traditionally, the charges have been assumed to be located on the surfaces of the electrodes. In adjusting the equipotentials to these surfaces one then has to take care about singularities. This difficulty is eliminated in the method of HOCH et al. by shifting the singularities into the interior of the electrodes. This combination of charges, apertures, and rings, which had not been studied before, makes this method very convenient.

An application to unsaturated magnetic lenses with practically infinite permeability μ is easily possible. One only has to use the analogy between $\underline{E} = -\mathrm{grad}\, V$, $\nabla^2 V = 0$ and $\underline{H} = -\mathrm{grad}\psi$, $\nabla^2\psi = 0$ in the vacuum part of the field. The formal "charges" in the interior of the pole pieces have no physical meaning, they only serve to satisfy the boundary conditions at the yoke surface. An application to a typical magnetic lens is shown in Fig.2.19. Reasonable accuracy is obtained with about 70 rings. The linear system (2.85) is then comparatively small and can easily be solved by means of the Gauss algorithm.

2.5.3 Linear Superposition of Magnetic Ring Fields

Some electron optical devices contain rotationally symmetric lenses built up of air-coils. We therefore discuss briefly the method of field calculation for such lenses. This will also be important in connection with the method described in Sect.2.5.4.

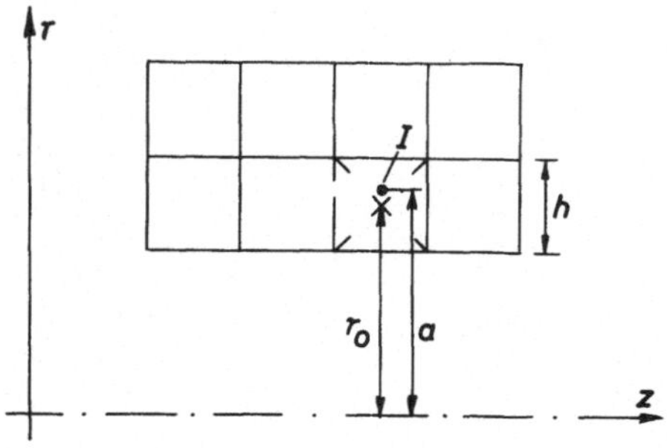

Fig.2.20. Simplified cross-section through the windings of an air-coil. The notation referring to one of the square-shaped elements is given explicitly

The whole body of the coil is notionally dissected into discrete toroidal parts with square-shaped cross-sections of the mesh width h. Each toroid is replaced by a ring conducting the same current I as the toroid and having the radius $a = (r_0^2 + h^2/12)^{1/2}$ [Ref.2.17, p.113], where r_0 is the mean radius of the toroid (Fig.2.20). The azimuthal component of the vector potential [2.30] is then given by

$$A(z,r) = \mu_0 Ia\{2[K(k) - E(k)]k^{-2} - K(k)\}(\pi s)^{-1} \quad ,$$

where K(k) and E(k) are elliptic integrals of the first and second kinds. The quantities s and k are defined by (2.84). The components of $\underline{B} = \nabla \times \underline{A}$ are calculated using (2.30) and (2.31). The corresponding differentiations can easily be performed by using the well-known derivatives of the elliptic integrals. The total field of the coils is obtained by appropriate summation of the contributions of all the rings.

In this way it is easy to compute the field of an air-coil everywhere in space. In the paraxial region, one has to take care to avoid formulae containing small differences of large quantities. This is not difficult, using the series expansions of the elliptic integrals.

The magnetic fields of coaxial circular ring currents can also be superimposed in order to approximate the magnetic field of a superconducting round lens. In this case the rings are to be located inside the superconducting body and a sequence of control points is to be defined at the surface. According to (2.37), the boundary problem to be solved is a Dirichlet problem for the flux potential F(z,r), related to A(z,r) by (2.29). The ring currents are regarded as unknowns to be obtained from the solution of a linear system of equations resulting from the condition F = const. at the control points. After solving this system, the field may be calculated at every point in the vacuum. Obviously, the field in the superconducting material is wrong, since in reality the screening currents form a two-dimensional continuous layer at the surface, and the field in the superconductor is practically zero. The vacuum field, however, is highly accurate, and only this is of practical importance. Apart from the replacement of the scalar potential V by the flux potential F and the omission of aperture fields, this method is analogous to that outlined in Sect.2.5.2 and seems to be much easier than that published by LUCAS [2.49].

2.5.4 The Integral Equation Method

The analytical methods described above may be regarded as special cases of a general procedure based upon the numerical solution of an integral equation. This method was first applied to electrostatic systems by CRUISE [2.9], LEWIS [2.45], SINGER and BRAUN [2.79], RAUH [2.63], READ et al. [2.64], HARRINGTON et al. [2.21], KURODA and SUZUKI [2.37]. Applications to magnetostatic systems have been given by HALACSY [2.19], ZAKY [2.87], JAWSON [2.33], ARMSTRONG et al. [2.2], NEWMAN et al. [2.57], SIMKIN and TROWBRIDGE [2.78], ISELIN [2.29], and LUCAS [2.49]. Here we will give a representation which is essentially a combination of Simkin's and Iselin's methods.

If a field calculation throughout the whole space is required, the scalar magnetic potential $\psi(\underline{r})$ introduced in Sect.2.1.1 is useless since this conflicts with (2.2). Therefore the rotational part of the total field is separated into

$$\underline{H}(\underline{r}) = \underline{H}_0(\underline{r}) - \mathrm{grad}\phi(\underline{r}) \quad , \tag{2.86}$$

$$\mathrm{curl}\ \underline{H}_0 = \underline{j} \quad , \quad \mathrm{div}\ \underline{H}_0 = 0 \quad . \tag{2.87}$$

The field $\underline{H}_0(\underline{r})$ is the field of the coil in the absence of any ferromagnetic material (for reasons of simplicity we assume $\mu = \mu_0$ for every nonferromagnetic material). This contribution $\underline{H}_0(\underline{r})$ can be calculated using Biot-Savart's law or, for rotationally symmetric coils, by means of the method described in Sect.2.5.3. It is thus well known everywhere.

First, the field calculation in unsaturated systems is described. We therefore assume μ = const. in the yoke material. From (2.1), we have div $\underline{H} = 0$ and in combination with (2.86,87) we finally obtain Laplace's equation, $\nabla^2\phi = 0$. This is valid everywhere except at the yoke surfaces, where μ is discontinuous.

For any solution $\phi(\underline{r})$ of Laplace's equation inside an arbitrary region R and in the limit, also on its closed boundary B, Green's integral theorem

$$\phi(\underline{r}) = \oint_B \left[G(\underline{r},\underline{r}')\,\frac{\partial\phi}{\partial n'} - \phi(\underline{r}')\,\frac{\partial G}{\partial n'}\right]dS' \tag{2.88}$$

is valid, where dS' is the surface element corresponding to the integration variable r', $\partial/\partial n'$ the derivative with respect to the surface normal $\underline{n}'$ in the outward direction and G(r,r') a Green's function. For reasons of simplicity G will be chosen as

$$G(\underline{r},\underline{r}') = \frac{1}{4\pi|\underline{r} - \underline{r}'|} \quad .$$

The first term in (2.88) may then be interpreted as a surface charge term, the local surface charge density being $\sigma = \partial\phi/\partial n'$. The second contribution is a surface dipole term, the corresponding dipole density being $-\phi(\underline{r}') \cdot \underline{n}'(\underline{r}')$. The integral dipole contribution vanishes for all closed parts of the boundary which are equipotentials.

2.5.4a Solution of Boundary-Value Problems

In principle the field calculation is straightforward if a general boundary condition of the third kind,

$$\alpha(\underline{r})\phi(\underline{r}) + \beta(\underline{r})\sigma(\underline{r}) = \gamma(\underline{r}) \quad , \quad (\sigma = \partial\phi/\partial n) \quad , \tag{2.89}$$

is given, which includes Dirichlet and Neumann conditions as special cases [for instance (2.36) or (2.37)]. In order to perform this calculation, the whole boundary B is dissected into N finite surface elements Δ_i. Here we describe only a very simple version of the theory where the elements Δ_i are chosen to be so small that the functions occurring in (2.89) may be regarded as constant. Equation (2.89) thus gives

$$\alpha_j\phi_j + \beta_j\sigma_j = \gamma_j \quad , \quad (j = 1,\ldots,N) \quad . \tag{2.90}$$

It is convenient to introduce the matrix elements

$$A_{ik} = \int_{\Delta_k} G(\underline{r}_i,\underline{r}_k')dS_k' \quad , \quad B_{ik} = \int_{\Delta_k} \frac{\partial G(\underline{r}_i,\underline{r}_k')}{\partial n_k'} dS_k' \quad , \tag{2.91}$$

where $\underline{r}_i$ refers to the centroid of the surface element Δ_i. Then (2.88) simplifies to a system of linear equations

$$\phi_i + \sum_{k=1}^{N} B_{ik}\phi_k = \sum_{k=1}^{N} A_{ik}\sigma_k \quad . \tag{2.92}$$

Together with (2.90) this forms a complete set of linear equations for the required boundary values, which can be solved by standard techniques. More details will be given at the end of the following section.

2.5.4b Solution of Problems with Interface Conditions

For reasons of simplicity we consider a system with only two contiguous regions, the vacuum domain (0) and the iron yoke (1) separated by the continuous surface B. An outer boundary enclosing the whole system gives no contribution to the integral (2.88) since asymptotically the field decreases as that of a dipole.

The boundary conditions to be satisfied at the yoke surface B are as follows. The potential ϕ must be continuous and hence the tangential component of $\underline{H}$ is continuous too. Furthermore, the normal component of $\underline{B}$ must be continuous. It is convenient to introduce the abbreviation

$$h(\underline{r}) = \underline{n}(\underline{r}) \cdot \underline{H}_0(\underline{r})$$

for the normal component of $\underline{H}_0$, the surface normal $\underline{n}$ being always directed outward from the yoke. Then for the derivatives $\sigma = \partial\phi/\partial n$, the condition

$$\mu_0[h(\underline{r}) + \sigma^{(0)}(\underline{r})] = \mu_1[h(\underline{r}) - \sigma^{(1)}(\underline{r})] \tag{2.93}$$

is satisfied, which is to be used in connection with (2.88). If the point of reference is located at the surface, this equation must be simultaneously valid with respect to both domains (0) and (1), which means that

$$\phi(\underline{r}) = \oint_B \left[G(\underline{r},\underline{r}')\sigma^{(i)}(\underline{r}') \pm \phi(\underline{r}') \frac{\partial G}{\partial n'}\right] dS' \quad , \quad (i = 0,1) \quad . \tag{2.94}$$

Eliminating the terms with $\sigma^{(i)}$ by use of (2.93) and introducing the abbreviation

$$\gamma = 2(\mu_1 - \mu_0)/(\mu_1 + \mu_0) \quad ,$$

we obtain the integral equation [2.70a]

$$\phi(\underline{r}) + \gamma \oint_B \phi(\underline{r}') \frac{\partial G}{\partial n'} dS' = \gamma \oint G(\underline{r},\underline{r}')h(\underline{r}')dS' \quad .$$

This is to be discretized in a manner similar to that described above, the result being

$$\phi_i + \gamma \sum_{k=1}^{N} B_{ik}\phi_k = \gamma \sum_{k=1}^{N} A_{ik}h_k \quad .$$

Since the terms on the right-hand side are well known, the complicated field calculation problem involving interface conditions is thus reduced to an ordinary boundary-value problem of Dirichlet type, which can be solved easily. The necessary surface charge densities are now obtained by solution of a linear system resulting from the discretization of (2.94),

$$\phi_i \mp \sum_{k=1}^{N} B_{ik}\phi_k = \sum_{k=1}^{N} A_{ik}\sigma_k^{(j)} \quad , \quad (j = 0,1) \quad ,$$

which is the same as (2.92) but now applied to both domains (0) and (1).

In order to calculate the magnetic field strength at an arbitrary point $\underline{r}$ in space, it is necessary to introduce and to evaluate the vector functions

$$\underline{a}_k(\underline{r}) = \nabla \int_{\Delta_k} G(\underline{r},\underline{r}_k')dS_k' \quad , \quad \underline{b}_k(\underline{r}) = \nabla \int_{\Delta_k} \frac{\partial G(\underline{r},\underline{r}_k')}{\partial n_k'} dS_k' \quad .$$

Then by differentiating and discretizing (2.94) and recalling (2.86), we finally obtain

$$\underline{H}(\underline{r}) = \underline{H}_0(\underline{r}) + \sum_{k=1}^{N} \left[-\sigma_k^{(i)}\underline{a}_k(\underline{r}) \mp \phi_k\underline{b}_k(\underline{r})\right] \quad ,$$

the superscript (i) and the signs to be chosen in accordance with the position of the point of reference. Care must be taken in the vicinity of singularities.

In this way, even true three-dimensional problems with interface conditions may be solved, since so far the assumption of rotational symmetry has not been required. A serious drawback will then be the very large memory required, since the matrices A_{ik} and B_{ik} are dense $N \times N$ matrices. In practice the method is more suitable for the solution of two-dimensional problems.

In systems with rotational symmetry the surface elements are conical frusta. For these the integration with respect to the azimuthal variable can be performed analytically and results in expressions containing the complete elliptic integrals of the first and second kinds. The integration over the arc length in any meridional section must be performed numerically. This will not be outlined here. Detailed investigations and more accurate approximations are given by HARRINGTON et al. [2.21], and by SINGER and BRAUN [2.79], for example.

2.5.4c Integral Equations Containing the Magnetization

The method described in the previous section is practically useless for the calculation of lenses with nonlinear media. For such lenses it is better to introduce the magnetization into the integral equation. This has been done by ZAKY [2.87], HALACZY [2.19], SIMKIN and TROWBRIDGE [2.78], ISELIN [2.29], LUCAS [2.49], and many other authors.

We still begin with (2.86) and (2.87) but now introduce the additional relation

$$\underline{B}(\underline{r}) = \mu_0\, \underline{H}(\underline{r}) + \underline{M}(\underline{r}) \quad , \tag{2.95}$$

where $\underline{M}(\underline{r})$ is the magnetization, a spatial vector function which is nonzero only inside the yoke. Combining (2.1,86,87,95), we obtain the Poisson equation

$$\nabla^2\phi(\underline{r}) = \frac{1}{\mu_0}\operatorname{div}\underline{M}(\underline{r}) \quad ,$$

which has the formal solution

$$4\pi\mu_0\phi(\underline{r}) = \int\limits_{\text{yoke}} \frac{-\nabla'\cdot\underline{M}(\underline{r}')dV'}{|\underline{r}-\underline{r}'|} = \int\limits_{\text{yoke}} \frac{\underline{M}(\underline{r}')\cdot(\underline{r}-\underline{r}')dV'}{|\underline{r}-\underline{r}'|^3} \quad , \tag{2.96}$$

dV' being the volume element corresponding to the variable $\underline{r}'$.

The evaluation of (2.96) is relatively easy for systems consisting of permanent magnetic materials where $\underline{M}(\underline{r})$ is a given function in space. This is usually not the case. Usually, the magnetization is related to the field strength by a nonlinear relation

$$\underline{M} = \underline{M}(\underline{H}) = \underline{M}(\underline{H}_0 - \operatorname{grad}\phi) \quad , \tag{2.97}$$

which only implicity gives $\underline{M}$ as a function of position. Introducing this into (2.96), one obtains a complicated nonlinear integro-differential equation for the scalar potential ϕ. This can only be solved iteratively. There are different solution techniques. The best one seems to be that of ISELIN [2.29], who replaces (2.97) by a locally linear relation to be evaluated in the neighbourhood of the preceding iteration guess. $\phi(\underline{r})$ is expanded with respect to a set of basis functions, and (2.96) is thus transformed into a linear system of equations for the coefficients which is then solved.

So far, we have only introduced integral equations for scalar functions $\phi(\underline{r})$. Some of the authors cited above have derived vector integral equations. We shall not treat these here, since they seem to be disadvantageous owing to their increased requirements for computing time and memory.

2.6 Field Calculation in Systems Without Rotational Symmetry

Though round electron lenses are usually used, some cases with lower symmetry are also of interest. Hence we shall briefly describe their calculation.

2.6.1 Lenses with Small Perturbations of the Rotational Symmetry

Since rotationally symmetrical lenses can never be built quite perfectly, the effects of small perturbations, such as shifts, tilts, or ellipticity, of the yokes on the field in the paraxial region are of interest. The method for studying them described below was introduced by JANSE [2.32]. Here it is slightly generalized.

First, as the zero-order approximation, the ideal round system is calculated, and then in a fine grid of points on the yoke surfaces the local field strength $\underline{H}(\underline{r})$ is determined. At each of these points, the local shift $\underline{s}(\underline{r})$ from the ideal surface to the real one is now defined. This shift may be in an arbitrary direction but must be very small. This is illustrated in Fig.2.21.

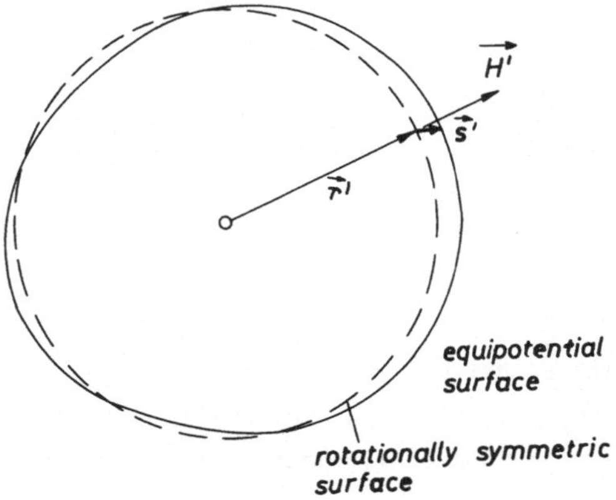

Fig.2.21. Simplified cross-section through a lens with perturbations of the rotational symmetry, showing the real and the ideal contour of a pole piece. The vectors $\underline{r}'$, $\underline{s}'$, $\underline{H}'$ are the projections of the position vector $\underline{r}$, the shift $\underline{s}(\underline{r})$, and the field strenght $\underline{H}(\underline{r})$

Since the desired equipotential is shifted to the real yoke surface, the scalar potential at the ideal surface is perturbed by a quantity

$$\delta\psi(\underline{r}) = \psi(\underline{r}) - \psi(\underline{r} + \underline{s}) \approx - \underline{s} \cdot \mathrm{grad}\ \psi = \underline{s}(\underline{r}) \cdot \underline{H}(\underline{r}) \quad . \tag{2.98}$$

This function now defines the boundary values for the field calculation in the region enclosed by the ideal rotationally symmetrical surface. These boundary values are to be introduced into the Fourier integrals (2.19). The procedure described in Sect.2.1.2 can then be applied. Usually only the terms with $|m| \leq 3$ are of practical interest. Finally, from the solutions of (2.18) only the axial values $\psi_m(z,0)$ are of importance. These are sufficient to study the influence of the perturbation on the electron trajectories.

2.6.2 Toroidal Deflection Systems

Systems of this kind are used in television tubes, in scanning microscopes, and in electron microrecorders for deflecting the electron beam. Two pairs of coils, rotated at 90° with respect to each other, are wound round a rotationally symmetrical ferrite yoke. This is performed in such a manner that each winding remains in a meridional plane. Figure 2.22a shows a cross-section through such a system and Fig.2.22b a meridional section.

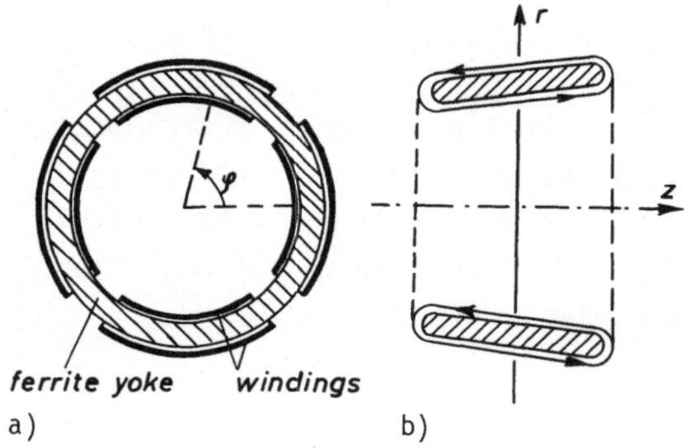

Fig.2.22a,b. Simplified representation of a toroidal deflection system: (a) cross-section; (b) meridional section. The outer-screening boundary is omitted for reasons of simplicity

The theory of such systems has been treated by SCHWERTFEGER and KASPER [2.71]. With the assumption that the contributions to the line integral $\int \underline{H} \cdot d\underline{s}$ can be neglected in the yoke, they derived the expression

$$\psi(z_s, r_s, \varphi) = - I \int_0^{\varphi} \nu(\varphi')d\varphi' \tag{2.99}$$

for the boundary values of the magnetic scalar potential at the surface of the system. Here the quantity I is the electric current in one pair of coils; $\nu(\varphi)$ is the density of windings, that is, $\nu(\varphi)d\varphi$ is the number of windings in the interval $\varphi \leq \varphi' < \varphi + d\varphi$. A more sophisticated investigation shows that the factor I should more exactly be replaced by $I\mu_r(1 + \mu_r)^{-1}$, μ_r being the relative permeability of the yoke material.

It is not necessary to perform the integration in (2.99). The Fourier components of ψ can be calculated directly from those of the function $\nu(\varphi)$. After obtaining these and including the factors $|r_s|^{-m}$ arising from (2.16), we know the inner-boundary values for (2.18). The outer-boundary values are simply zero on a sufficiently far distant surface enclosing the system or on any rotationally symmetrical boundary like the surfaces of round lenses. Equations (2.18) can now be solved numerically.

REGENAUER [2.64a] applied this method to magnetic deflecting systems with a round magnetic post-deflection lens. Devices of this kind are used in electron optical microrecorders and machines for electron lithography [2.79a]. The boundary conditions are as follows: at the yoke surface of the round lens the potential of the deflection field vanishes while that of the round lens has its appropriate values; conversely, at the yoke surface of the deflector the deflection potential has its appropriate values while that of the round lens is zero. In the field, obtained as the solution of this Dirichlet problem, the trajectories of electrons and, from these, the distortion coefficients of the system have also been computed.

2.6.3 Magnetic Multipole Systems

Electric or magnetic multipole systems are commonly used as stigmators or as strong focusing lenses in electron optical devices. Since these systems are not at all rotationally symmetric, the theory developed in this chapter cannot properly be applied to them. If, however, parts of the surfaces form a rotationally symmetrical face and the gaps between adjacent poles are comparatively narrow, as shown in Fig. 2.23a, the theory is approximately applicable. Systems of this kind are in use in the devices developed by SCHERZER, ROSE and their teams [2.69,70,36] in order to compensate the spherical aberration of third order and the axial chromatic aberration in an electron microscope of high resolution.

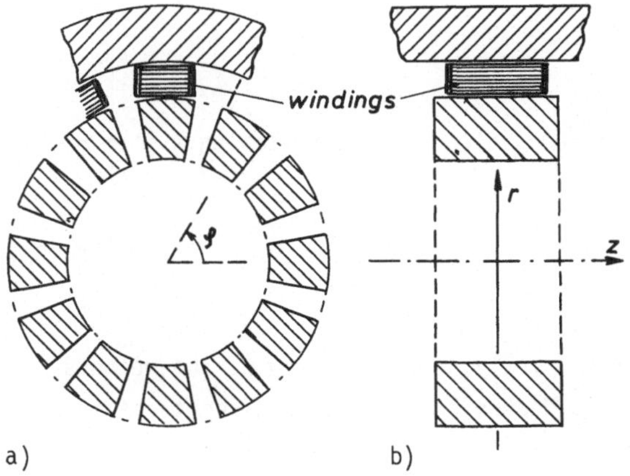

Fig.2.23a,b. Simplified representation of a magnetic multipole system: (a) cross-section, (b) meridional section. The figures show only parts of the coils exciting the pole pieces and of the outer screening ring

The necessary approximation for the treatment of such systems is such that in the gaps, the potential is linearly interpolated with respect to the azimuth φ. Then on a rotationally symmetrical surface consisting of the cylindrical bore, the gaps, the ring-shaped parts of the end-planes, and an outer surface enclosing the whole system, reasonable boundary values of the potential can be defined.

The axial section through this surface is shown in Fig.2.23b. Now the theory described in Sect.2.1.2 is again applicable. The errors in the incorrect boundary values do not seriously influence the field in the paraxial region. Hence this approximation is reasonable.

2.7 The Calculation of Electron Trajectories

The electron trajectories in round magnetic lenses are usually defined to be solutions of two types of ordinary differential equations. These types are the paraxial ray equation

$$\frac{d^2x}{dz^2} + \frac{\eta^2B^2(z)}{4U}\,x = 0 \quad , \quad \frac{d^2y}{dz^2} + \frac{\eta^2B^2(z)}{4U}\,y = 0 \quad , \tag{2.100}$$

valid only in the paraxial region and explained in Chap.1, and quite generally the Lorentz equation

$$\frac{d}{dt}\,\frac{m_0\underline{v}}{\sqrt{1-v^2/c^2}} = -e\underline{v}\times\underline{B}(\underline{r}) \quad . \tag{2.101}$$

Störmer's equation [Ref.2.17, p.126] may also be solved, but for non-meridional trajectories, this is numerically unfavourable; we shall therefore not consider it here.

The use of time as a parameter for the solution of (2.101) is rather inconvenient We therefore introduce the arc length s along the electron trajectory. Recalling that v = const. in magnetic fields, (2.101) is easily transformed into

$$\frac{d^2\underline{r}(s)}{ds^2} = -\frac{\eta}{\sqrt{U}}\,\frac{d\underline{r}}{ds}\times B(\underline{r}) \quad . \tag{2.102}$$

This second-order differential equation is equivalent to the following system of first-order equations:

$$\frac{d\underline{r}}{ds} = \underline{u} \quad , \quad \frac{d\underline{u}}{ds} = -\frac{\eta}{\sqrt{U}}\,\underline{u}\times\underline{B} \quad . \tag{2.103}$$

One constant of motion is given by $|\underline{u}| = 1$.

In the following sections we shall only briefly describe a few of the methods of solving ordinary differential equations. Modern comprehensive accounts of this subject are to be found, for example in LAMBERT [2.38] (initial-value problems) and WALSH [2.84] (boundary-value problems).

2.7.1 Solution of the Paraxial Ray Equation

The essential features of (2.100) are their linearity and the absence of terms containing the first derivatives. The following method is suitable for solving all linear differential equations having the form

$$\frac{d^2w(z)}{dz^2} + f(z) \cdot w(z) = 0 \quad , \tag{2.104}$$

w(z) being any real or complex solution. In order to derive an approximate difference equation for the solutions of (2.104), we start from a Taylor series expansion about an arbitrary point (z_m,w_m). By summation over the expansions for $w(z_m \pm h)$ all odd terms cancel out, the result being

$$w_h - 2w_m + w_{-h} = h^2w''_m + \frac{h^4}{12} w_m^{(4)} + O(h^6) \tag{2.105}$$

with $w_m = w(z_m)$, $w_{\pm h} = w(z_m \pm h)$. Differentiating (2.105) twice with respect to z at $z = z_m$ yields

$$w''_h - 2w''_m + w''_{-h} = h^2w_m^{(4)} + O(h^4) \quad .$$

This equation is used to eliminate the fourth derivative from (2.105), the result being

$$w_h - 2w_m + w_{-h} = \frac{h^2}{12} (w''_h + 10w''_m + w''_{-h}) \quad .$$

The second derivatives are now eliminated by means of (2.104), the result being the required difference equation

$$\left(1 + \frac{h^2}{12} f_h\right)w_h - 2\left(1 - \frac{5h^2}{12} f_m\right)w_m + \left(1 + \frac{h^2}{12} f_{-h}\right)w_{-h} = O(h^6) \tag{2.106}$$

with $f_m = f(z_m)$, $f_{\pm h} = f(z_m \pm h)$. Equation (2.106) was first derived by NUMEROV [2.59], MANNING and MILLMAN [2.50] and by FOX and GOODWIN [2.15]. It is a very favourable difference equation, since it is comparatively simple, numerically stable, and highly accurate for sufficiently small step-widths h. In each integration step, the function f(z) is evaluated only once and then stored. Application of (2.106) to the numerical solution of (2.104) is different for initial-value problems and eigenvalue problems.

2.7.1a Initial-Value Problems

We are confronted with an initial-value problem if initial conditions $w(z_0) = w_0$, $w'(z_0) = w'_0$ are given. Equation (2.106) cannot immediately be applied, because this requires a second value $w(z_0 + h) = w_1$, initially not known. This value can be calculated using a formula given by ZURMÜHL [Ref.2.90, p.395]:

$$w_1 = w_0 + hw'_0 + \frac{h^2}{36} (9.7w''_0 + 11.4w''_1 - 3.9w''_2 + 0.8w''_3) + O(h^6) \tag{2.107}$$

with $w''_j = w''(z_0 + j\,h)$. These quantities are to be eliminated by means of (2.104). Two additional equations for the variables w_1, w_2, w_3 are obtained by application of (2.106) to $z_m = z_0 + h$ and $z_m = z_0 + 2h$. The resulting system of linear equations is then solved for w_1. After this calculation has been performed, (2.106) can be used to find a numerical solution of (2.104) by successive integration with steps of equal step-width h. In each step a new value w_h is calculated from the preceding values of w_m and w_{-h} and stored, if necessary. After that the numbering is adjusted to the next step.

2.7.1b Boundary-Value Problems

These problems arise if an object plane $z = z_o$ and an image plane $z = z_i$ are given and the solution of (2.104) has to satisfy given boundary conditions with respect to these planes, usually $w(z_o) = w(z_i) = 0$. There is in general no solution to this problem, if f(z) is not properly adjusted to the boundary conditions. In order to perform this, it is necessary to replace f(z) by $\lambda \cdot f(z)$, the multiplicative eigenvalue λ being unknown initially.

An easy method for solving the boundary-value problem is the following. Starting with a rough guess λ' for λ, the initial-value problem with the initial conditions $w_o = 0$, $w_o' = 1$ is solved. From this solution the actual image plane $z_i'(\lambda')$ with $w(z_i') = 0$ is calculated, which will generally not be the correct one. Now λ' is changed and the solution of the initial-value problem is repeated, and so on. Starting from the values of $z_i'(\lambda')$ obtained from the first two iterations, the ordinary equation

$$(z_i - z_o)/(z_i'(\lambda') - z_o) = 1 \quad , \quad (\lambda' \to \lambda) \tag{2.108}$$

is now solved iteratively. This can be performed by applying the bisection method or the "regula falsi" or a reasonable combination of both. In each step a more accurate value of λ' is obtained which is then used for the next trajectory computati[on]. The procedure rapidly converges and can be terminated once sufficient accuracy has been achieved.

Finally, the corresponding lens excitation or relativistic accelerating voltage U can be determined from the equation $\lambda f(z) = \eta^2 B^2(z)/(4U)$.

2.7.1c Numerical Differentiation

The numerical calculation of aberration coefficients generally requires the solution of the paraxial ray equation and the calculation of the first derivative w'(z) of the solution. An appropriate formula for the required derivative is easily obtained from a Taylor series expansion. Adopting the notation introduced in connection with (2.105), subtraction of the corresponding Taylor series expansions yields the result

$$w_h - w_{-h} = 2hw_m' + \frac{h^2}{3} w_m''' + O(h^5) \quad . \tag{2.109}$$

Differentiating this twice with respect to z at $z = z_m$ yields

$$w_h'' - w_{-h}'' = 2\,h \cdot w_m''' + O(h^3) \quad .$$

Elimination of w_m''' by introducing this into (2.109) and elimination of $w_{\pm h}''$ by use of (2.104) finally results in

$$w_m' = \frac{1}{2h}\left[\left(1 + \frac{h^2}{6} f_h\right)w_h - \left(1 + \frac{h^2}{6} f_{-h}\right)w_{-h}\right] + O(h^4) \quad , \tag{2.110}$$

which is sufficiently accurate. Thus the calculation of the trajectory must be always one step in advance of the differentiation. The starting value w_0' is usually given. The last value w_i' can be calculated, if necessary, by integration over the last three values of $w'' = -f\,w$ by means of Simpson's rule.

2.7.1d Calculation of Aberration Coefficients and Optimization

All the aberration coefficients of round magnetic lenses can be expressed as integrals of the general form

$$C = \int_{z_0}^{z_i} F(z,f,f',g,g',h,h')\,dz \tag{2.111}$$

(Chaps.1 and 3), where f(z) denotes $\eta B(z)/(2\sqrt{U})$ and g(z), h(z) are two fundamental solutions of (2.104). The quantities z_0 and z_i are the coordinates of the object plane and the image plane, respectively. Here we only consider real imaging; formulation of the appropriate theory for asymptotic imaging presents no problem.

The solutions g(z), h(z) are computed and differentiated in the manner described in the previous sections. The function f(z) is differentiated by means of (2.77) or (2.78). Each expression of the form (2.111) can then be easily integrated by means of the extended trapezoidal rule or, more precisely, by means of the extended Simpson rule. This procedure is very straightforward, provided that reasonably simple integral expressions for the aberration coefficients can be derived. A more general method which is not based upon the evaluation of integrals, is given at the end of Sect.2.7.2b.

The determination of aberration coefficients plays an important role in optimization procedures, which have been used, for instance by CAMPBELL [2.6], BUTLER [2.5], TRETNER [2.82], MOSES [2.52]. The most advanced procedure seems to be that given by MOSES, who used the calculus of variations; this will now be briefly outlined. This procedure tries to find the best axial potential distribution, some constraints being given. The criterion for the best solution is the minimum of the spherical aberration coefficient. The constraints are fixed object and image plane, vanishing coefficient of anisotropic coma, and keeping the axial field strength below a reasonable limit. Together with the paraxial equation of motion these conditions are combined in one variational principle; the corresponding Euler equations are solved by Numerov's method.

This method is very elegant and works quite well but is not really satisfactory; its inherent difficulties are common to all optimization procedures of this kind. These are due to the attempt to define a whole lens field from its axial distribution. Then of course the difficulties described in Sect.2.1.3 will arise. Small, quite insignificant changes in the axial potential $\phi_0(z)$ will cause severe changes in the far off-axis region. In fact, the contours of yokes or pole pieces, determined by extrapolation of the paraxial fields, are very complicated and sometimes not well defined. Another difficulty is that optimizing procedures of this kind may run into uninteresting local extrema whereupon better and even simpler solutions will not be found by the program.

A realistic optimization procedure must start from a realistic and simple electrode or yoke configuration. All the image properties and aberration coefficients are then calculated. The geometry and the excitation are now slightly changed within realistic limits and the calculation repeated until it has converged to a minimum of the most important aberration coefficient. This procedure is, of course, very tedious and may also run into uninteresting extrema but has the advantage of always yielding realistic results. In its details the optimizing process strongly depends on the particular properties of the case in question and will therefore not be treated further in this chapter.

Image properties and aberration coefficients of lens combinations can be calculated by means of the matrix transfer formalism outlined in Chap.1 (see also [2.25]).

2.7.2 Solution of the Lorentz Equation

The best form of Lorentz's equation for a numerical solution is the system (2.103). This system of six coupled ordinary differential equations is a special case of the more general system

$$\underline{y}(s) = \underline{f}(s,\underline{y}) \quad , \tag{2.112}$$

the vector $\underline{y}$ being built up from the vectors $\underline{r}$ and $\underline{u}$ in (2.103).

2.7.2a The Runge-Kutta Method

The most familiar method for the solution of (2.112) is the usual Runge-Kutta procedure of fourth order ([2.68] or [Ref.2.90, p.407]), one step being defined by

$$\left.\begin{aligned}
\underline{k}_1 &= h\underline{y}'_n \quad , \\
\underline{k}_2 &= h\underline{f}(s_n + h/2,\ \underline{y}_n + \underline{k}_1/2) \quad , \\
\underline{k}_3 &= h\underline{f}(s_n + h/2,\ \underline{y}_n + \underline{k}_2/2) \quad , \\
\underline{k}_4 &= h\underline{f}(s_n + h,\ \underline{y}_n + \underline{k}_3) \quad , \\
s_{n+1} &= s_n + h \\
\underline{y}_{n+1} &= \underline{y}_n + (\underline{k}_1 + 2\underline{k}_2 + 2\underline{k}_3 + \underline{k}_4)/6 + O(h^5) \quad , \\
\underline{y}'_{n+1} &= \underline{f}(s_{n+1},\ \underline{y}_{n+1}) \quad .
\end{aligned}\right\} \tag{2.113}$$

In order to start the procedure, the initial vector $\underline{y}_0$ must be given, and $\underline{y}'_0 = \underline{f}(s_0, \underline{y}_0)$ must be calculated. Then, by repeating the sequence of calculations described above, the values of s, $\underline{y}(s)$, $\underline{y}'(s)$ are available at the end of each step. The step length may change from one step to the next.

There are procedures for changing the step length appropriately in order to keep the resulting truncation error small [Ref.2.90, p.410], but these procedures do not directly ensure that the resulting error remains smaller than a given tolerance limit. In doubtful cases, one should therefore repeat the calculation after halving the step length. A more precise Runge-Kutta procedure has been developed by GILL [2.16], (see also [2.68]). This, however, cannot be treated in the framework of this article.

Electron optical applications of the Runge-Kutta method are treated in detail in a series of papers by LAPEYRE [2.39] and LAPEYRE and LAUDET [2.40]. Often they are used in the third-order approximation [2.85]. Compared with Numerov's method, Runge-Kutta methods are less accurate and slower, for the same step length, since (2.113) requires four evaluations of $\underline{f}$ per step and has an intrinsic truncation error of fifth order. Runge-Kutta methods have, however, the advantage of general applicability.

2.7.2b A Predictor-Corrector Method

The Runge-Kutta procedure is necessary in order to compute the starting value required in more advanced predictor-corrector routines. Among these, HAMMING's modified predictor-corrector method (HPC) [2.20,61] is to be preferred to the classical MILNE method [2.51]. In electron optical calculations HPC was used by HAUKE [2.22], HOCH et al. [2.27], and KERN [2.35]. A still more accurate version of this method, called HPCD, has been developed by KASPER (so far unpublished).

In this procedure the accumulation of rounding errors is minimized by the use of a new numerically stable predictor formula which – compared with the classical formula – has practically the same discretization error but is more suitable for representation in incremental form. In the HPCD procedure a new set of variables having the index n + 1 is calculated from three preceding sets by

$$\begin{aligned}
\underline{p}_{n+1} &= \frac{8h}{3}\left[\frac{7}{8}\left(\underline{y}'_n + \underline{y}'_{n-2}\right) - \underline{y}'_{n-1}\right] - \underline{d}_{n-2} + \frac{29}{90} h^5 \underline{y}_n^{(5)} \quad ,\\
\underline{g}_{n+1} &= \underline{y}_n + \underline{p}_{n+1} + \frac{116}{125}\underline{\Delta}_n \,, \quad \underline{g}'_{n+1} = \underline{f}(s_n + h, \underline{g}_{n+1}) \quad ,\\
\underline{c}_{n+1} &= \frac{1}{8}\left[\underline{d}_n + \underline{d}_{n-2} + 3h\left(\underline{g}'_{n+1} + 2\underline{y}'_n - \underline{y}'_{n-1}\right)\right] - \frac{h^5}{40}\underline{y}_n^{(5)} \quad ,\\
\underline{\Delta}_{n+1} &= \underline{c}_{n+1} - \underline{p}_{n+1} \,, \quad \underline{d}_{n+1} = \underline{c}_{n+1} - \frac{9}{125}\underline{\Delta}_{n+1} \quad ,\\
s_{n+1} &= s_n + h \,, \quad \underline{y}_{n+1} = \underline{y}_n + \underline{d}_{n+1} \,, \quad \underline{y}'_{n+1} = \underline{f}(s_{n+1}, \underline{y}_{n+1}) \quad .
\end{aligned} \tag{2.114}$$

This sequence of statements is already a favourable representation with respect to programming. The vectors $\underline{p}_{n+1}$, $\underline{c}_{n+1}$, $\underline{d}_{n+1}$ are approximations for the increment $\underline{y}_{n+1} - \underline{y}_n$. This increment and the preceding ones are calculated without explicit use of the $\underline{y}$-vector differences. In this way the accumulation of rounding errors is minimized.

The meaning of the vectors appearing in (2.114) is as follows: $\underline{p}_{n+1}$ is the predictor in incremental representation, its discretizing error being given for information; $\underline{q}_{n+1}$ is the absolute modified predictor, the modification consisting in the fact that the discretization error is reduced to the sixth order in h. Thus $\underline{q}_{n+1}$ is already a highly accurate approximation for $\underline{y}_{n+1}$ as is $\underline{q}'_{n+1}$ for $\underline{y}'_{n+1}$. The vector $\underline{c}_{n+1}$ is the increment of Hamming's corrector, the discretization error being again given; $\underline{\Delta}_{n+1}$ is the difference vector serving for accuracy control; and $\underline{d}_{n+1}$ is the incremental representation of the modified corrector, the modification again being such that the discretization error is reduced to the sixth order in h. The last three statements are the computation of the final solution for the step with index n + 1.

The whole procedure starts with $\underline{y}'_0 = \underline{f}(s_0, \underline{y}_0)$. The first three sets with indices n = 1,2,3 are calculated by means of Gill's Runge-Kutta method. In order to have an accuracy control, each of these steps is repeated with half the step-size. If the accuracy is sufficient, the discretization error of the Gill procedure is reduced to the sixth order by means of extrapolation. The predictor-corrector routine now starts with n = 3, the statements are to be executed in the sequence given in (2.114). The step-length h is maintained if $\Delta = |\underline{\Delta}_{n+1}|$ is located within a given tolerance interval; h is halved if Δ exceeds the upper tolerance limit; then the sets with indices n, n-1, n-2 are recalculated by means of interpolation; and the integration step (2.114) is repeated with higher accuracy. If Δ is below the lower tolerance limit, h is doubled. This procedure is very similar to that described by RALSTON [2.61], the details not being treated here.

The main advantages of all predictor-corrector methods are the reliable control of accuracy and the automatic adjustment of the step-length to practically constant accuracy. In the case of a very low upper tolerance limit ($\Delta \leqq 10^{-7}$), the last statement – the final calculation of $\underline{y}'_{n+1}$ in (2.114) – can be omitted without significant loss of accuracy. Then $\underline{q}'_{n+1}$ has to be accepted as the final value of $\underline{y}'_{n+1}$. This was already noticed by KERN [2.35], when he tested his version of the predictor-corrector method. This is of great importance in electron optics, since the field calculation necessary for the evaluation of the function $\underline{f}(s,\underline{y})$ is the most time-consuming part of the procedure. This has to be done only once in each integration step.

It turns out that HPCD is the most economic method for the solution of systems of ordinary differential equations. It is much faster than all Runge-Kutta routines, since the latter require four evaluations of $\underline{f}(s,\underline{y})$ per step. Moreover, the Runge-Kutta routines are numerically unstable. HPCD is also superior to Numerov's method, since the former has a much wider range of applicability and the advantage of automatic adjustment of the step-length. HPCD is even superior to the extrapolation method described by STOER and BULIRSCH [2.80] if the field is obtained by interpolation in mesh grids. Practical tests show that extrapolation routines break down in the vicinity of the mesh lines, since there the higher-order partial derivatives of the potential are discontinuous, while HPCD works well in Hermite-interpolated fields. It is no problem to obtain nine valid digits in the solution with reasonable step-lengths and computing times. This is even of practical value if the field is less accurate but is at least unique and smooth within the same number of digits, and this is the case for interpolated fields.

By use of highly accurate methods like HPCD it is possible to compute aberration coefficients without using integral expressions for the required coefficients. In order to do this a sufficient number of Lorentz trajectories starting in the object plane are computed in full generality. Their coordinates and slopes in the image plane are then determined. These are now introduced into a series expansion with respect to the corresponding starting values in the object plane, the coefficients being unknown in the beginning. These coefficients are then determined by a least squares fit. Finally, the required aberration coefficients can be obtained by a mere scale transform. This procedure is generally applicable and works well in cases where the evaluation of integral expressions would be extremely complicated, for instance in calculations of the fifth-order spherical aberration of magnetic lenses. In this context it is essential to use field interpolations which are as smooth as possible.

2.8 Concluding Remarks

In this chapter we have discussed some methods for the computer-aided design of magnetic lenses. By appropriate use of these methods the electron optical properties of magnetic lenses can be calculated to a high degree of accuracy if the geometrical shape of the yokes, the distribution of windings, the magnetic properties of the yoke materials, and the lens excitation are completely specified. Optimization, however, remains a problem without a satisfactory solution.

It is the author's opinion that there is no general ideal method for electron optical design which works most favourably in every case. Since the choice of the most suitable method depends on the nature of the problem to be solved, a spectrum of different methods has been offered, and the reader is asked to select those which appear to be the best for his or her special demands.

References

2.1 W.F. Ames: *Numerical Methods for Partial Differential Equations* (Nelson, London 1969)
2.2 A.G.A.M. Armstrong, C.J. Collie, N.J. Diserens, M.J. Newman, J. Simkin, C.W. Trowbridge: *Proc. 5th Int. Conf. Magnet Technology, Rome, 1975*, ed. by N. Sacchetti, M. Spadoni, S. Stipcich (Laboratori Nazionali del CNEN, Frascati 1975) pp.168-182
2.3 K.A. Aziz: *The Mathematical Foundations of the Finite Element Method with Applications to Partial Differential Equations* (Academic, New York, London 1972)
2.4 O. Buneman: "A Compact Non-Iterative Poisson Solver", Report 294, Stanford University, California (1969)
2.5 J.W. Butler: *Proc. 6th Int. Cong. Electron Microscopy, Kyoto, 1966*, Vol.I, ed. by R. Uyeda (Maruzen Co., Tokyo 1966) pp.191-192
2.6 F.J. Campbell: Dissertation, University of Arizona (1967)
2.7 B.A. Carré: Comput. J. *4*, 73-78 (1961)
2.7a M.V.K. Chari, P.P. Silvester (eds.): *Finite Elements in Electrical and Magnetic Field Problems* (Wiley, New York, London, Sydney 1980)
2.8 J.S. Colonias: *Particle Accelerator Design: Computer Programs* (Academic, New York, London 1974)
2.9 D.R. Cruise: J. Appl. Phys. *34*, 3477-3479 (1963)
2.10 G.B. Denegri, G. Molinari, A. Viviani: In Ref.2.60, pp.104-110
2.11 W. Dommaschk: Optik *23*, 472-477 (1965/1966)
2.12 E. Durand: *Electrostatique*, Vol.2 (Masson, Paris 1966)
2.13 G.E. Forsythe, W.R. Wasow: *Finite Difference Methods for Partial Differential Equations* (Wiley, New York 1960)
2.14 L. Fox: In Ref.2.31, Chap.VII.1, pp.799-881
2.15 L. Fox, E.T. Goodwin: Proc. Cambridge Philos. Soc. *45*, 373-388 (1949)
2.16 S. Gill: Proc. Cambridge Philos. Soc. *47*, 96-108 (1951)
2.17 W. Glaser: *Grundlagen der Elektronenoptik* (Springer, Wien 1952)
2.18 T.N.E. Greville: In Ref.2.62, Vol.II, Part IV, Chap.8, pp.156-168
2.19 A.A. Halacsy: *Proc. 3rd Int. Conf. Magnet Technology, Hamburg 1970* (DESY, Hamburg 1972) pp.113-128
2.20 R.W. Hamming: J. Assoc. Comput. Mach. *6*, 37-47 (1959)
2.21 R.F. Harrington, K. Pontoppidan, P. Abrahamsen, N.C. Albertsen: Proc. Inst. Electr. Eng. *116*, 1715-1720 (1969)
2.22 R. Hauke: Dissertation, University of Tübingen, Germany (1977)
2.23 P.W. Hawkes: Comput. Aided Des. *5*, 200-214 (1973)
2.24 P.W. Hawkes (ed.): *Image Processing and Computer-aided Design in Electron Optics* (Academic, London, New York 1973) Part III, pp.229-433
2.25 P.W. Hawkes: In Ref.2.24, pp.230-248
2.26 M.B. Heritage: In Ref.2.24, pp.324-338
2.27 H. Hoch, E. Kasper, D. Kern: Optik *46*, 463-473 (1976)
2.28 H. Hoch, E. Kasper, D. Kern: Optik *50*, 413-425 (1978)
2.29 Ch. Iselin: In Ref.2.60, pp.15-18
2.30 J.D. Jackson: *Classical Electrodynamics* (Wiley, New York, London, Sydney 1962) p.141
2.31 D.H. Jacobs (ed.): *The State of the Art in Numerical Analysis* (Academic Press, London, New York 1977)
2.32 J. Janse: Optik *33*, 270-281 (1971)
2.33 M.A. Jawson: Proc. Roy. Soc. (London) A *275*, 23-32 (1963)
2.34 E. Kasper: Optik *46*, 271-286 (1976)
2.34a E. Kasper: unpublished (1980)
2.34b E. Kasper, F. Lenz: In *Electron Microscopy 1980*, Vol.1, Physics, ed. by P. Brederoo and G. Boom (North Holland, Amsterdam 1980) pp.10-15
2.35 D. Kern: Dissertation, University of Tübingen, Germany (1978)
2.36 H. Koops, G. Kuck, O. Scherzer: Optik *48*, 225-236 (1977)
2.37 K. Kuroda, T. Suzuki: Jpn. J. Appl. Phys. *11*, 1382 (1972)
2.38 J.D. Lambert: In Ref.2.31, Chap.IV.2, pp.451-500
2.39 R. Lapeyre: C.R. Acad. Sci. Paris *251*, 2144-2146 (1960); *252*, 3431-3433 (1961); *253*, 1315-1317 (1961); *254*, 237-239 (1962); *254*, 3825-3827 (1962); *256* , 1441-1443 (1963)

2.40 R. Lapeyre, M. Laudet: C.R. Acad. Sci. Paris *251*, 679-681 (1960); *251*, 863-865 (1960); *251*, 1874-1876 (1960); *253*, 1677-1678 (1961)
2.41 M. Laudet: C.R. Acad. Sci. Paris *241*, 1728-1730 (1955)
2.42 F. Lenz: Optik *7*, 243-253 (1950)
2.43 F. Lenz: Ann. Phys. (Leipzig) *19*, 82-88 (1956)
2.44 F. Lenz: In Ref.2.24, pp.274-282
2.45 H.R.J. Lewis: J. Appl. Phys. *37*, 2541-2550 (1966)
2.46 H. Liebmann: Sitzungsber. Bayer. Akad. Wiss. Munich, Heft 2, 385-416 (1918)
2.47 G. Liebmann: Brit. J. Appl. Phys. *1*, 92-103 (1950)
2.48 G. Liebmann, E.M. Grad: Proc. Phys. Soc. (London) B*64*, 956-971 (1951)
2.49 I. Lucas: J. Appl. Phys. *47*, 1645-1652 (1976)
2.50 M.F. Manning, J. Millman: Phys. Rev. *53*, 673 (1938)
2.51 W.E. Milne: Amer. Math. Monthly *38*, 14-17 (1931)
2.52 R.W. Moses: In Ref.2.24, pp.250-272
2.53 E. Munro: Dissertation, University of Cambridge (1971)
2.54 E. Munro: In Ref.2.24, pp.284-323
2.55 E. Munro: In Ref.2.60, pp.35-44
2.56 M.J. Newman: In Ref.2.60, pp.144-153
2.57 M.J. Newman, C.W. Trowbridge, L.R. Turner: *Proc. 4th Int. Conf. Magnet Technology, Brookhaven 1972*, ed. by Y. Winterbottom (U.S. Atomic Energy Commission, Washington 1973) pp.617-626
2.58 D.H. Norrie, G. de Vries: *The Finite Element Method* (Academic, New York, London 1973)
2.59 B. Numerov: Publ. Obs. Astrophys. Central Russie *2*, 188-288 (1923)
2.60 *Proceedings of the COMPUMAG, Conference on the Computation of Magnetic Fields, Oxford 1976* (Rutherford Laboratory, Chilton 1976)
2.61 A. Ralston: In Ref.2.62, Vol.I, Part III, Chap.8, pp.95-109
2.62 A. Ralston, H.S. Wilf (eds.): *Mathematical Methods for Digital Computers* (American ed.: Wiley, New York 1960, Vol.I, and 1967, Vol.II; German ed.: Oldenbourg, Munich 1969)
2.63 H. Rauh: Z. Naturforsch. *26a*, 1667-1675 (1971)
2.64 F.H. Read, A. Adams, J.R. Soto-Montiel: J. Phys. E *4*, 625-632 (1971)
2.64a G. Regenauer: Thesis, University of Tübingen (1980) (unpublished)
2.65 E. Regenstreif: Ann. Radioélectr. *6*, 51-83, 114-155 (1951)
2.66 J.K. Reid: In Ref.2.31, Chap.I.3, pp.85-146
2.67 C.H. Reinsch: Numer. Math. *10*, 177-183 (1967)
2.68 M.J. Romanelli: In Ref.2.62, Vol.I, Part III, Chap.9, pp.110-120
2.69 H. Rose: Optik *32*, 144-164 (1970)
2.70 H. Rose, E. Plies: In Ref.2.24, pp.344-369
2.70a W. Scherle: Private communication (1981), to be published. A factor 2 arises from the discontinuity of $\partial G/\partial n'$
2.71 W. Schwertfeger, E. Kasper: Optik *41*, 160-173 (1974)
2.72 A. Schuler: Dissertation, University of Vienna (1977)
2.73 U. Schumann: "Über die direkte Lösung der diskretisierten Poisson-Gleichung mittels zyklischer Reduktion", KFK-Ext. 8/75-6, Gesellschaft für Kernforschung, Karlsruhe, Germany (1976)
2.74 G. Shortley, R. Weller, P. Darby, E.H. Gamble: J. Appl. Phys. *18*, 116-129 (1947)
2.75 P. Silvester: Int. J. Eng. Sci. *7*, 849-861 (1969)
2.76 P. Silvester, A. Konrad: Int. J. Num. Meth. Eng. *5*, 481-497 (1973)
2.77 P. Silvester, A. Konrad: Comput. Phys. Commun. *5*, 437-455 (1973)
2.78 J. Simkin, C.W. Trowbridge: In Ref.2.60, pp.5-14
2.79 B. Singer, M. Braun: IEEE Trans. Electron. Devices *ED-17*, 926-934 (1970)
2.79a R. Speidel, G. Kilger, E. Kasper: Optik *54*, 433-438 (1979/1980)
2.80 J. Stoer, R. Bulirsch: *Einführung in die Numerische Mathematik II*, Heidelberger Taschenbücher, Band 114 (Springer, Berlin, Heidelberg, New York 1978)
2.81 H.L. Stone: SIAM J. Numer. Anal. *5*, 530-558 (1968)
2.82 W. Tretner: Optik *16*, 155-184 (1959)
2.83 R.S. Varga: *Matrix Iterative Analysis* (Prentice-Hall, Englewood Cliffs, New Jersey 1962)
2.84 J. Walsh: In Ref.2.31, Chap.IV, pp.501-533

2.85 C. Weber: In *Focusing of Charged Particles*, Vol.1, Chap.1.2, ed. by A. Septier (Academic, New York, London 1967) pp.45-99
2.86 D.M. Young: Trans. Am. Math. Soc. *76*, 92-111 (1954)
2.87 S.G. Zaky: Dissertation, University of Toronto (1969)
2.88 O.C. Zienkiewicz: *The Finite Element Method*, 3rd ed. (McGraw-Hill, London 1977)
2.89 O.C. Zienkiewicz, Y.K. Cheung: *The Finite Element Method in Structural and Continuum Mechanics* (McGraw-Hill, London 1967)
2.90 R. Zurmühl: *Praktische Mathematik für Ingenieure und Physiker*, 4th ed. (Springer, Berlin, Heidelberg, New York 1963)

3. Properties of Electron Lenses

F. Lenz

With 20 Figures

A magnetic lens is defined as an axisymmetric magnetic field existing in a finite axial interval between two field-free regions. The effect of saturation of the shielding material on the axial field distribution is discussed. Parameters describing lens strength and field form are defined, and scaling laws discussed. Results of numerical field calculations are given together with analytical models for symmetrical and asymmetrical lens fields. Field models are presented for which the paraxial trajectory equation can be solved analytically. A review is given of the paraxial properties, such as focal lengths, focal positions and chromatic aberration coefficients, and their dependence on lens strength and gap-bore ratio defining their field form, as well as unified representations in which each of these dependences is approximately described by one single curve in which the lens size and the gap-bore ratio occur only in the horizontal and vertical scales. A classification of third-order aberrations is given together with their unified representations for the limiting case of high magnification.

3.1 Concepts and Definitions

3.1.1 Definition of Lenses and Their Properties

A magnetic electron lens is an axially symmetric magnetic field which can be described by its axial flux density B(z). The concept of a lens implies that the extension of the field in axial direction is limited so that field-free regions exist on both sides of the lens. If the lens is used for imaging, these field-free regions are called "object space" and "image space", respectively.

The electron optical properties of magnetic electron lenses to be treated in the following contribution are the geometrical-optical properties relevant for focusing or for the formation of electron optical images or diffraction patterns. Such properties are focal length, focal positions, positions of principal planes, and coefficients of aberration. It is assumed throughout that the magnetic field in the lens is static or quasi-static and of rotational symmetry. While a static field is strictly time-independent, a quasi-static field varies sufficiently slowly with time for it to be regarded as practically constant during the time of flight of an electron through the lens field. Since in most electron optical instruments the electron velocity is much greater than 10^6 m/s, fields varying at frequencies less than 10 MHz are generally quasi-static. Whenever deviations from rotational symmetry are discussed, they will be assumed to be small. Magnetic "lenses" with a straight axis but with a two-fold or four-fold symmetry around this axis – i.e., "cylinder lenses", "quadrupole", or "octopole" lenses – will not be treated in this contribution. Since even-order aberrations do not occur in axially symmetric fields, the most important aberration coefficients of axially symmetric fields are those of first (paraxial) and third order.

3.1.2 Saturated and Unsaturated Lenses

We shall distinguish between "unsaturated" and "saturated" lenses. The concept of saturation refers to the magnetic saturation of the generally ferromagnetic pole pieces and shielding material of lenses in the form of iron-shielded magnetic coils or permanent magnets. A lens is said to be unsaturated if the flux density distribution $\underline{B}(\underline{r},I)$ created by a coil with I ampere-turns is proportional to I, i.e., if

$$\frac{\partial \underline{B}(\underline{r},I)}{\partial I} = \frac{\underline{B}(\underline{r},I)}{I} \quad . \tag{3.1}$$

This proportionality is true if the field equations are linear, e.g., if the flux density within the pole pieces and shielding material is sufficiently low, and if the magnetic permeability of these materials is sufficiently high. Then the magnetic field strength H within the material may be neglected and the scalar magnetic potential defined by

$$\psi(\underline{r}) = -\int_{\underline{r}_0}^{\underline{r}} \underline{H}(\underline{r}')\, d\underline{r}' \Rightarrow \underline{H}(\underline{r}) = -\mathrm{grad}\psi(\underline{r}) \tag{3.2}$$

can be treated as a constant on each pole-piece surface. In this case the potential difference between opposite pole pieces of an iron-shielded coil is equal to its number of ampere-turns ("lens excitation", denoted by NI in subsequent chapters)

$$I = \int H(z)\, dz = \mu_0^{-1} \int B(z)\, dz \quad . \tag{3.3}$$

In (3.3) the integration is extended along the axis from the field-free space on one side of the lens into the field-free space on the other one. The concept of a lens implies that such field-free regions exist on both sides. $\psi(\underline{r})$ is a unique function of position in every simply-connected region in which curl $\underline{H} = \underline{j}$ vanishes, i.e., a region not containing the windings of the coil. Actually it would be more precise to refer to this type of lens as a lens of "infinite permeability" [3.12] rather than calling it "unsaturated".

If the drop of scalar potential within the pole pieces and/or the shielding material is too large to be neglected, the lens is said to be saturated even if the magnetization within the material is below its saturation value M_s. Whenever the flux density in the gap or in the magnetic material is of comparable order of magnitude with the saturation magnetization, saturation effects can obviously not be neglected. As a practical rule, DUGAS et al. [3.10] find that saturation effects begin to play a noticeable role at some critical excitation I_c which amounts to 1100 S. A/mm for soft iron, i.e., $I_c \approx 0.6\ M_s S/\mu_0$, where S is the gap width.

In weak lenses, even if the coil current is switched off, assumption (3.1) may still not be justified. If the pole piece and/or shielding material show magnetic hysteresis, they may retain permanent magnetism, the contribution of which to the lens field may be noticeable if the lens is very weak or even switched off [3.2]. This may increase or decrease the focal length, depending on the sign of a previous stronger lens current determining the appropriate branch of the hysteresis curve. Definition (3.3) of the lens excitation I and the assumption of field-free regions on both sides of the lens are valid both for saturated and unsaturated lenses. Another example of an unsaturated lens is an iron-free coil, because its field strength is proportional to the lens excitation I while the shape of the field distribution does not depend on I. A lens showing typical effects of saturation in the shielding material is shown in Figs.2.11,12.

3.1.3 Lens Strength Parameters

GLASER [3.16] has introduced a dimensionless lens strength parameter

$$k^2 = 0.25\eta^2 B_0^2 a^2/U \qquad (3.4)$$

for his bell-shaped field. For practical purposes, (3.4) is not always a convenient parameter since sometimes neither the maximum axial magnetic flux density B_0 nor the half-width a of the axial flux density distribution B(z) is known. Other authors have used other length units to replace the half-width, such as the bore radius of the pole pieces (e.g., LIEBMANN [3.37]). Since the value of B_0 may be less readily available than the lens excitation I, it is often convenient to use a lens strength parameter κ^2 (denoted by g^2 in [3.30]) depending only on I and the relativistic accelerating voltage U, i.e.,

$$\kappa^2 = \eta^2\mu_0^2 I^2/(4\pi^2 U) = 0.00352\ I^2/U[A^{-2}\ V] \quad . \qquad (3.5)$$

The different dimensionless lens strength parameters are proportional to each other with factors depending, in the case of unsaturated lenses, only on the pole-piece dimensions such as bore diameter D and gap width S but not on I or U. In the case of saturated lenses, however, the field half-width generally increases with the excitation I. Some authors use, instead of κ^2, a "reduced excitation" $\varepsilon = I\ U^{-1/2}$ which is related to κ^2 by $\varepsilon^2 = 284.1\ \kappa^2\ [V^{-1}\ A^2]$ or $\varepsilon = 16.86\ \kappa[V^{-1/2}\ A]$.

3.1.4 Field-Form Factor

For GLASER's bell-shaped field, k^2 and κ^2 are identical because $\mu_0 I = \pi B_0\, a$. But for other fields B(z), even if they have the same conventional shape with one maximum and a monotonic decrease on both sides, the dimensionless "form factor"

$$I/(H_0 a) = \mu_0 I/(B_0 a) \tag{3.6}$$

may have values different from π, more often smaller than greater. For GLASER's generalized bell-shaped field

$$H(z) = H_0[1 + (z/d)^2]^{-\nu} \quad , \tag{3.7}$$

the form factor varies in the interval between π(for $\nu = 1$) and $(\pi/\ln 2)^{1/2} = 2.13$ (for $\nu \to \infty$). Most magnetic lenses have form factors within this interval. In unsaturated lenses, the form factor depends only on the pole piece dimensions. Saturation tends to increase the form factor of a lens with given dimensions.

3.1.5 Scaling of the Field

If in a physical equation all quantities with the dimension of length are multiplied by the same scaling factor, and all other quantities with other dimensions are multiplied by other scaling factors, one for each dimension, then the equation is called scale invariant if all scaling factors in the equation cancel each other. Let us, for example, consider the equation $\underline{v} = d\underline{r}/dt$. If the scale factors for length, time, and velocity are α_L, α_T, and α_V, respectively, then the equation is evidently scale invariant if $\alpha_L/(\alpha_T\alpha_V) = 1$. Let us apply this principle of scale invariance to the system of field equations for static magnetic fields:

$$\text{curl}\ \underline{H} = 0;\ \text{div}\ \underline{B} = 0;\ \underline{B} = \mu\underline{H} = -\mu\ \text{grad}\psi \quad , \tag{3.8}$$

where the permeability $\mu = B/H$ depends not only on position but also on B. Let α_I be the scaling factor for ψ and consequently also for the lens excitation I which, according to (3.2,3), is the difference between two different boundary values of ψ. If α_L is the scaling factor for length, then it follows from $\underline{H} = -\text{grad}\psi$ that the scaling factor for $\underline{H}$ is $\alpha_H = \alpha_I/\alpha_L$. If the system contains ferromagnetic material, then we have, within this material, $\underline{B} = \mu\underline{H} = \mu_0\underline{H} + \underline{M}$, where $\underline{M}$ is the magnetization. Consequently, $\underline{M}$ and $\underline{B}$ must have the same scaling factor α_H as $\underline{H}$, and the scaling factor for μ must be unity to make (3.8) scale invariant. It follows that magnetic fields containing magnetic materials can be scaled up or down only if the scaled lens model has either the same material composition as the original or if both materials have magnetization curves of the same shape, i.e., if the dependence of B/M_s on $\mu_0 H/M_s$, where M_s is the saturation magnification, is described by the same curve for both the model and the original material [3.29].

As a consequence of scaling, it is possible to measure the magnetic field distribution within a pole-piece configuration using scaled-up models if the real dimensions are so small and the field so strongly inhomogeneous that it is difficult to use probes of sufficiently small size. Such measurements are, in principle, even possible for saturated lenses provided that both in the scaled-up model and in the original the scale-invariant quantity $I/(M_s L)$ has the same value, where L stands for a typical length such as a gap width S or a bore diameter D. LANGNER and LENZ [3.29] have used a nickel model for a projector lens scaled up by a factor of four. Since nickel has a saturation magnetization four times less than that of the original pole-piece material consisting of an alloy of 35% Co and 65% Fe, the same lens excitation I could be applied to both model and original. The magnetization curves where the relative magnetization M/M_s is plotted against the relative magnetic field strength $\mu_0 H/M_s$ are very similar for both materials so that the scaling law can be applied even for the field inside the pole-piece material. A drawback of this method (apart from the high costs of big blocks of Ni or other pole-piece materials) is that the magnetic properties of Ni are very sensitive to small changes in temperature so that reliable measurements are rather expensive and hardly justified now that reliable methods for numerical computer calculation are available.

In the case of unsaturated lenses or rather "lenses of high permeability" the field equations are solved only in the vacuum ($\mu_r = 1$). They reduce to Laplace's equation $\nabla^2\psi = 0$ with the boundary condition that the pole-piece surfaces are equipotentials at $\psi = -I/2$ and $\psi = I/2$, respectively. In this case the only scaling law is $\alpha_H = \alpha_I/\alpha_L$, and such quantities as the field form factor $I/(H_0 a)$ or the dimensionless quantities $I/(H_0 S)$ and $I/(H_0 D)$ are scale invariant, where S and D are the gap width and bore diameter, respectively.

3.1.6 Scaling of Trajectories

Let us now apply the principle of scale invariance to Lorentz's equation of motion

$$m\, d^2\underline{r}/dt^2 = -e(d\underline{r}/dt) \times \underline{B} \quad . \tag{3.9}$$

This is scale invariant if the scaling factor for time $\alpha_T = \alpha_H^{-1} = \alpha_L/\alpha_I$. If we, in addition, wanted to scale charges and masses, i.e., compare the effect of a field on particles of different mass or charge, then we would also define appropriate scaling factors for mass and charge in order to make the equation of motion scale invariant. As long as we consider no other charged particles but electrons, however, m and e are treated as constants.

In electron optics we are often interested in the geometrical shape of the electron trajectories described by $\underline{r}(s)$, i.e., the dependence of the vector of position $\underline{r}$ on path length s rather than in its dependence $\underline{r}(t)$ on time. In other words we want to know where and from what direction, $\underline{t} = d\underline{r}/ds$, the electron hits a given recording plane or aperture but not when. If we eliminate time from (3.9) by replacing the differentiations with respect to time by differentiations with respect to path length, using the relation ds = v dt, we obtain

$$d^2\underline{r}/ds^2 = -(d\underline{r}/ds) \times (e\underline{B}/mv) = -(d\underline{r}/ds) \times (\eta\underline{B}U^{-1/2}) \tag{3.10}$$

since v is a constant in every magnetic field. Equation (3.10) is scale invariant if the scaling factor α_U for voltages is $\alpha_U = \alpha_I^2$. Since the modulus of the left-hand side of (3.10) is the curvature of the trajectory, it follows directly from (3.10) that it is equal to the component of $e\underline{B}/mv$ perpendicular to the tangent unit vector $\underline{t} = d\underline{r}/ds$.

If we identify the length unit L with either the field half-width a, the bore radius, the gap width, or any other characteristic dimension of the pole-piece configuration, and if we restrict ourselves to unsaturated lenses, then a change of the length unit is a similarity transformation for which the form factor $I/(H_0 a)$ defined by (3.6) is scale invariant. If we keep constant the dimensionless lens strength parameter k^2 or κ^2 defined by (3.4) or (3.5), respectively, then it follows from the scale invariance of $I/(B_0 a)$ that $\eta BU^{-1/2}a$ and hence (3.10) are scale invariant. Consequently, if the linear dimensions of the whole pole-piece configuration are all scaled up or down by the same constant scaling factor, then the same applies to the solutions of (3.10), i.e., the electron trajectories, which are scaled up or down by the same factor. The initial coordinates x_0 and y_0 of a given trajectory and all other trajectory coordinate values (e.g., in a recording plane or an aperture plane) having the dimension of a length are scaled by the same factor while the dimensionless parameters describing the direction of a trajectory such as initial directions are scale invariant [3.19].

As a general rule, all quantities having the dimension of length, such as the focal length f, the axial chromatic aberration coefficient C_c, and the coefficient C_s of third-order spherical aberration, are scaled together with the length unit while dimensionless quantities, such as f/a, C_c/a, and C_s/a, are scale invariant if the lens strength parameter k^2 or κ^2 is kept constant. The scaling laws are applicable to the geometric-optical characteristics of electron lenses but not to wave optical properties, unless one succeeds in solving the difficult problem of scaling the electron wavelength and the interatomic distances in scattering or diffracting media by the same scaling factor. There are other scaling laws for electron scattering and diffraction which will not be treated in this contribution.

3.2 Field Distribution in Unsaturated Lenses

For unsaturated lenses $\psi(\underline{r})$ is the solution of Laplace's equation $\nabla^2\psi = 0$ with the boundary condition that the pole-piece surfaces are equipotentials at $\psi = -I/2$ and $\psi = I/2$, respectively. $\underline{B}(z) = -\mu\ \mathrm{grad}\psi$ can then be obtained by differentiation of the potential $\psi(z)$ on the axis. As a consequence of the scaling law, the functions describing the dependence of ψ/I or HS/I on $\underline{r}/S$ do not depend on the excitation I or the gap width S but only on dimensionless shape parameters such as S/D. In the case of saturated lenses or when it is doubtful whether saturation effects are negligible or not, it may be easier and safer to measure B(z) using sufficiently small induction search coils or Hall probes. Under the assumption that the pole-piece bores are cylinders of infinite length and the pole-piece fronts defining the gap are unlimited parallel planes perpendicular to the axis, field distributions have been computed and/or measured by several authors (e.g., DOSSE [3.8], LENZ [3.31], LIEBMANN and GRAD [3.37], DURANDEAU et al. [3.13]). In unsaturated lenses this is a reasonable approximation because the electron optical properties of the lens depend mainly on the flux density distribution B(z) close to its maximum B_0, which in turn depends mainly on the shape of the pole-piece surface close to the edges where bores and gap intersect. Most practical lenses have cylindrical and plane pole-piece contours, at least in the close vicinity of these edges. The pole-piece contours further from the axis have less effect on the axial field distribution of an unsaturated lens, but they may be of great importance for practical design because the total flux determining the degree of magnetization in the magnetic material may depend strongly on the pole-piece shape at a greater distance from the axis. In order to avoid saturation, the pole pieces are often conically tapered [3.12,43,46]; see Sect.4.3.

3.2.1 Symmetrical Lenses

We call a lens "symmetrical" if a plane of symmetry z = 0 exists so that B(z) = B(-z). In this case we arrange the origin of the system of coordinates at r = 0; z = 0.

Figure 3.1 shows the dependence of HD/I on z/D for different values of the parameter S/D. Such normalized field distributions have been measured or computed by various authors (e.g., COSSLETT [3.6], HESSE [3.25], VAN MENTS and LE POOLE [3.42], LENZ [3.31]). More detailed, but for $S/D \ll 1$ and $S/D > 1$ not very accurate, tables may be found in the book by EL-KAREH and EL-KAREH [Ref.3.15, Vol.1]. All normalized field distributions have one maximum at z = 0 with H(z) decreasing monotonically on each side of it. Thus, for a rough approximation, the distributions can be described by three parameters: their excitation I, their maximum H_0, and their half-width a. In Fig.3.2 the dimensionless quantities H_0S/I and S/(2a) are plotted versus the gap-bore ratio S/D. It is a reasonable assumption that any two of these simple type of lens fields (which have the same values of I, H_0, and a) have approximately the same electron optical properties. These properties are independent of the manner in which the fields are created, i.e., whether they are formed by saturated or unsaturated configurations, or by iron-free, superconducting, or permanent magnet systems.

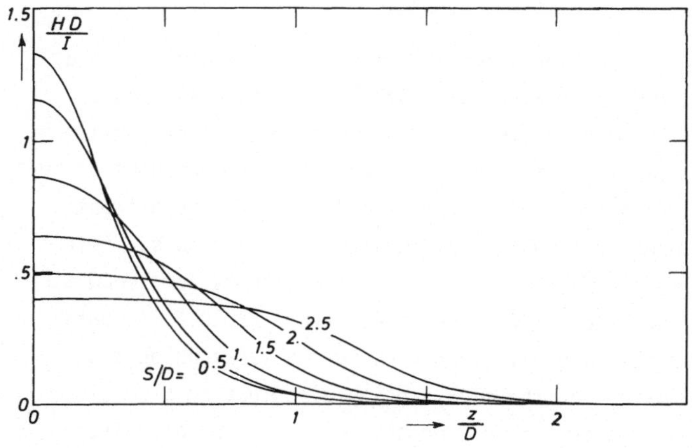

Fig.3.1. Normalized axial field distribution H(z)D/I in unsaturated symmetrical magnetic lenses for different values of the gap-bore ratio S/D

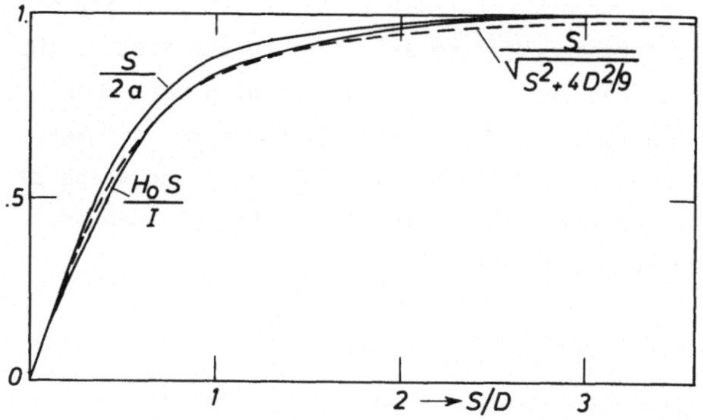

Fig.3.2. Quantities describing shape and half-width of the axial field distribution in unsaturated symmetrical magnetic lenses versus S/D. a is the field half-width, H_0 the maximum axial field strength, and $(S^2+4D^2/9)^{1/2}$ is DURANDEAU's characteristic length

Other authors have suggested other sets of three parameters to describe and compare empirical or computed axial field distributions H(z). OLLENDORFF [3.45] has characterised magnetic lenses by their excitation and two characteristic lengths $s_1 = I/H_0$ and $s_2 = \int_{-\infty}^{+\infty} [H(z)/H_0]^2 dz$. The latter integral occurs in the expression for the focal length of a weak lens, $f = s_1^2/(\pi^2\kappa^2 s_2)$. GLASER [3.18] has suggested using I, H_0, and a typical length $s_3 = [-2\,H_0/H''(z_m)]^{-1/2}$ so that the field curves H(z) of all lenses having these three parameters in common have the same apical curvature. For GLASER's bell-shaped field this typical length is equal to the half-width a. The idea behind GLASER's choice of parameters is that the field H(z) close to its maximum at $z = z_m$ has the greatest influence on the electron trajectories so that a good approximation is most important at $z \approx z_m$.

3.2.2 Asymmetrical Lenses

Again we assume that the pole-piece bores are cylinders and the pole-piece fronts are parallel planes, but in an asymmetrical lens the two bore diameters D_1 and D_2 differ from each other. Originally, a typical reason why an asymmetrical lens was used instead of a symmetrical one was that in the objective lens of an electron microscope the specimen holder had to be inserted in a sufficiently wide upper bore while the lower bore was designed narrower in order to concentrate the field, i.e., to make its half-width as small as possible. In many modern electron microscopes, however, the specimen is introduced sideways through the pole-piece gap so that there is no need to make one bore wider than the other. There may be other reasons for making a lens asymmetrical. An example is the "pinhole lens" which, according to LIEBMANN [3.40], has the advantage of combining low spherical aberration with the requirement for a relatively small lens excitation, provided the pinhole lens is located on the incident side as seen by the electron beam.

In an unsaturated cylindrical bore with diameter D the decrease in field strength for $|z| \gg S$ is proportional to $\exp(-4.8097|z|/D)$ [3.33], where the numerical factor in the exponent is twice the first zero α_1 of the Bessel function J_o ($\alpha_1 = 2.40483$). Consequently the decrease of the field in an asymmetrical lens is steeper on the side where the bore diameter is smaller. The value of H_oS/I depends in approximately the same manner on the gap-bore ratio S/D as in symmetrical lenses if D stands for the mean bore diameter, $D = (D_1 + D_2)/2$. The axial position z_m of the field maximum, however, and the two half-widths a_1 and a_2 defined by $H(z_m - a_1) = H(z_m + a_2) = 0.5\, H(z_m) = 0.5\, H_o$ depend on the asymmetry ratio D_1/D_2. Figure 3.3 illustrates the quantities S, D_1, D_2, H_o, z_m, a_1, and a_2. In Fig.3.4 z_m/S, $(z_m - a_1)/S$, and $(z_m + a_2)/S$ are plotted versus D_1/D_2 for different values of $S/D = 2\,S/(D_1 + D_2)$. The values for z_m/S are in good agreement with those given by DUGAS et al. [3.10], and the values of a_1, a_2, and z_m with those by LIEBMANN [3.39], but Fig.3.4 covers a wider range of parameters. The relative full half-width $(a_1 + a_2)/S$ and the relative maximum axial field strength H_oS/I do not vary by more than 2% with varying D_1/D_2.

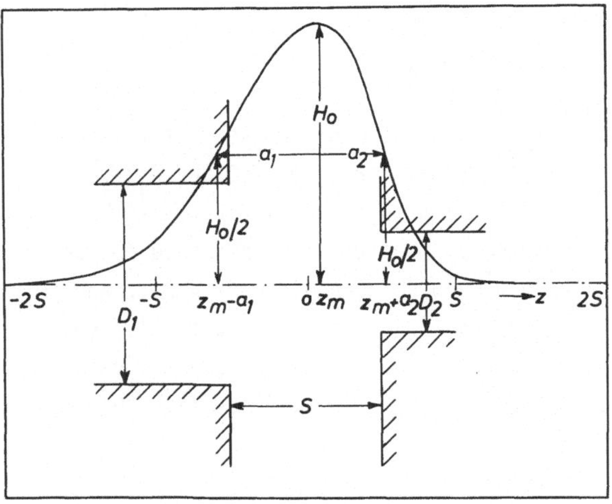

Fig.3.3. Illustration of field maximum H_o, maximum position z_m, half-widths a_1 and a_2, and bore diameters D_1 and D_2 for an asymmetric lens

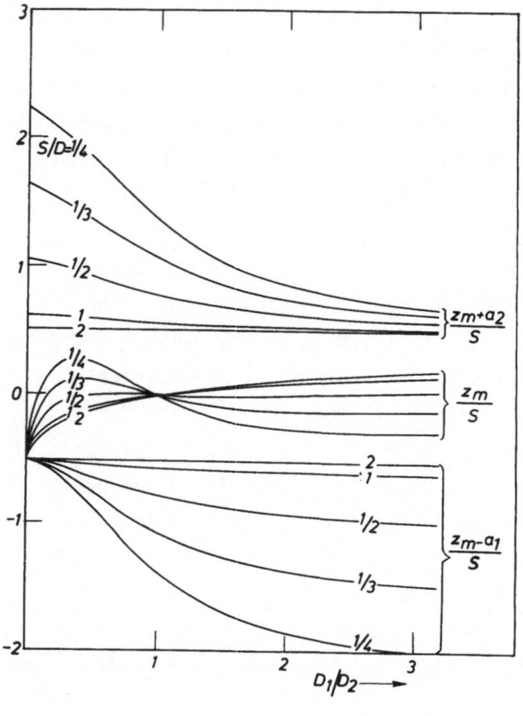

Fig.3.4. Maximum position z_m and half-widths a_1 and a_2 in unsaturated asymmetrical lenses versus asymmetry parameter D_1/D_2

It is sufficient to compute the quantities shown in Fig.3.4 for $D_1 \geqq D_2$ because of an obvious symmetry of this figure with respect to an exchange of the indices 1 and 2 and a simultaneous change of the sign of the z coordinate. If we distort the horizontal scale by plotting against $\log(D_1/D_2)$ or $(D_1 - D_2)/(D_1 + D_2)$ rather than D_1/D_2, the curves in Fig.3.4 become centrosymmetric with respect to the origin ($D_1 = D_2$; $z = 0$). In other words, we have $z_m(D_1/D_2) = -z_m(D_2/D_1)$; $a_1(D_1/D_2) = a_2(D_2/D_1)$; $a_1(D_1/D_2) + a_2(D_1/D_2) = a_1(D_2/D_1) + a_2(D_2/D_1)$; $H_0(D_1/D_2) = H_0(D_2/D_1)$. If the values of z_m, a_1, a_2, and H_0 are known for the symmetrical lens ($D_1/D_2 = 1$), then they are also known for $D_1/D_2 \to 0$ and $D_1/D_2 \to \infty$. With $D_1 = 0$ and $D_2 = 2\,D \neq 0$, the axial field distribution $H(z)$ for $z > -S/2$ is identical with the axial field distribution $H(z)$ that would be obtained for a symmetrical lens with bore radii $D_1^* = D_2^* = D^* = D_2$, a gap width $S^* = 2\,S$, and an excitation $I^* = 2\,I$, i.e., with $S^*/D^* = S/D$ and $I^*/S^* = I/S$, but with $H(z)$ vanishing for $z < -S/2$. Hence for $D_1/D_2 \to 0$ we have $a_1 \to 0$, $z_m \to -S/2$; and the value of a_2/S for $D_1/D_2 \to 0$ is twice the value of a_2/S for $D_1 = D_2$ and the same S/D ratio.

3.3 Analytical Field Models

3.3.1 GLASER's Bell-Shaped Field

The bell-shaped field [3.16]

$$H(z) = H_0[1 + (z/a)^2]^{-1} \quad , \tag{3.11}$$

exhibits the qualitative behaviour of the field on the axis of a typical electron lens in so far as it has a maximum at $z = 0$ and drops monotonically to zero on both sides. It is, however, not typical of an unsaturated lens with cylindrical bores because its decrease for large $|z|$ is slower than exponential. Its field-form factor, $I/(H_0 a) = \pi$, is considerably greater than the form factors of the unsaturated lenses treated in Sect.3.2, which range between 2.0 and 2.3. Saturated lenses, however, sometimes have form factors comparable to or even greater than that of GLASER's bell-shaped field.

A very important feature of the bell-shaped field is that it permits not only an analytic solution of the paraxial trajectory equation but also an explicit description of all paraxial and third-order aberration coefficients in terms of elementary analytic functions of the lens strength parameter k^2 and the object position z_0. Since the image position z_i and the magnification M depend on k^2 and z_0, any two of the four quantities z_0, z_i, k^2, and M can be used as the two independent parameters defining the imaging conditions. GLASER's bell-shaped field has contributed much to a qualitative understanding of the dependence of electron-optical properties of magnetic lenses on the imaging parameters such as k^2 and z_0, especially at a time when powerful modern computers were not available and numerical field and trajectory calculations were very tedious.

Another property of the bell-shaped field model, which GLASER considered to be most important, is that the relations between object position z_0, image position z_i, and magnification M satisfy Newton's imaging equation

$$M = -(z_i - z_{fi})/f = -f/(z_{fo} - z_o) \quad , \tag{3.12}$$

where z_{fo} and z_{fi} are the focal positions on the object and image side, respectively. For other lenses, Newton's imaging equation generally holds only if both object and image position are in the field-free regions, while in the case of GLASER's model (3.11) it is valid even if the object position and/or the image position are situated within the lens field. The definition of principal planes makes sense only if Newton's imaging equation holds. Otherwise, the position of the principal planes depends on object and image position, and the classical construction of conjugate planes for collinear imagery fails. GLASER and BERGMANN [3.20] have even questioned the physical meaning of such concepts as focal point and focal length if the image equation is not Newtonian.

Since the field model (3.11) contains only two independent parameters, viz. B_0 and a, it cannot be adapted to a given field-form factor or to a given bore ratio D_1/D_2 of an asymmetric lens. In order to make his model more flexible, GLASER has suggested two generalizations. One of them, which is restricted to symmetrical lenses, is given by (3.7). With increasing ν, the field decrease in the bores becomes steeper, and the form factor

$$I/(H_0 a) = \Gamma(\nu - \tfrac{1}{2})\Gamma(\tfrac{1}{2})(2^{1/\nu} - 1)^{-1/2}\Gamma^{-1}(\nu) \tag{3.13}$$

decreases. The half-width is $a = d(2^{1/\nu} - 1)^{1/2}$. For $\nu = 3/2$, (3.7) describes the field of an iron-free lens formed by a linear circular current with radius d. For $\nu \to \infty$, (3.7) can be written as

$$H(z) = H_0 \exp[-(z/a)^2 \ln 2] \quad . \tag{3.14}$$

An analytical solution of the paraxial trajectory equation for $\nu \neq 1$ is not known. The other generalization suggested by GLASER is applicable to asymmetrical lenses and will be discussed in Sect.3.3.4.

3.3.2 Analytical Models for a Symmetrical Lens with S << D

The axial field of a symmetrical unsaturated lens with S << D [3.1,9,22,24] is given by

$$H(z) = \frac{2I}{\pi D}\int_0^\infty \frac{\cos(2tz/D)}{I_0(t)}\,dt = \frac{2I}{D}\sum_{k=1}^{\infty} \exp(-2\alpha_k|z|/D)/J_1(\alpha_k) \quad . \tag{3.15}$$

This solution of Laplace's equation was first derived to describe the electrostatic field formed by two coaxial cylinders of equal diameter D separated by a distance S << D. In (3.15), $I_0(t) = J_0(it)$ is the modified Bessel function, and the α_k are the zeros of the Bessel function J_0. For a numerical evaluation of (3.15) the integral is more convenient for small $|z|$ while the sum converges more rapidly for large $|z|$. For $|z| \ll D$, (3.15) yields $H(z) = 1.326\,(I/D)[1 - 7.53\,(z/D)^2 + O(z^4)]$. GRAY [3.22] recommended

$$H(z) = 1.318\,(I/D)\cosh^{-2}(2.636\,z/D) \tag{3.16}$$

as a very good approximation for (3.15). Both (3.15) and (3.16) decrease exponentially for large $|z|$, the exact solution (3.15) with the correct factor $\alpha_1 = 2.40483$ in the exponent of $\exp(-2\alpha_1|z|/D) = \exp(-4.80966|z|/D)$, while (3.16) is proportional to $\exp(-5.272|z|/D)$ instead. Another field model [3.23,34] combines the correct exponential decrease with the existence of an analytical solution to the paraxial trajectory equation:

$$H(z) = [2\alpha_1 I/(\pi D)]/\cosh(2\alpha_1 z/D) \quad . \tag{3.17}$$

The field-form factors are $I/(H_0 a) = 2.306$ for the exact solution (3.15), 2.269 for (3.16), and 2.385 for (3.17). The axial magnetic potentials $\psi(z)$ from which the axial field distributions (3.15-17) can be derived are

$$\psi(z) = -\frac{I}{\pi}\int_0^\infty \frac{\sin(2tz/D)}{tI_o(t)}\,dt = I\left[\sum_{k=1}^{\infty} \frac{\exp(-2\alpha_k|z|/D)}{\alpha_k J_1(\alpha_k)} - \frac{1}{2}\right]\mathrm{sign}\, z \tag{3.18}$$

for the exact solution (3.15),

$$\psi(z) = -(I/2)\tanh(2.636\, z/D) \tag{3.19}$$

for GRAY's approximation (3.16), and

$$\psi(z) = -(2I/\pi)\arctan[\tanh(\alpha_1 z/D)] \tag{3.20}$$

for the field model (3.17). The arbitrary additive constant in the potential is chosen so that $\psi(0) = 0$ in all three cases.

3.3.3 Analytical Models for Symmetrical Lenses with Finite S/D Ratios

For symmetrical lenses it is a reasonable approximation that the boundary values of $\psi(D/2,z)$ on the cylindrical surface $r = D/2$ form a linear function of z within the gap, i.e., $\psi(D/2,z) = -Iz/S$ for $|z| < S/2$. Actually the field is stronger close to the edges $r = D/2$, $|z| = S/2$ where the cylindrical bore and the plane pole-piece fronts intersect and weaker in the middle of the gap, but this deviation from linearity has only little effect on the field on the axis. The advantage of the assumed linearity is that the field H(z) for finite S/D ratios can then be obtained by convoluting the field models $H^{(o)}(z)$ used in Sect.3.3.2 for S/D = 0 with a uniform rectangular function which is equal to 1/S for all $|z| < S/2$ and vanishes for $|z| > S/2$:

$$H(z) = S^{-1}\int_{-S/2}^{+S/2} H^{(o)}(z - t)dt = S^{-1}\left[\psi^{(o)}(z - \tfrac{S}{2}) - \psi^{(o)}(z + \tfrac{S}{2})\right] \quad . \tag{3.21}$$

In (3.21), $H^{(o)}(z)$ and $\psi^{(o)}(z)$ denote the axial field strength and axial magnetic potential, respectively, obtained for $S/D \to 0$. Using (3.21), H(z) can easily be determined for arbitrary values of S/D and z by simple subtraction if $\psi^{(o)}(z)$ is tabulated for a sufficient number of z/D values [3.3]. Table 3.1 lists the values of $\psi^{(o)}(z)$ according to (3.18).

The convoluted GRAY model is obtained by replacing $\psi^{(o)}(z)$ in (3.21) by (3.19) for $\psi(z)$:

$$\begin{aligned} H(z) &= (I/2S)\{\tanh[1.318(S/D)(1+2z/S)]+\tanh[1.318(S/D)(1-2z/S)]\} \\ &= (I/S)\sinh(2.636\, S/D)/[\cosh(2.636\, S/D)+\cosh(5.272\, z/D)] \quad . \end{aligned} \tag{3.22}$$

If in this expression the factor 2.636 is replaced by $\alpha_1 = 2.4048$ and 5.272 by $2\alpha_1$, it becomes identical with a model suggested by LENZ [3.33] and found by intuition rather than by convolution:

$$H(z) = (I/2S)\left\{\tanh\left[\frac{\alpha_1 S}{2D}\left(1 + \frac{2z}{S}\right)\right] + \tanh\left[\frac{\alpha_1 S}{2D}\left(1 - \frac{2z}{S}\right)\right]\right\}$$

Table 3.1. Scalar magnetic potential according to (3.18)

z/D	$-\psi(z)/I$	z/D	$-\psi(z)/I$	z/D	$-\psi(z)/I$
0	0	0.50	0.4298	1	0.4935
0.05	0.0659	0.55	0.4443	1.1	0.4960
0.10	0.1294	0.60	0.4560	1.2	0.4975
0.15	0.1885	0.65	0.4653	1.3	0.4985
0.20	0.2417	0.70	0.4726	1.4	0.4990
0.25	0.2882	0.75	0.4785	1.5	0.4994
0.30	0.3281	0.80	0.4830	1.6	0.4996
0.35	0.3615	0.85	0.4866	1.7	0.4998
0.40	0.3891	0.90	0.4895	1.8	0.4999
0.45	0.4116	0.95	0.4917	∞	0.5

$$= (I/S)\sinh(\alpha_1 S/D)/[\cosh(\alpha_1 S/D) + \cosh(2\alpha_1 z/D)]$$

$$= N\left\{1 + \tanh\left[\frac{\alpha_1 S}{2D}\left(1 - \frac{2z}{S}\right)\right]\right\}\left\{1 + \tanh\left[\frac{\alpha_1 S}{2D}\left(1 + \frac{2z}{S}\right)\right]\right\} \tag{3.23}$$

where N is a normalizing factor, $N = (I/4S)[1 - \exp(-2\alpha_1 S/D)]$. This model can be interpreted as a product of two hyperbolic-tangent-steps at $z = \pm S/2$. The maximum axial field strength is

$$H(0) = (I/S)\tanh(\alpha_1 S/2D) \quad . \tag{3.24}$$

The convoluted field model (3.17) is obtained by inserting $\psi(z)$ from (3.20) for $\psi^{(o)}(z)$ into (3.21):

$$H(z) = \frac{2I}{\pi S} \arctan \frac{\sinh(\alpha_1 S/D)}{\cosh(2\alpha_1 z/D)} \quad . \tag{3.25}$$

The maximum axial field strength becomes

$$H(0) = \frac{2I}{\pi S} \arctan[\sinh(\alpha_1 S/D)] \quad . \tag{3.26}$$

The principle of convolution can also be applied to the field of a solenoidal coil with radius R and length S. The axial field of a linear circular current is

$$H(z) = (IR^2/2)(R^2 + z^2)^{-3/2} \quad , \tag{3.27}$$

and its axial potential

$$\psi(z) = -(I/2)z(R^2 + z^2)^{-1/2} \quad . \tag{3.28}$$

Convolution according to (3.21) yields for a coil of length S

$$H(z) = \frac{I}{2S}\left[\frac{z + S/2}{[R^2 + (z + S/2)^2]^{1/2}} - \frac{z - S/2}{[R^2 + (z - S/2)^2]^{1/2}}\right] \quad . \tag{3.29}$$

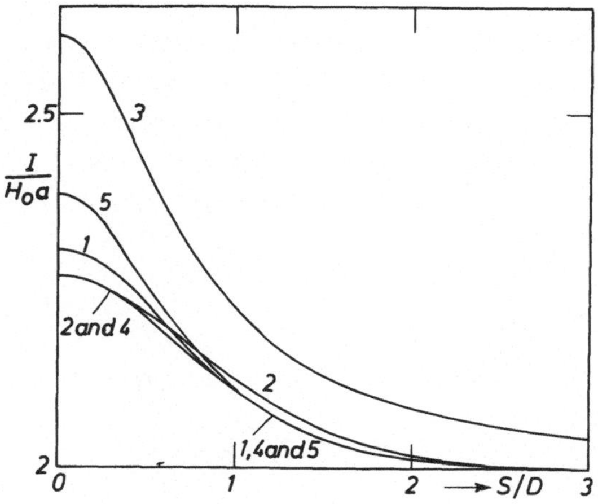

Fig.3.5. Field-form factors $I/(H_0 a)$ versus S/D for different unsaturated lens models. Curve 1: Precise numerical calculation, 2: LENZ's model (3.23), 3: DURANDEAU's model (3.29), 4: convoluted GRAY's model (3.22), 5: convoluted LENZ-GRIVET model (3.25)

According to DURANDEAU [3.12], this is a good approximation for the field of an unsaturated lens with gap width S and bore diameter D if $R = D/3$ and if $0.5 < S/D < 2$. This approximation is called the "equivalent solenoid lens" by MULVEY and WALLINGTON [3.44]. The maximum axial field strength is

$$H(0) = I(S^2 + 4D^2/9)^{-1/2} \quad .$$

A comparison of the broken curve in Fig.3.2 with $H_0 S/I$ allow us to estimate the accuracy of this approximation. In Fig.3.5 the field-form factors $I/(H_0 a)$ are plotted versus S/D for the different field models. One might conclude from Fig.3.5 that DURANDEAU's model is the least accurate as far as the prediction of field half-widths and field-form factors is concerned, but it has a compensating advantage. From a practical point of view it is desirable to have a unified representation of electron optical properties depending on only one characteristic length instead of two independent parameters S and D. DURANDEAU [3.12] has introduced the length unit

$$L = (S^2 + (2D/3)^2)^{1/2} \approx (S^2 + 0.45D^2)^{1/2} \qquad (3.30)$$

so that, with (3.30), he obtains $H_0 = I/L$. Another quantity which, to a tolerable approximation, depends on L rather than on S and D separately is the field half-width which, according to DURANDEAU, is

$$a \approx 0.485(S^2 + 0.45D^2)^{1/2} = 0.485L \quad , \qquad (3.31)$$

at least for $0.5 < S/D < 2$ (compare Fig.3.2). Together with $H_0 = I/L$, this implies that the form factor $I/(H_0 a) = 2.062$ does not depend on S/D. Curve 3 in Fig.3.5 shows that the accuracy of this assumption is limited. The introduction of a length unit (3.31) is plausible, considering that the field distribution for finite S/D ratios can be approximately described by the convolution of a function $H^{(o)}(z)$ for $S/D \to 0$ with a uniform rectangular ("top-hat") function equal to 1/S for $|z| < S/2$ and vanishing for $|z| \geqq S/2$. The half-width of $H^{(o)}(z)$ is proportional to D, viz.

$a = 0.327D \approx D/3$ according to (3.15), while the half-width of the rectangular function is $S/2$. For Gaussian distributions it is a general rule that the square of the half-width of the convolution is the sum of the squares of the half-widths of the convoluted distributions. For non-Gaussian distributions this is not correct but often a good approximation. In the present case, we would obtain $a^2 = (S/2)^2+(D/3)^2$, in almost perfect agreement with (3.31).

3.3.4 Analytical Models for Asymmetrical Lenses

GLASER's bell-shaped field model (3.11) can be generalized to describe asymmetrical lenses [3.7,16]:

$$H(z) = H_0[1 + (z/a_1)^2]^{-1} \quad \text{for} \quad z < 0 \;;$$

$$H(z) = H_0[1 + (z/a_2)^2]^{-1} \quad \text{for} \quad z > 0 \;. \tag{3.32}$$

Again we have three parameters (H_0, a_1, and a_2) which can be used to adapt the field model to the real lens field. The half-widths are a_1 and a_2, and the field form factor $I/(H_0 a) = 2I/[H_0(a_1 + a_2)] = \pi$ does not depend on a_1 and a_2. The paraxial trajectory equation can be solved for $z \leq 0$ as well as for $z \geq 0$, and the two solutions can be fitted together so that they are continuous and have a continuous derivative. Thus analytical relations are obtained in which the object positions, image positions, magnification, and lens strength are dependent on each other. For a calculation of third-order aberrations, the field model (3.32) is of doubtful value because the second and higher derivatives of $H(z)$ are discontinuous at $z = 0$. Thus the off-axis field

$$H_z(r,z) = H(0,z) - (r^2/4)H''(0,z) + O(r^4)$$

becomes undefined and ambiguous in the plane $z = 0$. In other words $H(z)$ from (3.32) is compatible with Laplace's equation for $z < 0$ and $z > 0$ but not for $z = 0$. In the derivation of the integral expressions for the third-order aberrations, integration by parts is used so that one has to be careful when applying the results to fields with discontinuous derivatives.

DURANDEAU [3.12] has modified his field model (3.24) as follows to make it applicable to asymmetric fields:

$$H(z) = \frac{I}{2S}\left[\frac{z + S/2}{[(D_1/3)^2 + (z + S/2)^2]^{1/2}} - \frac{z - S/2}{[(D_2/3)^2 + (z - S/2)^2]^{1/2}}\right] \;. \tag{3.33}$$

A similar modification for asymmetric fields [3.33] is possible for the field model (3.23) leading to

$$H(z) = N\left\{1 + \tanh\left[\frac{\alpha_1 S}{2D_1}\left(1 + \frac{2z}{S}\right)\right]\right\}\left\{1 + \tanh\left[\frac{\alpha_1 S}{2D_2}\left(1 - \frac{2z}{S}\right)\right]\right\} \;. \tag{3.34}$$

The normalizing factor N is chosen so that the integral over H(z) from $-\infty$ to $+\infty$ is equal to the lens excitation I. The semi-empirical formula

$$N = \frac{I}{2S} \frac{1 - \exp(-2\alpha_1 S/D)}{1 + 0.379[(D_1 - D_2)/(D_1 + D_2)]^2 \exp(-\alpha_1 S/D)} \tag{3.35}$$

approximates the normalizing factor to within 0.6% over the whole range of parameters, i.e., $0 \leqq D_1/D_2 < \infty$ and $0 \leqq S/D < \infty$. Another, even simpler modification of the field model (3.23) is

$$H(z) = \frac{I}{2S}\left\{\tanh\left[\frac{\alpha_1 S}{2D_1}\left(\frac{2z}{S} + 1\right)\right] - \tanh\left[\frac{\alpha_1 S}{2D_2}\left(\frac{2z}{S} - 1\right)\right]\right\} \quad . \tag{3.36}$$

It is not identical with (3.34), and it does not contain a complicated normalizing factor. Figure 3.6 shows, for comparison, the normalized axial field strength $H(z)S/I$ according to (3.32) (Curve 1), (3.33) (Curve 2), and (3.34) (Curve 3) for $S/D = 1$, $D_1/D_2 = 1.5$. While (3.33) and (3.34) contain S, D_1 and D_2 as parameters, the parameters H_o, a_1, and a_2 of Curve 1 were chosen so that the maximum position z_m and the half-widths a_1 and a_2 were the same as in curve 3 and $H_o = 2I/[\pi(a_1+a_2)]$. Hence all three curves refer to the same lens excitation I. The dots denote values from an accurate finite-difference calculation. The values according to (3.36) were also computed, but they are so close to those of Curve 3 that they cannot be represented as a separate curve in Fig.3.6.

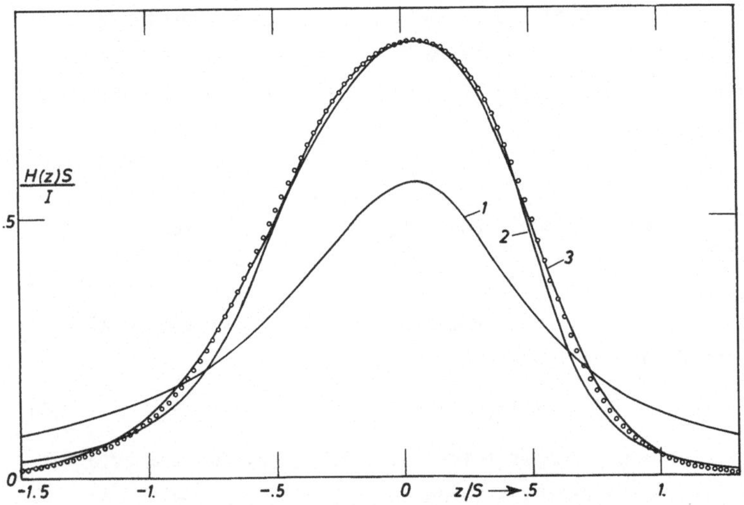

Fig.3.6. Comparison of field models for an asymmetrical lens with $S/D = 1$, $D_1/D_2 = 1.5$. Curve 1: GLASER's model for the same half-widths and excitation, 2: DURANDEAU's model (3.33), 3: LENZ's model (3.34). The dots show results of a precise numerical calculation

Evidently, DURANDEAU's approximation (3.33) is very good close to the field maximum, while approximation (3.34) fits better for large $|z|$ and is more suitable for determining the half-widths a_1 and a_2. GLASER's approximation is poor for unsaturated fields at equal lens excitation. Of course, it is better if not only a_1 and a_2, but also H_0 are taken from other models or from measured values. However, it then refers to a quite different lens excitation.

3.3.5 Analytical Solutions of the Paraxial Trajectory Equation

The paraxial trajectory equation

$$d^2x/dz^2 + \eta^2 B^2(z)x/(4U) = 0 \tag{3.37}$$

can be solved for GLASER's bell-shaped field (3.11) [3.16]. Substituting

$$z = a\cot\varphi \tag{3.38}$$

and defining

$$\omega = (1 + k^2)^{1/2} = [1 + \eta^2 B_0^2 a^2/(4U)]^{1/2} \quad , \tag{3.39}$$

one finds the general solution

$$x(\varphi) = C\sin[\omega(\varphi - \gamma)]/\sin\varphi \tag{3.40}$$

with C and γ as arbitrary constants of integration. If z varies from $-\infty$ via 0 to $+\infty$, φ varies from π via $\pi/2$ to 0. The solution with

$$\gamma = \pi/\omega \qquad \text{and} \qquad C = -1/\omega \tag{3.41}$$

describes the paraxial trajectory $\bar{G}(z) = \sin(\omega\varphi)/(\omega\sin\varphi)$ whose asymptote in image space ($\varphi \to 0$, i.e., $z \to \infty$) is parallel to the axis at unit distance. Its zero $\varphi = \pi/\omega$ defines the focal position z_{fo} (called "midfocal length" by EL-KAREH and EL-KAREH [3.15]) on the object side

$$z_{fo} = a\cot(\pi/\omega) \quad . \tag{3.42}$$

Its slope at $z = z_{fo}$ determines the focal length f_o on the object side

$$f_o = a/\sin(\pi/\omega) \quad . \tag{3.43}$$

The symmetrical solution $G(z)$ with $\gamma = \pi - \pi/\omega$ and $C = 1/\omega$ defines the focal position z_{fi} and focal length f_i on the image side

$$z_{fi} = -a\cot(\pi/\omega) = -z_{fo} \quad , \quad f_i = a/\sin(\pi/\omega) = f_o \quad . \tag{3.44}$$

For $\omega = 2$, i.e., $k^2 = 3$, the lens becomes telescopic, i.e., any two trajectories entering the lens with parallel asymptotes in object space leave it with parallel asymptotes in image space. Every positive integer value of ω defines a telescopic lens strength $k^2 = \omega^2 - 1$. For $\omega > 2$ the trajectory $\sin(\omega\varphi)/(\omega\sin\varphi)$ has more than one zero in the interval $0 \leqq \varphi \leqq \pi$, so there is more than one focal point; instead of (3.42-44), we have

$$f = f_o = f_i = a/\sin(n\pi/\omega) \quad , \quad z_{fo} = -z_{fi} = a \cot(n\pi/\omega) \quad , \tag{3.45}$$

where n may be any integer in the interval $1 \leq n \leq \omega$.

Let us now assume that an object is arranged in an object plane $z = z_o = a \cot\varphi_o \leq z_{fo}$. Then the solutions g(z) and h(z) of the paraxial trajectory equation defined by

$$g(z_o) = 1 \quad , \quad g'(z_o) = 0 \quad , \quad h(z_o) = 0 \quad , \quad h'(z_o) = 1 \tag{3.46}$$

are

$$h(\varphi) = -\frac{a}{\omega} \frac{\sin[\omega(\varphi - \varphi_o)]}{\sin\varphi \sin\varphi_o} \tag{3.47}$$

and

$$g(\varphi) = \frac{\omega\cos[\omega(\varphi - \varphi_o)]\sin\varphi_o + \sin[\omega(\varphi - \varphi_o)]\cos\varphi_o}{\omega\sin\varphi} \quad . \tag{3.48}$$

The coordinates x(z) and y(z) of a trajectory defined by the initial conditions $x_o = x(z_o)$, $y_o = y(z_o)$, $x_o' = x'(z_o)$, $y_o' = y'(z_o)$ in the object plane $z = z_o$ are then

$$x(z) = x_o g(z) + x_o' h(z) \quad , \quad y(z) = y_o g(z) + y_o' h(z) \quad . \tag{3.49}$$

In an image plane $z = z_i$ for which $h(z_i)$ vanishes, $x_i = x(z_i)$ and $y_i = y(z_i)$ depend only on x_o and y_o but not on x_o' and y_o' in paraxial approximation. In other words all paraxial trajectories starting from the same object point (x_o, y_o) are reunited at the same image point (x_i, y_i). The image position $z_i = a \cot\varphi_i$ defined by $h(z_i) = 0$ follows from (3.47)

$$\varphi_i = \varphi_o - \pi/\omega \quad , \quad z_i = a \cot(\varphi_o - \pi/\omega) \quad . \tag{3.50}$$

For the magnification M we obtain

$$M = x_i/x_o = y_i/y_o = g(z_i) = -\sin\varphi_o/\sin\varphi_i = 1/h'(z_i) \quad . \tag{3.51}$$

Using (3.42-44,50,51), one can easily verify that Newton's imaging equations (3.12) are satisfied.

For $\omega > 2$, two-step, three-step, or multi-step images can be defined by

$$\varphi_i = \varphi_o - n\pi/\omega \quad , \quad z_i = a \cot(\varphi_o - n\pi/\omega) \quad , \tag{3.52}$$

for which Newton's imaging equations are satisfied as well if the corresponding paraxial characteristics defined by (3.45) are used. Again n may be any integer in the interval $1 < n \leq \omega$. Since the imaging equations are Newtonian, it makes sense to define principal planes

$$z_{po} = z_{fo} + f_o = a \cot[n\pi/(2\omega)] \quad , \quad z_{pi} = z_{fi} - f_i = -z_{po} \quad . \tag{3.53}$$

Measuring object and image planes from the principal planes, Newton's imaging equations can be written as

$$(z_{po} - z_o)^{-1} + (z_i - z_{pi})^{-1} = f^{-1} \quad . \tag{3.54}$$

From (3.40,41) we can also derive asymptotic paraxial characteristics such as asymptotic (or "virtual") focal lengths v_o and v_i and focal positions z_{vo} and z_{vi}. The symbol v recalls RUSKA's [3.46] original definition of the asymptotic focal length as "Vergrößerungsweite", which has become obsolete. v_o is defined as the inverse of the slope in object space of the paraxial solution (3.41), and z_{vo} as the coordinate of the point of intersection of its asymptote in object space with the axis. One obtains

$$v_o = -\omega a/\sin(\pi\omega) \quad , \quad z_{vo} = -a\cot(\pi\omega) \quad . \tag{3.55}$$

The corresponding quantities on the image side follow from the symmetrical trajectory with $\gamma = \pi - \pi/\omega$ and $C = 1/\omega$:

$$v_i = v_o = -\omega a/\sin(\pi\omega) \quad , \quad z_{vi} = -z_{vo} = a\cot(\pi\omega) \quad . \tag{3.56}$$

Since every trajectory has only one asymptote on either side, there is no ambiguity even for $\omega > 2$.

Another field model [3.23,34] for which the paraxial trajectory equation (3.37) can be solved analytically is [cf. (3.17)]

$$H(z) = H_o/\cosh(z/d) \quad , \tag{3.57}$$

where $d = a/\ln(2 + \sqrt{3}) = 0.7593a$ may or may not be related to the bore diameter of an unsaturated lens as in (3.17). Substituting

$$\tanh(z/d) = u \quad , \tag{3.58}$$

the paraxial trajectory equation becomes

$$(1 - u^2)\frac{d^2x}{du^2} - 2u\frac{dx}{du} + \kappa^2 x = 0 \quad .$$

Its general solution can be expressed in terms of the solutions $P_\nu(u)$ and $Q_\nu(u)$ of Legendre's differential equation

$$x(z) = AP_\nu[\tanh(z/d)] + BQ_\nu[\tanh(z/d)] \quad , \tag{3.59}$$

where $\nu(\nu + 1) = \kappa^2$. The solution with $A = 1$ and $B = 0$ describes a paraxial trajectory whose asymptote in image space is parallel to the axis at unit distance. For integer values $\nu = n$ the lens becomes telescopic with telescopic lens strength parameters $\kappa^2 = n(n + 1) = 2, 6, 12, 20, 30$, etc. For half-integer values $\nu = n + 1/2$, the solutions P_ν and Q_ν can be expressed as complete elliptical integrals. Simple analytic expressions for the dependence of the paraxial characteristics on the lens strength parameter are not known except for the asymptotic focal length v and focal position z_v:

$$v_o = v_i = \pi d/[2\sin(\pi\nu)] \quad ; \quad z_{vi} = -z_{vo} = \frac{\pi d}{2}\cot(\pi\nu) + d(\psi(\nu) + \gamma) \quad , \tag{3.60}$$

where $\psi(\nu) = d\ln\Gamma(\nu + 1)/d\nu$ and $\gamma = 0.5772$ is Euler's constant.

3.4 Paraxial Properties

The paraxial properties, which can be directly derived from the paraxial equation (3.37) and its general solution, are the real and asymptotic cardinal elements, i.e., focal lengths, focal positions, and principal plane positions, as well as the real and asymptotic paraxial chromatic aberration coefficients C_c, C_D, and C_Θ.

Numerical values of the paraxial quantities mentioned will be given for symmetrical lenses only. For asymmetrical lenses it is, however, a good approximation to define $D = (D_1 + D_2)/2$ as in Sect.3.2.2 and to use the curves and tables calculated for symmetrical lenses. This approximation is good not only for the field form parameters but also for the paraxial properties [3.14]. Focal positions should, in this approximation, be counted from the field maximum.

3.4.1 Weak Lens Approximation

Introducing the normalized axial induction

$$b(z) = \pi B(z)/(\mu_0 I) \Rightarrow \int_{-\infty}^{+\infty} b(z)dz = \pi \qquad (3.61)$$

and the lens strength parameter κ^2 defined by (3.5), the paraxial trajectory equation (3.37) may be rewritten as

$$x'' + \kappa^2 b^2 x = 0 \quad . \qquad (3.62)$$

In using (3.62) instead of (3.37), one is making the tacit assumption that it makes physical sense to split the axial field distribution $B(z)$ up into one factor κ^2 that depends only on lens strength and another $b(z)$ that depends only on the shape of the field but not the lens strength. As we have seen, this is a reasonable assumption for unsaturated lenses. Indeed, it is a necessary assumption if one intends to describe the dependence of lens properties on lens strength κ^2.

One way of solving the paraxial trajectory equation (3.62) for given initial conditions [3.42] is the iterative method of Picard and Lindelöf. A sequence of approximate solutions $x_n(z)$ is defined by

$$x''_{n+1}(z) = -\kappa^2 b^2(z) x_n(z) \quad ; \quad n = 0, 1, 2, \ldots, \qquad (3.63)$$

where $x_0(z)$ is a linear function of z satisfying the given initial conditions. Once $x_n(z)$ is known, the right-hand side of (3.63) is a known function of z which can be directly integrated twice to obtain $x_{n+1}(z)$. After each integration, the constant of integration is determined so that the given initial conditions are satisfied for every $x_n(z)$. With increasing n, the sequence $x_n(z)$ converges towards the solution $x(z)$. For sufficiently weak lenses, $x_1(z)$ is already a good approximation.

If we apply this method to the solution $G(z)$ with the initial conditions $G(-\infty) = 1$, $G'(-\infty) = 0$, we have $G_0(z) \equiv 1$, and

$$G_1'(z) = -\kappa^2 \int_{-\infty}^{z} b^2(\zeta)d\zeta \quad , \tag{3.64}$$

$$G_1(z) = 1 - \kappa^2 \int_{-\infty}^{z} \int_{-\infty}^{\xi} b^2(\zeta)d\zeta d\xi = 1 - \kappa^2 \int_{-\infty}^{z} (z - \zeta)b^2(\zeta)d\zeta \quad . \tag{3.65}$$

If the lens is short, i.e., if the image focal point is situated in the practically field-free image space, both the real and the asymptotic focal length are determined by the slope of the outgoing asymptote in image space. Thus we have

$$f_i^{-1} = v_i^{-1} = -G'(+\infty) \approx -G_1'(+\infty) = \kappa^2 \int_{-\infty}^{+\infty} b^2(z)dz = \frac{\eta^2}{4U} \int_{-\infty}^{+\infty} B^2(z)dz \quad . \tag{3.66}$$

The equation of the outgoing asymptote in image space is

$$G(z) \approx G_1(z) = 1 - (z/f_i) + \kappa^2 \int_{-\infty}^{+\infty} \zeta b^2(\zeta)d\zeta \quad . \tag{3.67}$$

The last term on the right-hand side vanishes for symmetrical lenses; it is small compared to unity for all short lenses because it is equal to $\langle z\rangle/f_i$, where

$$\langle z\rangle = \int_{-\infty}^{+\infty} b^2(z)\, z\, dz \Big/ \int_{-\infty}^{+\infty} b^2(z)\, dz \tag{3.68}$$

is the average value of the z coordinate, averaged with b^2 as a weight function. It is understood that the origin of the z coordinate is arranged somewhere in the lens centre. It is even possible to arrange it at the "centre of gravity" of the lens so that $\langle z\rangle$ vanishes by definition. We must further assume that b(z) decreases sufficiently strongly as $|z| \to \infty$ for the integrals occurring in the definition of $\langle z\rangle$ (3.68) to exist. GLASER and BERGMANN [3.20] have shown that b(z) must decrease more strongly than $|z|^{-1}$ or $|z|^{-3/2}$ in order to ensure the existence of a focal point or focal plane, respectively. In the later case, even $\langle z^2\rangle$ exists. We shall henceforward include the assumption that both conditions are satisfied in the definition of the concept of a lens. This is a more precise definition than the verbal statement in the introduction of Sect.3.1.1 that "field-free regions" exist on both sides of the lens.

Since the image focal position z_{fi} is defined as the zero of G(z), we have, for weak lenses, $z_{fi} \approx z_{vi} \approx f_i \approx v_i$. The same reasoning for a solution $\bar{G}(z)$ defined by the initial conditions $\bar{G}(+\infty) = 1$, $\bar{G}'(+\infty) = 0$ leads to analogous expressions for the object focal lengths and focal positions so that we finally obtain for weak lenses [3.5]

$$\begin{aligned} z_{fi} &= z_{vi} = f_i = v_i = -z_{fo} = -z_{vo} = f_o = v_o = \left[\kappa^2 \int_{-\infty}^{+\infty} b^2(z)dz\right]^{-1} \\ &= \frac{4U}{\eta^2}\left[\int_{-\infty}^{+\infty} B^2 dz\right]^{-1} \quad . \end{aligned} \tag{3.69}$$

In the weak lens approximation we may consequently omit the indices i and o of the focal lengths and write $f = f_o = f_i$ and $v = v_o = v_i$. We shall, however, for the rest of this section denote the focal length by f_o in order to avoid confusion with the coefficient of anisotropic coma, which is conventionally denoted by the same letter f.

As mentioned in Sect.3.2.1 the various focal lengths and positions of a weak lens can be expressed by the lens strength parameter κ^2 using OLLENDORFF's [3.45] characteristic lengths:

$$s_1 = I/H_o \quad ; \quad s_2 = \int_{-\infty}^{+\infty} (H(z)/H_o)^2 \, dz \quad . \tag{3.70}$$

With (3.61) and (3.69) we obtain for weak lenses

$$f_o^{-1} = \pi^2\kappa^2 s_2/s_1^2 \quad . \tag{3.71}$$

For the convoluted GRAY model (3.22), which is a good approximation for unsaturated symmetrical lenses, the factor on the right-hand side of (3.71) can be evaluated analytically:

$$\pi^2 s_2/s_1^2 = \frac{\pi^2}{S}\left[\frac{1 + \tanh^2(\alpha S/2D)}{2\tanh(\alpha S/2D)} - \frac{D}{\alpha S}\right] \quad , \quad \alpha = 2.636 \quad . \tag{3.72}$$

Figure 3.7 shows $f_o\kappa^2/D = s_1^2/(\pi^2 s_2 D)$ for weak lenses versus S/D. For $S/D \rightarrow 0$, $f_o\kappa^2/D \rightarrow 3/\pi^2\alpha = 0.1153$, while for $S/D \rightarrow \infty$, $f_o\kappa^2/D \rightarrow [(S/D) + (1/\alpha)]/\pi^2 =$
$= 0.0384 + 0.1013\ S/D + 0.0168\ D/S$

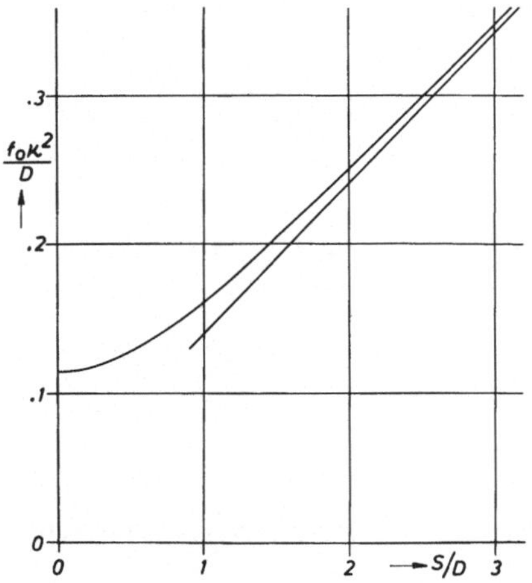

Fig.3.7. The ratio of f_o/D and $1/\kappa^2$ in the weak lens approximation as a function of S/D

The application of the weak lens approximation is not restricted to the solution G(z). Applying it to the solution h(z) defined by the initial conditions $h(z_o) = 0$, $h'(z_o) = 1$ at the object plane $z = z_o$, we have $h_o(z) = z - z_o$, and (3.63) becomes for n = 0

$$h_1''(z) = -\kappa^2 b^2(z)(z - z_0) \quad . \tag{3.73}$$

Integrating over z, we obtain

$$h_1'(z) = 1 - \kappa^2 \int_{z_0}^{z} b^2(\zeta)(\zeta - z_0)\, d\zeta \tag{3.74}$$

and

$$h_1(z) = z - z_0 - \kappa^2 \int_{z_0}^{z} \int_{z_0}^{\xi} b^2(\zeta)(\zeta - z_0)\, d\zeta d\xi$$

$$= z - z_0 - \kappa^2 \int_{z_0}^{z} b^2(\zeta)(\zeta - z_0)(z - \zeta) d\zeta \quad . \tag{3.75}$$

If the object position $z = z_0$ is in the practically field-free object space, we may replace the lower limit in the integrals by $- \infty$. We then obtain for the asymptote in image space

$$h'(+\infty) \approx h_1'(+\infty) = 1 - (\langle z\rangle - z_0)/f_0 \approx 1 + z_0/f_0 \tag{3.76}$$

and

$$h(z) \approx h_1(z) = z - z_0 + (zz_0 - z\langle z\rangle - z_0\langle z\rangle + \langle z^2\rangle)/f_0 \approx z - z_0 + zz_0/f_0 \quad . \tag{3.77}$$

Since the zero of h(z) defines the image position z_i, we obtain Newton's lens equation for weak lenses:

$$z_i = z_0 f_0/(z_0 + f_0) \quad . \tag{3.78}$$

The image is real if $z_0 \leq -f_0$, and virtual if $z_0 > -f_0$. The magnification becomes

$$M = h'(z_0)/h'(z_i) \approx 1/h'(+\infty) = f_0/(f_0 + z_0) = z_i/z_0 \quad . \tag{3.79}$$

It is negative for a real image and positive for a virtual one.

The weak lens approximation can also be applied to estimate the coefficients of chromatic aberration:

$$C_C = \kappa^2 \int_{z_0}^{z_i} b^2 h^2 dz \quad , \quad C_D = \kappa^2 \int_{z_0}^{z_i} b^2 ghdz \quad , \quad C_\theta = \frac{\kappa}{2} \int_{z_0}^{z_i} bdz \quad . \tag{3.80}$$

In the weak lens approximation we may replace the lower and upper limits of the integrals by $- \infty$ and $+ \infty$, respectively. Further, we may replace h in the integrands by $-z_0$, and g by 1, so that (3.80) become

$$C_C = \frac{z_0^2}{f_0} = \left(\frac{1 - M}{M}\right)^2 f_0 \quad , \quad C_D = -z_0/f_0 = \frac{M - 1}{M} \quad , \quad C_\theta = \pi\kappa/2 \quad . \tag{3.81}$$

For high magnification ($M \to -\infty$; $z_0 \to -f_0$) we have in the weak lens approximation

$$C_C = f_0 \quad , \quad C_D = 1 \quad , \quad C_\theta = \pi\kappa/2 \quad . \tag{3.82}$$

The third-order aberration coefficients can be treated correspondingly. As an example, the third-order spherical aberration coefficient may be written as

$$C_S \equiv B = (\kappa^2/12) \int_{z_0}^{z_i} (16\kappa^2 b^4 + 5b'^2 - bb'')h^4\, dz \quad . \tag{3.83}$$

Replacing the limits of integration by $-\infty$ and $+\infty$, and $h(z)$ in the integrand by $-z_o$, one obtains

$$B = \frac{\kappa^2 z_o^4}{12} \int_{-\infty}^{+\infty} (16\kappa^2 b^4 + 5b'^2 - bb'')dz = \frac{\kappa^2}{6}\left(\frac{1 - M}{M}\right)^4 f_o^4 \int_{-\infty}^{+\infty} (8\kappa^2 b^4 + 3b'^2)\ dz\ . \quad (3.84)$$

One might be tempted to neglect the term $8\kappa^2 b^4$ in the integrand since we are studying the behaviour for small values of κ^2. A comparison of the weak lens approximation (3.84) with the analytic solutions available for the model fields mentioned in Sect.3.3.5 shows, however, that the range of lens strengths for which (3.84) is a good approximation is wider if the term $8\kappa^2 b^4$ is retained. Since $f_o \propto \kappa^{-2}$, B is proportional to κ^{-6} for $\kappa^2 \to 0$. For high magnification ($M \to -\infty$) the factor $[(1 - M)/M]^4$ may be omitted.

Treating the other third-order aberration coefficients in the same way, we obtain for the coefficient of isotropic astigmatism:

$$C = \frac{\kappa^2}{6}\left(\frac{1 - M}{M}\right)^2 f_o^2 \int_{-\infty}^{+\infty} (8\kappa^2 b^4 + 3b'^2)dz = B\left(\frac{M}{1 - M}\right)^2 f_o^{-2}\ . \quad (3.85)$$

The coefficient D of field curvature has the same asymptotic behaviour as C for $\kappa^2 \to 0$ because C and D are related to each other by Petzval's theorem,

$$C - D = -2\kappa^2 \int_{z_o}^{z_i} b^2 dz \approx -2/f_o \propto \kappa^2\ , \quad (3.86)$$

while C and D are proportional to κ^{-2} for $\kappa^2 \to 0$. For the coefficient E of isotropic distortion we have in the weak lens approximation

$$E = B[M/(M - 1)]^3 f_o^{-3}\ , \quad (3.87)$$

and for the coefficient F of isotropic coma,

$$F = Bf_o^{-1} M/(M - 1)\ . \quad (3.88)$$

For $\kappa^2 \to 0$, E has a finite limit while F is proportional to κ^{-4}. For the coefficients of anisotropic third-order aberrations, the weak lens approximation yields

$$c = 2\kappa^3 (1 - M)M^{-1} f_o \int_{-\infty}^{+\infty} b^3\ dz\ , \quad e = \kappa^3 \int_{-\infty}^{+\infty} b^z\ dz\ ,$$

$$f = [(1 - M)/M]^2 f_o^2 \kappa^3 \int_{-\infty}^{+\infty} b^3\ dz\ . \quad (3.89)$$

For $\kappa^2 \to 0$, c is proportional to κ, e to κ^3, and f to κ^{-1}. Some authors call e "spiral distortion".

3.4.2 Strong Lens Approximation

In most practical applications, magnetic electron lenses are operated at lens strengths smaller than the first telescopic one, so that their behaviour for $\kappa^2 \to 0$ is of more practical interest than that for $\kappa^2 \to \infty$. We can, however, obtain some interesting information from an investigation of lens properties for this other limiting case.

Due to saturation effects, the assumption that the axial field distribution $B(z) = \kappa b(z)$ can be separated into one factor κ depending only on lens strength and another factor $b(z)$ depending only on field shape is not applicable to a lens containing ferromagnetic material in which the lens strength is increased by raising the lens excitation beyond reasonable limits where κ^2 can be varied without affecting $b(z)$. It does, however, make physical sense to apply the strong lens approximation to a lens in which the exitation remains small enough to justify the neglect of saturation effects, and the high values of κ^2 are attained by using a sufficiently low value of the accelerating voltage.

The solution $G(z)$ of the paraxial trajectory equation (3.62) which remains finite for $z \to -\infty$ can for large values of the lens strength parameter κ^2 be approximated using the WKBJ method (WENTZEL [3.47], KRAMERS [3.28], BRILLOUIN [3.4], JEFFREYS [3.27]), i.e., by expanding ln G into a generally semi-convergent series of powers of κ^{-1}:

$$G(z) = \exp[\kappa\gamma_0(z) + \gamma_1(z) + \kappa^{-1}\gamma_2(z) + O(\kappa^{-2})] \quad . \tag{3.90}$$

Introducing (3.90) into the paraxial trajectory equation (3.62) and comparing terms of equal power of κ, we obtain a sequence of recurrence formulae for the $\gamma_i(z)$, the first few of which can be integrated:

$$\gamma_0(z) = \pm i \int_{-\infty}^{z} b(\zeta)d\zeta \quad ; \quad \gamma_1(z) = -\frac{1}{2} \ln b(z) \quad . \tag{3.91}$$

There are two independent solutions $G_1(z)$ and $G_2(z)$:

$$G_{1,2}(z) = [b(z)]^{-1/2} \exp\left[\pm i\kappa \int_{-\infty}^{z} b(\zeta)d\zeta + O(\kappa^{-1})\right] \quad . \tag{3.92}$$

We neglect the term of order κ^{-1} in the exponent and form a linear combination $\tilde{G}(z)$ which remains finite for $z \to -\infty$:

$$\tilde{G}(z) = [b(z)]^{-1/2} \sin\left[\kappa \int_{-\infty}^{z} b(\zeta)d\zeta\right] \tag{3.93}$$

so that

$$\tilde{G}'(z) = -\frac{1}{2} b^{-3/2} b' \sin\left[\kappa \int_{-\infty}^{z} b(\zeta)d\zeta\right] + \kappa b^{1/2} \cos\left[\kappa \int_{-\infty}^{z} b(\zeta)d\zeta\right] \quad .$$

$\tilde{G}(z)$ is a solution of the differential equation

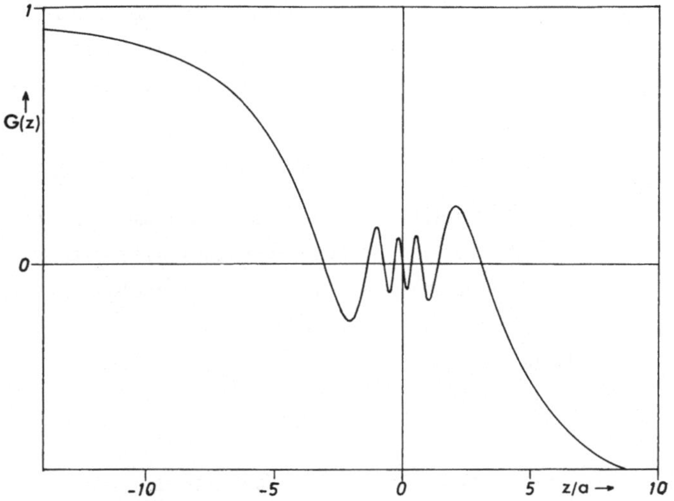

Fig.3.8. Example of a telescopic trajectory in a strong lens (GLASER's bell-shaped field, $k^2 = 99$)

$$\tilde{G}'' + \kappa^2 b^2 \tilde{G} = \left(\frac{3}{4}\frac{b'^2}{b^2} - \frac{b''}{2b}\right)\tilde{G} \tag{3.94}$$

instead of (3.62). It may be a good approximation for G(z) if κ^2 is large and if b'/b and b"/b are small enough for the terms on the right-hand side to be negligible compared to $\kappa^2 b^2 \tilde{G}$.

The lens is telescopic if $\tilde{G}(z)$ remains finite for $z \to +\infty$. In the strong-lens approximation this occurs for integer values of κ, because for $z \to \infty$ the argument of the sine function in (3.93) becomes $\kappa\pi$. Indeed, we have for GLASER's bell-shaped field telescopic lens strengths for $\kappa = (n^2 - 1)^{1/2} = n - O(n^{-1})$, and for the analytic model lens (3.57) $\kappa = (n^2 + n)^{1/2} = n[1 + O(n^{-1})]$.

$\tilde{G}(z)$ has zeros at $z = z_f$ if

$$\kappa \int_{-\infty}^{z_f} b(z)dz = n\pi \tag{3.95}$$

and extrema at $z = z_M$ if

$$\tan\left[\kappa \int_0^{z_M} b(z)dz\right] = \frac{2\kappa b^2(z_M)}{b'(z_m)} \quad . \tag{3.96}$$

As an example, Fig.3.8 shows for GLASER's bell-shaped field, $G(z)=\sin[\omega(\pi-\varphi)]/(\omega\sin\varphi)$ for $\omega = 10$, i.e., $\kappa^2 = 99$, with $\cot\varphi = z/a$, according to (3.38). The same curve represents, for $\kappa = 10$, i.e., $\omega^2 = 101$,

$$\tilde{G}(z)/\tilde{G}(-\infty) = \kappa^{-1}[1 + (z/a)^2]^{1/2} \sin\{\kappa[\pi - \text{arc}\cot(z/a)]\} = \frac{\sin[\kappa(\pi - \varphi)]}{\kappa\sin\varphi} . \tag{3.97}$$

If the asymptotic behaviour of $b(z)$ for $|z| \to \infty$ is known, e.g., $b(z) \approx A|z|^{-\nu}$ or $b(z) \approx A \exp(-C|z|)$, the dependence of f and z_f on κ for large κ can be estimated using approximation (3.93). It is, however, possible in these simple cases to solve the paraxial ray equation (3.62) analytically, replacing $b(z)$ by its asymptotic expression. For $b(z) \approx A|z|^{-\nu}$ one obtains for large κ

$$z_f \approx -\beta_n(\nu)(A\kappa)^{1/(\nu-1)} \quad ; \quad f \approx \delta_n(\nu)(A\kappa)^{1/(\nu-1)} \tag{3.98}$$

with factors $\beta_n(\nu)$ and $\delta_n(\nu)$ depending on the power exponent ν and the number n of the zero considered but not on lens strength. $\beta_1(\nu)$ increases monotonically with increasing ν from $1/\pi$ for $\nu = 2$ to 1 for $\nu \to \infty$, while $\delta_1(\nu)$ decreases monotonically from $1/\pi$ for $\nu = 2$ to 0 for $\nu \to \infty$. For an exponential field decrease $b(z) \approx A \exp(-C|z|)$, one obtains for large κ

$$z_f \approx -\frac{1}{C} \ln \frac{\kappa A}{C\alpha_n} \quad ; \quad f \approx [C\alpha_n J_1(\alpha_n)]^{-1} \quad , \tag{3.99}$$

where the α_n's are, as in Sect.3.3.2, the zeros of the Bessel function J_0. According to field model (3.22), we have $A/C\alpha_1 = 0.496(D/S) \sinh(2.636\, S/D)$ and $C = 5.272/$ The focal length f, which decreases with increasing lens strength κ^2 in the weak lens approximation, increases again for stronger lenses if the field falls off with a power law $b(z) \propto |z|^{-\nu} (\nu > 3/2)$ for large $|z|$. Hence there must be a minimum focal length f_{min} for some finite value of the lens strength parameter κ^2. An example is GLASER's bell-shaped field with $\nu = 2$, where $f_{min} = a$ for $\kappa^2 = 3$. The focal position z_f decreases monotonically with increasing lens strength, changing its sign at some finite value. In GLASER's bell-shaped field this again occurs for $\kappa^2 = 3$, i.e., the first telescopic lens strength.

In the case of an exponential field decrease, both z_f and f decrease monotonically with increasing lens strength κ^2, but f has a finite limit for $\kappa^2 \to \infty$, while z_f changes sign, and $|z_f|$ increases logarithmically for $\kappa^2 \to \infty$. If the field decreases more strongly than exponentially, e.g., $b(z) \propto \exp(-Cz^2)$, f decreases monotonically towards zero for $\kappa^2 \to \infty$.

We shall not attempt to derive analytic expressions for the dependence of the asymptotic cardinal elements v and z_v on the lens strength κ^2 for $\kappa^2 >> 1$. It is, however, evident that for each telescopic lens strength value both v and z_v become infinite and have finite values in between. In the interval between two telescopic lens strength values, $v = v_o = v_i$ does not change sign but has an extremum somewhere close to the centre of the interval. In the same interval z_{vi} decreases monotonically from $+\infty$ to $-\infty$, going through zero close to its centre. In addition we shall refrain from investigating the aberration coefficients for $\kappa^2 >> 1$ with two exceptions. In the case of an exponential decrease, $b(z) \approx A \exp(-C|z|)$, it is possible to calculate the limits of the spherical aberration coefficient B and the axial chromatic aberration coefficient C_c for $|M| \to \infty$ and $\kappa^2 \to \infty$ [3.34]:

$$B = (4C)^{-1} \quad , \quad C_c = (2C)^{-1} \quad . \tag{3.100}$$

This result is independent of the number n of the zero of $\bar{G}(z)$ chosen as the focal and object plane.

3.4.3 Focal Length and Focal Position

The object focal length f_o of a magnetic lens can be experimentally determined by measuring the magnification M and the image position z_i [3.46]:

$$f_o = \frac{z_{vi} - z_i}{M} + O(M^{-2}) \quad , \tag{3.101}$$

where $z_i - z_{vi}$ is the distance of the image plane from the asymptotic image focal position z_{vi}. In the case of one-step imaging, M is negative. Even if z_{vi} is not known, the focal length f_o can be determined from (3.101) because for strong lenses and high magnification ($|M| >> 1$) z_{vi} is small compared to z_i while for weak lenses $z_{vi} \approx f_o$ so that $f_o = z_i/(1 - M)$. If the axial field distribution B(z) is known from measurements or computer calculations, f_o and f_i can be derived from the solutions G(z) and $\bar{G}(z)$ of the paraxial trajectory equation defined by the initial conditions $G(-\infty) = 1$, $G'(-\infty) = 0$, $\bar{G}(+\infty) = 1$, $\bar{G}'(+\infty) = 0$. After defining the focal positions z_{fo} and z_{fi} by

$$0 = \bar{G}(z_{fo}) \quad , \quad 0 = G(z_{fi}) \quad , \tag{3.102}$$

the focal lengths are defined by

$$\bar{G}'(z_{fo}) = f_o^{-1} \quad , \quad G'(z_{fi}) = -f_i^{-1} \quad . \tag{3.103}$$

Correspondingly, the asymptotic focal length v is defined by

$$\bar{G}'(-\infty) = -G'(+\infty) = v^{-1} \quad , \tag{3.104}$$

and the asymptotic focal positions by

$$z_{vi} = \lim_{z\to\infty} [z + v\, G(z)] \quad ; \quad z_{vo} = \lim_{z\to-\infty} [z - v\, \bar{G}(z)] \quad . \tag{3.105}$$

The result of calculations and measurements for unsaturated symmetrical lenses is shown in Fig.3.9 for the object focal length f_o, Fig.3.10 for the objective focal position z_{fo}, Fig.3.11 for the asymptotic focal length v, and Fig.3.12 for the asymptotic object focal position z_{vo}. For symmetrical lenses, additional figures for the corresponding quantities in image space are not required because $f_o = f_i$, $z_{fo} = -z_{fi}$, $v_o = v_i = v$, and $z_{vo} = -z_{vi}$. The horizontal scale shows the "reduced excitation",

$$\varepsilon = I\, U^{-1/2} \doteq 16.86\,\kappa \; [A\, V^{-1/2}] \tag{3.106}$$

because this seems to be more convenient for the user than plotting against κ^2. As an example, a lens operated at I = 4000 A and U = 100 kV is working at a reduced excitation $\varepsilon = 4000\,(10^5)^{-1/2}\, A\, V^{-1/2} = 12.65\, A\, V^{-1/2}$ or $\kappa^2 = 0.563$.

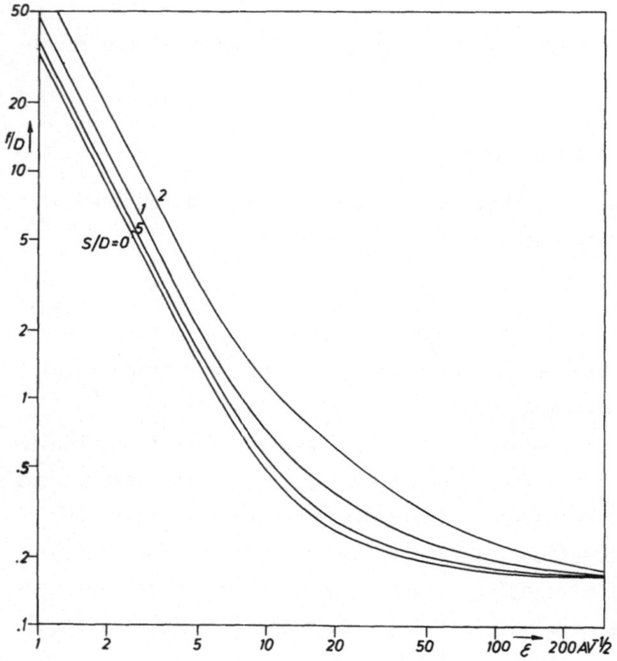

Fig.3.9. Focal length f in unsaturated symmetrical lenses versus reduced excitation $\varepsilon = I\,U^{-1/2} = 16.86\kappa\,[\mathrm{A\,V^{-1/2}}]$ for different values of the gap-bore ratio S/D. For $\varepsilon \to \infty$, f/D approaches the limit $[2\alpha_1^2 J_1(\alpha_1)]^{-1} = 0.1665$ for all S/D

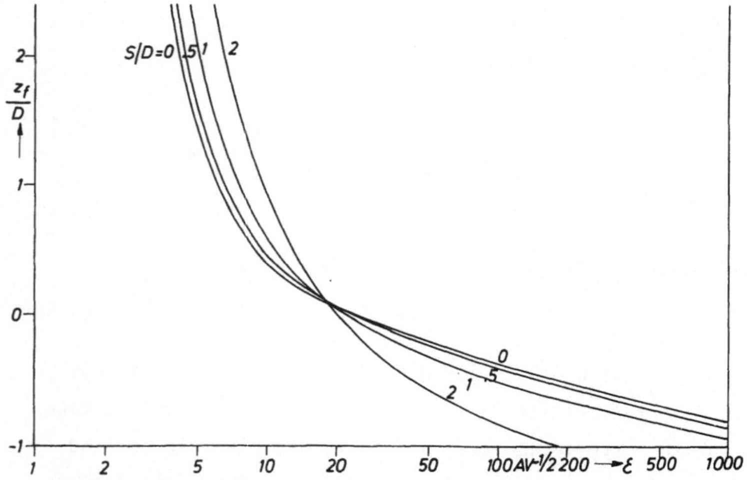

Fig.3.10. Focal position z_f in unsaturated symmetrical lenses versus reduced excitation ε

DURANDEAU et al. [3.10a,14] have calculated the real and asymptotic focal length and focal position as well as the chromatic aberration and image rotation of asymmetric lenses for $1/3 \leq D_1/D_2 \leq 3$; D_1 is the bore diameter on the object side and D_2 that on the image side. With $D = (D_1 + D_2)/2$ their values are close to those found for symmetrical lenses with the same S/D value, especially in the case of the asymptotic focal length. With increasing D_1/D_2, the object focal length decreases

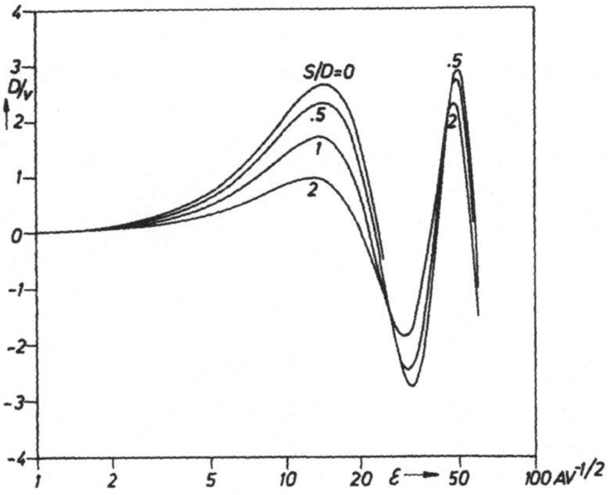

Fig.3.11. Inverse asymptotic focal length in unsaturated symmetrical lenses versus reduced excitation ε. Since v has poles but no zeros, D/v is plotted rather than v/D. The zeros of D/v characterize telescopic lens strength values

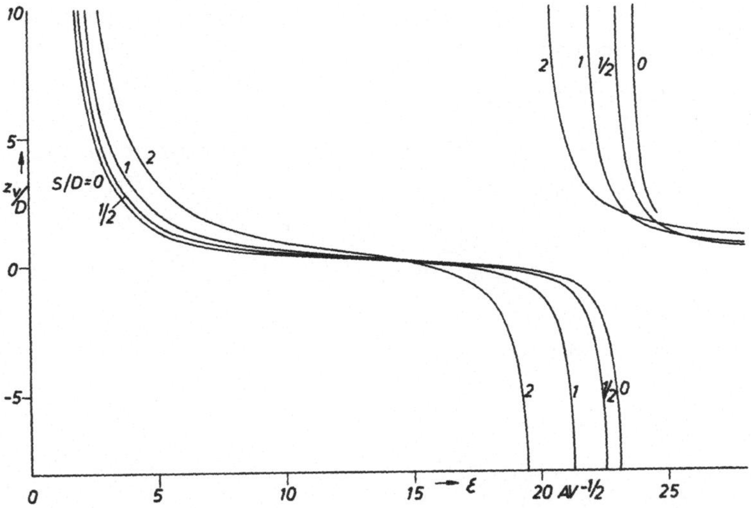

Fig.3.12. Asymptotic focal position z_v in unsaturated symmetrical lenses versus reduced excitation ε for subtelescopic and slightly supertelescopic lens strengths

while the image focal length increases, and all focal positions, i.e., z_{fo}, z_{fi}, z_{vo}, and z_{vi}, are shifted towards the side where the bore is narrower. The coefficient of chromatic aberration for high magnification varies with the same tendency as the object focal length. The relative deviation of the paraxial quantities for asymmetrical lenses from those for the corresponding symmetrical ones increases with increasing lens strength.

3.4.4 Chromatic Aberration

The three coefficients of chromatic aberration, which can be expressed by the integrals in (3.80), depend not only on lens strength but also on magnification. Their dependence on magnification in weak lenses is given by (3.81). An alternative, but equivalent, way of expressing the coefficients of chromatic aberration is the following.

We introduce complex coordinates $w(z)$ in the rotated system and $u(z)$ in the fixed system,

$$w(z) = x(z) + iy(z) \quad , \quad u(z) = X(z) + iY(z) = w(z)\exp[i\Theta(z)] \quad , \tag{3.107}$$

with

$$\Theta(z) = \frac{\eta}{2} U^{-1/2} \int_{z_o}^{z} B(\zeta)d\zeta \quad ,$$

in agreement with the notation in Hawkes' chapter (1.25). The complex coordinate of the general trajectory with the relativistic acceleration voltage U is then

$$w(z,U) = w_o g(z,U) + w_o' h(z,U) \quad , \tag{3.108}$$

where the paraxial solutions $g(z,U)$ and $h(z,U)$ are again defined by their initial conditions at $z = z_o$ as in (3.46). A trajectory with the relativistic acceleration voltage $U + \Delta U$ has the complex coordinate

$$w(z,U + \Delta U) = w_o g(z,U + \Delta U) + w_o' h(z,U + \Delta U) \quad . \tag{3.109}$$

In the image plane $z = z_i$ defined by $h(z_i,U) = 0$, we have the complex coordinate

$$\begin{aligned} w(z_i,U + \Delta U) &= w_o g(z_i,U + \Delta U) + w_o' h(z_i,U + \Delta U) \\ &= w_o\left[g(z_i,U) + \frac{\partial g_i}{\partial U}\Delta U\right] + w_o' \frac{\partial h_i}{\partial U}\Delta U \quad , \end{aligned} \tag{3.110}$$

neglecting terms of higher than linear order in ΔU. With $M = g(z_i,U)$ we obtain for the complex coordinate u in the fixed system

$$\begin{aligned} u(z_i,U + \Delta U) &= w(z_i,U + \Delta U)\exp\left[\frac{i\eta}{2}(U + \Delta U)^{-1/2}\int_{z_o}^{z_i} B(z)dz\right] \\ &= \exp(i\Theta_i)\left[w_o\left(M + \frac{\partial g_i}{\partial U}\Delta U\right) + w_o' \frac{\partial hi}{\partial U}\Delta U - \frac{i\Theta_i M}{\partial U}\Delta U w_o\right] \quad . \end{aligned} \tag{3.111}$$

Comparing this with the definition of the chromatic aberration coefficients C_c, C_D, and C_Θ,

$$u(z_i,U + \Delta U) = M\exp(i\Theta_i)\left\{w_o - [C_c w_o' + (C_D + iC_\Theta)w_o]\frac{\Delta U}{U}\right\} \quad , \tag{3.112}$$

we obtain for the coefficients of chromatic aberration:

$$C_c = -M^{-1}\frac{\partial h_i}{\partial U}U \quad , \quad C_D = -M^{-1}\frac{\partial g_i}{\partial U}U \quad , \quad C_\Theta = \Theta_i/2 \quad . \tag{3.113}$$

The physical meaning of the quantities $\partial h_i/\partial U$ and $\partial g_i/\partial U$ can be interpreted as follows. We form the linear combination of $g(z,U + \Delta U)$ and $h(z,U + \Delta U)$,

$$t(z) = g(z_i,U + \Delta U)\,h(z,U + \Delta U) - g(z,U + \Delta U)\,h(z_i,U + \Delta U) \quad , \tag{3.114}$$

which vanishes for $z = z_i$, i.e., in the image plane conjugate to $z = z_o$ for $\Delta U = 0$. This linear combination $t(z)$ has another zero in object space at $z_o + \Delta z$ if

$$\Delta z = M^{-1}\frac{\partial h_i}{\partial U}\Delta U \Rightarrow C_c = -U\frac{\Delta z}{\Delta U} \quad . \tag{3.115}$$

The magnification $M + \Delta M$ in this case, i.e., if the object plane $z = z_o + \Delta z$ is imaged into the image plane $z = z_i$ with the relativistic acceleration voltage $U + \Delta U$, is

$$M + \Delta M = \frac{t'(z_o + \Delta z, U + \Delta U)}{t'(z_i, U + \Delta U)} = M + \frac{\partial g_i}{\partial U}\Delta U \Rightarrow C_D = -\frac{U}{M}\frac{\Delta M}{\Delta U} \quad . \tag{3.116}$$

In the case of high magnification ($M \to -\infty$), we have $z_i \to \infty$ and

$$\Delta z = \frac{\partial z_{fo}}{\partial U}\Delta U \quad , \quad \Delta M = -\frac{M}{f_o}\frac{\partial f_o}{\partial U}\Delta U \quad , \tag{3.117}$$

so that, for high magnification, we have

$$C_c = -U\frac{dz_{fo}}{dU} \quad , \quad C_D = \frac{U}{f_o}\frac{df_o}{dU} \quad . \tag{3.118}$$

As an example, since the dependence of f_o and z_{fo} on lens strength is known for GLASER's bell-shaped field, we obtain from (3.42) and (3.43) by differentiation

$$C_c = \frac{\pi\kappa^2 a}{2\omega^3 \sin^2(\pi/\omega)} \quad , \quad C_D = -(\pi\kappa^2/2\omega^3)\cot(\pi/\omega) \quad . \tag{3.119}$$

Of course, we would have obtained the same results by evaluating the integrals in (3.80). Another example is the unsaturated lens for $S \gg D$ ("top-hat field") for which

$$B(z) = \begin{cases} B_o & \text{for} \quad |z| \leqq S/2 \\ 0 & \text{for} \quad |z| > S/2 \quad . \end{cases} \tag{3.120}$$

In this case we have for $0 < \kappa < 1/2$

$$f_o = \frac{S}{\pi\kappa\sin(\pi\kappa)} \quad , \quad z_{fo} = -\frac{S}{2} - \frac{S\cot(\pi\kappa)}{\pi\kappa} \quad , \tag{3.121}$$

$$C_c = \frac{\pi\kappa + \sin(\pi\kappa)\cos(\pi\kappa)}{2\pi\kappa\sin^2(\pi\kappa)} S \quad , \quad C_D = \frac{1}{2} + \frac{\pi\kappa}{2}\cot(\pi\kappa) \quad , \tag{3.122}$$

and for $\kappa \geqq 1/2$

$$f_o = \frac{S}{\pi\kappa} \quad , \quad z_{fo} = \frac{S}{2} - \frac{S}{2\kappa} \quad , \quad C_c = \frac{S}{4\kappa} \quad , \quad C_D = 1/2 \quad . \tag{3.123}$$

If not only the relativistic voltage U but also the lens excitation I is subject to variations, we have instead of (3.110)

$$w(z_i, U + \Delta U, I + \Delta I) = w_o\left[g(z_i, U) + \frac{\partial g_i}{\partial U}\Delta U + \frac{\partial g_i}{\partial I}\Delta I\right] + w_o'\left[\frac{\partial h_i}{\partial U}\Delta U + \frac{\partial h_i}{\partial I}\Delta I\right] \quad . \tag{3.124}$$

In unsaturated lenses, g_i and h_i depend only on the lens strength parameter κ, and we have

$$\frac{\Delta\kappa}{\kappa} = \frac{\Delta I}{I} - \frac{\Delta U}{2U} \quad , \quad U\frac{\partial}{\partial U} = -2\kappa\frac{\partial}{\partial\kappa} \quad , \quad I\frac{\partial}{\partial I} = \kappa\frac{\partial}{\partial\kappa} \quad , \tag{3.125}$$

so that we only have to replace $\Delta U/U$ by $\Delta U/U - 2\Delta I/I$ in (3.112). In saturated lenses, however, in which not only does κ^2 depend on U and I but the normalized axial induction b(z) also depends on I, $\Delta U/U$ and $\Delta I/I$ have to be considered separately and independently, $\Delta U/U$ must be replaced by $\Delta U/U - f(I)\Delta I/I$ with a function f(I), the value of which equals two for $I \to 0$ but may considerably differ from two for greater I when saturation effects may become important.

Figures 3.13,14 show the dependence of C_c and C_D on the reduced excitation ε for one-step high magnification in unsaturated lenses. As a practical rule it is found that the ratio C_c/f_o decreases slowly and monotonically from unity for $\varepsilon \to 0$ to 0.62 for $\varepsilon \to \infty$. Figure 3.13 shows that C_c/D has no extremum but decreases monotonically with increasing ε. Some authors (LIEBMANN [3.38], BROOKES et al. [3.4a], KAMMINGA et al. [3.27a]) have, however, studied the minimum of C_c which occurs if ε is varied while the field strength $H_p = I/S$ in the gap, U, and S/D are kept constant. Under these conditions, $C_cH_pU^{-1/2}$ is proportional to $\varepsilon C_c/D$ and has a minimum close to $\varepsilon = 15\ \mathrm{A\ V^{-1/2}}$ for all values of S/D. The minimum value of $C_cH_pU^{-1/2}$ decreases monotonically with increasing S/D but there is no significant decrease beyond S/D = 2 where $C_cH_pU^{-1/2} = 4.5\ \mathrm{A\ V^{-1/2}}$.

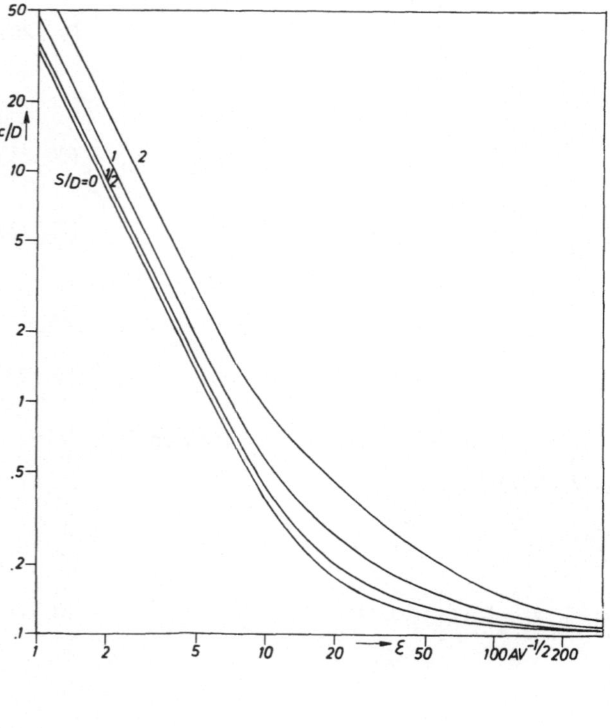

Fig.3.13. Coefficient C_c of axial chromatic aberration for high magnification in unsaturated symmetrical lenses versus reduced excitation ε. For $\varepsilon \to 0$, C_c/D approaches the limit $(4\alpha_1)^{-1} = 0.104$

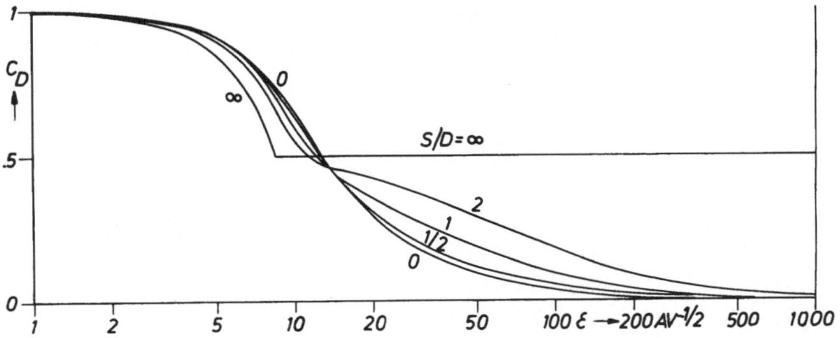

Fig.3.14. Coefficient C_D of "chromatic aberration of magnification" for high magnification in unsaturated symmetrical lenses versus reduced excitation ε

3.4.5 . Unified Representations

The similar shape of the curves representing the dependence of some of the focal properties and aberration coefficients on lens strength for different values of S/D has led some authors (LIEBMANN [3.41], DUGAS et al. [3.10]) to suggest "unified representations". The idea consists in introducing appropriate scaling factors depending on S and D for both the vertical scale representing the various focal lengths and positions, and the horizontal scale representing lens strength or relative excitation. Thus the family of curves for different values of S/D can, to some approximation, be described by one universal curve. Since the real and asymptotic focal length and focal position and the coefficients of axial chromatic aberration and spherical aberration have the dimension of a length, the reference unit for vertical scaling should also be a length, e.g., the gap width S times some function of S/D. LIEBMANN suggested using the sum S + D, but better approximations by unified curves seem to result from using DURANDEAU's length unit L defined by (3.30). Another possible vertical scaling unit is the minimum asymptotic focal length v_{min} which, according to MULVEY and WALLINGTON [3.44], is equal to $0.55\,(S^2 + 0.56\,D^2)^{1/2}$ for $S/D \leqq 3$. The lens strength κ_0^2 or reduced excitation ε_0, for which this minimum is attained, may be used as the horizontal scaling unit. If v/v_{min} is plotted against $\varepsilon/\varepsilon_0$, then all curves with different parameters S/D will by definition go through the point $v/v_{min} = 1$, $\varepsilon/\varepsilon_0 = 1$. However, the user of the plot also needs to know how v_{min}/S and ε_0 depend on S/D. Another possible horizontal scaling unit, introduced by JANDELEIT and LENZ [3.26], is the telescopic lens strength κ_t^2 or the telescopic reduced excitation ε_t. In Table 3.2 the values of L/D, v_{min}/D, ε_0, κ_0^2, ε_t, and κ_t^2 are given for a number of S/D values.

Figures 3.15,16 show the unified representation of f_0/L, v_0/L, C_c/L, z_{fo}/L, and z_{vo}/L against $\varepsilon/\varepsilon_t$. The curves for all unsaturated lenses with $0 \leqq S/D < \infty$ fall within the shaded areas. Since for $S/D \to \infty$ ("top-hat field") we have analytic expressions, all the curves can be approximated by

Table 3.2. Scaling factors needed for the unified representation of the gap-bore ratio S/D

S/D	v_{min}/D	ε_0	κ_0^2	ε_t	κ_t^2	L/D
0	0.377	14.35	0.724	23.3	1.91	0.667
0.25	0.391	14.29	0.718	23.2	1.89	0.712
0.5	0.433	14.10	0.699	22.8	1.83	0.833
1	0.591	13.54	0.645	21.6	1.65	1.202
2	1.034	12.68	0.565	20.0	1.40	2.108
$\to \infty$	0.550 S/D	10.89	0.417	16.9	1.00	S/D

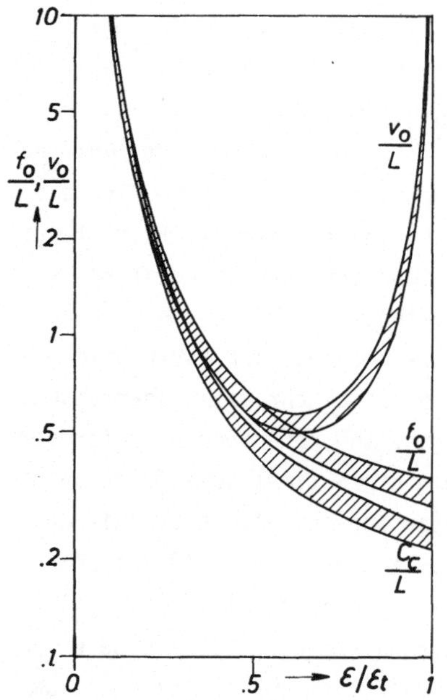

Fig.3.15. Unified representation of f_0/L, v_0/L, and C_c/L versus $\varepsilon/\varepsilon_t$ in unsaturated symmetrical lenses for subtelescopic lens strengths ($\varepsilon \leqq \varepsilon_t$)

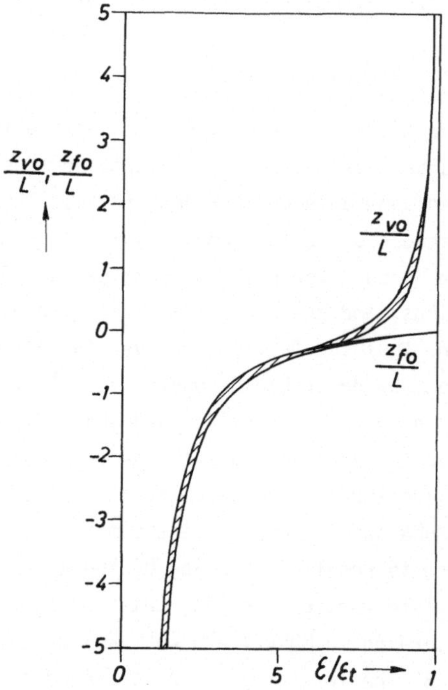

Fig.3.16. Unified representation of z_{f0}/L and z_{v0}/L versus $\varepsilon/\varepsilon_t$ in unsaturated symmetrical lenses for subtelescopic lens strengths ($\varepsilon \leqq \varepsilon_t$)

$$\frac{v_0}{L} \approx \left[\frac{\pi\varepsilon}{\varepsilon_t} \sin\left(\frac{\pi\varepsilon}{\varepsilon_t}\right)\right]^{-1} , \quad \frac{f_0}{L} \approx \begin{cases} v_0/L & \text{for} \quad \varepsilon \leqq \varepsilon_t/2 \\ \varepsilon_t/(\pi\varepsilon) & \text{for} \quad \varepsilon > \varepsilon_t/2 \end{cases} ,$$

$$\frac{z_{v0}}{L} \approx -\frac{1}{2} - \left[\frac{\pi\varepsilon}{\varepsilon_t} \tan\left(\frac{\pi\varepsilon}{\varepsilon_t}\right)\right]^{-1} , \quad z_{f0}/L \approx \begin{cases} z_{v0}/L & \text{for} \quad \varepsilon \leqq \varepsilon_t/2 \\ \frac{1}{2}(1 - \varepsilon_t/\varepsilon) & \text{for} \quad \varepsilon > \varepsilon_t/2 \end{cases} , \qquad (3.126)$$

$$\frac{C_c}{L} \approx \frac{(2\pi\varepsilon/\varepsilon_t) + \sin(2\pi\varepsilon/\varepsilon_t)}{(2\pi\varepsilon/\varepsilon_t)\ [1-\cos(2\pi\varepsilon/\varepsilon_t)]} \quad \text{for} \quad \varepsilon < \varepsilon_t/2 \ , \quad \frac{C_c}{L} \approx \frac{\varepsilon_t}{4\varepsilon} \quad \text{for} \quad \varepsilon \geqq \varepsilon_t/2$$

It is evident from Fig.3.14 that the shapes of the curves $C_D(\varepsilon)$ are so very different from each other that they cannot be unified by scaling.

3.5 Third-Order Aberrations

3.5.1 Classification of Third-Order Aberrations

The aberration coefficients B,C,D,E,F,c,e,f in the "aperture-free system" [3.21] can be defined using the complex coordinates introduced in (3.107):

$$\begin{aligned} w(z_i) = M[w_o &+ Bw_o'^2\bar{w}_o' + 2(F+if)w_ow_o'\bar{w}_o' + (F-if)w_o'^2\bar{w}_o \\ &+ (D+C)w_o\bar{w}_ow_o' + (C+ic)w_o^2\bar{w}_o' + (E+ie)w_o^2\bar{w}_o] \\ &+ \text{terms of 5th and higher order} \ , \end{aligned} \qquad (3.127)$$

where $w_o = w(z_o)$, $w_o' = w'(z_o)$, etc., and conjugate complex quantities are denoted by bars.

Another set of aberration coefficients can be defined in the "system with aperture" using an equation with the same structure as (3.127) but replacing w_o' and $\bar{w}_o'$ by $w_B = w(z_B)$ and $\bar{w}_B = \bar{w}(z_B)$ on the right-hand side, where $z = z_B$ is the aperture plane. In this contribution, this system will not be used, and therefore no symbols will be defined for the corresponding coefficients. If the coefficients of third-order aberrations are known in the aperture-free system, however, the coefficients in the system with aperture can be directly derived from them since

$$w_o = w_Bh'(z_B) - w_B'h(z_B); \quad w_o' = w_B'g(z_B) - w_Bg'(z_B) \quad . \qquad (3.128)$$

Expressing w_o and w_o' in (3.127) by w_B and w_B' from (3.128), rearranging and comparing the coefficients of the third-order terms, the relations between the coefficients in the aperture-free system and those in the system with aperture can easily be derived. In the practically important case that the aperture is situated in the plane of the diffraction image, i.e., if $g(z_B) = 0$ and $h(z_B)g'(z_B) = 1$, the relations between the two sets of coefficients become even simpler than in the general case.

The asymptotic aberration coefficients describe the relation between the asymptotes in object and image space. They do not depend on object or image position or on magnification. The asymptote of a general trajectory in object space,

$$w(z) = w_o + w_o'(z - z_{vo}) \quad , \qquad (3.129)$$

is characterized by two complex parameters w_o and w_o' describing its position and direction in the asymptotic object focal plane $z = z_{vo}$. The asymptotic aberration coefficients as defined by LENZ [3.35] describe its asymptote in image space:

$$w(z) = w_o(z_{vi} - z)/v + w_o'v - \frac{z - z_{vi}}{v}\Big[B_a w_o'^2\bar{w}_o'$$

$$+ 2(F_a+if_a)w_o w_o'\bar{w}_o' + (F_a-if_a)\bar{w}_o w_o'^2 + (D_a+C_a)w_o\bar{w}_o w_o' + (C_a+ic_a)w_o^2\bar{w}_o'$$

$$+ (E_a+ie_a)w_o^2\bar{w}_o\Big] - v\Big[A_a w_o^2\bar{w}_o + (F_a+1/2-if_a)w_o'^2\bar{w}_o' + (C_a + D_a)w_o w_o'\bar{w}_o'$$

$$+ (C_a-ic_a)w_o'^2\bar{w}_o + (2E_a-v^{-2}-2ie_a)w_o\bar{w}_o w_o' + (E_a-1/2v^{-2}+ie_a)w_o^2\bar{w}_o'\Big] \quad . \qquad (3.130)$$

The asymptotic aberration coefficients are related to the coefficients (pqrs) in HAWKES' contribution (1.175-182) by

$$(2000) = -(v^4/4)A_a \ , \quad (0200) = -B_a/4 \ , \quad (0020) = -v^2C_a \ ,$$

$$(1100) = -(v^2/2)D_a \ , \quad (1010) = -v^3E_a + v/2 \ , \quad (0110) = -vF_a \ ,$$

$$(1001) = -v^3e_a \ , \quad (0101) = -vf_a \ , \quad (0011) = -v^2c_a \ . \qquad (3.131)$$

In (3.131), v is the asymptotic focal length denoted by f in HAWKES' contribution. The asymptotic aberration coefficients $B_a,C_a,D_a,E_a,F_a,c_a,e_a$, and f_a are identical with the corresponding magnification-dependent coefficients B,C,D,E,F,c,e, and f in Hawkes' contribution for the case of high magnification ($m \to 0$).

For GLASER's bell-shaped field (3.11) all third-order aberration coefficients can be expressed as analytic functions of object position z_o and lens strength κ^2. This has been done by GLASER and LAMMEL [3.17] for a system with an aperture, by GLASER [3.19] for an aperture-free system, and by JANDELEIT and LENZ [3.26] for the asymptotic third-order aberration coefficients.

3.5.2 Unified Representation in the Aperture-Free System

As in the case of paraxial characteristics, it is possible to give unified representations by scaling the coordinates. Again we are using the telescopic excitation as a horizontal scaling unit, i.e., we plot against $\varepsilon/\varepsilon_t$. Since the dimensions of the aberration coefficients in the aperture-free system are powers of the length unit, they can be made dimensionless by dividing or multiplying them by an appropriate power of DURANDEAU's length unit (3.30). In Fig.3.17 the reduced isotropic coefficients B/L, F, LC, LD, and L^2E are plotted against $\varepsilon/\varepsilon_t$, while Fig.3.18 shows the reduced anisotropic coefficients f, Lc and L^2e. The data in both figures are restricted to the case of high magnification ($M \to -\infty$).

Figure 3.17 shows that, among other things, the coefficient B/L has no extremum but decreases monotonically with increasing ε. B is the coefficient of spherical aberration for which many authors use the symbol C_s. As in the case of chromatic aberration, a minimum of C_s occurs if ε is varied while H_p = I/S, U and S/D are

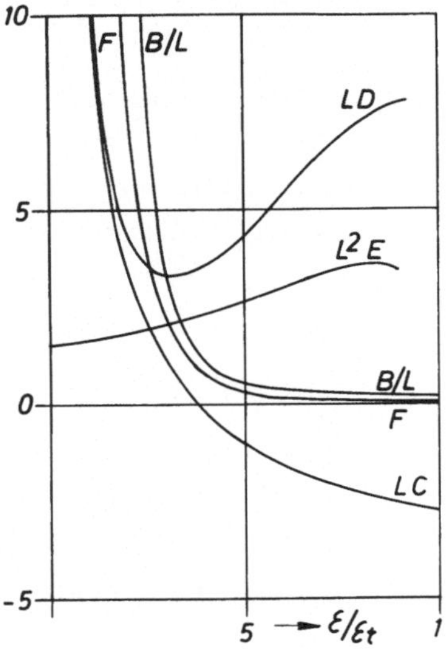

Fig.3.17. Unified representation of the isotropic aberration coefficients B/L, F, LC, LD, and L^2E versus $\varepsilon/\varepsilon_t$ in unsaturated symmetrical lenses for subtelescopic lens strengths and high magnification. This figure is not applicable if S/D > 2

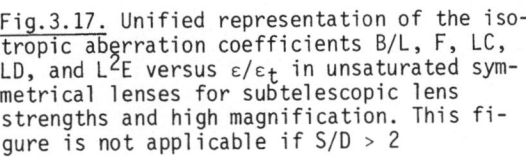

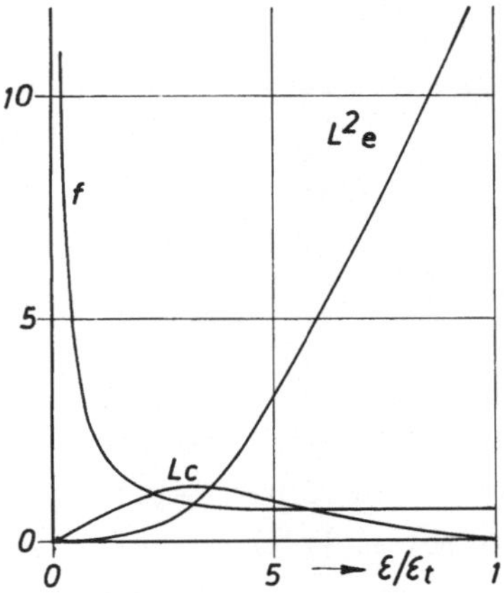

Fig.3.18. Unified representation of the anisotropic aberration coefficients f, cL, and eL^2 over $\varepsilon/\varepsilon_t$ in unsaturated symmetrical lenses for subtelescopic lens strengths

kept constant [3.4a,27a,37a]. Under these conditions, $C_sH_pU^{-1/2}$ is proportional to $\varepsilon C_s/L$ and has a minimum for every value of S/D. The same is true for $C_sH_oU^{-1/2}$, where H_o is the maximum field strength on the axis, because $H_o/H_p = H_oS/I$ is a constant depending on S/D but not on ε. A minimum of $C_sH_pU^{-1/2}$ slightly above 4 A $V^{-1/2}$ is found for all S/D $\geqq$ 1, and a slightly smaller value for the minimum of $C_sH_oU^{-1/2}$ for all S/D. The lowest minimum of $C_sH_oU^{-1/2}$ = 3.2 A $V^{-1/2}$ is obtained for $S \ll D$ at a lens strength close to the first telescopic one, i.e., under conditions similar to those of the single-field condenser-objective lens [3.28a,45a]. KUNATH et al. [3.28a] found that the coefficient of spherical aberration can be made even smaller than Liebmann's minimum values if the lens is operated at highly supertelescopic lens strengths and with saturated pole pieces.

BARTHERE et al. [3.0] and DUGAS et al. [3.10a] have calculated the coefficient C_s of spherical aberration at high magnification for a series of asymmetrical lenses in the range $1/3 \leqq D_1/D_2 \leqq 3$. If D_1/D_2 is varied while ε, S, and $D = (D_1 + D_2)/2$ are kept constant, C_s increases with increasing D_1/D_2; D_1 and D_2 are the bore diameters on the object and the image side, respectively. Hence, if D_1 is made greater than D_2, e.g., in order to gain space for the specimen stage, C_s is greater than the value for a symmetrical lens taken from Fig.3.17. On the other hand, if it is possible to make D_1 much smaller than D_2 ("pinhole lens"), C_s is smaller than in the case of a symmetrical lens with the same S/D ratio. For D_1/D_2 = 3, C_s may become up to 60% greater, and for $D_1/D_2 = 1/3$ up to 25% smaller than for the symmetrical lens $D_1/D_2 = 1$.

While the unified representations of the paraxial characteristics can be used for all values of S/D from 0 to ∞, Fig.3.17 is not applicable to the case S >> D because the isotropic aberration coefficients increase beyond all limits for $S/D \to \infty$. This follows from the unlimited increase of B'^2 at $z \approx \pm S/2$. Thus for S >> D the integral

$$\int_{-S/2-\Delta}^{-S/2+\Delta} B'^2(z)dz = \int_{S/2-\Delta}^{S/2+\Delta} B'^2(z)dz \propto (S^2D)^{-1}$$

increases beyond all limits for $D \to 0$ at constant S. The curves in Fig.3.17 are drawn for $S/D = 1$. They are a good approximation for $0 < S/D < 3/2$, a tolerable approximation for $S/D = 2$, but they should not be used for $S/D > 2$. This restriction does not apply to the anisotropic aberration coefficients shown in Fig.3.18 because their aberration integrals exist even for $S/D \to \infty$.

For $S/D \to \infty$ ("top-hat field") one obtains

$$f = \begin{cases} (\pi\varepsilon/2\varepsilon_t)\,\sin^{-2}(\pi\varepsilon/\varepsilon_t) + (1/4)\,\cot(\pi\varepsilon/\varepsilon_t) & \text{if} \quad \varepsilon < \varepsilon_t/2 \;, \\ \pi/4 & \text{if} \quad \varepsilon > \varepsilon_t/2 \;. \end{cases}$$

$$cL = \begin{cases} (\pi\varepsilon/2\varepsilon_t) + (\pi\varepsilon/\varepsilon_t)^2\,\cot(\pi\varepsilon/\varepsilon_t) & \text{if} \quad \varepsilon < \varepsilon_t/2 \;, \\ \pi\varepsilon/(4\varepsilon_t) & \text{if} \quad \varepsilon > \varepsilon_t/2 \;, \end{cases}$$

$$eL^2 = \begin{cases} (\pi\varepsilon/\varepsilon_t)^3/2 + (\pi\varepsilon/\varepsilon_t)^2\,\sin(2\pi\varepsilon/\varepsilon_t)/8 & \text{if} \quad \varepsilon < \varepsilon_t/2 \;, \\ (\pi/4)(\pi\varepsilon/\varepsilon_t)^2 & \text{if} \quad \varepsilon > \varepsilon_t/2 \;. \end{cases}$$

These analytic expressions can be used as approximations even if S is not large compared to D. For the anisotropic coma f, this approximation is tolerable for the whole range $0 < \varepsilon < \varepsilon_t$, while the anisotropic astigmatism c and distortion e are well approximated mainly for $\varepsilon < \varepsilon_t/2$.

3.5.3 Unified Representation of the Asymptotic Aberration Coefficients

Unified representations are also possible for the asymptotic aberration coefficients [3.26]. Again, they are not applicable for the isotropic coefficients if S >> D because they increase beyond all limits for $S/D \to \infty$.

Since B_a, C_a, D_a, and F_a become infinite for $\varepsilon \to 0$, and B_a, C_a, D_a, E_a, and F_a for $\varepsilon \to \varepsilon_t$, we have plotted in Fig.3.19 the quantities A_aL^3, B_aL^3/v^4, C_aL^3/v^2, D_aL^3/v^2, $-E_aL^3/v$, $-F_aL^3/v^3$ which remain finite in the interval $0 \le \varepsilon \le \varepsilon_t$.

The integral expressions for the anisotropic aberration coefficients do not diverge for S >> D, so the unified representation can be used for all values of S/D. Since f_a becomes infinite for $\varepsilon \to 0$, and f_a and c_a for $\varepsilon \to \varepsilon_t$, we have plotted

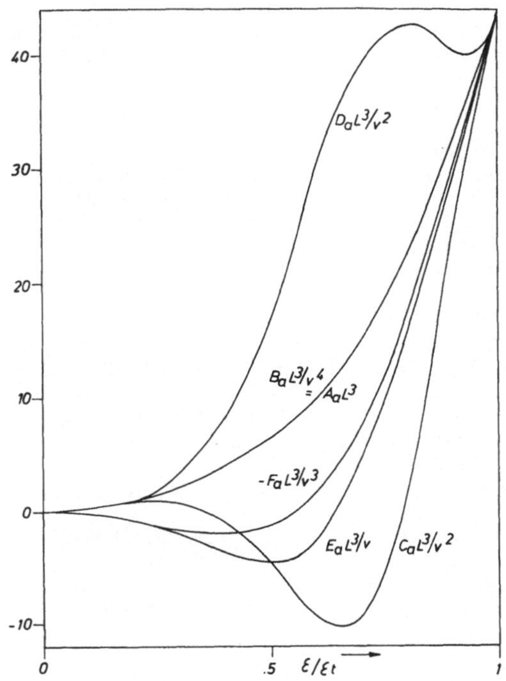

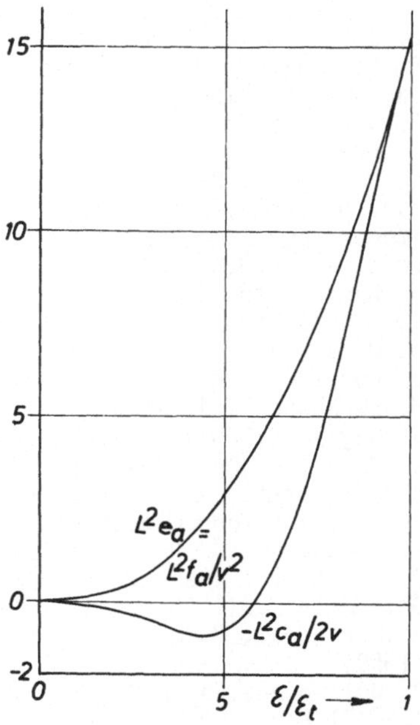

◄ Fig.3.19 Fig.3.20 ▲

Fig.3.19. Unified representation of the isotropic asymptotic aberration coefficients. Since all except A_a have poles at either $\varepsilon = 0$ or $\varepsilon = \varepsilon_t$, the figure shows products of the aberration coefficients with an appropriate power of the asymptotic focal length v to make them finite at both ends and with an appropriate power of DURANDEAU's length unit L to make them dimensionless. The figure is not applicable if $S/D > 2$

Fig.3.20. Unified representation of the anisotropic asymptotic aberration coefficients, again multiplied by appropriate powers of v and L to make them finite and dimensionless

in Fig.3.20, the quantities L^2e_a, L^2f_a/v^2, and $-L^2c_a/(2v)$ which remain finite in the interval $0 \le \varepsilon \le \varepsilon_t$.

Again, it is possible to express the asymptotic anisotropic aberration coefficients as analytic functions for the top-hat field $S/D \to \infty$. One obtains

$$f_a = (\pi\varepsilon/2\varepsilon_t)\,\sin^{-2}(\pi\varepsilon/\varepsilon_t) + (1/4)\,\cot(\pi\varepsilon/\varepsilon_t) \quad ,$$

$$c_aL = (\pi\varepsilon/2\varepsilon_t) + (\pi\varepsilon/\varepsilon_t)^2\,\cot(\pi\varepsilon/\varepsilon_t) \quad ,$$

$$e_aL^2 = \frac{1}{2}(\pi\varepsilon/\varepsilon_t)^3 + (\pi\varepsilon/\varepsilon_t)^2\,\sin(2\pi\varepsilon/\varepsilon_t)/8 \quad .$$

These analytic expressions are good approximations in the whole range $0 < \varepsilon < \varepsilon_t$ and $0 < S/D < \infty$ for f_a and c_a. For e_a the approximation is poorer, yielding slightly high values.

References

3.0 J. Barthère, J. Dugas, P. Durandeau: C.R. Acad. Sci. Paris *250*, 3461-3463 (196

3.1 S. Bertram: Proc. Inst. Radio Eng. N.Y. *28*, 418-420 (1940); J. Appl. Phys. *13*, 496-502 (1942)

3.2 B. von Borries, F. Lenz, G. Opfer: Optik *10*, 132-136 (1953)

3.3 H. Bremmer: Optik *10*, 1-4 (1953)

3.4 L. Brillouin: C.R. Acad. Sci. Paris *183*, 24-26 (1926)

3.4a K.A. Brookes, T. Mulvey, M.J. Wallington: *Proc. 4th Eur. Reg. Conf. Electron Microscopy, Rome 1968*, ed. by D.S. Bocciarelli (Tipografia Poliglotta Vaticana, Rome 1968) Vol.I, pp.165-166

3.5 H. Busch: Ann. Phys. (Leipzig) *81*, 974-993 (1926)

3.6 V.E. Cosslett: J. Sci. Instrum. *17*, 259-264 (1940)

3.7 J. Dosse: Z. Phys. *117*, 316-321 (1941)

3.8 J. Dosse: Z. Phys. *117*, 437-443, 722-753 (1941)

3.9 M. Duchesne: C.R. Acad. Sci. Paris *228*, 1407-1408 (1949)

3.10 J. Dugas, P. Durandeau, Ch. Fert: Rev. Opt. Theor. Instrum. *40*, 277-305 (1961)

3.10a J. Dugas, P. Durandeau, B. Fagot: *Proc. 2nd Eur. Reg. Conf. Electron Microscopy, Delft, 1960*, Vol.I, ed. by A.L. Houwink, B.J. Spit (De Neederlandse Vereiniging voor Electronenmicroscopie, Delft 1960) pp.35-40

3.11 P. Durandeau: J. Phys. Radium *16*, 72S (1955)

3.12 P. Durandeau: Thesis, University of Toulouse; Ann. Fac. Sci. Univ. Toulouse Sci. Math. Sci. Phys. *21*, 1-88 (1957)

3.13 P. Durandeau, B. Fagot, J. Barthère, M. Laudet: J. Phys. Radium *20a*, 80A-90A (1959)

3.14 P. Durandeau, Ch. Fert, P. Tardieu: C.R. Acad. Sci. Paris *246*, 79-81 (1958)

3.15 A.B. El-Kareh, J.C.J.El-Kareh: *Electron Beams, Lenses and Optics* (Academic, New York, London 1970)

3.16 W. Glaser: Z. Phys. *117*, 285-315 (1941)

3.17 W. Glaser, E. Lammel: Arch. Elektrotech. *37*, 347-356 (1943)

3.18 W. Glaser: Ann. Phys. (Leipzig) *7*, 213-227 (1950)

3.19 W. Glaser: *Grundlagen der Elektronenoptik* (Springer, Wien 1952)

3.20 W. Glaser, O. Bergmann: Z. Angew. Math. Phys. *1*, 363-379 (1950); *2*, 159-188 (1951)

3.21 W. Glaser, H. Grümm: Österr. Ing. Arch. *6*, 360-372 (1952)

3.22 F. Gray: Bell Syst. Techn. J. *18*, 1-31 (1939)

3.23 P. Grivet: C.R. Acad. Sci. Paris *233*, 921-923 (1951); *234*, 73-75 (1952)

3.24 W.W. Hansen, D.L. Webster: Rev. Sci. Instrum. *7*, 17-23 (1936)

3.25 M.B. Hesse: Proc. Phys. Soc. London B*63*, 386-401 (1950)

3.26 O. Jandeleit, F. Lenz: Optik *16*, 87-107 (1959)

3.27 H. Jeffreys: Proc. London Math. Soc. (2), *28*, 81-90 (1928); *33*, 246-252 (1932)

3.27a W. Kamminga, J.L. Verster, J.C. Francken: Optik *28*, 442-461 (1968/1969)

3.28 H.A. Kramers: Z. Phys. *39*, 828-840 (1926)

3.28a W. Kunath, W.D. Riecke, E. Ruska: *Proc. 6th Int. Cong. Electron Microscopy, Kyoto, 1966*, Vol.I, ed. by R. Uyeda (Maruzen, Tokyo 1966) pp.139-140

3.29 G. Langner, F. Lenz: Optik *11*, 171-180 (1954)

3.30 F. Lenz: Z. Angew. Phys. *2*, 337-340 (1950)

3.31 F. Lenz: Optik *7*, 243-253 (1950)

3.32 F. Lenz: Z. Angew. Phys. *2*, 448-453 (1950)

3.33 F. Lenz: Ann. Phys. (Leipzig) *8*, 124-128 (1950)

3.34 F. Lenz: Ann. Phys. (Leipzig) *9*, 245-258 (1951)

3.35 F. Lenz: Optik *14*, 72-82 (1957); *Proc. 1st Eur. Reg. Conf. Electron. Microscopy, Stockholm, 1956*, ed. by F.S. Sjöstrand, J. Rhodin (Almqvist and Wiksell Stockholm, 1957) pp.48-51

3.36 G. Liebmann: Proc. Phys. Soc. London B*62*, 753-772 (1949)

3.37 G. Liebmann, E.M. Grad: Proc. Phys. Soc. London B*64*, 956-971 (1951)

3.37a G. Liebmann: Proc. Phys. Soc. London B*64*, 972-977 (1951)

3.38 G. Liebmann: Proc. Phys. Soc. London B*65*, 188-192 (1952)

3.39 G. Liebmann: Proc. Phys. Soc. London B*68*, 679-686 (1955)

3.40 G. Liebmann: Proc. Phys. Soc. London B*68*, 682-685 (1955)

3.41 G. Liebmann: Proc. Phys. Soc. London B*68*, 737-745 (1955)

3.42 M. van Ments, J.B. Le Poole: Appl. Sci. Res. B*1*, 3-17 (1947)
3.43 T. Mulvey: Proc. Phys. Soc. London B*66*, 441-447 (1953)
3.44 T. Mulvey, M.J. Wallington: Repts. Prog. Phys. *36*, 347-421 (1973)
3.45 F. Ollendorff: *Berechnung magnetischer Felder* (Springer, Wien 1952)
3.45a W.D. Riecke, E. Ruska: *Proc. 6th Int. Cong. Electron Microscopy, Kyoto, 1966*, Vol.I, ed. by R. Uyeda (Maruzen, Tokyo 1966) pp.19-20
3.46 E. Ruska: Arch. Elektrotech. *38*, 102-130 (1944)
3.47 G. Wentzel: Z. Phys. *38*, 518-529 (1926)

4. Practical Lens Design

W. D. Riecke

With 150 Figures

The physical background and many practical considerations concerning the design and construction of magnetic electron lenses are explained. In a short historical introduction the evolution of lens design principles is outlined. General rules for establishing a suitable working point are discussed for each of the fundamental operating modes: as an objective, as a projector, and as a condenser. Optimization of magnetic lenses with respect to geometric optical aberrations is also described. The design of the magnetic circuit of the lens is explained, starting with the pole-piece system and completing it with a suitably shaped casing. Turning to other essential components of magnetic lenses, the design of lens coils is treated as well as that of stigmators, of aperture alignment and exchange systems, and of mechanical lens alignment devices. Measures are described that reduce the sensitivity of the magnetic lens to environmental vibrations and stray magnetic fields. The general layout of electron optical systems composed of permanent magnet lenses is treated, and means for varying the focal length of these lenses are described. The design aspects arising from the mutual interaction of the permanent magnet and its associated magnetic circuit are discussed: pole-piece systems, lens casing, and means of varying the focal length. In conclusion, the design of superconducting electron lenses is treated, starting with lenses featuring conventional ferromagnetic soft-iron pole pieces or rare-earth pole pieces fabricated from dysprosium or holmium. The particular requirements that arise when a lens is to be operated in the persistent current mode or when its field is charged by using a flux pump are explained. Finally, the design of the superconducting shielding lens is considered.

The whole chapter has been written with the aim of providing a solid basis for the design of magnetic electron lenses and for the reliable prediction of the appropriate shape and the properties of its components. Useful technological and design information is contained in a number of cross sections of whole magnetic lenses and of their components, which illustrate the text. Moreover, both the text and the rather elaborate captions contain many design hints which should help to avoid the worst pitfalls.

4.1 Introduction

The art and science of magnetic electron lens design goes back to the late 1920s and early 1930s when BUSCH [4.25] discovered that rotationally symmetric magnetic fields can be used to focus electrons in a way that is analogous to the focusing action of glass lenses on light beams. From BUSCH's work it had become clear almost from the outset that in order produce a lens of high refractive power, the field along the axis of rotational symmetry had to be confined to a narrow region, comparable in length to the focal length required, and that the magnetic field strength within this region must be sufficiently high.

This important aspect of the physics of the magnetic lens was clearly realized by RUSKA [4.175] when he designed his first electromagnetic objective lens, in which he employed high-permeability pole pieces with conically tapered surfaces in order to achieve the desired concentration of the magnetic field. Cylindrical bores were provided on the common axis of the pole pieces in order to open a passage for the electron beam to enter and leave the region of intense magnetic field. The magnetic potential required between the pole pieces for building up the concentrated magnetic field in the pole-piece gap was produced by a current-carrying coil. The axis of the coil coincided with the axis of the pole pieces, so that all disturbance of the rotational symmetry was avoided. In order to prevent the magnetic field of the coil from reaching out into space and thereby causing unwanted deflection of the electron beams, the coil was completely shrouded in a cylindrical iron casing with flat upper and lower lids, and on the axis short thick-walled tubular pieces connected these lids ferromagnetically to the pole pieces. Excessive temperature rise of the lens due to the Joule heat generated in the coil was prevented by incorporating water cooling within the lens casing. A lens of this construction is shown diagrammatically in Fig.4.1, and it should be noted that all of its relevant details are still part of the today's magnetic lens design. Nevertheless, some vital features remained to be added and much technological experience had to be gained before the modern objective lens was created, the performance of which comes close to its theoretical electron optical limits. The most important of these additional features are the stigmator and provisions for mutually aligning the pole pieces.

Apart from the universally prevalent lens design of Fig.4.1, some other types of lenses have been studied for which different magnetic circuit configurations have been used. For example, in the early years of electron microscopy VON ARDENNE [4.3] developed an objective lens based on the design principles for strong laboratory electromagnets. Here, in order to increase the magnetic field strength in the pole-piece gap as much as possible, two field-producing coils were used and arranged symmetrically with respect to the pole-piece system and as close to it as possible (Fig.4.2). Another novelty was that the coils were not enclosed in the usual can-shaped casing. Instead, a thick-walled yoke was added to complete the magnetic

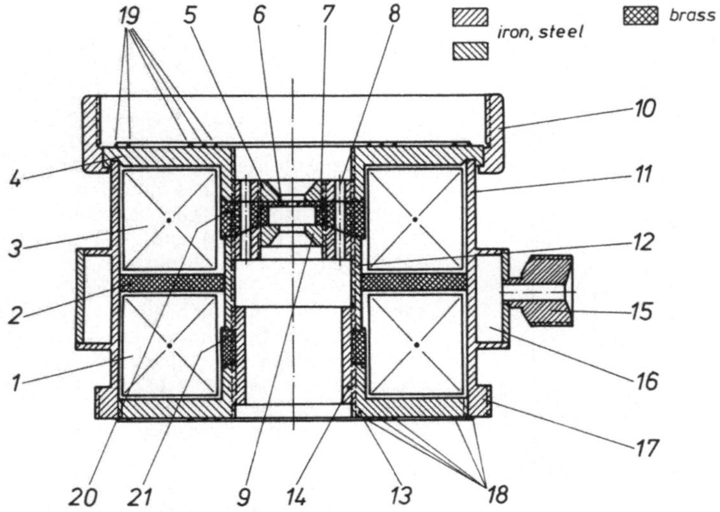

Fig.4.1. Cross-section of early electromagnetic electron objective lens designed by RUSKA (adapted from [4.175]). The lens core (12) was interrupted by two ring-shaped nonmagnetic spacers (20), (21). The upper spacer (20) was placed within the seat of the objective pole-piece system and was an integral part of the objective lens gap. A magnetic field generated across the lower spacer (21) could be used as an intermediate lens, but in the present drawing a ferromagnetic ring (14) is shown which short circuits the lower gap and eliminates the second lens. Magnetic circuit: the two gaps (20) and (21) were in series and energized by a lens coil divided into two parts (1) and (3), separated by a thick brass plate (2) to improve the cooling provided by the water chamber (16). Objective lens pole-piece system: upper (5) and lower (9) soft-iron pole pieces, aperture (6), brass spacer (7), pumping ducts (8). Upper (4) and lower (13) end flanges, lens casing (11) with thread (17) and screw cap (10) for connecting the lens to adjacent part of the column. (18), (19) rims for improving vacuum sealing action of flat rubber seals, (15) cooling-water connection

circuit, again a feature common with electromagnets. A particular advantage of using a yoke was that free access was left to the roomy pole-piece area. Space was available for the specimen stage controls and for the alignment mechanism of a small aperture, positioned before and close to the specimen, which was used to limit the area of illumination to a few micrometres only. The entire pole-piece system could be removed from the column without dismantling the microscope, a particularly welcome feature at a time when the objective apertures were mounted on small tubes, which had to be inserted into the bore of the lower objective pole piece and had frequently to be taken out for cleaning. The obvious disadvantage of the yoke lens was that the rather open layout of the lens coils gave rise to magnetic stray fields which caused a decentring with respect to the remainder of the microscope column. This is probably the reason why this particular lens design did not survive in electron microscopy.

With another still more extreme yoke lens design (Fig.4.3), developed by KINDER and PENDZICH [4.90] at about the same time at the AEG-Forschungslaboratorium, an attempt was made to exploit the magnetic stray fields generated by using them to

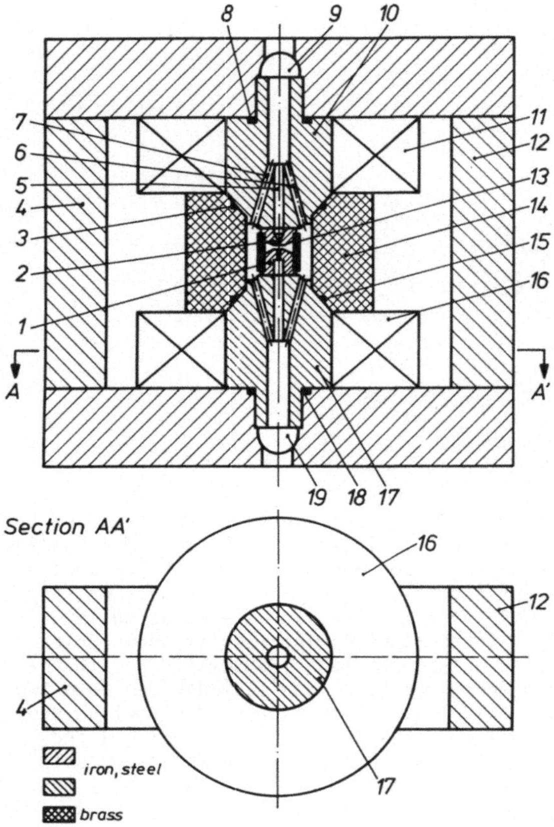

Fig.4.2. Cross-section of VON ARDENNE's yoke-type electron objective lens (adapted from [4.3]). The pole-piece system consisted of the pole pieces (1) and (2), which were joined by a nonmagnetic spacer (13); this could be inserted into its seat between the lens cores (10) and (17) or extracted from it through a door in the vacuum jacket (14). When carrying out this procedure, the remainder of the vacuum space of the microscope was separated from the vacuum space of the objective lens by sliding lid-valves into the chambers (9) and (19) and sealing off the central bores of the lens cores (10) and (17). (An analogous technique is still employed today for specimen interchange with some simplified commercial electron microscopes, which do not dispose of a specimen airlock.) The lens cores are pierced by additional vacuum ducts (6), (7). They bypass the central bore (5) which is very inefficient for pumping because of the narrowness of the pole-piece bores. (3), (8), (15), (18) are O-rings, (11), (16) the lens coils and (4), (12) ferromagnetic connecting bars to keep the yokes and the whole assembly together

create an additional lens in the microscope column [4.91]. Here again, two coils were used, but in this case they were placed entirely outside the microscope column and surrounded the vertical connecting pieces of the yoke. As in the case of VON ARDENNE's lens, the purpose of this design was to provide optimal access to the pole-piece region. The construction of what is now called a side-entry stage would thus be comparatively easy. Another advantage was expected from the fact that the Joule heat of the coils had been removed from the interior of the microscope column. It can easily be seen from basic physical reasoning that a strong magnetic stray field must inevitably appear on the axis of the microscope. This is a consequence of the fact that the closed line integral $\oint H ds$ over the magnetic field strength H is equal to the ampere turns NI encircling the path of integration. In the present case, if the path of integration extends along the microscope axis z and up to infinity, the encircling current is zero, and consequently

$$\int_{-\infty}^{+\infty} H\,dz = 0 \quad , \quad \text{i.e.} \quad \int_{\text{lens field}} H\,dz = \int_{\text{stray field}} H\,dz \quad .$$

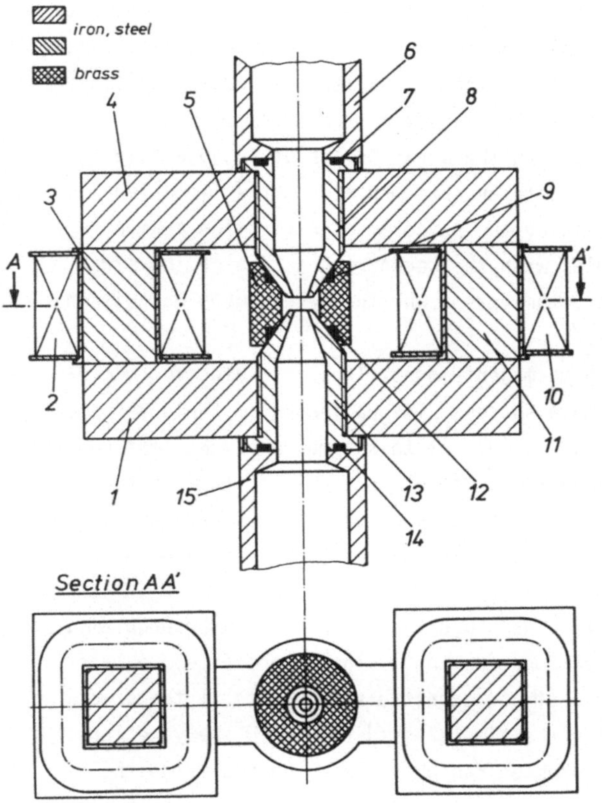

← Fig.4.3. Magnetic yoke lens after KINDER and PENDZICH. The yoke-type iron circuit consists of the two yokes 1, 4 which are connected by the cores 3, 11 of the coils 2, 10. The pole pieces 8, 13 are screwed into the yokes against a nonmagnetic spacer 5, which contains the rubber vacuum seals 9, 12. Two ferromagnetic screening tubes 6, 15 protect the electron beam from the magnetic stray field of the yokes and provide a vacuum wall for the vacuum space around the beam path. Rubber rings 7, 14 provide a vacuum seal between the tubes and the pole pieces. Magnetically, the coils 2 and 10 are operated in parallel (adapted from [4.90]; parts 6 and 8 and parts 13 and 15 were fabricated as single pieces and rings 7 and 14 removed in a later version)

Fig.4.4. (a) Schematic drawing of the magnetic stray field produced by a yoke lens. (b) Schematic drawing showing the superposed stray fields of two yoke lenses used as an additional intermediate lens ↓ (adapted from [4.91])

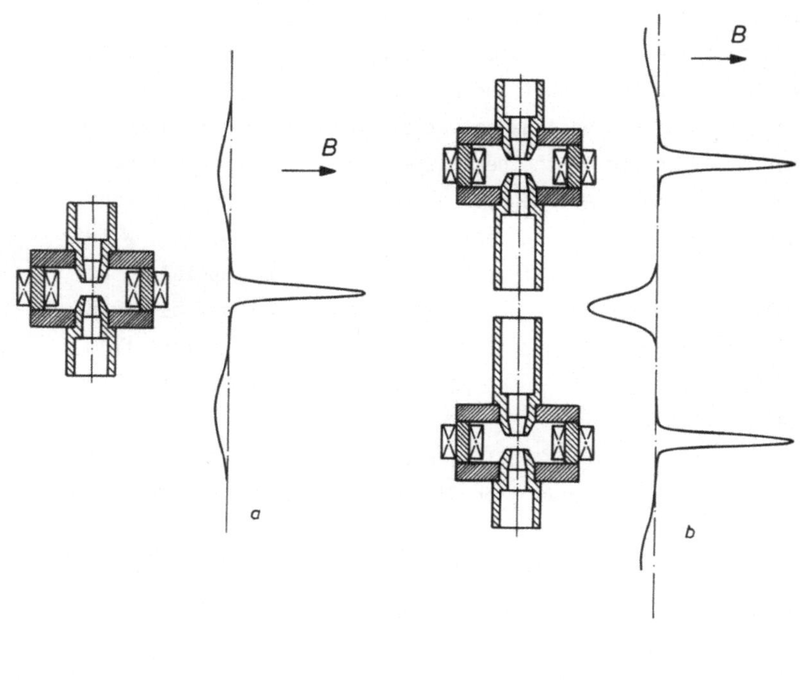

Thus, the magnetic potential in the stray field is the same (although of opposite sign) as the magnetic potential available for the lens field proper. This is shown schematically in Fig.4.4a for the case of a single yoke-lens. By employing long ferromagnetic shielding tubes to concentrate the magnetic stray fields into a narrow gap, KINDER [4.91] succeeded in superposing the stray fields of an objective "yoke lens" and a projector "yoke lens" to create an intermediate "stray field lens" (Fig.4.4b) which was useful for increasing the range of magnification available. For this type of yoke lens difficulties with the decentring action of the stray fields were also experienced, probably the reason why the development was discontinued.

Efforts to introduce the so-called permanent-magnet lens into electron microscope design have met with a little more success. The beginning of the development of these lenses goes back to about 1940 [4.19]. At this time, when the lens currents were supplied by batteries and consequently suffered in stability from thermal and discharging drift and from electrical instabilities of the setting resistors, the highly constant lens fields generated by permanent magnets appeared to be most attractive.

On the other hand, permanent-magnet lenses had their own particular problems, one of them being the impossibility of constructing a single lens (i.e., a lens with one gap only) without producing a strong magnetic stray field at the same time (Fig.4.5). This situation recalls the similar situation with the yoke lens, and it was again discovered that the problem can be remedied by employing the stray field to energize a second pole-piece gap [4.20,142] (Fig.4.6). In the years that followed, considerable engineering ingenuity has been devoted to designing permanent-magnet lens systems, which have been used both in laboratory microscopes [4.179,23,24,121] and commercial instruments [4.144,145,89,123].

Nevertheless, commercial electron microscopes employing permanent magnets had comparatively little success in the market and were sooner or later withdrawn, perhaps because of the limited beam voltage compatible with permanent-magnet lenses. This is due to another disadvantage of the permanent magnet for constructing electron lenses. For in practice, permanent magnets do not provide magnetic potentials higher than about 3000 Oe cm, which is equivalent to about NI $\approx$ 2400 A-t. With a lens energized by a coil this is just sufficient for a microscope operating at 50 kV acceleration voltage. Nevertheless, permanent-magnet lenses may be useful for special applications, e.g., for lenses operating at a potential different from ground. In Sect.4.6, therefore, the design of permanent-magnet lenses will be considered in some detail.

In the course of time, the iron-shrouded lens became the universally adopted magnetic electron lens, and its design will be discussed thoroughly in the following sections.

More recently some new concepts have been introduced into magnetic electron lens design which look rather promising. Among these "unconventional" electron lenses,

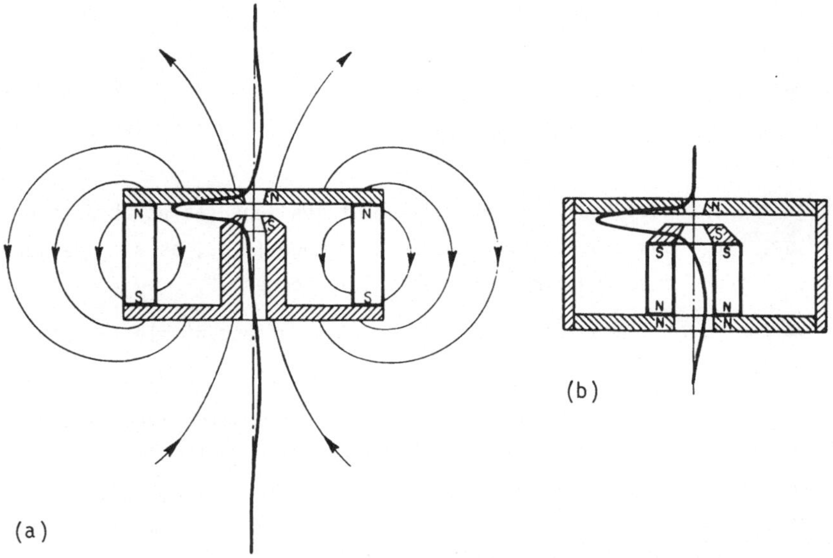

Fig.4.5a,b. Schematic drawing of single-gap permanent-magnet lenses which have a strong magnetic stray field on the lens axis; (a) with an external magnet which produces stray fields before and after the lens casing; (b) with an internal magnet which generates a stray field within the casing and the magnet [4.179]

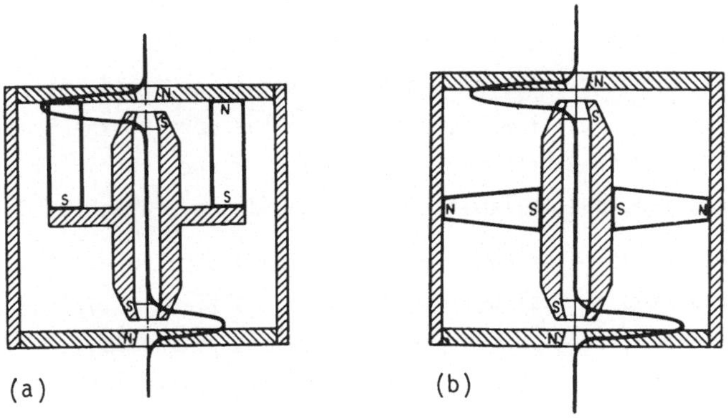

Fig.4.6a,b. Schematic drawing of double-gap permanent-magnet lenses. (a) With axial magnets; (b) with radial magnets [4.179]

the iron-free miniature lens [4.104], the "single-pole" snorkel lens [4.130] and the highly compact coil with forced cooling [4.129] seem to be on the brink of being introduced into electron optical instrument design on a larger scale. They will be discussed later in this volume [4.133]. The superconducting electron lens, which will be treated in Sect.4.7, is only just emerging from the laboratory development stage.

4.2 General Rules for the Electron Optical Design of the Magnetic Lens

The art of constructing an electron lens involves a considerable number of rather different requirements that have to be combined in such a way that they are compatible with each other and merge into a well-balanced design. The most important of these requirements may be listed as follows:

1) Specification of the intended field of application of the lens (as an objective lens, a projector, etc.) and consequently its working mode (e.g., as a condenser-objective lens, a second zone lens, etc. for objective lenses).

2) Specification of the focal length required and the limit of acceptable values for the geometric optical aberrations.

3) Selection of the gap width S and the bore diameters D of the pole-piece system and of the number of ampere-turns NI which will generate a magnetic field of such a shape and strength that at a given beam voltage Φ the working mode selected according to 1) and the optical requirements as specified under 2) are realized. Of course, the possibility of attaining the desired lens qualities is subject to the basic physical laws and the properties of the materials employed in constructing the lens.

4) Establishing the shape of a magnetic circuit, which encloses the lens coil and concentrates almost all of the magnetic potential generated by the current in the coil at the pole-piece gap. The magnetic circuit should have a sufficiently large cross-section, so that magnetic saturation does not set in and noticeable stray fields do not appear on the optical axis, both within the bore of the lens core and in neighbouring parts of the instrument. The design of the lens is often subject to additional restrictions due to limitations on the space available.

5) Usually, provision has to be made for the additional electron optical elements required to safeguard the expected optical performance, such as fixed or alignable apertures, stigmators, and deflection systems. With objective lenses, specimen stages and airlocks often have to be blended into the lens design proper.

6) It goes almost without saying that the space immediately surrounding the optic axis must be sealed off vacuum tight against air and must be connected to the pumping system of the instrument. With the high state of development of vacuum technology today, this usually presents no particular difficulties and will be treated here at most only marginally.

7) Finally, the construction of the lens should be guided by some rules, which aim at reducing the image deterioration caused by environmental disturbances. Thus the lens has to provide adequate shielding against fluctuating and AC magnetic stray fields generated by nearby electrical installations. The pole-piece

system and also the specimen stage should be mechanically decoupled from the microscope column or be rigid enough to resist vibrations, so that their shape and their position with respect to the optic axis are stable to a fraction of the resolving power of the lens, at least for the time required for one exposure.

In what follows, we discuss how allowance is made in practical lens design for the points raised in 1) to 3), whereas the aspects mentioned in 4) to 7) will be treated at a later stage.

Fortunately, the electron optical properties of rotationally symmetric magnetic fields have been studied in considerable detail and are also treated in Chaps.1-3 of this book [4.66,85,102]. For the actual design, use can be made of a wealth of electron optical data published by many authors. What is required for the purposes of lens design are the focal lengths, the focal positions, and often the aberration coefficients, data that are usually given as functions of beam voltage Φ, pole-piece geometry and number of ampere turns NI of the lens. Although some minor differences sometimes show up between values published by different authors, the data are mostly accurate enough for practical purposes, and which to use is often a matter of personal preference.[1] By way of example, some are listed as references: [4.36,39,41,83,106,108]; see also Appendix A.

The study of the physical laws determining the electron optical properties of magnetic fields and the generation and shaping of these fields has revealed some *scaling rules which are fundamental for practical lens design*. They are expressed below under the assumption that the ferromagnetic materials employed in the lens are not fully saturated and their permeability is at least of the order of 100. These rules are then as follows:

1) With a given constant configuration of ferromagnetic material, the electron beam paths remain unaffected if the beam voltage Φ and the field-generating current I are changed simultaneously in such a way that the characteristic parameter $I/\sqrt{U}$ remains constant. [Here $U = \Phi(1 + 0.978 \cdot 10^{-6}\, \Phi/V)$ is the so-called relativistically corrected beam voltage.]

2) The magnetic field strength $H(x,y,z)$ at all points x,y,z outside the ferromagnetic material is proportional to the lens current I for a given constant configuration of the magnetic circuit:

$$H(x,y,z) = H_0(x,y,z)\,\frac{I}{I_0} \quad ,$$

where $H_0(x,y,z)$ is the field distribution due to an arbitrary value of current I_0.

1 The notation employed in this chapter is not always quite the same as that adopted in the remainder of this book, mainly to harmonize with published information. For the same reason, many quantities and formulae are given both in SI and cgs units.

3) If all the dimensions of the magnetic circuit are scaled up by a factor m (conservation of similarity), and if at the same time the parameter $NI/\sqrt{U}$ characterizing the working mode of the lens remains unchanged, then the original beam paths described by the equations $x = x(z)$; $y = y(z)$ are also scaled up correspondingly by m to $x = m \cdot x(z/m)$; $y = m \cdot y(z/m)$. Here, N is the number of turns of the lens coil.

4) If all the dimensions of a magnetic circuit are scaled up by a factor m (conservation of similarity) and if the original lens current I_0 is increased m times to $I = m\, I_0$, then the magnetic field is scaled up correspondingly and the same value of magnetic field strength is conserved at corresponding points: $H(x,y,z) = H_0(x/m,y/m,z/m)$. (This rule is sometimes known as KELVIN's rule [4.203].)

By employing these rules, data given for one particular combination of pole-piece dimensions, lens current, and accelerating voltage can be readily converted to another set of experimental conditions. In lens design, therefore, the required pole-piece dimensions and ampere-turns can be predicted rather reliably on the basis of the specified electron optical conditions: operating mode, focal length, and beam voltage.

With the lenses used for electron microscopes, three basic types can be distinguished: objective lenses, condenser lenses, and projectors. This subdivision is complete in the sense that by employing magnetic lenses with the properties typical of these, most of the other electron beam instruments can be created, scanning microscopes and electron beam lithography machines for example, and it does not matter that different names are sometimes used for these basic lens types. Some typical aspects of the construction of these lenses will now be discussed, especially in regard to points 1) to 3) of the more general aspects listed at the beginning of this section.

4.2.1 Objective Lenses

4.2.1a Objective Lens Aberrations, Optimum Apertures

Objective lenses are usually designed to generate a highly resolved image of the specimen when used with a conventional electron microscope or to produce a small probe when employed with a scanning instrument. In practice, the resolving power or the smallness of the probe is of the order of only a few Å and is limited by geometric optical aberrations, such as spherical aberration or chromatic aberration, on the one hand, and by diffraction aberration, on the other hand.

It is well known that the radii of the aberration discs for each of these aberrations taken separately and at the imaging aperture angle α are given by

$$\delta_S = C_S\alpha^3 \qquad \text{for spherical aberration} \quad , \tag{4.1}$$

$$\delta_c = C_c\alpha \frac{\Delta U}{U} \qquad \text{for chromatic aberration} \quad , \tag{4.2}$$

$$\delta_d = 0.6 \frac{\lambda}{\alpha} \qquad \text{for diffraction} \quad . \tag{4.3}$$

Here, C_s is the coefficient of spherical aberration, C_c the coefficient of chromatic aberration, ΔU the deviation of the relativistically corrected beam voltage from its nominal value

$$U = \Phi\left(1 + \frac{e\Phi}{2m_0c^2}\right) \quad , \tag{4.4}$$

and

$$\lambda/\text{Å} = \frac{12.56}{\sqrt{U/V}} \tag{4.5}$$

is the wavelength of the electron. In these equations, Φ is the acceleration voltage, e and m_0 the electron charge and rest mass, c the velocity of light.

Consequently, the best resolving power is obtained as an optimum compromise between geometric optical and diffraction aberrations. Here, two cases should be distinguished according to the field of application in which the objective lens will be mainly employed: imaging at high resolution and imaging of thicker specimens. In both cases, with decreasing imaging aperture angle α the geometric optical aberrations become smaller while the diffraction aberration increases. An optimum of the angle α can thus be found for which the total aberration becomes smallest. It is found that

1) for high-resolution work (which implies using thin specimens so that inelastic scattering losses and the corresponding chromatic aberration are tolerably small), the compromise is between spherical aberration and diffraction. According to SCHERZER [4.186], the best resolving power which can be expected in this case is

$$\delta_{s,d} = 0.43 \sqrt[4]{C_s\lambda^3} \quad . \tag{4.6}$$

It is assumed here that incoherent dark-field imaging is used at the corresponding optimum imaging aperture

$$\alpha_{s,d} = 1.14 \sqrt[4]{\frac{\lambda}{C_s}} \quad . \tag{4.7}$$

For other imaging modes (e.g., for bright-field phase contrast and for coherent dark field) the values of the numerical proportionality factors are a little different but remain of the order of 0.5 and 1, respectively.

2) With thicker specimens, for which the chromatic aberration caused by inelastic scattering and its accompanying electron energy spread largely exceeds the spherical aberration, the compromise is established between chromatic aberration and diffraction. The optimum resolving power obtained in this way is [4.173]

$$\delta_{c,d} = 1.1\sqrt{\lambda \cdot C_c \cdot \frac{\Delta U}{U}} \quad , \tag{4.8}$$

with the corresponding optimum aperture

$$\alpha_{c,d} = 0.77\sqrt{\frac{\lambda}{C_c \frac{\Delta U}{U}}} \quad . \tag{4.9}$$

Here

$$\Delta U = \Delta\Phi(1 + e\Phi/m_0c^2) \tag{4.10}$$

is the relativistically corrected deviation of the beam voltage corresponding to the expected spread $e\Delta\Phi$ of the electron energy $e\Phi$.

Before we can make use of (4.6-9) for the overall planning of the lens, a reasonable estimate for $\Delta U/U$ is needed. To a large extent, this value also depends on the specific kind of specimen to be investigated. A formula useful for a preliminary estimate of the energy loss in the specimen has been given by VON BORRIES [4.22] and can be cast into the form

$$e\Delta U_s[\mathrm{eV}] \approx 10^2 \cdot \gamma[\mathrm{g/cm^3}] \cdot t[\mu\mathrm{m}] \quad . \tag{4.11}$$

Here, γ is the mass density of the specimen [g/cm^3] and t its thicknesss [μm]. To this must be added the energy spread due to fluctuations of the acceleration voltage, the Maxwellian distribution of the electron emission velocity and, at higher currents, possibly the BOERSCH effect, as well as a virtual voltage fluctuation $\Delta\Phi_I/\Phi_I = -2\Delta I/I$, which corresponds to the chromatic aberration caused by fluctuations ΔI of the lens current I and is equivalent in its electron optical effect.

Now, comparing (4.6) and (4.8) by putting $\delta_{c,d} > \delta_{s,d}$ yields a limit criterion for establishing whether a design with an especially low C_c or a design with an especially low C_s can be expected to be more advantageous for obtaining a high resolving power. The criterion may be expressed as follows: chromatic aberration supersedes spherical aberration as the dominant geometric optical aberration if

$$\frac{\Delta U}{U} > \left(\frac{0.43}{1.1}\right)^2 \frac{\sqrt{C_s\lambda}}{C_c} = 0.15\sqrt{\frac{\lambda}{C_c}}\sqrt{\frac{C_s}{C_c}} \quad . \tag{4.12}$$

This statement has consequences for both practical lens design and operation; "operation" means that the optimum imaging aperture should be determined by employing (4.9) rather than (4.7).

In the actual design of objective lens pole-piece systems, the ratio of the gap width S to the bore diameter D will almost always lie in the range $0.5 < S/D < 2$, and within this range C_s and C_c differ by less than a factor of 2. Thus $\sqrt{C_s/C_c}$ will not differ too much from unity, and the "critical" voltage fluctuation can be readily determined with an accuracy sufficient for practical purposes.

By way of example, putting $\lambda = 3.7\cdot10^{-2}$ Å (which corresponds to Φ = 100 kV) and $C_c \approx 1$ mm yields $\Delta U/U > 10^{-5}$. Consequently, if the objective lens has to be designed for high-resolution phase-contrast work where the effect of spherical aberration must show up clearly and should not be masked by chromatic aberration, both beam voltage and lens current fluctuation have to be kept well within a few ppm at most, and from (4.11), the specimen should be only a few nanometres thick.

With regard to magnetic electron lens design, the stability of the lens current and of the beam voltage must be regarded as crucial design parameters which to a large degree determine the actual lens performance. It is interesting to note that the chromatic aberration of magnetic electron lenses that is caused by fluctuations of the lens current has no exact counterpart in light optics, where it would be equivalent to small fluctuations in the size of the glass lenses.

A discussion of the techniques that are employed to stabilize lens currents and beam voltage is beyond the scope of this book. But in the present context some guide rules will be discussed for establishing the design data for electron lenses and especially for those having particularly small spherical and chromatic aberrations.

4.2.1b Electron Optical Properties Fundamental for Objective Lens Design

In the following it will be assumed that the magnetic field strength within the pole-piece system is low enough so that the pole pieces are not yet saturated. Some *electron optical properties* of the lens, relevant in the present context, can then be encapsulated in the following *guide rules*:

1) With a high-magnification objective lens the object position very nearly coincides with the front focal point.

2) The position of this object-side focal point, taken with respect to the pole faces, completely determines the lens strength, which can be characterized by the parameter

$$k^2 = \frac{eB_0^2h^2}{8m_0U} \quad , \tag{4.13}$$

which has been introduced by GLASER [4.57] for the special field distribution

$$B(z) = \frac{B_0}{1 + \frac{z^2}{h^2}} \quad , \tag{4.14}$$

or by

$$k^2 = 0.022\,\frac{H_0^2R^2}{U} = \beta\,\frac{(NI)^2}{U} \quad , \tag{4.15}$$

a notation preferred by LIEBMANN [4.106]. Here, R is the radius of the bore of the (symmetrical) pole-piece system, H_0 the peak field, B_0 the peak flux den-

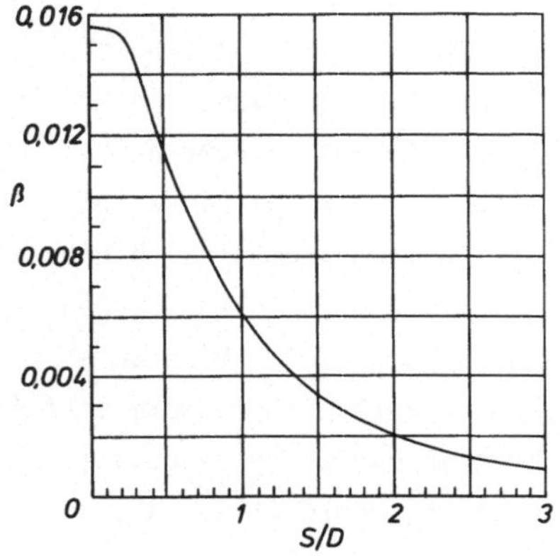

Fig.4.7. Dependence of LIEBMANN's lens strength parameter β on the gap to bore diameter ratio S/D [4.106]

sity of the magnetic lens field having axial half-width h, and β is LIEBMANN's lens-strength parameter, which is a function of the geometry of the pole-piece system. So, in order to predict the electron optical properties of the lens when employing LIEBMANN's notation, k^2 has to be used in combination with an additional parameter indicating the aspect ratio of the pole-piece system, such as the ratio of gap width S to bore diameter D = 2R, S/D, or the ratio h/R of the field distribution half-width h to the bore radius R. β is likewise a function of S/D (Fig.4.7).

3) For a given lens strength and aspect ratio, the electron optical cardinal elements of the lens (e.g., focal length, distance of the focal points from the centre of the field) and those aberration coefficients that are expressed in units of length (such as the coefficients C_s and C_c of spherical and axial chromatic aberration) can be scaled up and down in proportion to any characteristic length in the pole-piece system. By way of example, S, R, or D are well suited to serve as the characteristic "unit of length" for the pole-piece system. This is a consequence of the scaling laws explained above.

4) Aberration coefficients that are given as numbers without a dimension of length are not affected in value by scaling the pole-piece system up or down. (Examples for these are the coma coefficient and the coefficients of chromatic aberration of magnification and of rotation.)

5) The coefficient of Seidel (third-order) astigmatism is expressed in units of reciprocal length. On scaling the pole-piece system down, it increases accordingly in inverse proportion to the characteristic length.

For electron microscopes where only the object field region close to the axis must be imaged at high resolution, it will be advantageous to diminish C_s and C_c as far as possible. As a consequence of 3) the gap width S should be made so small that a further compression of the gap would lead to oversaturation of the pole-piece tips. This critical gap width,

$$S_{min} = \mu_0 \frac{NI}{B_s} \quad , \qquad (4.16)$$

can be determined by employing in this equation the number NI of ampere-turns at which the objective lens will be operated and the saturation flux density B_s of the pole-piece material, expressed in tesla (T) = V s/m^2; $\mu_0 = 4\pi \cdot 10^{-7}$ T m/A is the magnetic permeability of vacuum. If the saturation flux density B_s^* of the pole-piece material is expressed in gauss, as is often the case in commercial data sheets, the critical gap width can be determined from

$$S_{min} = 4\pi \frac{NI}{B_s^*} \text{ mm} \quad . \qquad (4.17)$$

It should be clearly understood that (4.16,17) will yield the "optimum" gap width S_{min} for a lens of predetermined ampere-turns NI. The lens thus established will not necessarily be the "absolutely" optimum lens in the sense of having the lowest possible values of spherical and/or chromatic aberration that can be obtained at a given accelerating voltage Φ and pole-piece material saturation flux density B_s.

4.2.1c Optimizing the Objective Lens with Regard to C_s and C_c

The problem of determining the objective lens of minimum spherical or chromatic aberration has recently been studied in some detail by MULVEY and WALLINGTON [4.127]. Why this minimum comes about can be understood as follows. To start with, we assume that a specific class of pole-piece systems is selected which is characterized by a constant value of the gap width S to bore diameter D ratio S/D. Further, we assume that the pole-piece tips are just saturated magnetically so that the actual gap width S and the corresponding number of ampere-turns NI at which the lens will be operated are connected by (4.16) or (4.17). If the electron optical properties of the lens are studied at constant beam voltage Φ, it is found that with increasing ampere-turns NI the ratios C_s/S and C_c/S of the values of the coefficients C_s and C_c of spherical and chromatic aberration to the gap width S decrease steadily. The decrease is rather pronounced at first up to a lens strength $NI/\sqrt{U}$ of a little over 5 A/V$^{\frac{1}{2}}$ and then becomes progressively slower, the higher the lens strength becomes. On the other hand, the gap width S of the pole-piece system must be enlarged in proportion to NI if the pole-piece tips are to be kept just at the point of saturation. In order to obtain the absolute values of C_s and C_c, the relative values C_s/S and C_c/S have to be multiplied by S. Whereas the decrease of C_s/S and C_c/S outweighs the increase of S at lower values of $NI/\sqrt{U}$, still yielding a corresponding decrease in C_s and C_c, this trend is reversed at the higher values of the lens strength $NI/\sqrt{U}$. This is because the unchanged rate of enlargement of the pole-piece gap S finally overtakes the effect of the decrease of C_s/S and C_c/S, which is now slow, and C_s and C_c then increase together with $NI/\sqrt{U}$. The points C_{smin} and C_{cmin} of minimum spherical and chromatic aberration and the corresponding lens strength values $(NI/\sqrt{U})_{min\ C_s}$ and $(NI/\sqrt{U})_{min\ C_c}$ separate the two ranges.

It is important to realize that the following two factors completely determine the actual values of C_{smin} and C_{cmin}:

1) *One pole-piece system shape parameter*, e.g., S/D. This is equivalent to a specification of the field shape and at the same time determines the values of $(NI/\sqrt{U})_{min\ C_s}$ and $(NI/\sqrt{U})_{min\ C_c}$ corresponding to the minimum spherical and chromatic aberrations. It also determines the relative values $(C_s/S)_{min\ C_s}$ and $(C_c/S)_{min\ C_c}$.

2) *The beam voltage Φ (or U) together with the saturation flux density B_s* of the pole-piece material. The combination of Φ(or U) and B_s determines the minimum gap widths $S_{min\ C_s}$ and $S_{min\ C_c}$ corresponding to the minima of the aberrations C_s and C_c. The value of $S_{min\ C_s}$ can be obtained by substituting $NI = (NI/\sqrt{U})_{min\ C_s} \cdot \sqrt{U}$ into (4.16):

$$S_{min\ C_s} = \frac{\sqrt{U}}{B_s}\left[\mu_o \cdot \left(\frac{NI}{\sqrt{U}}\right)_{min\ C_s}\right] , \qquad (4.18)$$

and likewise, mutatis mutandis, for $S_{min\ C_c}$. The expression in the square bracket is "universal" in the sense that it is a function of S/D only and does not depend on the accelerating voltage Φ.

Employing (4.18), the minimum spherical aberration can be written in the form

$$C_{s\ min} = \left(\frac{C_s}{S}\right)_{min\ C_s} \cdot S_{min\ C_s}$$

$$= \frac{\sqrt{U}}{B_s}\left[\mu_o\left(\frac{C_s}{S}\right)_{min\ C_s}\left(\frac{NI}{\sqrt{U}}\right)_{min\ C_s}\right] , \qquad (4.19)$$

where again the expression in square brackets depends on S/D only, so that the same is true of

$$\frac{C_{s\ min}}{\sqrt{U}/B_s} = \mu_o\left(\frac{C_s}{S}\right)_{min\ C_s}\left(\frac{NI}{\sqrt{U}}\right)_{min\ C_s} . \qquad (4.20)$$

Similarly, for chromatic aberration:

$$\frac{C_{c\ min}}{\sqrt{U}/B_s} = \mu_o\left(\frac{C_c}{S}\right)_{min\ C_c}\left(\frac{NI}{\sqrt{U}}\right)_{min\ C_c} . \qquad (4.21)$$

These "universal" functions depend on S/D only and are shown as graphs in Fig.4.8. Also drawn as curves are the values which C_c assumes if C_s is a minimum and vice

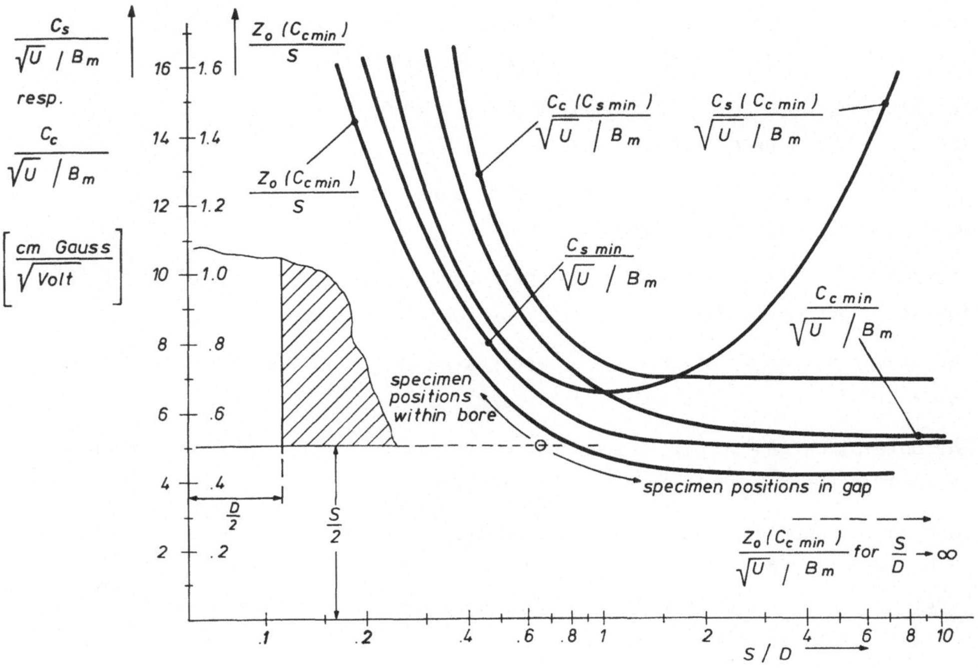

Fig.4.8. Dependence of the coefficients of minimum spherical aberration $C_{s\ min}$, minimum chromatic aberration $C_{c\ min}$, and corresponding specimen position on the aspect ratio S/D. The data are shown in universal units with B_m as saturation flux density, U as relativistically corrected beam voltage, and S as gap width of the pole-piece system [4.41,106,127]

versa. The corresponding "universal" functions can be written as

$$\frac{C_s(C_{c\ min})}{\sqrt{U}/B_s} = \mu_o\left(\frac{C_s}{S}\right)_{min\ C_c}\left(\frac{NI}{\sqrt{U}}\right)_{min\ C_c} , \tag{4.22}$$

$$\frac{C_c\ (C_{s\ min})}{\sqrt{U}/B_s} = \mu_o\left(\frac{C_c}{S}\right)_{min\ C_s}\left(\frac{NI}{\sqrt{U}}\right)_{min\ C_s} . \tag{4.23}$$

Here $(C_s/S)_{min\ C_c}$ corresponds to the lens strength at which C_c is a minimum and $(C_c/S)_{min\ C_s}$ to the lens strength of minimum C_s.

Concerning their relevance for practical lens design, the curves in Fig.4.8 show that:

1) $C_{s\ min}$ and $C_{c\ min}$ become steadily smaller with increasing aspect ratio S/D but almost all the possible advantage of choosing a larger S/D is already achieved at about S/D ≈ 1;
2) within 0.7 < S/D < 1.5, all the coefficients vary by less than about 10% from their mean;

3) for the more extended range $0.5 < S/D < 2$ the variation from the mean is still within 20%, except for $C_{c\ min}$, which may deviate up to 30%;

4) for $S/D > 2$, $C_s(C_{c\ min})$ increases strongly with S/D, while below $S/D < 0.5$ all the aberrations become very large;

5) the mean of $C_s(C_{c\ min})$ is only about 20% larger than the mean of $C_{s\ min}$, and the mean of $C_c(C_{s\ min})$ rises above the mean of $C_{c\ min}$ by about 15%.

From these observations and additional data contained in [4.106] and[4.127], the following *rules* emerge *for the design of an "optimum" objective lens.*

1) The aspect ratio S/D of the pole-piece system should be chosen within the range $0.5 < S/D < 2$ or, better still, $0.7 < S/D < 1.5$.

2) The pole-piece gap should be made so small that the pole tips are just close to saturation. This gap width can be predicted by using (4.16) or (4.17).

3) Concerning spherical and chromatic aberration, there is no pronounced advantage to be gained by designing an objective lens to have $C_{s\ min}$ rather than $C_{c\ min}$ or vice versa, or by operating the lens at a working point between these two salient points of operation.

4) The specimen position for the $C_{s\ min}$ operation mode is close to the centre of the pole-piece gap so that a $C_{s\ min}$ lens is virtually a single-field condenser-objective lens. The lens strength is about $NI/\sqrt{U} \approx 20\text{-}25\ A/V^{\frac{1}{2}}$.

Based on the above considerations the design of the symmetric objective lens can be optimized with respect to its aberrations. A similar analysis was carried out earlier by LIEBMANN [4.107] and showed basically the same trend as the more recent results of MULVEY and WALLINGTON [4.127] for which these authors claim a better accuracy.

In the above, it has been assumed that the pole-piece systems in question are symmetrical, i.e., both pole-piece bores have the same diameter. Concerning the more general *asymmetric lenses*, where the two pole-piece bores have different diameters, data have been published [4.39,110,111] on the paraxial imaging properties and on the spherical aberration. Unfortunately, there are no data available on the optimization of magnetic lenses in this more general case. Many objective lenses of electron microscopes are constructed asymmetrically with a larger upper bore, thus providing access for introducing the specimen from above, along the beam path, into the lens field proper within the pole-piece gap.

4.2.1d Basic Objective Lens Operating Modes

There is one final point of prime importance for the objective lens design: the *operating mode* in which the lens will be employed. This aspect can be treated only rather summarily here because evidently selecting a specific operating mode is closely connected with deciding on the whole electron optical layout of the instrument concerned. Figure 4.8 shows that it is not possible to select a pole-piece system and operating mode (specimen position) which is markedly superior to all others with respect to C_s or C_c. The only point that really counts in this context is to make the gap width S as small as possible, going at least up to the onset of saturation [cf. (4.16)]. Thus, when considering the advantages and drawbacks of specific operating modes, other electron optical and design criteria prevail.

It will usually be an advantage to dispose of a gap space S as large as possible in order to accommodate elaborate specimen handling and treating equipment. From (4.16) we see that high-ampere-turn objective lenses are favourable in this respect, i.e., lenses with a specimen position well immersed in the high-field-strength region.

On the other hand, not every specimen position is compatible with an objective lens which can be operated conveniently. The problem here is the electron optical effect of the strong pre-field: that part of the lens field which – if seen in the direction of the beam – precedes the specimen.

Now it is well known that in order to obtain a satisfactory electron optical performance as regards the aberrations of the electron objective, the specimen should be illuminated by a parallel bundle of rays, and this may become difficult to control if a strong lens action is caused by the pre-field just before the specimen.

The electron optical behaviour of different objective lens modes may be characterized by introducing the notion of the entrance pupil (e.g., [4.154]) as follows: The points within the specimen plane are illuminated by conical bundles of electron rays, the tips of these cones coinciding with the respective specimen points. Those rays which travel on the cone axes in the space close to the specimen pass through and intersect one another afresh in the axial point of the entrance pupil plane, and the diameter of the electron beam within this plane governs the cone angle, i.e., the illumination aperture. The actual axial position of the entrance pupil plane determines whether the cone axes intersect in a point before the specimen (divergent illumination) or after it (convergent illumination), or if they are parallel to each other and to the lens axis (parallel illumination).

Thus, for every lens strength $NI/\sqrt{U}$, there is a specific axial position of the entrance pupil plane which produces parallel illumination of the specimen. (In this context we may disregard the helical twist of the electron paths around the field axis, so that 'parallel illumination' refers to beams parallel in the rotated coordinate system (e.g., [4.57].) In real space, with increasing off-axis distance, 'parallel' beams become increasingly skew about the lens axis. In practice, complications arise if, with increasing lens strength, the entrance pupil position comes close to the lens field or even enters its front part, which happens for about $NI/\sqrt{U} \approx 20\ A/V^{\frac{1}{2}}$. These difficulties stem from restrictions to which the

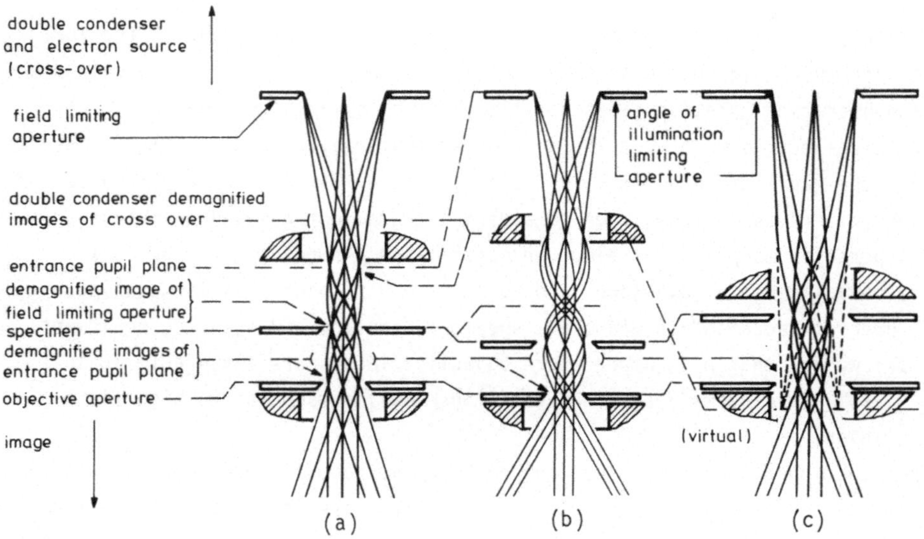

Fig.4.9a-c. Schematic drawing of electron beam paths for basic operation modes of objective lenses. (a) Condenser-objective mode; (b) second-zone mode; (c) normal-objective mode

beam diameter in the entrance pupil must then be subjected, a diameter of the order of 1 μm or less if the illumination aperture of the specimen is to be kept to some 10^{-4} rad as is necessary for phase contrast imaging.

Such small beam diameters cannot be successfully produced in practice by employing physical apertures. Even if the latter can be fabricated at all, they contaminate rapidly and give rise to severe optical disturbances. The obvious remedy is to utilize the condenser system to project a sufficiently and correspondingly demagnified image of the electron source into the plane of the entrance pupil. But this may interfere, as regards overall electron optical system design, with another requirement of practical microscopy, which is to restrict the illumination of the specimen to an area only a few micrometres in diameter at most. This is important in order to avoid excessive specimen heating and damage as well as charging up. In addition, it is most desirable for practical microscopy that the illumination aperture and the size of the illuminated area can be adjusted and varied independently.

It turns out that for each field shape there are only three special specimen positions and corresponding lens strengths where these somewhat conflicting requirements can be met. The respective ray paths are shown schematically in Fig. 4.9 and illustrate a *classification of objective lenses*, each class having its specific properties, operation modes and merits.

1) In the *normal-objective-mode lens*, the specimen is positioned before the lens field or within the comparatively weak field before the half-width point, so

that, in contrast to 2) and 3) below, no strong converging action is generated close before the specimen plane (Fig.4.9c). The lens strength amounts to about $NI/\sqrt{U} \approx 10$ to 15 $A/V^{\frac{1}{2}}$. The entrance pupil is situated a fair distance before the lens field, compared to the field extension, and a comparatively large physical aperture can be placed there in order to determine the angle of specimen illumination. The area of illumination is restricted by employing the double condenser to project a considerably demagnified image of the electron source into the specimen plane. Speaking in terms of Sect.4.2.1c, the normal-mode objective lens is practically a $C_{c\ min}$ lens if the gap width is made as small as possible, while respecting the magnetic saturation limit. The axial position of neither the specimen nor the entrance pupil aperture is critical. Most electron microscope objective lenses are operated in this mode (e.g. [4.4,93,141,207]).

2) In the *condenser-objective mode*, the ray paths are essentially telescopic and the specimen is just at the optical centre of the lens field (Fig.4.9a). In order to run the lens in this mode the lens strength $NI/\sqrt{U}$ and hence the ampere-turns NI of the lens coil have to be adjusted precisely so that the image side focal point of the pre-field (that part of the field which precedes the specimen plane) lies a little way in front of the object side focal point of the objective lens field proper (downstream from the specimen plane, as seen in the direction of the beam). The corresponding lens strength ranges from $NI/\sqrt{U} \approx 20$ to 24 $A/V^{\frac{1}{2}}$ for lenses having equal diameter pole-piece bores and aspect ratios from S/D = 2 to 0.5 [4.137]. The pre-field lens acts as a short-focal-length condenser and Fig.4.9a shows that it can be employed to project a strongly demagnified image of a field (of illumination) limiting aperture onto the specimen plane [4.156]. Extremely small electron probes can be produced in this way, easily below 0.1 μm in diameter and even down to only a few nanometres in size [4.164]. The probes may be used for microanalysis purposes, by electron diffraction or X-ray emission for example, and they allow the effects of specimen heating and charging up to be reduced to the utmost. The small-probe generating capability is certainly one of the prime benefits offered by this type of lens, but it has to be purchased at the price of some disadvantages:

a) Illumination apertures α below 10^{-4} rad cannot be attained in practice because with pre-field-condenser focal lengths f of a few millimetres this would require a beam diameter αf of considerably less than 0.1 μm in the entrance pupil plane, a nearly impossible task for the long-focal-length second-condenser lens.

b) The specimen and the small-diameter probe should be in focus at the same time but the focusing of the pre-field condenser and of the objective lens proper are coupled because these lenses are parts of one and the same magnetic field. Therefore, the simultaneous focusing of specimen and probe can only succeed if the specimen is situated exactly in the specific telefocus plane with a possible deviation of a few micrometres at most, the amount of defocus which can be tolerated so as not to cause too much blurring and spreading of the probe when the specimen is exactly focused. In order to accomplish this despite mechanical tolerances and 'waviness' of specimens, a specimen stage with a precise axial fine adjustment is the best remedy [4.164] (Sect.4.4.4). Another corrective measure to counteract inaccuracies in specimen position is to run the condenser-objective lens at a little less than telefocal lens strength and add an additional condenser lens with long focal length at a position rather close to the pre-field. The specimen is now focused exactly using the objective-lens current controls, and we can employ the additional condenser to supply the deficit in refractive power of the pre-field lens required to put the whole system exactly into the telecentric mode [4.155]. Speaking in the terms of Sect.4.2.1c, the condenser-objective lens virtually becomes a $C_{s\,min}$ lens if the gap width is made so small that magnetic saturation of the pole pieces just sets in. An experimental electron microscope with a condenser-objective lens has been described in [4.157]. Its alignment procedure, which is fundamentally different from the procedure used with the normal-objective-mode lens, has been treated in [4.163].

3) In the *second-zone-mode* lens (sometimes also called the SUZUKI lens after its inventor [4.196]), the pre-field is telecentric (Fig.4.9b). Here, a bundle of electron rays that is parallel to the optic axis on entering the lens field will be again parallel to the lens axis in the region close to the specimen plane. An interesting property of this special pre-field is that the optical conjugate of the specimen plane lies in the region of incipient field. The conjugate plane is imaged into the specimen plane at a demagnification of about 3 to 5 times, the actual value being determined by the pole-piece system aspect ratio S/D and hence by the field shape. The angle of illumination in the specimen plane is increased correspondingly by the same factor if compared to the angle of illumination prevailing in the conjugate plane. The alignment procedure for the second-zone lens is rather similar to the habitual straightforward alignment with the normal-objective-mode lens. It can be seen by comparing Fig.4.9b,c that in both cases the angle of specimen illumination is determined by the size of the aperture situated in the entrance pupil plane, and only the moderate aperture-increasing action of the second-zone lens pre-field has to be taken into careful consideration. Again in both cases, the illumination of the specimen can be restricted to an area of the order of about 1 μm in diameter by employing the double condenser to generate a correspondingly demagnified image of the electron source (the crossover). There, the additional demagnifying action of the second-zone lens pre-field may be of advantage, allowing electron probes with diameters down to about 0.1 μm to be projected into the specimen plane [4.197].

In the first stages of the development of the second-zone lens it had been hoped that a lens with an especially low coefficient C_s of spherical aberration would emerge. This is in fact true if C_s is considered relative to the objective focal length f or the gap width S because C_s/f and C_s/S decrease with increasing lens strength [4.106], and the lens strength required for the second-zone mode is about $NI/\sqrt{U} \approx 30$ to 35 $A/V^{1/2}$, i.e., it is more than double that of the normal-objective mode and roughly 50% higher than that of the condenser-objective mode. But correspondingly, in order to avoid oversaturation of the pole pieces, the gap width has to be enlarged too, and the net result is a lens which is practically the $C_{s\,min}$ lens of Sect.4.2.1c. The main advantage of the second-zone lens design is thus the comparatively large gap space, which may be useful for introducing a tilting stage or elaborate specimen handling equipment [4.87]. A second-zone lens is employed in some commercial electron microscopes [4.1,213].

In Fig.4.9, pole-piece systems with equal diameter bores have been indicated, but the same (functional) classification also applies to pole-piece systems where the bore diameters are different from each other, at least as long as the changes in focal length of pre-field and objective field proper (which of course follows the specimen) remain moderate and do not completely change the electron optical behaviour. Likewise, the general layout of lens coils and lens casing does not (if designed properly) influence the electron optical performance of the objective lens. It therefore does not matter for the classification whether the whole lens design, including coils and casing, is mirror symmetric with respect to a mid-plane S'S" (Fig.4.10a) (e.g. [4.157]), or whether the casing is constructed asymmetrically with the pole-piece gap close to one of the end flanges (Fig.4.10b), corresponding to RUSKA's first lens design (Fig.4.1).

The term "condenser-objective" should thus be reserved for an objective lens for which the illumination-side electron optical conjugate of the specimen plane is at a large distance in front of the lens field. Using a side-entry stage or introducing the specimen into the higher-field-strength region within the gap does not necessarily create a condenser-objective lens.

4.2.1e The Pin-Hole Objective Lens

In conclusion, a strongly asymmetric objective lens has to be mentioned which does not fit into the above classification and has become known as LIEBMANN's pin-hole lens (Fig.4.11). The idea behind this lens was nearly to eliminate the pre-field and its ensuing complications for the specimen illumination without seriously affecting the objective lens part of the field following the specimen. As a recipe for achieving this, LIEBMANN [4.107] has proposed replacing the illumination-side pole piece with an extended ferromagnetic pole plate, its surface passing through the point where the maximum of the axial magnetic field

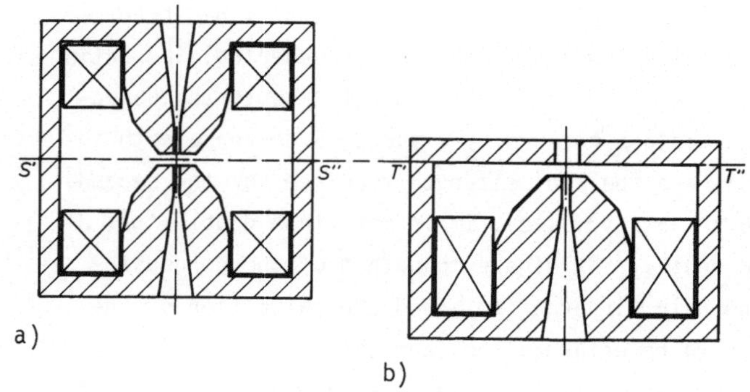

Fig.4.10a,b. Classification of objective lenses according to the design of the magnetic circuit: (a) Symmetrical lens; (b) asymmetrical lens

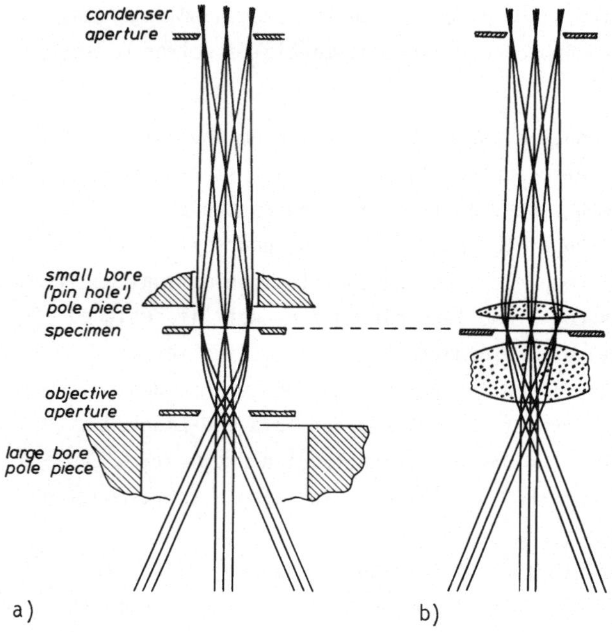

Fig.4.11a,b. Schematic drawing of the beam paths in LIEBMANN's pin-hole lens. (a) Electron beam path; (b) light optical analogue

would have been in the corresponding 'conventional' symmetric lens. The field distribution between the plate and the remaining pole piece is not changed at all and is obtained at half the expense of ampere-turns. In fact, the field distribution may be considered to be generated by the remaining pole piece and its mirror image with respect to the plate surface, with the mirror-image pole piece assumed to be of reverse magnetization. (This approach for understanding the field distribution is in exact analogy to the method of images well known in electrostatics, e.g. [4.81].) In order to provide a passage through which the electrons can enter the lens field, the pole plate is pierced by a tiny bore, the 'pin-hole', which is of course on the same axis as the bore in the pole piece, and the field distribution is not changed seriously if the pin-hole remains narrow in comparison to the pole-piece bore. For this, it is sufficient to keep the diameter 2 R_p of the pin-hole within 10 to 20% of the diameter of the pole-piece bore [4.110].

Although even a commercial electron microscope [4.136] has been equipped with a pin-hole objective lens, there are few quantitative data available from which reliable predictions can be made about details of the electron optical performance of the pin-hole lens, such as focal lengths, position of focal points, and/or principal planes given as functions of lens strength $NI/\sqrt{U}$ and aspect ratio. For some time, the prospects of the pin-hole lens were handicapped by misgivings that small enough pin-holes could not be machined with sufficiently good rotational symmetry to keep the axial astigmatism within reasonable limits. Rather recently, YADA and KAWAKATSU [4.216] have shown in actual practice that pin-hole bores with diameters down to at least 0.4 mm can be machined with sufficient precision for high-resolution imaging, and that – in agreement with theoretical predictions – coefficients C_s of spherical aberration can be obtained of about 0.5 mm at moderate lens strengths of little more than $NI/\sqrt{U} \approx 10\ A/V^{\frac{1}{2}}$. Evidently, only specimen positions in the gap can be employed, and so, in the terms of Sect.4.2.1d, the working modes will correspond to and cover the range from condenser-objective to second-zone lens. The ampere-turns required are about half those necessary for the corresponding working modes of the symmetrical objective lens. Because the effect of the pre-field is greatly reduced with the pin-hole lens, specimen positions between those for the condenser-objective and the second-zone modes are also practicable. For a first estimate of the refractive power $1/f_{ph}$ or focal length f_{ph} of the pin-hole lens pre-field, a formula given by LIEBMANN [4.107] can be used:

$$\frac{1}{f_{ph}} \approx k_s^2 \frac{z_0}{R^2} \quad . \tag{4.24}$$

Here, k_s^2 is LIEBMANN's lens strength for the corresponding symmetrical lens (4.15), z_0 the distance of the specimen from the pole plate with the pin-hole, and R the bore radius of the image-side pole piece. It can easily be seen that (4.24) results from BUSCH's well-known equation for the focal length of a thin lens [4.25],

$$\frac{1}{f} = \frac{e}{8m\Phi} \int_0^{z_0} B^2\, dz \quad ,$$

by putting $B = B_0$ = const. with B_0 as the maximum axial flux density of the corresponding symmetrical lens. Equation (4.24) thus describes the refractive power of a short piece of homogeneous field B_0 extending over the length z_0 and is an upper limit for the actual refractive power of the pin-hole lens pre-field.

4.2.2 Projector Lenses

4.2.2a Basic Design Considerations for Projector Lenses

Design and operation-mode requirements for projector lenses are distinctly different from those corresponding to a good objective lens. As its name indicates, the task of the projector is not to generate an image of a real specimen but rather to be used as the second or consecutive step of an imaging system and to project an intermediate image already generated by an objective lens or a preceding projector into a further plane. For this, projection at high magnification will be mainly employed in order to increase the total magnification of the image and make small highly resolved detail visible. On the other hand, projection at about unit magnification may be required to transfer an intermediate image from an inaccessible place in the electron optical column into the object plane of a subsequent (magnifying) projector lens. Sometimes both of these functions have to be performed alternatively by a single lens, for example, in the selected-area microdiffraction technique introduced by LE POOLE [4.103] where often one and the same lens is employed as both intermediate projector and diffraction lens [4.58].

With a lens to be designed as an intermediate or final projector, aperture-dependent aberrations can be disregarded because the intermediate image forming the object for the final projector is already at considerable magnification M_i and the aperture $\alpha_i = \alpha_0/M_i$ of the electron beams passing through the individual intermediate image points is correspondingly small in comparison with the aperture α_0 present within the objective lens. In practice, final projector lenses do not differ from objective lenses as regards magnetic flux density and field extension, and the same is true of the coefficients C_s and C_c of spherical and chromatic aberration. For comparison, the aberrations of the final projector can be referred back from the intermediate image plane to the specimen plane by dividing them by M_i. Applying (4.1,2) it is then easily seen that the effect of the spherical aberration of the projector lens is smaller by $1/M_i^4$ and its axial chromatic aberration smaller by $1/M_i^2$, both compared to the corresponding aberrations of the objective lens. M_i being of the order of 10^2 to 10^4, the contributions of the projector lens to the total spherical and chromatic aberration are extremely small and can be disregarded, and there would be no benefit to be gained from trying to optimize the projector lens in this respect.

Concerning lens design it is important to note that analogous reasoning also applies to axial astigmatism, an aberration which is generated by ellipticity and/or corrugation of the pole pieces [4.2] and thus must be expected to arise also with projector-lens pole-piece systems. As this aberration is also proportional to the imaging aperture $\alpha_i = \alpha_0/M_i$, it scarcely impairs the resolution at all and only as $1/M_i^2$ compared to the effect of a similar deformation of the objective-lens pole pieces. Stigmators are thus not required for projector lenses, and in the workshop it is not necessary to go to extremes in trying to machine exactly rotationally symmetric pole-piece bores, whereas this is compulsory for objective lenses (Sect.4.4.2).

As a consequence of the scaling laws explained in Sect.4.2, a shorter focal length may also be obtained with projector lenses if, for constant aspect ratio, the dimensions of the pole-piece system are made smaller. This will be advantageous for obtaining high magnification.

The question of what is the minimum projector focal length has been investigated by several authors (e.g. [4.36,108,127]). The answer to this question depends on the point of view from which the design problem is seen. Is the absolutely minimal value of the focal length required, with no restrictions assumed as regards pole-piece bore diameter, gap width, and lens strength? Or is a specific pole-piece system to be operated at its minimum focal length while other 'fringe' benefits are derived by selecting a suitable S/D value?

As a consequence of the scaling rules described in Sect.4.2, one condition for attaining the minimum focal length is that the pole-piece gap should be made as small as possible. For lenses not oversaturated magnetically this means that the gap width S_{min} to be employed will depend both on the ampere-turns NI of the lens coil and on the saturation flux density B_s; S_{min} can thus be determined from (4.16) or (4.17).

It is important to note that minimizing the pole-piece spacing S in this way is the basis for obtaining the minimum focal length both in absolute value and for a projector lens having other favourable electron optical properties. As in the case of the optimized objective lens treated in Sect.4.2.1c, the minimum possible projector focal length is completely determined by:

1) a pole-piece-system shape parameter, usually the aspect ratio S/D;
2) the beam voltage Φ (or U) and the magnetic saturation flux density B_s of the pole-piece material.

For the case of the projector lens of minimum absolute focal length investigated by MULVEY and WALLINGTON [4.127], the minimum possible focal length can be predicted to be

$$f_{p\ abs\ min} \approx 0.55\sqrt{S^2 + 0.56D^2} \quad . \tag{4.25}$$

This simple expression provides an accurate estimate, sufficient for practical applications throughout the aspect ratio range $0.5 \lesssim S/D \lesssim 2$. The corresponding lens strength required is nearly independent of S/D and increases slightly from $NI/\sqrt{U} \approx 10\ A/V^{1/2}$ at $S/D \approx 2$ with decreasing aspect ratio to about $NI/\sqrt{U} \approx 11\ A/V^{1/2}$ at small S/D.

It should be clearly understood that to obtain the minimum absolute value of the projector focal length, it is assumed that the minimum possible value S_{min} of the gap width is employed for every value of the lens coil excitation NI, according to (4.16) and (4.17). So, with a given pole-piece material and magnetic saturation flux density B_s, a different pole-piece system results for every combination of ampere-turns NI and aspect ratio S/D.

On the other hand, it is well known that for every pole-piece system of fixed dimensions a minimum projector focal length arises for a corresponding specific value of the coil exitation NI. With magnetically non-saturated pole-piece systems, the projector focal length f_p can be represented as a universal curve throughout the useful range of lens strengths if both the projector focal length f_p and the corresponding ampere-turns $NI(f_p)$ are shown as relative values $f_p/f_{p\,min}$ and NI/NI_{min}; the scaling factors are the particular minimum values of focal length $f_{p\,min}$ and corresponding ampere-turns NI_{min} of the actual pole-piece system employed [4.108]. It is most convenient for projector-lens design that this universal relative projector focal length curve, which is shown as a continuous curve in Fig.4.12, changes very little within the range of lens strengths and aspect ratios employed in practice. If the relative values $f_p/f_{p\,min}$ are multiplied by $f_{p\,min}$, absolute values f_p of the projector focal length can be predicted. The minimum projector focal length $f_{p\,min}$ employed in this context is a function of the aspect ratio S/D and – as a consequence of the scaling laws described in Sect. 4.2 – proportional to a characteristic length of the pole-piece system, e.g., the bore radius R. Thus, the normalized minimum projector focal length $f_{p\,min}/R$ is a function of S/D only and can be represented as the universal curve shown in Fig.4.13. Finally, the ampere-turns NI_{min} required to obtain the minimum focal length $f_{p\,min}$ can be predicted from

$$\frac{NI_{min}}{\sqrt{U}} \approx 13.5\ A/V^{\frac{1}{2}} \quad , \tag{4.26}$$

a formula which is good to a few percent accuracy within the aspect ratio range $0 < S/D \lesssim 2$.

It should be called to mind that compressing the pole-piece gap up to saturation magnetization is not necessary for the validity of the continuous curves $f_p/f_{p\,min}$ of Fig.4.12 and $f_{p\,min}/R$ of Fig.4.13. On the other hand, in order to understand better the interrelationship between absolute minimum projector focal length $f_{p\,abs\,min}$ and "relative" minimum projector focal length $f_{p\,min}$, we shall for the moment assume that the pole-piece system is just saturated magnetically at NI_{min}, so that from (4.16)

$$S_{min}(NI_{min}) = \mu_0 \frac{NI_{min}}{B_s} \quad . \tag{4.27}$$

Since $R = 0.5S/(S/D)$, the smallest possible focal length of a projector lens having the lens strength defined by (4.26) is given by

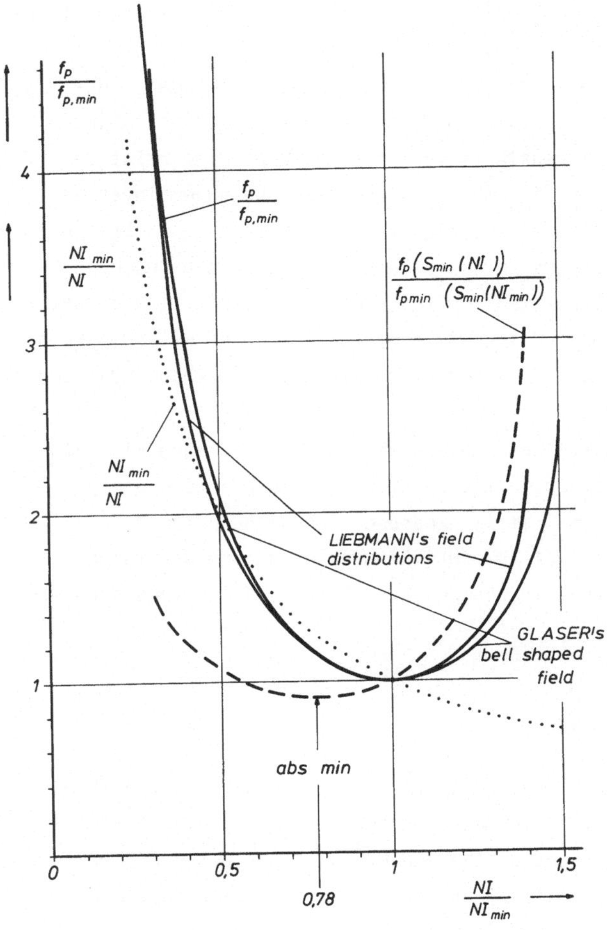

Fig.4.12. The dependence of projector focal length f_p and corresponding reciprocal ampere-turns, shown in relative units, on the relative lens strength NI/NI_{min}; f_p and NI_{min} are the minimum attainable focal length and the corresponding ampere-turns for a pole-piece system of fixed gap width S and bore diameter D. The values f_p (S_{min}) refer to a series of pole-piece systems with constant aspect ratio S/D where, for every value NI of ampere-turns, the gap width is made so small that the pole-piece tips are just saturated magnetically [4.108, 127]

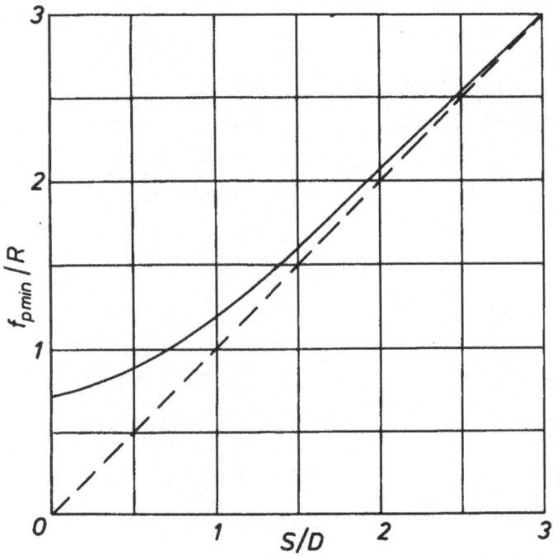

Fig.4.13. Dependence of the minimum projector focal length $f_{p\,min}$ on the pole-piece aspect ratio S/D, shown in units of bore radius R = D/2. The minimum $f_{p\,min}$ is attained for a given pole piece with fixed dimensions [4.108]

$$f_{p\,min}(S_{min}(NI_{min})) = \frac{f_{p\,min}}{R} \cdot \frac{1}{2} \frac{S_{min}(NI_{min})}{S/D} \quad . \tag{4.28}$$

Consequently, the continuous curve of Fig.4.12 has now to be understood as describing the projector focal length f_p for a lens with a pole-piece system just saturated magnetically at the lens strength (4.26), the corresponding relative minimum projector focal length $f_{p\,min}\,(S_{min}(NI_{min}))$ of (4.28) being employed as the unit length of reference. Speaking more strictly, only that part of the continuous curve that belongs to the ampere-turns range $NI < NI_{min}$ can be used in this way. Outside this range, as stipulated for the present argument, the magnetization of the pole pieces would go beyond the limit of saturation of $NI > NI_{min}$, so that the actual projector focal length f_p would be somewhat larger than indicated by the continuous curve of Fig.4.12.

For the next step in our considerations, we assume that in the range $NI < NI_{min}$ the pole-piece system is shrunk just up to the point where the pole-piece tips become saturated magnetically, and that it is expanded correspondingly if $NI > NI_{min}$. It is immediately clear from (4.16) that the shrinking or expansion has to be performed by the factor NI/NI_{min}, so that in consequence of the scaling laws a change of projector focal length results:

$$\frac{f_p(S_{min}(NI))}{f_{p\,min}(S_{min}(NI_{min}))} = \frac{NI}{NI_{min}} \cdot \frac{f_p}{f_{p\,min}} \quad . \tag{4.29}$$

Here, the left-hand side denominator is given by (4.28). Expression (4.29) is therefore a universal function; it is shown as a broken curve in Fig.4.12 and is again (within a few percent) independent of the aspect ratio for $0.5 < S/D < 2$. The curve represents the smallest (relative) values of the projector focal length that can be obtained without magnetic oversaturation over the whole range of lens strength values employed in practice. Its lowest point indicates the absolute minimum projector focal length $f_{p\,abs\,min}$ described above by (4.25).

Finally, the hyperbola NI_{min}/NI has been plotted in Fig.4.12, as a dotted line; this is the reciprocal of the scaling factor NI/NI_{min} employed in (4.29). It can easily be seen that as long as this hyperbola remains above the $f_p/f_{p\,min}$ curve, the projector focal length can be made smaller than $f_{p\,min}$ or $f_{p\,min}\,(S_{min}(NI_{min}))$ by suitably shrinking the pole-piece system.

4.2.2b Optimization of Projector Lenses

It has been explained in Sect.4.2.2a that aperture-dependent aberrations can be disregarded with projector lenses. On the other hand and in contrast to the conditions prevailing with objective lenses where only small specimen regions close to the axis are imaged, projector lenses must admit as objects comparatively large intermediate image fields, which may well cover a noticeable fraction of the cross-sectional area of the pole-piece bore. The dominant aberrations are therefore the so-called field aberrations, and it is desirable to optimize the design of projector lenses with respect to these.

The field aberrations of projector lenses can be grouped as follows [4.171, 172]:

1) First-order chromatic field aberrations:

a) Chromatic aberration of the projector magnification M_p

$$\frac{\Delta M_p}{M_p} = C_{c,mag}\left(\frac{\Delta U}{U} - 2\,\frac{\Delta I}{I}\right) \tag{4.30}$$

causes a radial blurring of the off-axis image points if either the beam voltage U or the current I in the lens coil fluctuates by ΔU or ΔI, respectively.

b) Chromatic aberration of image rotation

$$\frac{\delta_\rho}{r_i} = \frac{r_i \Delta\rho}{r_i} = C_{c,rot}\left(\frac{\Delta U}{U} - 2\,\frac{\Delta I}{I}\right) \tag{4.31}$$

is the origin of a circumferential blurring of the off-axis image points and is caused by the corresponding fluctuations $\Delta\rho$ of the angle ρ of image rotation.

The extension $\delta_r = (\Delta M_p/M_p)r_i$ and $\delta_\rho = r_i\,\Delta\rho$ of the aberration blurs increases in proportion to the off-axis distance r_i of the respective image points.

2) Third-order (or Seidel) distortion aberrations (of monochromatic beams):

a) Radial distortion

$$\frac{\Delta r_i}{r_i} = C_d\left(\frac{r_o}{R}\right)^2 \tag{4.32}$$

generates a radial displacement Δr_i of an image point originally situated at an axial distance r_i and corresponding to an axial distance $r_o = r_i/M_p$ in the intermediate image serving as the object for the projector.

b) Spiral distortion

$$\frac{\Delta t_i}{r_i} = C_{sp}\left(\frac{r_o}{R}\right)^2 \tag{4.33}$$

causes an analogous displacement Δt_i of the image point at r_i in the tangential (circumferential) direction, i.e., at right angles to Δr_i.

Both Δr_i and Δt_i increase as the third power r_i^3 of the off-axis distance r_i.

In the above $C_{c,mag}$ and $C_{c,rot}$ are the coefficients of the chromatic field aberrations. From [Ref.4.158, Eq.(38)], it can be seen that for a projector lens used at high magnification, the coefficient of chromatic aberration of magnification can be expressed as

$$C_{c,mag} = -\frac{U}{f_p}\frac{\partial f_p}{\partial U} \tag{4.34}$$

or

$$C_{c,mag} = \frac{1}{2}\frac{NI}{f_p}\frac{\partial f_p}{\partial (NI)} = \frac{1}{2}\frac{NI/NI_{min}}{f_p/f_{p\,min}}\cdot\frac{\partial f_p/f_{p\,min}}{\partial NI/NI_{min}} \quad . \tag{4.35}$$

As has been explained above, $f_p/f_{p\,min}$, which is shown in Fig.4.12, can to a good approximation be considered as a universal function of NI/NI_{min} in the range of aspect ratios used in practice $0.5 \leq S/D \leq 2$, the same being true for $C_{c,mag}$, too.

The coefficient $C_{c,rot}$ of chromatic aberration of image rotation is just equal to half the angle ρ of image rotation (cf. [4.158]):

$$C_{c,rot} = -\frac{1}{2}\rho = -0.0932\,\frac{NI}{\sqrt{U}}\frac{\sqrt{V}}{A}\ \text{rad} \quad , \tag{4.36}$$

which can be written in a more universal form as

$$C_{c,rot} = -\frac{1}{2}\rho_{min}\frac{NI}{NI_{min}} \quad . \tag{4.37}$$

Here, ρ_{min} is the angle of image rotation (in rad) corresponding to NI_{min}, the ampere-turns required to give the minimum focal length of a given fixed pole-piece system.

It has been mentioned above that the lens strength $NI_{min}/\sqrt{U}$ required for the minimum projector focal length f_p is about 13.5 $A/V^{\frac{1}{2}}$ for aspect ratios $0.5 \leq S/D \leq 2$ and changes little within this range. Hence the same is true for the corresponding angle of image rotation, which is given by

$$\rho_{min} = 0.1863\,\frac{NI_{min}}{\sqrt{U}}\frac{\sqrt{V}}{A}\ \text{rad} \tag{4.38}$$

$$\approx 2.5\ \text{rad} \quad .$$

To a good approximation, therefore, $C_{c,rot}$ can also be expressed as a universal function

$$C_{c,rot} = -1.25\,\frac{NI}{NI_{min}} \quad . \tag{4.39}$$

$C_{c,mag}$ and $C_{c,rot}$ are shown as graphs in Fig.4.14.

The coefficients C_d and C_{sp} of the third-order field aberrations as well as the corresponding equations (4.32) and (4.33) have been used above in a form originally introduced by LIEBMANN [4.108]. On the other hand, the equations written as above do not readily yield a straightforward answer to the question

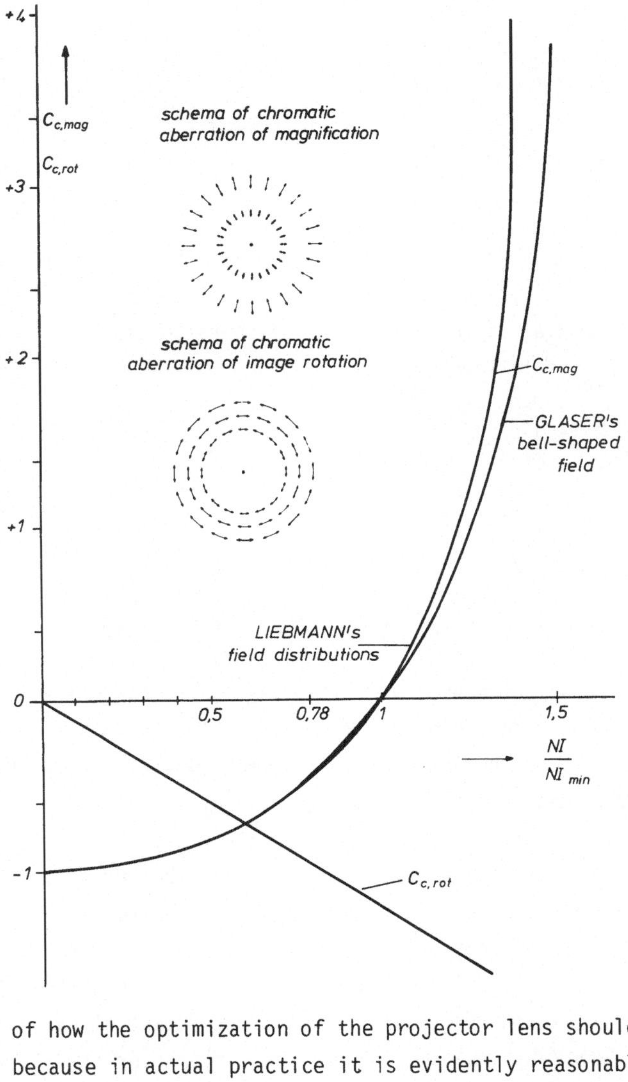

Fig.4.14. Dependence on the relative lens strength NI/NI_{min} of the coefficients $C_{c,mag}$ of the chromatic aberration of magnification and $C_{c,rot}$ of the chromatic aberration of image rotation

of how the optimization of the projector lens should be brought about. This is because in actual practice it is evidently reasonable to compare the distortions of images which have the same external diameter $2r_i$.

If, therefore, we begin the optimization procedure by stipulating the image radius r_i, the distance L of the centre of the projector pole-piece system from the image plane, and the magnification $M_p = r_i/r_o$, R becomes a rather involved function of the projector focal length $f_p = L/M_p = Lr_o/r_i$, the pole-piece system aspect ratio S/D, and the lens strength $NI/\sqrt{U}$ (cf. Figs.4.12,13). C_d and C_{sp} are likewise functions of S/D and $NI/\sqrt{U}$ but do not depend on f_p. Thus, S/D and $NI/\sqrt{U}$ are the only design parameters which – in all possible combinations reasonable for practical applications – are actually available for the lens optimization.

Following these lines of reasoning, (4.32) and (4.33) can be modified into

a) radial distortion

$$\frac{\Delta r_i}{r_i} = C_r\left(\frac{r_i}{L}\right)^2 \tag{4.40}$$

and

b) spiral distortion

$$\frac{\Delta t_i}{r_i} = C_t\left(\frac{r_i}{L}\right)^2 \quad . \tag{4.41}$$

Here, the "new" coefficients C_r of radial distortion and C_t of spiral distortion are related to LIEBMANN's C_d and C_{sp} as follows:

$$C_r = C_d\left(\frac{f_p}{R}\right)^2 = C_d\left(\frac{f_p}{f_{p\,min}}\right)^2 \cdot \left(\frac{f_{p\,min}}{R}\right)^2 \tag{4.42}$$

and

$$C_t = C_{sp}\left(\frac{f_p}{R}\right)^2 = C_{sp}\left(\frac{f_p}{f_{p\,min}}\right)^2 \cdot \left(\frac{f_{p\,min}}{R}\right)^2 \quad . \tag{4.43}$$

C_r and C_t are shown as graphs in Fig.4.15. Like C_d and C_{sp} they are functions of $NI/\sqrt{U}$ and S/D only.

Equations (4.40,41) show that the actual amount of (relative) radial and spiral distortion present in the image does not depend on M_p, R, and f_p but – besides $NI/\sqrt{U}$ and S/D – only on the ratio r_i/L of the instrument's column-geometry parameters r_i and L, or in other words, on the image projection (half) angle β, given by $\tan\beta = r_i/L$.

To make this point still more intelligible, we shall consider for the moment a projector design characterized by a set of design parameters r_i, L, $NI/\sqrt{U}$, and S/D. Then, for a magnification M_p, a focal length $f_p = L/M_p$ is required, and from this the absolute values of bore radius R and gap width $S = 2R \cdot S/D$ can be determined from Figs.4.12,13. It is essential in this context that for constant $NI/\sqrt{U}$ and S/D, we have $f_p = \kappa R$ with $\kappa = \kappa(NI/\sqrt{U}, S/D)$ as a proportionality factor that is constant here. Then, by using $r_i = M_p r_0$ and $L = M_p \cdot f_p$, we have

$$\frac{r_i}{L} = \frac{r_0}{\kappa R} \quad , \tag{4.44}$$

which means that for a given set of r_i, L, $NI/\sqrt{U}$ and S/D values, the same fraction of the pole-piece bore is always covered by the intermediate image serving as the object for the projector, regardless of the actual value of projector magnification M_p or focal length f_p.

Now, from Figs.4.12,14,15, the following conclusions can be drawn as regards the optimization of the projector lens design.

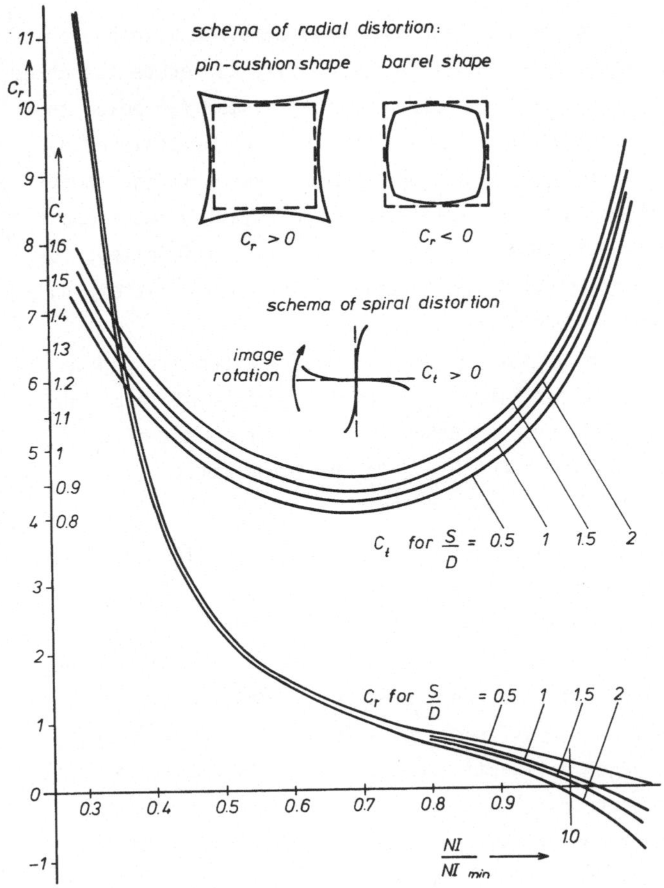

Fig.4.15. Dependence on the relative lens strength NI/N_{min} of the coefficients C_r of radial distortion and C_t of spiral (tangential) distortion. (Curves calculated from values given in [4.108])

1) Throughout the lens strength range $0.5 \leqslant NI/NI_{min} \lesssim 1.1$, the minimum obtainable projector focal length $f_p(S_{min}(NI))$ is particularly small and varies by less than about 10%.

2) The absolute minimum $f_{p\ abs\ min}$ (4.25) of the projector focal length that can be attained at all is smaller by only about 5% than the minimum focal length $f_{p\ min}$ provided by a projector which is operated at $NI/NI_{min} = 1$ and has a pole-piece system just at magnetic saturation; it will thus rarely be justified to barter the advantages [cf. 5) and 6)] gained by operating the projector at $NI/NI_{min} \approx 1$ against a rather minor decrease in focal length and increase in magnification.

3) As regards the coefficient $C_{c,rot}$, it is seen from Fig.4.14 that chromatic aberration of image rotation is the dominant chromatic field aberration around $NI/NI_{min} \approx 1$. $C_{c,rot}$ decreases in direct proportion to NI. It is important to realize that $C_{c,rot}$ is completely determined by the angle of image rotation, i.e., by $NI/\sqrt{U}$, and does not depend either on S/D or on the absolute pole-piece size, i.e. on R.

In actual instrument design, the projector lens is always combined with one or several other lenses, and so it is often possible drastically to reduce the chromatic error of image rotation of the whole electron optical system far below the value for an individual lens, or even to eliminate it. This can be achieved by judiciously balancing the value and the direction of the currents in the individual lens coils in such a manner that the individual image rotations counteract and the total image rotation of the lens system is more or less eliminated, together with the chromatic aberration of image rotation of the whole system. An example of such a compensation is described in [4.158].

In conclusion, for optimizing a projector lens only a moderate benefit can be expected from considerations concerning the $C_{c,rot}$ of the individual lens. Moreover, as can be seen from Fig.4.14, only a rather limited improvement of $C_{c,rot}$ can be achieved by decreasing NI from about NI_{min} to the lower limit $NI/NI_{min} \approx 0.5$ of the practical lens strength range.

4) As can be seen from Fig.4.15, the coefficient C_t of spiral distortion does not change much throughout the useful range of lens strengths $0.5 \lesssim NI/NI_{min} \lesssim 1.1$. There is little room for an optimization that would really count in terms of image quality. Whereas the spiral distortion is usually not particularly evident in the images of the common types of electron microscopical specimens, it may become rather conspicuous in electron diffraction patterns. With patterns containing reflection spots of larger diffraction angles especially, this may well lead to noticeable measurement errors.

   With a high-magnification electron lens system, a mutual compensation of the spiral distortion [analogous to the procedure mentioned in 3)] cannot be performed successfully because the imaging (half) angles β of the intermediate projectors and of the objective are smaller by orders of magnitude than the final imaging angle. Thus, for specimen imaging, spiral distortion has only to be taken into account with final projectors. On the other hand, with the objective, a spiral distortion appears in its back focal plane if the diffraction pattern of the specimen is generated there. Therefore, with LE POOLE's microdiffraction mode [4.103], compensation of this distortion by means of the spiral distortion of the projector is feasible for special combinations of pole-piece systems and lens strengths [4.152].

5) It is immediately obvious from the $f_p/f_{p\,min}$ curve of Fig.4.12 that its minimum indicates a lens strength NI/NI_{min} for which the chromatic aberration of magnification vanishes. In Fig.4.15 this corresponds to the zero of the universal $C_{c,mag}$ curve, obviously an optimum operating point for projector lenses; it is characterized by $NI = NI_{min}$.

6) A similar situation arises with regard to radial distortion. Here though, as can be seen in Fig.4.15, C_r is a universal function of NI/NI_{min} only for its larger

values, which correspond to the lower lens strength range $NI/NI_{min} < 0.7$. On approaching NI_{min}, the C_r curve splits into separate branches, each branch belonging to a different pole-piece aspect ratio S/D. The NI values corresponding to the zero values of C_r straddle around NI_{min}, $C_r = 0$ coinciding with NI_{min} for about $S/D \approx 1.6$. This would obviously be the best choice, because C_r and $C_{c,mag}$ will then vanish simultaneously at NI_{min}.

The main detrimental effect caused by radial distortion in high-magnification images is a falsification of the magnification and of the area of the images of particles that are small in comparison to the whole image field. HILLIER has drawn attention to the fact that completely erroneous results can be caused in the determination of particle size distributions by a rather moderate radial distortion [4.76].

For a small particle imaged at a distance r_i from the center of the image plane, the change of magnification caused by radial distortion is $\Delta M_r/M_r \approx 3C_r(r_i/L)^2$ if seen in the radial direction, $\Delta M_t/M_t = C_r(r_i/L)^2$ if seen in the tangential (circumferential) direction, and $\Delta A/A = 4C_r(r_i/L)^2$ for the change in area. By employing the data presented in Fig.4.15, the measurement accuracy for a specific design and operating point can be predicted.

If a high degree of freedom from radial distortion is required, as in electron beam lithographic systems for example, a mutual compensation of this aberration can be achieved for two lenses working in the telefocus mode by driving them with the same number of ampere-turns NI and employing pole-piece systems having the same aspect ratio S/D although different in size [4.112].

From the above, the following *guide rules* emerge for the design of a high quality projector lens.

1) The pole-piece aspect ratio should be close to $S/D \approx 1.6$.

2) The lens should be operated at or close to $NI/\sqrt{U} \approx 13.5\ A/V^{\frac{1}{2}}$ because that means killing three birds with one stone: employing the minimum focal length $f_{p\,min}$ and putting $C_{c,mag} \approx 0$ and $C_r \approx 0$ simultaneously.

Finally, it should be mentioned that a special two-gap projector lens design has been introduced [4.76] with the purpose of correcting radial distortion (Fig.4.16). This correction was based on the observation that with a radial distortion-free projector lens the image side half-angles β between the rays and the lens axis should increase exactly in proportion to the off-axis distance r_0 of the respective rays of the parallel bundle entering the lens.

Now, for the correction, the aspect ratio S/D and the size of the two gaps have to be the same. They are energized by the same coil and are magnetically in series, so that the paraxial focal length f_{p0} of both lenses is the same. Moreover, this focal length must be just equal to the spacing d of the two gaps: $f_{p0} = d$. Lens 1 focuses the close-to-axis rays exactly into the centre of lens 2, which they pass

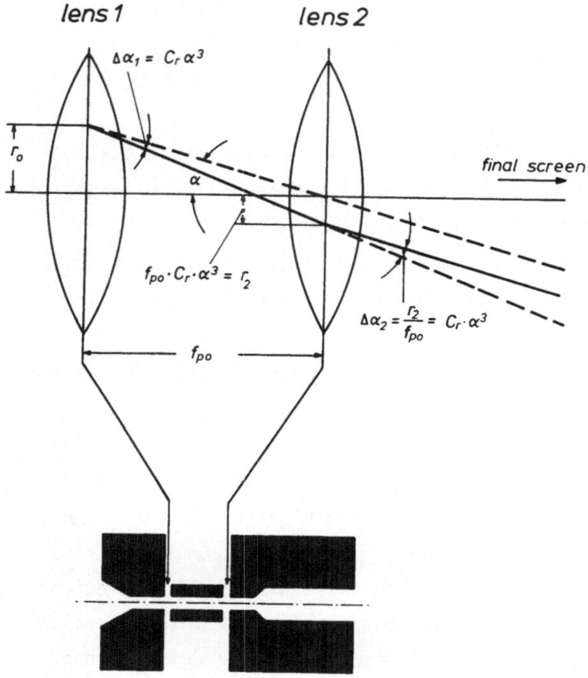

Fig.4.16. Pole-piece system of HILLIER's two-gap projector len and light optical analogue of its beam paths. The gaps are magnetically in series and ener gized by one and the same coil [4.76]

unaffected. The off-axis rays are subjected to an additional deflection $\Delta\alpha_1 = C_r\alpha^3$ caused by the radial distortion aberration of lens 1, and they enter lens 2 at an axial distance $r_2 = f_{p0}\,C_r\alpha^3$. There they are deflected back by $\Delta\alpha_2 = -r_2/f_{p0} = -C_r\alpha^3$ the ray now has the proper direction which an undistorted ray would have, whereas its parallel displacement $r_2 = f_{p0}\,C_r\alpha^3$ is so small that it does not matter in practice.

Although it was experimentally demonstrated that a projector made according to these ideas could quite successfully remove radial distortion, the design was not adopted widely in instrumental practice. Only one radial distortion-free focal length can be used with a lens of specific construction: in comparison with the $f_{p\,min}$ design recommended above, which also provides radial distortion-free operation at a specific focal length, the two-gap projector does not offer any particular advantage. On the contrary, the radial distortion correction can be realized only if the image-side focal point of the first lens lies not too close behind the magnetic field of the first gap so that the optic action of the second lens can be clearly separated from the deflection in the first lens. For a corrected system therefore, short focal lengths that are comparable to $f_{p\,min}$ and the field extension of a single gap are not feasible.

4.2.2c Practical Realization of a Large Magnification Range for Projector Lenses

To have a large range of magnifications available for imaging the specimen is imperative with an electron microscope. It is, however, now quite evident from the considerations of the preceding section that with a projector lens of fixed dimensions a large range of focal lengths and magnifications cannot be covered without running into excessive aberrations.

A slight improvement in performance was obtained by WEGMANN [4.209], again by employing two closely spaced gaps but with a separate coil and magnetic circuit for each of the lenses. The general correction idea behind this arrangement was quite similar to that employed with HILLIER's double-gap lens [4.76], but now the current in the coils and the focal lengths of the individual lenses could be chosen independently. Thus, an additional degree of freedom was provided for keeping radial distortion within reasonable limits while at the same time permitting some variation of the projector focal length. The magnification could be changed continuously by a factor of about three, the radial distortion remaining within a few percent.

A still larger magnification range can be obtained by employing a projector equipped with a set of suitably different pole-piece systems and some mechanical means for exchanging these pole-piece systems that does not impair the high vacuum of the microscope column. The most successful of these designs is the pole-piece system revolver introduced by RUSKA [4.177] in 1944 and since then employed in numerous commercial and experimental designs [4.178,182,198,199,157]. Here, four different pole-piece systems are inserted into four corresponding bores of a rotatable nonmagnetic turret (Fig.4.17). The pole-piece axes are equidistant from the rotation axis of the turret, and this rotation axis is at just this distance from the axis of the microscope. Thus, by suitable rotation of the turret, any one of the pole-piece axes can be made to coincide with the instrument axis and the lens is then available as a final projector. The remainder of the magnetic circuit of the projector is quite conventional. The pole-piece system actually in use connects up with the lens core at its lower pole piece and with the upper lens casing flange at the upper one.

Evidently, in order to allow exchange of the pole-piece systems, the latter cannot be screwed or clamped into place, but when the lens coil is energized, considerable magnetic attraction forces lock the appropriate pole-piece system into place. A small gap or a little looseness has to be provided between the surfaces in contact in order to avoid too much friction when turning the turret. It has proved useful for this purposes to equip the upper flange with a cylindrical central core which can move freely to some extent in the axial direction. Further, it is prudent not to hold the pole-piece systems rigidly within their seats in the turret bores but to allow a little tilting so that the connecting surfaces of the

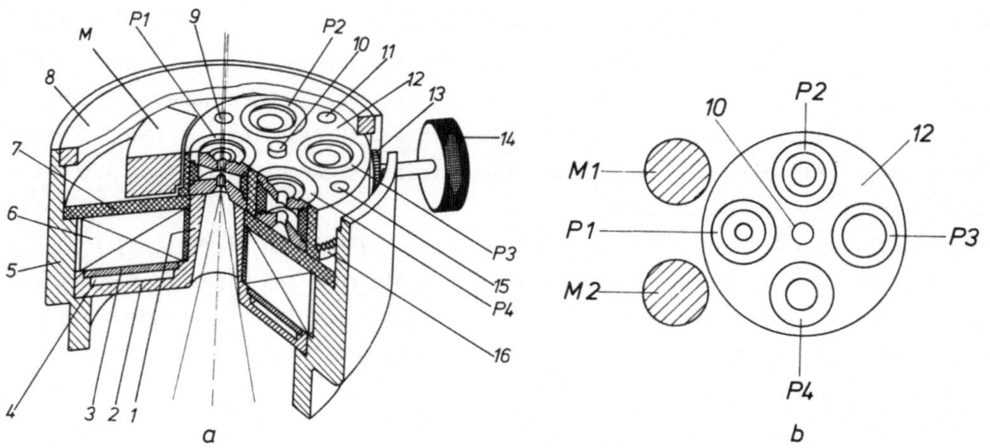

Fig.4.17. (a) Projector lens with pole-piece revolver by RUSKA [4.177]. The revolver drum 12 is made of brass and holds four different pole-piece systems P1 to P4. The revolver is rotated around its axis 10 by turning the knob 14 and with it the toothed wheel 13 which locks into a toothed rim 16 on the circumference of the revolver drum 12. The drum rests on the brass plate 7, and for better evacuation of the narrow gap between the drum 12 and the plate 7, bores 9, 10, 15 are provided. The remaining parts of the lens are essentially conventional. The pole-piece system in use (P1 in the drawing) rests on the lens core 1. The magnetic circuit is completed by the lower flange 2, the external casing 5 and the upper flange 8. For removing the heat from the lens coil 6, a water chamber 4 is provided in the lower flange and closed by an iron lid 3. A magnetic balancing mass M is provided to reduce the asymmetry of the magnetic stray field caused by the pole pieces not in use. (b) Shows another possibility for placing balancing masses. The two pole-piece systems not in use, P2 and P4 together with the balancing chunks M1, M2, form a four-fold symmetric arrangement with respect to the pole-piece system in use P1, a symmetry which is only slightly disturbed by the more distant pole-piece system P3. This design has been used successfully in the electron microscope described in [4.157]

magnetic circuit come flat against each other and magnetically resistive gaps with their resulting stray fields are avoided.

When the turret is rotated by small angles only, a nearly linear movement of the pole-piece system results, which is at right angles to the instrument axis and in the circumferential direction of the pole-piece turret. If an additional micrometer drive is available for traversing the turret in the direction connecting the microscope axis and the turret axis (i.e., at right angles to the first movement), a transverse alignment of the pole-piece system with respect to the optic axis of the microscope can be effected [4.182]. Since the progress of the alignment has to be appraised by observing the electron optical image, the pole piece has to be moved while the lens coil is energized. This gives rise to considerable friction between the pole pieces and the adjoining cores as they are magnetically attracted against each other. In actual practice, therefore, the movement of the pole pieces is often observed to be jerky and sometimes rather

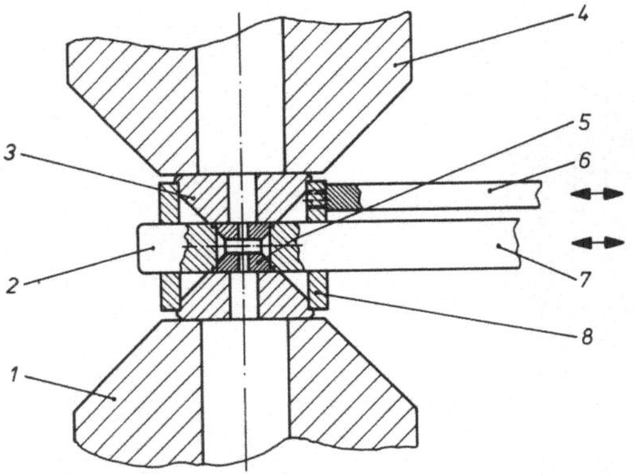

Fig.4.18. Pole-piece system exchange device after VON ARDENNE [4.3]. The lens cores 1,4 act as pole pieces for the largest focal length after both other pole-piece systems have been retracted by means of the rods 6 and 7. In order to obtain the intermediate focal length, the pole-piece system 3 had to be moved into place by pushing the rod 6 inwards, whereas the short focal length pole-piece system 5 still remains retracted. Then, only the tip 2 of the rod 7 still rests within the guiding channel bored into the brass spacer 8 which connects the pole pieces 3. The drawing shows the third operating mode with the short focal length pole-piece system 5 in place and ready for operation

erratic. A much better design for lens alignment is to provide for a transverse movement of the whole lens body, as will be described below in Sect.4.2.3.

One of the practical problems with the construction of a projector lens having a pole-piece system revolver is caused by the presence of the unemployed pole pieces, which drastically endanger the rotational symmetry of the magnetic stray field within the lens casing. This can easily give rise to a peculiar elliptical or even egg-shaped distortion of the final image of several percent, especially with pole-piece systems having large gaps so that the asymmetrical stray field may penetrate into the space around the instrument axis and so into the lens field proper. By adding a ferromagnetic balancing mass in a position opposite the turret with respect to the instrument axis, the magnetic stray field can be made roughly rotationally symmetric so that the corresponding distortion is avoided [4.178]. The proper size and position of the balancing mass must be determined experimentally by trial and error (an example is sketched in Fig.4.17), because a theoretical treatment of the problem is so far not available.

There are certainly still other feasible designs for interchanging the projector pole-piece systems. A linear arrangement of three pole-piece systems on a slider has been proposed by VON BORRIES and RUSKA [4.21] but has apparently never been put into practice. A similar design has been employed by VON ARDENNE [4.3] with an experimental electron microscope. Here, the pole-piece systems of a set of three are not all the same size but fit into each others' gaps (Fig.4.18). For a given pole-piece system the next larger one in terms of gap and bore size provides the tips of the lens cores.

A quite different method of varying the focal length of the projector at constant lens strength $NI/\sqrt{U}$ is to enlarge or decrease the gap width of the pole-piece system continuously. It is immediately clear from Fig.4.13 that at constant bore radius R, the minimum focal length $f_{p\,min}$ is roughly proportional to the aspect ratio S/D, tending to $f_{p\,min} \approx 0.5S$ for larger values of S/D and to

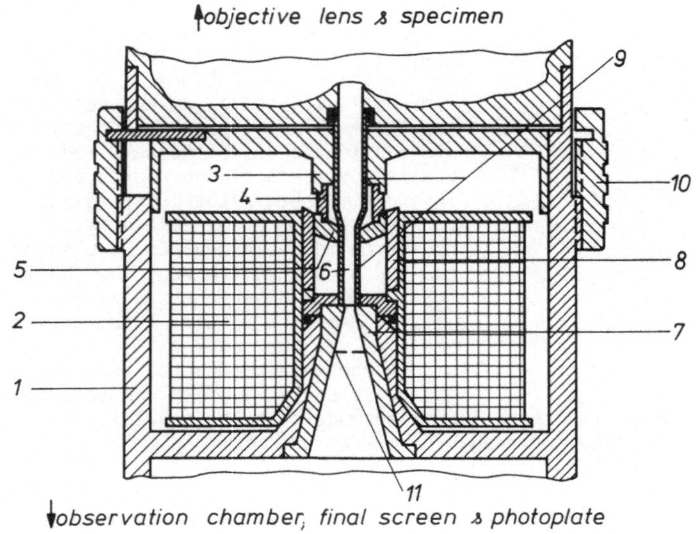

Fig.4.19. Variable focal length projector lens having two gaps, employed with the PHILIPS electron microscope EM 75 [4.35]. The upper pole piece 5 of the variable gap is mounted outside the vacuum space and can be moved over about 15 mm by turning a threaded ring 10 which encircles the microscope column. Between the sliding pole piece 5 and the upper flange 3, a non-magnetic spacer 4 is interspersed so that a second gap is formed; this weakens the field in the lower gap 6 between the pole piece 5 and the lower pole piece, which sits on the lens core 7. The magnetic circuit is completed in the conventional way by the lower flange and the external casing 1 and is energized by the coil 2. The vacuum wall around the beam is formed by a brass tube 9. When the upper gap slides down with the pole piece 5, the gap 4 is gradually short-circuited magnetically by the ferromagnetic sleeve 8, and so the magnetomotive force is transferred more and more to the lower gap. An aperture 11 limits the image field

$f_{p\,min} \approx 0.7$ R for the very small ones. A 1:5 variation of the projector focal length and magnification can be achieved in this way. It is interesting to note that because of the universal character of the $f_p/f_{p\,min}$ curve of Fig.4.12, i.e. of its very weak dependence on S/D, the same variation can be attained for any value of NI/NI_{min} although working at NI_{min} and hence $f_{p\,min}$ will usually be preferred in order to minimize the field aberrations. The variable gap design has been successfully employed with a commercial electron microscope [4.35] but with the additional artifice of providing a second gap of constant gap length S which is magnetically in series with the variable gap and connected to the sliding pole piece (Fig.4.19). This generates an additional weak lens, at the same time reducing the field strength in the variable gap by using up a part of the magnetic potential NI. The net effect is to reduce the magnification of the whole two-gap combination by a factor of two in comparison to the variable gap alone. On the other hand, when making the variable gap smaller and thus its focal length shorter, the additional gap is gradually short-circuited magnetically by sliding into a ferromagnetic sleeve. This arrangement increases the focal length and magnification range up to 1:9.

4.2.3 Condenser Lenses

Condenser lenses are required to concentrate the illuminating electron beam into the investigated area of the specimen plane. If the diameter of this area is not too small, say of the order of several 1/100 mm, designing a condenser lens is not at all difficult. Gap and bore can then be of the order of 1 cm and the required ampere-turns would correspond to a lens strength around $NI/\sqrt{U} \approx 5\ A/V^{1/2}$ or even lower, producing a focal length in the 10 cm range,

Axial astigmatism and spherical aberration can be disregarded here because of the smallness of the specimen illumination angles required, which are of the order of 10^{-3} rad or less. In the specimen plane, the beam deflection caused by the aberrations remains of the order of about 1 μm. This is smaller than the beam diameters required there, and there is practically no broadening of the spot size, even though the coefficients characterizing the aberrations may become rather large for such weak lenses, of the order of 10 m to 100 m for the coefficient C_s of spherical aberration.

Moreover, because of the smallness of the imaging (half) angles, field aberrations are irrelevant too. When designing a weak condenser lens, therefore, there is no optimization procedure, analogous to those described above for objective and projector lenses, to be followed. Here, only the condenser focal length f_c is determined by the general layout of the whole electron optical system. The designer is at liberty to base his choice of pole-piece dimensions on other criteria, e.g. the need to have a large enough gap S to insert and align a condenser aperture easily, or the desire to make the bore diameter D large enough to avoid machining problems in the workshop. Having selected S,D and consequently R = D/2 and S/D, $f_{c\,min}$ can be taken from Fig.4.13; $f_c/f_{c\,min}$ is calculated, and the required ampere-turns are determined from Fig.4.12 using $NI_{min} \approx 13\sqrt{U}\ A/V^{1/2}$. For $f_c/f_{c\,min} > 4$ the ampere-turns cannot be determined from Fig.4.12. With such very weak lenses, GLASER's bell-shaped field is a satisfactory model and leads to

$$\frac{f_c}{f_{c\,min}} = -\frac{1}{1.47} \cdot \frac{\sqrt{1 + k^2}}{\sin(\pi\sqrt{1 + k^2})} \quad , \tag{4.45}$$

k^2 being defined by (4.13). From this an approximate formula can be derived for establishing the required ampere-turns:

$$\frac{NI}{NI_{min}} \approx 0.65\sqrt{\frac{f_{c\,min}}{f_c}} \quad . \tag{4.46}$$

A more careful design of the condenser lens is required if electron beams of micron or submicron diameter are to be projected into the specimen plane or into the entrance pupil plane of a condenser-objective lens [4.164]. For this, the electron source has first to be demagnified rather strongly, and one or several short-focal-length condenser lenses are employed for this purpose [4.151,182].

Essentially, a short-focal-length condenser lens is like a projector operated with the beam direction reversed. Expressions (4.40) and (4.41) for radial and spiral distortion therefore remain valid if the imaging (half) angle, arctan $(r_i/L) \approx r_i/L$ is replaced by the (half) angle under which the source is seen from the lens. As this angle is orders of magnitude smaller than the projector imaging angle, third-order field aberrations are negligibly small with short-focal-length condensers. Chromatic field aberration can be disregarded too, because the electron source and its image are rather close to or even on the lens axis. Finally, the beam spot diameter spreading due to spherical aberration and axial astigmatism will in practice always remain so small as to be of no account.

As in the case of the weak long-focal-length condenser lens, therefore, there is no actual optimization to be carried out when designing a short-focal-length condenser. In fact, there is no particular point in operating the lens at NI_{min}, and for any specific focal length f_c chosen according to the general layout of the illuminating system, a corresponding pair of values of bore radius R and ampere-turns can be selected from Figs.4.12,13.

Often a variation of the refractive power of the short focal length condenser is required during operation of an electron microscope, in order to change the demagnification of the source. There is no fundamental objection from the straightforward electron optical point of view to effecting this focal length change by a corresponding variation of the ampere-turns. On the other hand, a variation of the ampere-turns also means that the magnetic potential distribution in the iron circuit of the lens changes, giving rise to a change of the magnetic stray field emanating from lens core and lens casing, causing a corresponding deviation of the electron beam path and thus a misalignment within the illuminating system or even within the imaging system. Such a disturbance can be avoided by equipping the short-focal-length condenser with a pole-piece system revolver (as described in Sect.4.2.2c) and operating the lens always with the same number of ampere-turns [4.159].

We now consider the long focal length condenser lens to be employed in a double [4.182] or triple [4.153] condenser lens system; its design requires careful consideration to ensure that this long-focal-length condenser does not become the performance-limiting factor of the whole system. As regards aberrations, the minimum beam spot size attainable is essentially limited by the spherical error of this lens, because it can safely be assumed that its axial astigmatism can be compensated by means of a stigmator. If the beam cross-section is not to be deteriorated by an increase of its size and a decrease of its current density, the aberration disc must evidently remain small in comparison with the spot diameter.

In this context, it is important to bear in mind that quite generally the coefficient C_s of spherical aberration is strongly magnification-dependent [4.62,63]. If both the object- and the image-side focal points are located outside the magnetic field of the lens so that asymptotic imaging prevails, the coefficient of spherical aberration referred to the image plane of a lens producing a magnification M_c can be expressed as

$$C_s(M_c) = C_s(0) \cdot (1 + M_c)^4 \quad . \tag{4.48}$$

Here, $C_s(0)$ is the coefficient of spherical aberration that would apply to the object plane if the lens was used as an objective lens at very high magnification, i.e., with the specimen plane close to the object-side focus. Thus, $C_s(0)$ is the coefficient termed C_s in most of the theoretical papers and for which extensive graphical and tabulated values are available. With the illuminating aperture α in the specimen plane, the radius of the circle of least confusion for spherical aberration can hence be written as

$$\delta_c = \frac{1}{4} C_s(M_c) \cdot \alpha^3 = \frac{1}{4} C_s(0) \cdot (1 + M_c)^4 \alpha^3 \quad . \tag{4.49}$$

In practice, the maximum diameter $2\delta_c$ of the circle of least confusion that is still compatible with the smallest beam diameter Ψ to be generated in the specimen plane by the condenser system should not be too large a fraction of this diameter, say about $2\delta_c \approx 0.2\Psi$. Now, what we want to know when designing the long-focal-length condenser lens is its focal length, of course, and also its coefficient $C_s(0)$ of spherical aberration. However, $C_s(0)$ is determined by δ_c as given by (4.49) only if both M_c and α have been specified beforehand. For M_c this is usually the case because it is determined by the position of the final long-focal-length condenser between the demagnified image of the electron source and the specimen plane. On the other hand, α is an 'experimental' quantity in the sense that it has to be chosen according to the particular investigation in question.

In order to estimate the largest value of the coefficient of spherical aberration that is still compatible with the various electron optical requirements, the largest actual aperture has to be used with (4.49). This critical aperture α_c is about one fifth of the imaging aperture of the objective lens, i.e. about $\alpha_c \approx 2 \times 10^{-3}$ rad. Then for a spot size Ψ and $2\delta_c \approx 0.2\Psi$, (4.49) yields the guide rule

$$C_s(0) \approx 5 \cdot \frac{10^7 \Psi}{(1 + M_c)^4} \quad . \tag{4.50}$$

In Table 4.1, values of $C_s(0)$ are shown for some typical magnifications M_c and beam diameters Ψ.

Table 4.1.

	$C_s(0)$	
M_c	for $\Psi = 0.1\ \mu m$	for $\Psi = 1\ \mu m$
0.5	1 m	10 m
1	300 mm	3 m
2	60 mm	600 mm

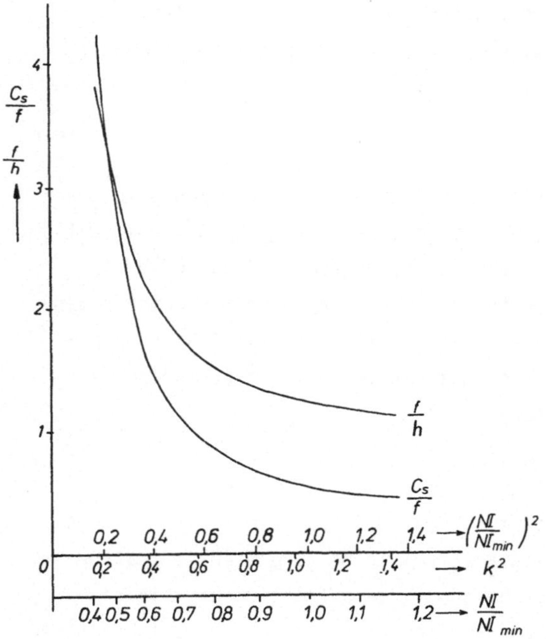

Fig.4.20. Dependence of the coefficient C_s of spherical aberration and of the focal length f on the relative lens strength NI/NI_{min}. C_s is shown in units of f, while f is in units of the field half width h. GLASER's bell-shaped field has been assumed as the lens field distribution

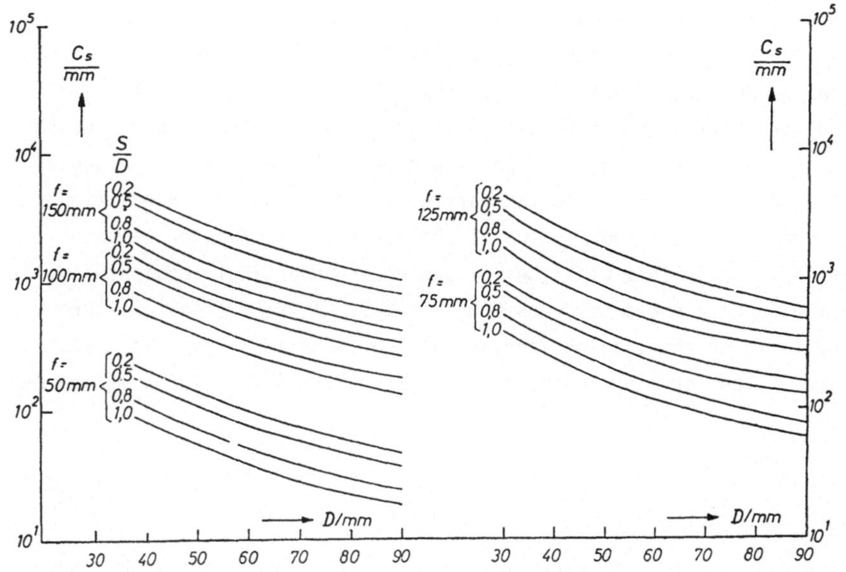

Fig.4.21. Coefficient C_s of spherical aberration for a series of electron lenses having LIEBMANN's field distribution. The dependence of C_s on the bore diameter D is shown. The focal length f and the pole-piece system aspect ratio S/D are used as parameters to distinguish the different curves (values after [4.41,106])

The focal length of a long-focal-length condenser lens is mostly between 50 mm and 100 mm, so that with the smaller beam diameters $\Psi \approx 0.1$ µm and with magnifications $M_c \gtrsim 1$, the coefficients of spherical aberration $C_s(0)$ should be of the

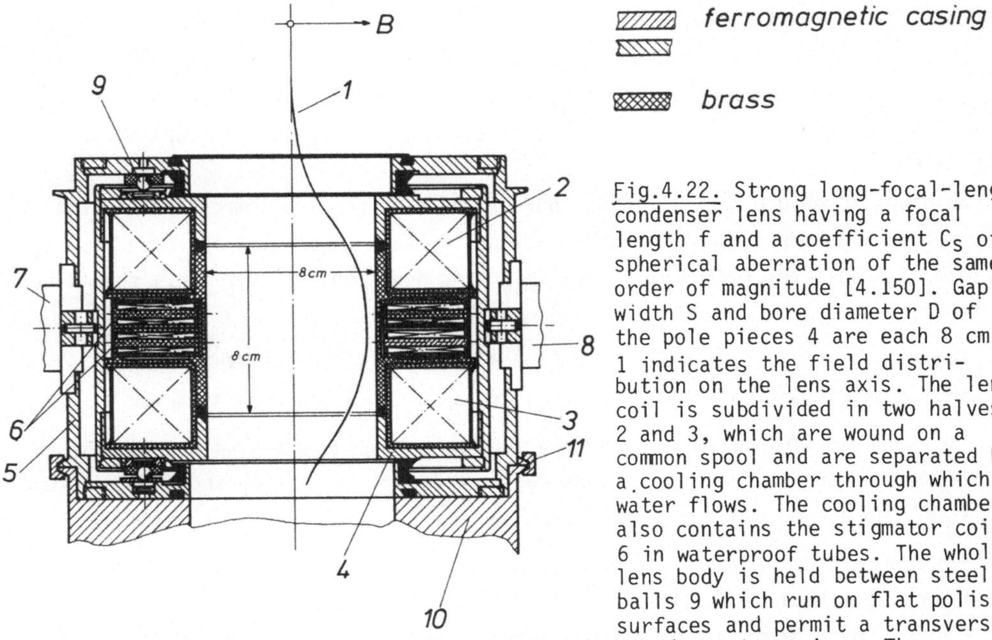

Fig.4.22. Strong long-focal-length condenser lens having a focal length f and a coefficient C_s of spherical aberration of the same order of magnitude [4.150]. Gap width S and bore diameter D of the pole pieces 4 are each 8 cm. 1 indicates the field distribution on the lens axis. The lens coil is subdivided in two halves 2 and 3, which are wound on a common spool and are separated by a cooling chamber through which water flows. The cooling chamber also contains the stigmator coils 6 in waterproof tubes. The whole lens body is held between steel balls 9 which run on flat polished surfaces and permit a transverse alignment of the whole lens by pairs 7, 8 of drives and countersprings. These drives and countersprings are connected to the steel tube 5, which constitutes the wall of the electron optical column proper. The whole assembly is connected to the adjacent elements 10 of the column by clamping rings 11

same order of magnitude as the focal length, a typical property of a strong magnetic lens (Fig.4.20). It is also seen from Fig.4.20 that with such a strong lens the half-width of the field distribution is not very different from the focal length. Compared to the usual short-focal-length lenses and to the weak condenser, the extension of the magnetic field and consequently the gap width required are comparatively large. From Fig.4.21 it is seen that a coefficient of spherical aberration $C_s(0)$ safely below 1 m can be actually attained with a sufficiently large bore diameter between about 5 to 10 cm and a high enough aspect ratio. A condenser lens designed according to these guide rules is reproduced in Fig.4.22 [4.150] and has been shown to generate beam diameters of about 0.1 μm [4.153].

In conclusion, when designing a long-focal-length condenser lens it is always good policy to employ a bore diameter of at least about 2 cm: in the case of a weak lens ($NI \lesssim 0.4\ NI_{min}$), C_s increases extremely fast as NI decreases and might inadvertently overrun the limit beyond which its effect on the condenser performance is no longer acceptable.

4.3 Design of the Magnetic Circuit

What we require for a magnetic electron lens is essentially a short stretch of strong rotationally symmetric field. Ideally, it should be more or less confined to a volume surrounding the lens axis (axis of rotation of the field) and spread from a few mm to some cm at most along this axis. In practice, to produce such a field, a rather extended magnetic circuit has to be employed, the shape of which has been indicated in Fig.4.1.

In order to ensure that the behaviour of the lens remains comprehensible within a complex, high-performance, electron optical system, especially in regard to alignment procedures [4.158,163], the following additional requirements should be satisfied.

1) The pole-piece system should be designed so that the actual field shape can be reasonably well predicted in terms of gap width S and bore diameter D = 2R.

2) The magnetic flux density within the lens core should remain well below saturation in order to prevent spurious magnetic fields from appearing on the column axis, where they might act as an unwanted lens or as deflecting fields.

3) The drop of the magnetic potential along the lens casing should be kept as low as possible because it gives rise to magnetic stray fields, which – by penetrating into the microscope column above or below the lens – might produce misaligning deflection fields.

Usually, meeting conditions 1) and 3) also means that nearly the whole magnetic potential NI supplied by the lens coil is available at the lens gap. As a rule of thumb, the magnetic potential drop along lens core and lens casing together should not exceed about 5% of the ampere-turns provided by the coil.

Depending on the basic design of their respective magnetic circuits, two types of magnetic lenses may be distinguished (Fig.4.10):

1) The asymmetric lens which has a single lens coil and the pole-piece system close to one of the flanges of the casing (Fig.4.10b).

2) The symmetric lens which has two coils and the pole-piece system in its centre, so that the whole lens is essentially symmetric with respect to its normal-to-axis midplane (Fig.4.10a).

Since the two-coil design has the advantage of providing a higher total number of ampere-turns, the symmetrical design is often employed for condenser-objective lenses and in high-voltage electron microscopes.

In practice, for designing a magnetic electron lens, the logical approach is to begin with the pole-piece system because its shape will determine an important part of the magnetic flux that has to be accommodated by the lens core. Employing this knowledge and having a rough idea of the configuration of lens coil and lens casing, the shape of the lens core can be determined. Having established the magnetic flux entering the top and bottom flanges from the cores (with a symmetric lens as in Fig.4.10a) or from one core and one pole piece (with an asymmetric one as in Fig.4.10b), the cross-sectional area of the flanges can be determined and finally the thickness of the external cylindrical casing, which completes and closes the magnetic circuit. The consecutive steps of this design routine will now be considered in detail.

4.3.1 Magnetic Design of the Pole-Piece System

4.3.1a Calculation of the Magnetic Flux Density Distribution within the Pole-Piece System

For the following we shall assume that we have already decided which electron optical design parameters to employ, i.e., which bore diameter, gap width and number of ampere-turns in the coil. These data could have been obtained by considerations such as those indicated in Sect.4.2.

In most of the theoretical work on obtaining the field shape for a given pole-piece geometry, the relaxation method is used [4.85]. For this, the pole faces adjoining the gap are usually assumed to be planes extending orthogonally to and far out from the lens axis up to a region where the potential drop across the gap can be stipulated to be linear to a good approximation.

In practice, the pole faces cannot be extended out thus far without pushing the lens diameter beyond reasonable limits. A satisfactory compromise is to make the pole-face diameter about three to four times larger than the bore diameter [4.126]. The validity of this rule was tested by carefully measuring the focal lengths of a set of projector lenses and comparing them to calculated values. Generalizing from these results, it can be expected that MULVEY's recipe [4.126] will secure an electron optical performance of the lens in line with precalculated values to within about 1%.

In Fig.4.23 typical configurations of pole-piece systems are shown. Just as in the construction of the lens body, symmetrical and asymmetrical pole-piece systems can be distinguished. The symmetrical pole-piece system (Fig.4.23a) consists of a pair of truncated cones with the (half) apex angle around $\psi_0 \approx 50^\circ$ to 60°. With the asymmetric pole-piece system (Fig.4.23b) one truncated conical pole piece faces a flat extended pole piece, which is usually part of or identical with the upper or lower flange of the lens casing. Asymmetrical pole-piece systems are thus mostly used in conjunction with asymmetrical lens bodies, whereas symmetrical pole-piece systems are employed both with symmetrical and asymmetrical lenses.

In order to function satisfactorily within the framework of the magnetic circuit, the pole-piece system has to be designed according to the following guidelines.

1) It has to concentrate the magnetic flux into the parallel gap between the pole faces, thereby generating a short field of high strength on the lens axis, and

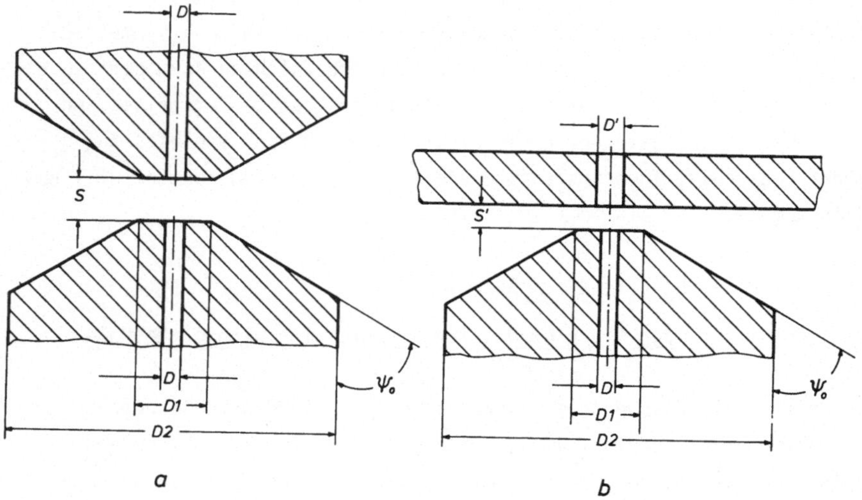

Fig.4.23a,b. Typical configurations of pole-piece systems. (a) Symmetrical pole-piece system. (b) Asymmetrical pole-piece system

it has to make available over the short space between the pole faces and with as little loss as possible the magnetic potential provided by the lens coil.

2) Within the conical parts of the pole pieces, the flux density should decrease with increasing distance from the pole faces as rapidly as possible and down to a value that can be easily accommodated by the lens core.

It will become evident from what follows that the bulk of the magnetic flux enters the pole pieces across their conical surfaces. A useful "introductory" model for predetermining the flux distribution with an actual design and for furnishing design guidelines is thus provided by the double-cone pole-piece system shown in Fig.4.24 [4.74].

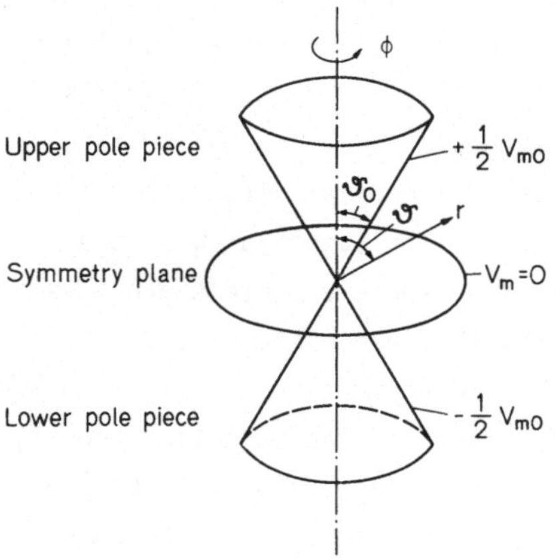

Fig.4.24.
The double-cone pole-piece system model

Spherical polar coordinates r, ϑ, ϕ are obviously well adapted to the study of the magnetic potential distribution V_m, which is determined within the space between the conical surfaces by Laplace's equation:

$$\nabla^2 V_m = \frac{1}{r^2}\cdot\frac{\partial}{\partial r}\left(r^2\cdot\frac{\partial V_m}{\partial r}\right) + \frac{1}{r^2\sin\vartheta}\cdot\frac{\partial}{\partial\vartheta}\left(\sin\vartheta\cdot\frac{\partial V_m}{\partial\vartheta}\right) + \frac{1}{r^2\sin^2\vartheta}\cdot\frac{\partial^2 V_m}{\partial\phi^2} = 0 \quad .(4.51)$$

This equation can be simplified considerably if we observe that:

1) because of the rotational symmetry of the pole-piece surfaces about the polar axis, the potential distribution V_m cannot depend on the azimuthal angle ϕ, so that

$$\frac{\partial^2 V_m}{\partial\phi^2} \equiv 0 \quad ; \qquad (4.52)$$

2) the pole-piece surfaces may, to a good approximation, be assumed to be equipotentials; then, the double-cone pole-piece system does not contain a natural unit of length and the value V_m of the potential must be insensitive to any change of scale of the radius vector r,

$$\frac{\partial V_m}{\partial r} \equiv 0 \quad . \qquad (4.53)$$

Using (4.52,53), the Laplace equation (4.51) reduces to

$$\frac{\partial}{\partial\vartheta}\left(\sin\vartheta\cdot\frac{\partial V_m}{\partial\vartheta}\right) = \cos\vartheta\cdot\frac{\partial V_m}{\partial\vartheta} + \sin\vartheta\cdot\frac{\partial^2 V_m}{\partial\vartheta^2} = 0 \quad . \qquad (4.54)$$

A solution of this equation can be obtained following the usual mathematical procedure of first substituting $u = \partial V_m/\partial\vartheta$ to reduce the second-order differential equation (4.54) to first order. The resulting first-order equation for $u = u(\vartheta)$ can be readily integrated with respect to ϑ to yield

$$u = \frac{\partial V_m}{\partial\vartheta} = \frac{c_1}{\sin\vartheta} \quad . \qquad (4.55)$$

A second integration yields

$$V_m(\vartheta) = c_1 \ln\tan(\vartheta/2) + c_2 \quad . \qquad (4.56)$$

The constants of integration c_1 and c_2 have to be determined from the boundary conditions, and it is appropriate to assign the zero value of the potential to the symmetry plane $\vartheta = \pi/2$: $V_m(\pi/2) = 0$. With $V_{mo} = NI$ as the total magnetic potential difference between the conical pole-piece surfaces, it follows that these surfaces have the potentials $V_m(\vartheta_o) = +V_{mo}/2$ and $V_m(\pi - \vartheta_o) = -V_{mo}/2$; the potential distribution in the free space between the cones can be expressed as

$$V_m(\vartheta) = \frac{1}{2} V_{mo} \frac{\ln\tan(\vartheta/2)}{\ln\tan(\vartheta_o/2)} \quad . \qquad (4.57)$$

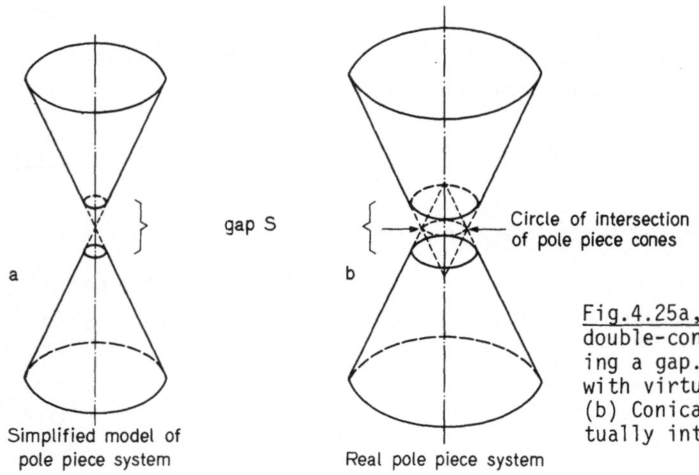

Fig.4.25a,b. Schematical drawing of double-cone pole-piece system having a gap. (a) Conical pole pieces with virtually coincident tips. (b) Conical pole pieces with virtually intersecting cones

The equipotential surfaces are conical surfaces too and the magnetic field lines are arcs of circles orthogonal to them.

The magnetic flux density

$$B = -\mu_o \, \mathrm{grad}_\vartheta V_m = -\frac{\mu_o}{r}\frac{\partial}{\partial\vartheta}V_m$$

$$= -\frac{\mu_o}{r\sin\vartheta}\cdot\frac{V_{mo}}{2\cdot\ln(\tan(\vartheta_o/2))} \qquad (4.58)$$

is largest at the pole-piece surfaces and smallest in the symmetry plane. It decreases in inverse proportion to the distance r from the centre of the pole-piece system.

In the calculation of the magnetic flux entering the conical pole-piece surfaces, assuming circular lines of force leads to rather simple and straightforward equations. However, this simplification evidently calls for some additional justification because firstly, it might appear rather crude to neglect the presence of the pole-piece gap (Fig.4.25a), and secondly, the two pole-piece cones do not have a common tip. Instead, their surfaces would intersect around a circle as shown in Fig.4.25b. (Of course, the pole pieces do not actually touch each other because of the gap: the intersection is "virtual".)

It can be shown [4.74] that with an intersecting double-cone pole-piece model like that of Fig.4.25b, the magnetic field lines in the wedge close to the circle of intersection are shaped like arcs of circles too. On the other hand, at distances from the intersection circle large compared with its radius, the magnetic field lines will again be arcs of circles because, as seen from there, the way the cone tips actually come together does not matter.

It can therefore be expected that even between these two regions the magnetic field lines will not deviate too much from a circular path and that it is legiti-

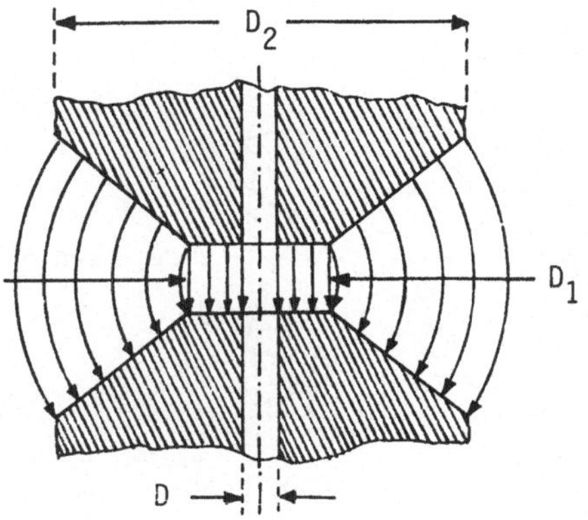

Fig.4.26. Simplified first-approximation representation of the path of the magnetic field lines in the space between the pole pieces

mate to employ this model of circular field lines throughout the space between the conical surfaces for the purpose of calculating the flux entering the pole pieces. The actual calculation will be outlined here only rather briefly so as to provide some understanding of how the results can be applied to the design of the pole-piece system (for details of the calculation see [4.74]).

Figure 4.26 shows that it is appropriate to consider the magnetic field separately in two regions: the essentially parallel field in the gap proper between the plane pole faces and the arc-of-circle shaped field between the conical pole-piece slopes. The magnetic flux ω traversing each of these two regions can then be expressed with the aid of HOPKINSONs' law [4.78], which may be written as

$$\omega = \mu_0 \frac{NI}{R_m} \tag{4.59}$$

if measured in volt seconds, or as

$$\omega^* = \frac{4\pi}{10} \frac{NI}{R_m^*} \tag{4.60}$$

with ω^* in Gauß centimetre squared. Here, R_m is the magnetic resistance which in free space is exactly analogous to the electrical resistance of a substance with a specific conductivity $\rho = 1$ (dimensionless), and so R_m is measured in reciprocal units of length, e.g., in 1/m. In (4.60), R_m^* has to be expressed in 1/cm.

The gap resistance R_s and the resistance R_c of the space between the conical surfaces are magnetically in parallel, so that the total resistance R_p of the pole-piece system can be obtained from

$$\frac{1}{R_p} = \frac{1}{R_s(D_1)} + \frac{1}{R_c(D_1, D_2)} \quad . \tag{4.61}$$

When determining the resistance of the gap, the bore can be neglected, because in a well-designed pole-piece system we have $D_1 \approx 3...4D$, and so the bore area is only about 10% or less of the pole-face area. Moreover, although the flux density within the bore area is somewhat less than that within an equal area between the actual

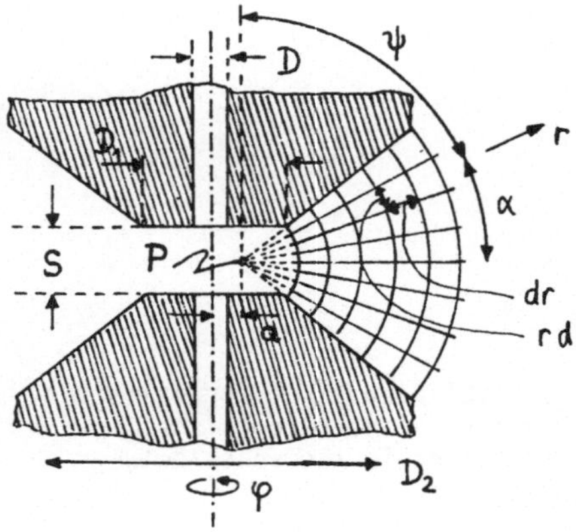

Fig.4.27. Illustration of the quantities employed for the determination of the magnetic resistance between the pole pieces

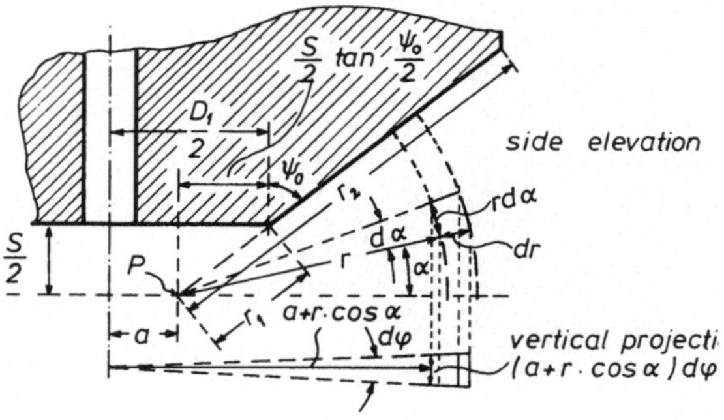

Fig.4.28. Illustration of quantities used for the calculation of the magnetic flux traversing between the conical pole-piece surfaces

pole faces, it still contributes to the total flux within the pole pieces. The error caused by neglecting the presence of the bore leads to an underestimation of the gap resistance by only a few percent, which does not matter in practice. The gap resistance can hence be written down directly as

$$R_s = \frac{4S}{\pi D_1^2} \quad . \tag{4.62}$$

How the resistance between the conical pole-piece slopes is calculated has been indicated in Fig.4.27. The equipotential surfaces are conical surfaces which intersect around a ring of radius

$$a = \frac{D_1}{2} - \frac{S}{2} \cdot \tan\psi_0 \quad .$$

They subdivide the magnetic tubes of force which are formed and limited by the arc-of-circle-shaped magnetic field lines. Thus, small box-like volume elements are partitioned off; the lengths of their sides are dr, r.dα and (a + r·cosα)·dϕ and — in the direction of the magnetic field lines — their magnetic resistance is r · dα/{(a + r·cosα) · dϕ · dr} (Fig.4.28).

Now, for calculating the magnetic resistance R_c between the conical pole-piece slopes, the resistance between two closely spaced equipotential surfaces is first established as a parallel arrangement of the volume elements. The resulting flattish and nearly two-dimensional resistors are then added up in a series extending from one pole piece to the other. (For details of this calculation and the following discussion, see [4.166].) It turns out that in order to obtain numerical values for R_c three different formulae have to be employed, each formula being applicable to a specific range of pole-piece design parameters.

For specifying the respective ranges of applicability, some characteristic functions are useful:

$$A = 2\pi a(\ln r - \ln r_1) \tag{4.63}$$

$$C = 2\pi(r - r_1) \tag{4.64}$$

$$W(r,a) = r/a - \ln(r/a) \quad . \tag{4.65}$$

Here, $r_1 = S/2 \cos\psi_0$ is the distance of the corner of the pole face from the virtual intersection circle of radius a, r is the distance of a volume element or the corresponding tube of force, and

$$r_2 = (D_2 - 2a)/2 \sin\psi_0$$

is the distance of the exterior corner of the pole-piece slope (Fig.4.28).

With these definitions, the magnetic resistance between the conical pole-piece slopes can be written as follows:

1) For A = C:

$$R_C^{(1)} = \frac{1}{\pi(\hat{r}_1 - r_1)} \frac{\cos\psi_0}{1 + \sin\psi_0} \quad ; \tag{4.66}$$

2) for $|A| < |C|$:

$$R_C^{(2)} = \frac{2}{\sqrt{C^2 - A^2}} \ln \frac{C + A \cdot \sin\psi_0 + \sqrt{C^2 - A^2} \cdot \cos\psi_0}{A + C \cdot \sin\psi_0} \quad ; \tag{4.67}$$

3) for $|A| > |C|$:

$$R_C^{(3)} = \frac{4}{\sqrt{A^2 - C^2}} \arctan \frac{\sqrt{A^2 - C^2} \cdot \cos\psi_0}{(A + C) \cdot (1 + \sin\psi_0)} \quad . \tag{4.68}$$

Here, A and C are functions of r, as are $R_C^{(2)}$ and $R_C^{(3)}$, which represent the magnetic resistance of a part of the conical pole-piece slope extending between the corner of the pole face and a circle on its surface which is parallel to the circle of virtual intersection of the cones and at a distance r from it.

The expression for $R_C^{(1)}$ describes the magnetic resistance for a specific value $\hat{r}_1$ of r, which for given a and $r_1 < a$ is completely determined by the equation

$$\frac{\hat{r}_1}{a} - \ln\left(\frac{\hat{r}_1}{a}\right) = \frac{r_1}{a} - \ln\left(\frac{r_1}{a}\right) \quad , \tag{4.69}$$

an immediate consequence of A = C and (4.63,64).

When employing the function W(r/a) defined by (4.65), the ranges of applicability of (4.66-68) can be easily related to actual design problems. For this, W(r,a) is shown as a function of r/a in Figs.4.29a-c.

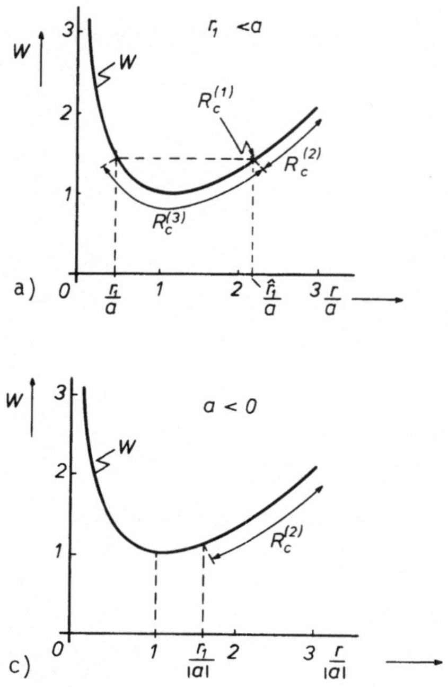

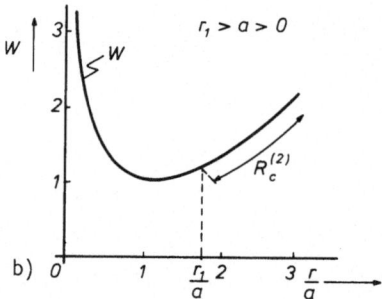

Fig.4.29a-c. Representation of the characteristic function W (4.65) used to specify the ranges of applicability of (4.66-68). (a) For $r_1<a$. (b) For $r_1>a>0$. (c) For $a<0$

It is now immediately obvious that (apart from the trivial case $r_1 = \hat{r}_1$) (4.69) refers to a configuration where r_1/a relates to the branch of the curve to the left of the minimum $r = a$ and $\hat{r}_1/a$ to the branch to the right of it. Here, corresponding couples of r_1/a and $\hat{r}_1/a$ have the same values $W(r_1,a) = W(\hat{r}_1,a)$ of the ordinates. As the branch to the left of the minimum of the W curve relates to $r/a < 1$, it follows that the case A = C can only occur if $a > r_1$, i.e., a pole-piece system with a pole face that extends rather far and a (half) cone apex ψ_0 remaining well below 90°. This condition can be expressed by the inequality

$$D_1 > S \cdot \frac{1 + \sin\psi_0}{\cos\psi_0} \quad , \tag{4.70}$$

so that D_1 must become very large as ψ_0 approaches $\pi/2$. It can also be shown easily that (4.70) automatically excludes the cases $a = 0$ and $a < 0$, which would not be compatible with the definition of the function W in (4.65).

We repeat here and it must be clearly realized that $\hat{r}_1$ is not a free variable but is completely determined by r_1 and a, through (4.69). Equations (4.66,69) are both subject to the relations $r_1 < a$ and $a > 0$. For other pole-piece cone extensions r, (4.67,68) have to be applied.

If $a > 0$ and $r_1 < a$ are still true, and if the pole pieces or those parts of the pole-piece cones just considered do not extend up to the distance $\hat{r}_1$ from the circle of virtual intersection of the cones, (4.68) has to be employed to calculate the magnetic resistance $R_C^{(3)}(r)$. It can be seen immediately from Fig.4.29 that $W(r,a) < W(\hat{r}_1,a)$ for $r_1 < r < \hat{r}_1$, a relation equivalent to $A > C$ for $a > 0$. On the

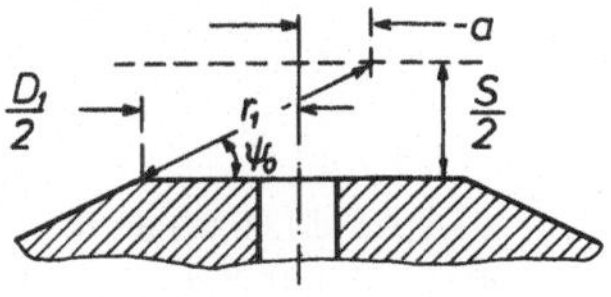

Fig.4.30. Illustration of essential geometric data of a pole-piece system with a < 0

other hand, for $r > \hat{r}_1$ and again a > 0, we have $W(r,a) > W(\hat{r}_1,a)$, a relation which can be shown to be equivalent to A < C (Fig.4.29b). For the more extended pole-piece cones, then, the magnetic resistance has to be calculated by employing (4.67) for $R_C^{(2)}$.

Finally, we come to the case a < 0, which corresponds to the situation characterized in Fig.4.30. Here, the (virtual) tips of the pole-piece cones are moved apart in comparison to the case a = 0 of Fig.4.25a, where the tips coincide. For a < 0 it is immediately evident from Fig.4.30 that r_1 is always greater than $|a|$. Only the branch of the W curve to the right of the minimum applies here, therefore, as indicated in Fig.4.29c where W is defined, using absolute values of a < 0, by

$$W(r,a) = \frac{r}{|a|} - \ln \frac{r}{|a|} \quad . \tag{4.71}$$

For the calculation of the magnetic resistance across the space between the pole-piece cones, it is always true that $r > r_1$ or $r_2 > r_1$ and $W(r,|a|) > W(r_1,|a|)$ or $W(r_2,|a|) > W(r_1,|a|)$. These relations are equivalent to $|A| < |C|$, so that in this case also the magnetic resistance $R_C^{(2)}$ is given by (4.67).

As an aid for practical lens design and as a summary of the above results, the respective ranges of application of (4.66-68) have been indicated in Figs. 4.29a-c [4.166].

The total magnetic resistance R_p of pole-piece systems can now be calculated by employing (4.61-68). From this and using HOPKINSONs' law (4.59) or (4.60), the magnetic flux traversing the free space between the pole pieces can be determined from

$$\omega_p = \mu_0 \frac{NI}{R_p} \tag{4.72}$$

in volt seconds with R_p in metres, or

$$\omega_p^* = \frac{4\pi}{10} \frac{NI}{R_p^*} \tag{4.73}$$

in Gauß centimetre squared with R_p^* expressed in centimetres. By dividing ω_p or ω_p^* by the corresponding cross-sectional area of the pole piece taken orthogonal to the axis of rotation, the average flux density (induction) $\overline{B}$ within these respective areas can be determined: $\overline{B} = 4\omega_p/(\pi D_3^2)$.

Numerical results obtained using the above procedure strictly represent the magnetic flux density in the connecting surface between a pole piece of (fixed) external diameter D_2 and the adjacent lens core. In fact, they can also be under-

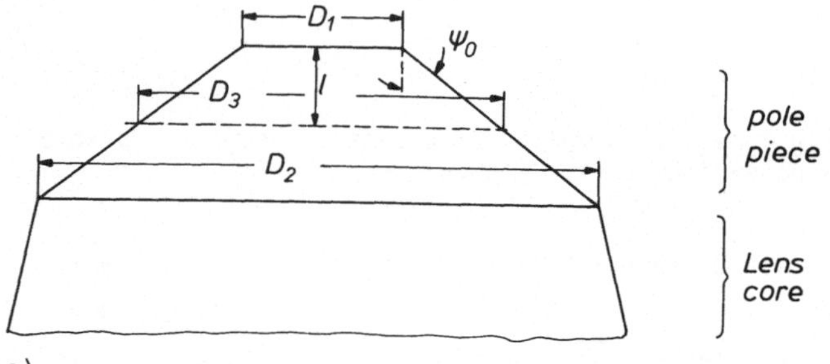

a)

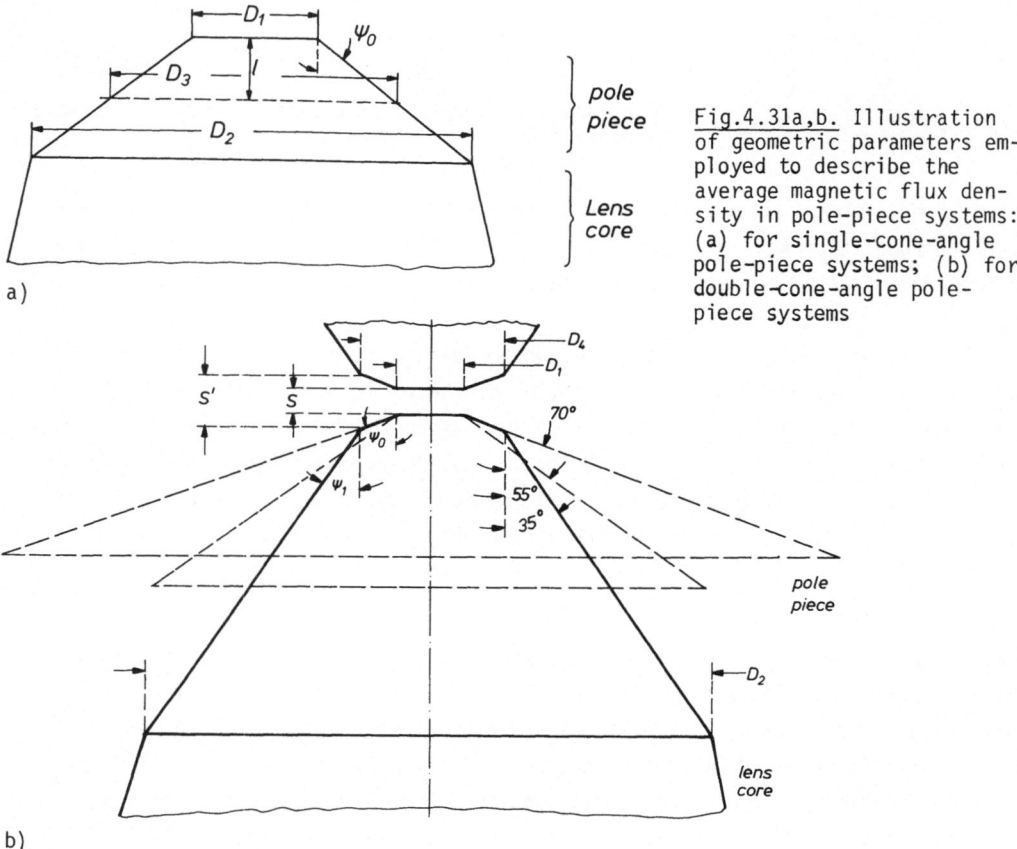

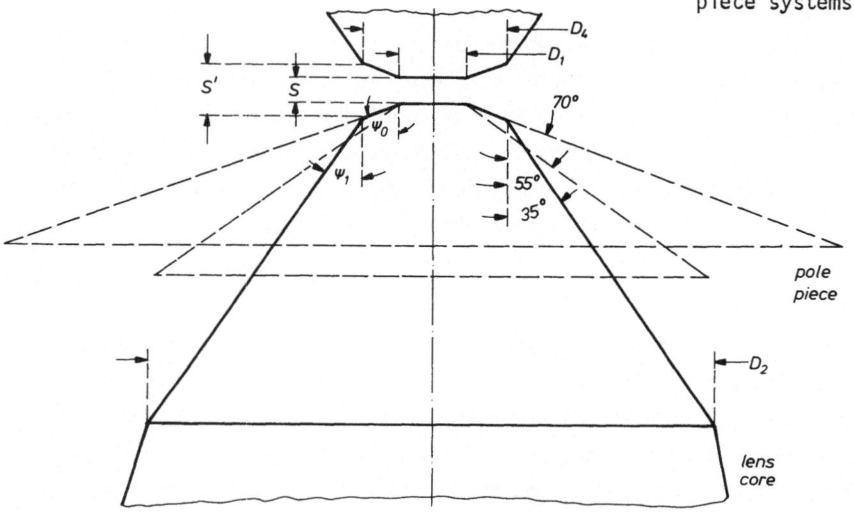

b)

Fig.4.31a,b. Illustration of geometric parameters employed to describe the average magnetic flux density in pole-piece systems: (a) for single-cone-angle pole-piece systems; (b) for double-cone-angle pole-piece systems

stood as describing the magnetic flux density within the pole pieces themselves at a distance

$$l = \frac{1}{2}(D_3 - D_1)\cot\psi_0 \qquad (4.74)$$

from the flat pole surface adjacent to the gap (Fig.4.31). In (4.63,64), the following expression for r,

$$r = (D_3 - 2a)/2\sin\psi_0 \quad ,$$

has then to be used.

4.3.1b Applications and Comparison with Experimental Results

The distance l can now be employed as an independent variable for plotting the average flux density to be expected in the pole pieces. In order to facilitate the application of the results to pole-piece systems of different size, it is appropriate to make use of the scaling rules of Sect.4.2 and normalize l with respect to the pole-face diameter D_1 as the "natural" unit of length. The specific number NI of ampere-turns can likewise be eliminated from the calculated results by normaliz-

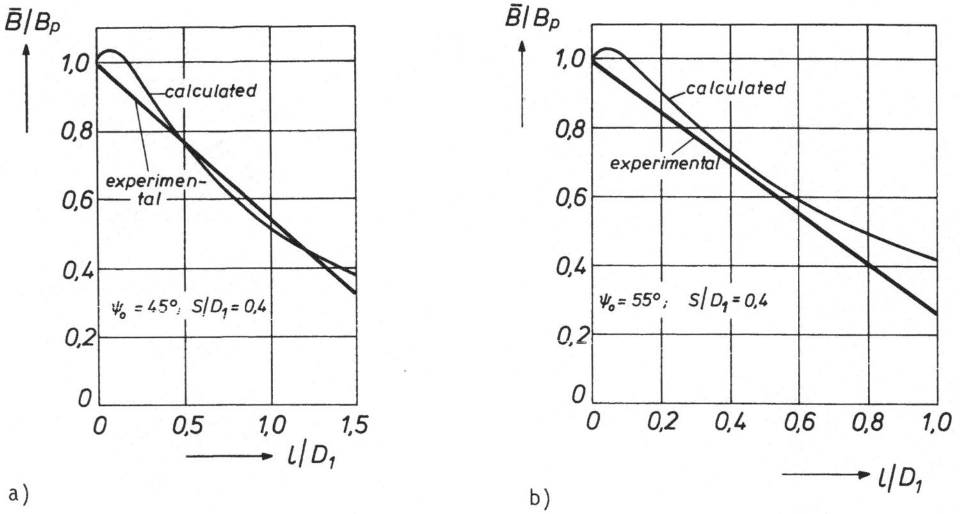

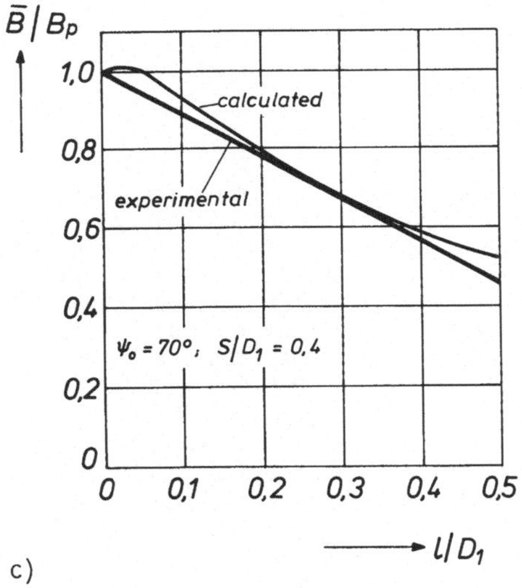

Fig.4.32a-c. Calculated variation of the average magnetic flux density $\bar{B}$ of a series of pole-piece systems having the same ratio $S/D_1 = 0.4$ of gap width S to pole-face diameter D_1 but different cone taper angles ψ_0. $\bar{B}$ is shown in units of the magnetic flux density B_p present in the parallel gap. Experimental values from [4.126] are included for comparison

ing $\bar{B}$ with respect to $B_p = \mu_0 NI/S$, which is the magnetic flux density that would prevail in the homogeneous field between pole faces extending much farther than their mutual spacing S. By way of example and in order to provide data for practical pole-piece design, several curves representing calculated flux density distributions are shown in Figs.4.32-34. On comparing these curves with experimental values [4.74,126] also shown in the figures, it becomes evident that for pole-piece design the magnetic flux density distribution can be reliably predicted to within about 5% or even better, provided that the pole pieces are not oversaturated.

The specific pole-piece systems for which the magnetic flux density distributions are shown in Figs.4.32-34 have been selected to bring out some details in the optimal design of the pole pieces. Thus, in Figs.4.32a-c the effect of choosing different

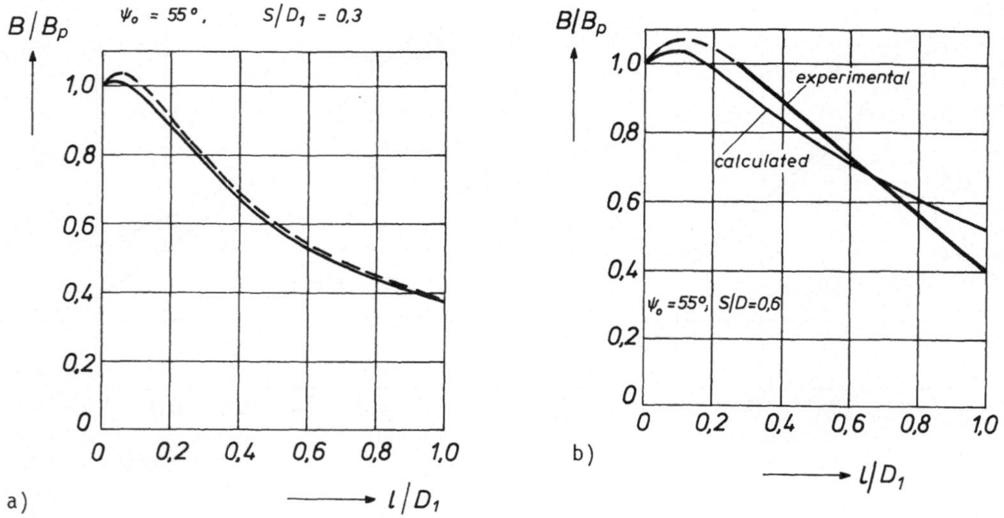

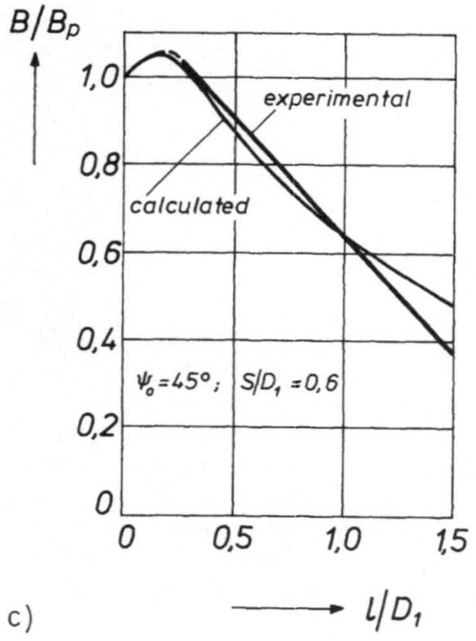

Fig.4.33a-c. Calculated variation of the average magnetic flux density $\bar{B}$ in a set of pole-piece systems having the same taper angle but different aspect ratios S/D_1. $\bar{B}$ is shown in units of the magnetic flux density B_p present between the parallel pole faces. Experimental values after [4.126] are included for comparison. (a) $S/D_1 = 0.3$, $\psi_0 = 55°$. (b) $S/D_1 = 0.6$, $\psi_0 = 55°$. Figure 4.32b, which refers to $\psi_0 = 55°$, $S/D_1 = 0.4$, should be regarded as the third case of the present series. (c) Here $S/D_1 = 0.6$ and $\psi_0 = 45°$ so that on taking this together with Fig.4.32a another pair of pole pieces is obtained having the same taper angle $\psi_0 = 45°$ but different aspect ratios S/D_1

cone angles ψ_0 for the pole-piece slopes becomes evident for a set of pole-piece systems having the same aspect ratio S/D_1 of the gap width S to the pole-face diameter D_1. It is seen immediately that in order to decrease the flux density $\bar{B}$, a shorter pole piece can be employed with the larger cone angles ψ_0.

On the other hand, it is often more desirable to try and restrict the external diameter of the pole pieces as far as possible because this will also allow a small inner diameter and electric resistance of the lens coil to be employed, resulting in a lower electrical power dissipation for a given number of ampere-turns. Thus, in Fig.4.35, for pole-piece systems having the same aspect ratio S/D_1 of the gap space but different cone-slope angles ψ_0, the external diameters D_3/D_1 of the pole

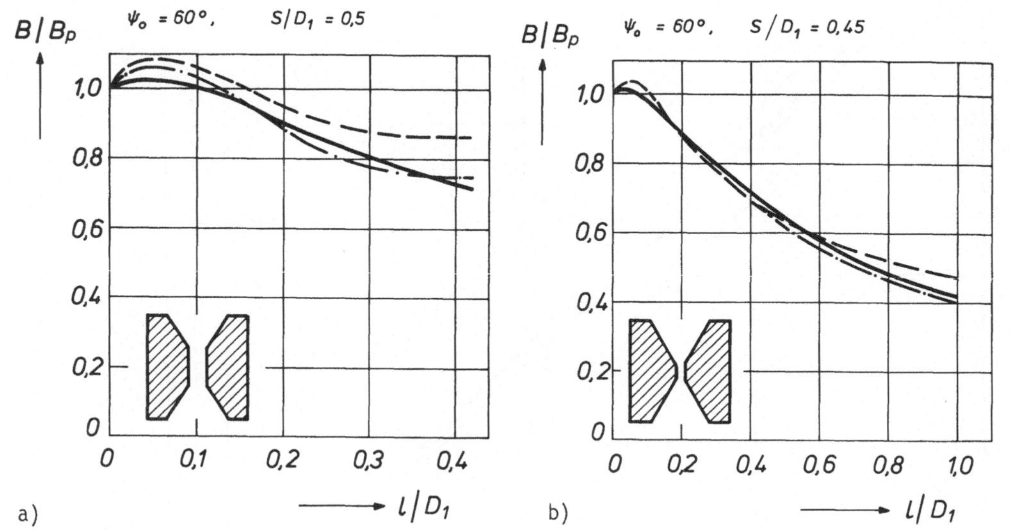

Fig.4.34a,b. Illustration of the effect on the flux density distribution in pole pieces caused by asymmetrical positioning of a lens coil with respect to the pole-piece system. Calculated values are compared to experimental data, obtained with the projector lens represented in Fig.4.38 [4.74]. ——— calculated values of flux density; ----- flux density in the pole piece connected to the lens core; -.-.-.-.- flux density in the pole piece opposite the lens core. (a) For a rather "open" pole-piece system with a comparatively short conical part; the asymmetry of the field distribution is evident. (b) For a pole-piece system with a more extended conical part; the asymmetry of the field distribution is nearly suppressed

pieces are shown as a function of the decrease of the average flux density $\bar{B}/B_p$, both quantities normalized to the respective values D_1 and B_p at the pole face.

These calculated values predict that when still close to the gap, the magnetic flux density $\bar{B}$ does not decrease but at first even becomes somewhat larger than the flux density B_p between the pole faces, a behaviour which has been confirmed experimentally (Figs.4.33,34) and may be understood as being caused by a field concentration produced by the sharp wedge at the junction between the pole face and the conical slope. As is to be expected, the effect becomes more noticeable with a stronger taper of the pole-piece slopes, i.e., with decreasing ψ_0. Consequently, if the pole-piece system is to be used with a gap flux density B_p close to magnetic saturation, it will be advisable to avoid slope angles ψ_0 below about 45°.

Moreover, Fig.4.35 provides an answer to the question of which is the optimum slope angle ψ_0, and in this context it is instructive to discuss the same question for the double-cone pole-piece model of Fig.4.24. For this, the magnetic flux density distribution between the cones is described by (4.58), after which the magnetic flux penetrating into the pole-piece cones and the corresponding average flux density $\bar{B}$ can be calculated straightforwardly. If the magnetic potential difference V_{mo} between the cones is eliminated by introducing a hypothetical gap flux

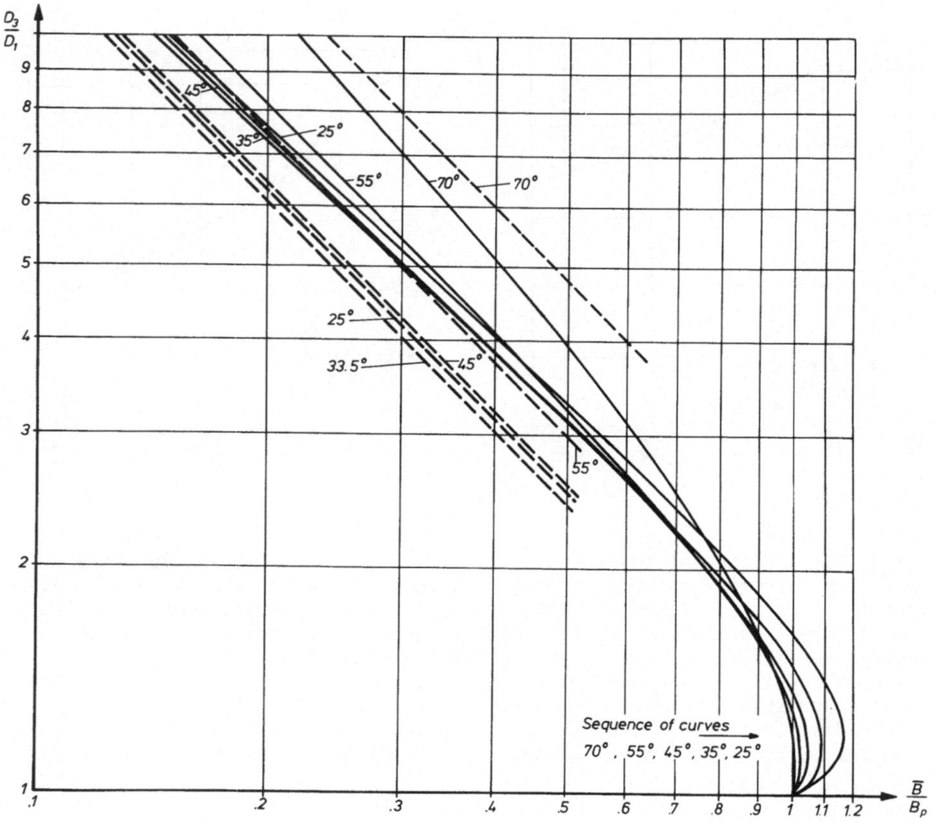

Fig.4.35. Diameter D_3 of the pole piece required for a specified diminution of the relative magnetic flux density $\bar{B}/B_p$. D_3 is shown in units of the pole-face diameter D_1, and the cone-taper angle ψ_0 is employed as a parameter for the different curves. For all the pole-piece systems the same aspect ratio $S/D_1 = 0.4$ has been assumed. The continuous curves refer to pole-piece systems of the type shown in Fig.4.27, whereas the broken lines refer to the double-cone pole-piece model (for details, see text)

density $B_p = \mu_o V_{mo}/S$, we obtain

$$\frac{\bar{B}}{B_p} = -\frac{2S/D_1}{\sin\psi_o \ln(\tan(\psi_o/2))} \cdot \frac{1}{D_3/D_1} \quad . \tag{4.75}$$

Here, D_1 is just a unit of length and has no immediate physical meaning. Above all, it does not in general describe the diameter of the cones at a distance S/2 from the centre of the system (but this would be the case if $S/D_1 = \cot\psi_o$, i.e., for $\psi_o = 68.9^\circ$ at $S/D_1 = 0.4$ in the example used for Fig.4.35). Now, if $\bar{B}/B_p$ is drawn as a function of D_3/D_1 on a doubly logarithmic scale, a family of parallel straight lines results which are inclined at -45°; the logarithm of the first fraction in (4.75) represents the parameter indicating the individual lines. They have been drawn as broken lines in Fig.4.35.

In regions far from the gap, i.e. with Fig.4.35 in regions where the diameter D_3 of the pole piece is large compared to the diameter D_1 of the pole face, and for a common angle ψ_0 of the pole-piece taper, the continuous curves of Fig.4.35 which have been calculated from (4.61-68) can be expected to approach the corresponding straight broken lines obtained by employing the double-cone pole-piece model.

This is rather obvious for $\psi_0 = 70^\circ$, which for $S/D_1 = 0.4$ nearly corresponds to the truncated double-cone pole-piece system with coincident tips shown in Fig. 4.25a. It can be readily understood that in this case the continuous curve approaches the corresponding broken straight line from below, for the magnetic flux actually traversing the gap is smaller by a factor of two than the flux which would have penetrated into the short conical caps now removed in order to generate the gap. With the same pole-piece diameter D_3, therefore, the average flux density $\bar{B}$ is smaller for the actual pole-piece system with a gap than for the double-cone system without a gap.

Conversely, the other continuous curves shown in Fig.4.35 indicate an average flux density $\bar{B}$ somewhat larger than for the double-cone system with coincident cone tips. This is quite plausible because here, the pole face area is definitely larger and the magnetic resistance of the gap correspondingly smaller than would be the case for the same gap width and a double-cone pole-piece system with coincident (virtual) cone tips.

Now, the optimum slope-taper angle ψ_0 can be understood to be the one that produces the largest decrease of the average flux density $\bar{B}$ relative to a given value B_p or, equivalently, the one that requires the smallest pole-piece diameter D_3 for a specified diminution of the average flux density $\bar{B}/B_p$. In the case of the pole-piece model with coincident cone tips, this optimum can be obtained easily by employing (4.75) and putting its first derivative with respect to ψ_0 equal to zero. This yields $\psi_{0,opt} = 33.5^\circ$, and it is seen from the numerical results based on (4.61-68) and shown as continuous curves in Fig.4.35 that the optimum slope angle here is about 35°. Hence, both calculations indicate nearly the same optimum angle. The optimum is rather flat in both cases. Optimizing the slope angle in the sense discussed here is therefore not particularly critical.

We can now draw a general conclusion from Fig.4.35, taking into account both the results concerning the optimization of the slope angle ψ_0 and the findings regarding the increase of the flux density near the pole faces. It is seen that for pole-piece design, a slope angle ψ_0 between 55° and 45° would usually be a good choice. Nevertheless, for pole pieces operated close to saturation, a larger slope angle, say around $\psi_0 \approx 70^\circ$, might be advantageous in order to avoid oversaturation and to decrease the flux density as quickly as possible.

As the pole-piece diameter becomes comparatively large for ψ_0 going up to about 70°, one might wonder whether a pole-piece shape as shown in Fig.4.31b could be of advantage. Here, the pole piece is assumed to be composed of two cones truncated so that their junction surfaces have equal sizes and the pole-piece cross-sectional area does not change abruptly when going from one cone to the other. At the junction surface the rather large taper angle ψ_o of the first conical part which is adjacent to the gap changes drastically to the more slender taper of the second conical part, which is connected to the lens core. This shape was intuitively employed long ago (e.g. [4.210]) for strong electromagnet design and was assumed to be of advantage for achieving a particularly high field strength in the gap. In the present context the question is whether by employing a pole-piece system with two taper angles as in Fig.4.31b a smaller overall diameter D_2 of the pole-piece base will suffice than that required for the more straightforward pole pieces with uniform taper of Fig.4.31a.

Pole pieces of this type can be treated by extending the methods of calculating the flux density explained above. The average magnetic flux density $\bar{B}_4$ in the junction surface between the cones can be obtained straightaway, by treating the first conical part like a normal pole piece. The average magnetic flux density $\bar{B}_4$ in the junction surface D_4 between the two parts is thus known. The flux density in the second pole-piece part can be established by employing the following artifice. Both first pole-piece parts adjacent to the gap S are assumed to be removed, so that the second pole parts would form a "normal" pole-piece system which has the taper angle ψ_1, the pole-face diameter D_4 and the gap width $S' = S + (D_4 - D_1) \cot\psi_o$, and is of course operated at NI ampere-turns too. Now, when applying our standard procedure to the calculation of the magnetic flux density in this "virtual" pole-piece system, the proper amount of flux density entering the pole piece via its conical surface is taken into account. On the other hand, a density $\bar{B}_4' = \mu_o NI/S'$ of the parallel magnetic flux is implicity assumed to be present in the "virtual" gap S' when we apply our procedure, and this flux density $\bar{B}_4'$ is smaller than the actual average flux density $\bar{B}_4$ in the junction surface between the two conical pole parts. Therefore, a correction has to be applied to the standard procedure by adding the corresponding flux difference $(\bar{B}_4 - \bar{B}_4')D_4^2\pi/4$ to the flux previously calculated for the second pole part by employing the methods described in Sect.4.3.1a. The actual magnetic flux distribution in the second pole part is thus established and the distribution of the average flux density can be calculated readily.

Numerical investigations have shown that a tangible advantage arises from the use of a double-cone-angle pole-piece system as shown in Fig.4.31b only if the pole-face region is operated quite close to magnetic saturation. Then, a cone taper-angle $\psi_o \approx 70^\circ$ becomes imperative, whereas $\psi_o \approx 55^\circ$ might already lead to disadvantages.

By way of example, in Fig.4.31b, a double-cone-angle pole-piece system is shown which has been specified by the aspect ratio S/D_1 and the cone angles $\psi_o = 70^\circ$ and $\psi_1 = 35^\circ$, as well as by the condition that at the exit planes which connect the pole pieces to the lens core, the average magnetic flux density B_2 should have dropped to $\bar{B}_2/B_p = 0.2$ (i.e., from a gap flux density a little over 20 kG = 2 T down to about 4 kG = 0.4 T, a reasonable diminution as will be explained in Sect.4.3.2).

It turns out that, quite generally, the external diameter D_2 of the pole-piece system becomes smaller if the junction surface between the two conical parts is moved towards the pole face. On the other hand, the average flux density $\bar{B}_4$ in the junction surface should have dropped sufficiently to avoid magnetic oversaturation which might otherwise be caused by the rather pronounced rise of the flux density to be expected for $\psi_1 = 35^\circ$ near the virtual pole face (Fig.4.35). There is thus no optimum design in the rigorous sense of the word, but $\bar{B}_4/B_p = 0.8$ is a sound compromise and has been used as the example for the pole-piece design outlined in Fig.4.31b.

For comparison, the corresponding single-cone-angle pole pieces are also indicated in Fig.4.31b for $\psi_0 = 70^\circ$ and $\psi_0 = 55^\circ$. The following observations may be made.

1) The external diameter $D_2 = 8.5\ D_1$ of the double-cone-angle pole-piece system is markedly smaller than the diameter $D_2 = 11.1\ D_1$ required for the corresponding single-cone-angle $\psi_0 = 70^\circ$ pole pieces.

2) The external diameter $D_2 = 8.5\ D_1$ of the double-cone-angle pole-piece system is nearly the same as (strictly speaking even a little larger than) the diameter $D_2 = 8.1\ D_1$ already obtained with a single-cone-angle $\psi_0 = 55^\circ$ geometry.

3) The amount of magnetic flux fed into the lens core from the pole-piece system is nearly doubled for the single-cone-angle $\psi_0 = 70^\circ$ pole-piece system in comparison with both other geometries. Consequently, the single-cone-angle $\psi_0 = 70^\circ$ lens will turn out to be rather bulky compared to the other two designs studied here, because firstly, it must be equipped with a lens coil of a rather extended inner diameter, and secondly, a large cross-section is required for the lens casing in order to accommodate the heavier flux.

4) Finally, high-permeability material suitable for pole-piece construction is rather expensive, and so the quantity of material required may be of interest when deciding on a specific design. Here, the single-cone-angle pole-piece systems turn out to be definitely superior, requiring for a $\psi_0 = 55^\circ$ design only about half the material necessary for a double-cone-angle system and about two-thirds of the material required for $\psi_0 = 70^\circ$.

Taking all these points into account, the merits of the single-cone-angle $\psi_0 = 55^\circ$ pole-piece system become still more convincing. Nevertheless, employing a double-cone-angle design might be of advantage in cases of particularly high magnetic saturation.

Another essential design parameter for pole-piece systems is the aspect ratio S/D_1 of the gap. In Fig.4.33, the flux diminution in the pole pieces is shown for some aspect ratios other than $S/D_1 = 0.4$, used for Fig.4.32. Calculated and experi-

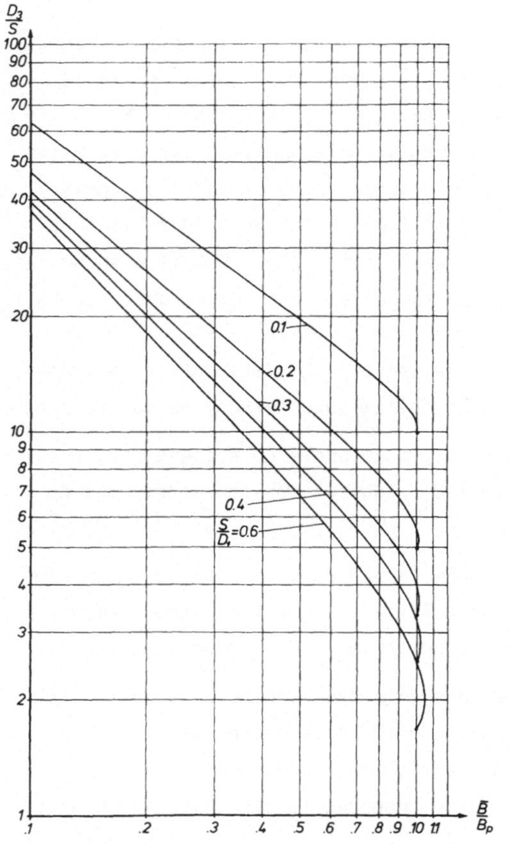

Fig.4.36. Diameter D_3 of the pole piece required for a specified diminution of the relative magnetic flux density $\bar{B}/B_p$. Here, in contrast to Fig.4.35, D_3 is shown in units of the gap width S. The aspect ratio S/D_1 is employed as the parameter characterizing the different curves, which refer to pole-piece systems of the type illustrated in Fig.4.27 having the common cone taper angle $\psi_0=55°$

mental values are shown for comparison. Again the agreement between the two sets of values is quite satisfactory.

For the design of the pole-piece system, the choice of gap width S is governed by electron optical considerations as explained in Sect.4.2. It is therefore valuable to know how the external diameter D_3 of the pole pieces can be varied by employing a suitable diameter of the pole face, both cone taper angle ψ_0 and diminution $\bar{B}_2/B_p$ of the average magnetic flux density having been specified previously. Since S is constant here, it is appropriate to represent D_3 in the normalized form D_3/S (see Fig.4.36 which shows this dependence for the favourable cone angle $\psi_0 = 55^\circ$).

Evidently, D_3/S does not change much for aspect ratios $0.2 \lesssim S/D_1 \lesssim 0.6$, whereas for a rather narrow but extended gap with $S/D_1 = 0.1$, the required pole-face diameter increases up to about twice that value. MULVEY's rule $D_1 \approx 3...4D$, for the minimum suitable size of the pole-face diameter D_1 with a pole-piece bore D [4.126] can now be expressed in the form

$$\frac{S}{D} \approx 3 \ldots 4 \frac{S}{D_1} \quad . \tag{4.76}$$

It is seen that for lens aspect ratios $S/D \gtrsim 1$, which include nearly all the pole-piece geometries used in practice, the external diameters D_3 required for the pole pieces do not differ much. Only with lens aspect ratios $S/D \ll 1$, which have been employed sometimes with projector lenses [4.176], does a larger relative pole-piece diameter D_3/D_1 become necessary.

Finally, it should be mentioned that the flux density distribution within the pole pieces may be somewhat modified by the position of the lens coil and the shape of the lens casing. This effect is quite noticeable if asymmetric lens designs are employed as with the projector lens casing shown in Fig.4.38. It is seen that the flux distribution asymmetries are more striking for rather "open" pole-piece systems with comparatively narrow conical surfaces (Fig.4.34a) than for a more "shielded" construction in which the conical surfaces extend rather far out in comparison to the height of the gap (Fig.4.34b).

In conclusion and with reference to Fig.4.23b it can be seen that the lower surface of the flat pole plate here acts essentially as a magnetic mirror surface. Therefore, the magnetic flux distribution within a conical pole piece opposite a flat pole plate can be readily calculated by adapting the methods explained above. For this calculation the flat pole plate is replaced by a "virtual" mirror image of the conical pole piece, so that the proper flux distribution is calculated if $2S'/D_1$ is used as the aspect ratio and 2NI as the magnetic potential difference between the conical pole piece and its mirror image.

4.3.2 Magnetic Design of the Lens Core

4.3.2a Some General Considerations

It is clear from the foregoing discussion that the magnetic behaviour of the pole-piece system can be predicted reliably and that its design may be based on sound principles. Therefore, the magnetic flux emanating from the pole piece and entering the lens core via the connecting surface is also a known quantity.

The designer is now faced with the problem of determining the additional flux that penetrates into the lens core through its cylindrical walls. By adding this to the flux coming from the pole piece, the total magnetic flux within the lens core and its variation in the direction of the lens axis can be established. After deciding what flux density is permissible within the lens core, the cross-section required for the core and hence its shape can be derived immediately.

To determine the field distribution around the core is a rather involved problem because of the presence of the current-carrying lens coil. Unlike the analogous problem with the pole-piece system, where the magnetic field between the pole

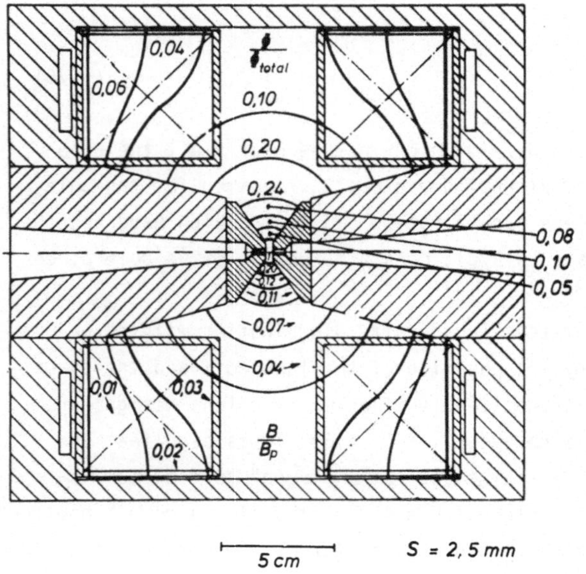

Fig.4.37. Simplified cross-section of a symmetric electron lens which is employed as a condenser-objective lens [4.157]. The continuous curves trace the path of the magnetic field lines. In the upper half of the drawing the numbers indicate the fraction ω/ω_{total} of the total magnetic flux ω_{total} passing through the respective partial volumes of the space enclosed by the casing. In the lower half of the drawing, the numbers represent the relative magnetic flux density B/B_p, where B_p denotes the nominal flux density $B_p = \mu_0 NI/S$ in the gap

pieces is determined by LAPLACE's equation, MAXWELL's equations now apply within the coil-filled space:

$$\text{curl}\ \vec{H} = \vec{j} \quad , \tag{4.77}$$

$$\text{div}\ \vec{B} = 0 \quad . \tag{4.78}$$

Here, $\vec{j}$ is the current density and $\vec{H} = \vec{B}/\mu_r\mu_0$ the magnetic field strength.

An approximate idea of the form of the field and flux distribution to be generally expected is furnished by Figs.4.37,38. Figure 4.37 shows the case of a symmetric magnetic circuit, which could typically be employed for a lens operated at comparatively high lens strength, say about NI ≈ 10000 A (e.g., a condenser-objective lens to be used at 100 to 150 kV or a lens of medium strength for a high voltage electron microscope). Figure 4.38 refers to the case of an asymmetric lens, designed to be operated at about NI = 5000 A as a projector with a pole-piece exchange turret (Sect.4.2.2c) for a 100 kV electron microscope.

In both figures, a series of solid curves indicates the path of magnetic field lines. If these curves are assumed to be rotated around the lens axis, they subdivide the volume enclosed by the lens casing into partial volumes in each of which a constant amount of magnetic flux is transported between the lens cores or between the pole pieces or between the lens cores and the external casing.

The numbers appearing in the upper half of Fig.4.37 represent the fraction ω/ω_{tota} of the total magnetic flux ω_{total} that is conveyed by the corresponding partial volume. Also, in the lower half of Fig.4.37 some values of the corresponding relative flux density B/B_p are indicated for illustration. The normalization has been effected with respect to the flux density $B_p = \mu_0 NI/S$ in the parallel part of the

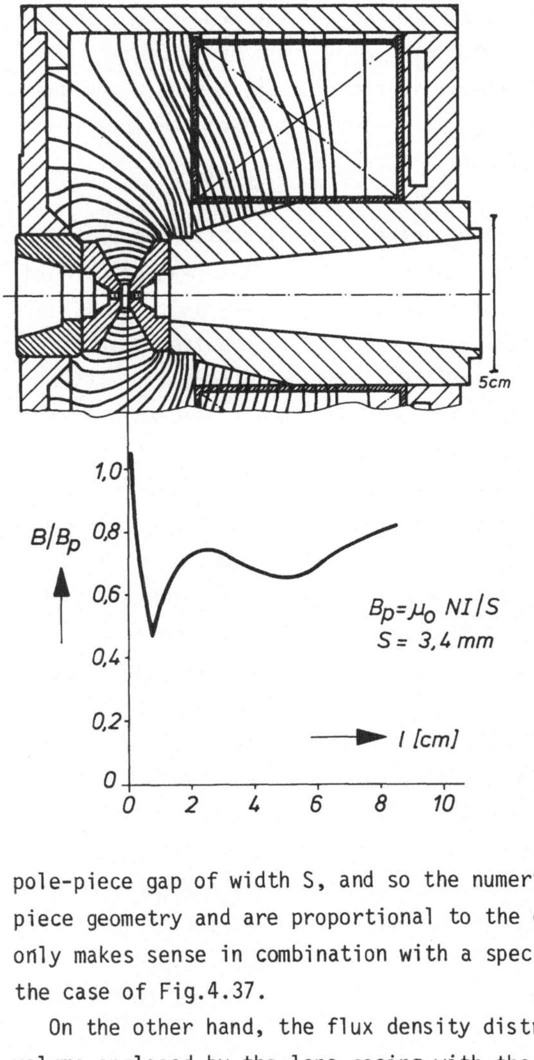

Fig.4.38. Simplified cross-section of an asymmetric electron lens which is actually equipped with a pole-piece revolver turret (Fig.4.17) and employed as a final projector [4.157]. The curves in the space enclosed by the casing indicate the path of the magnetic field lines. The rotationally symmetric partial volumes limited by two consecutive field lines each contain about 3% of the total magnetic flux generated within the lens casing. The curve below the cross-section shows the average magnetic flux density in the lens core (measured values) [4.73]

pole-piece gap of width S, and so the numerical values of B/B_p depend on the pole-piece geometry and are proportional to the gap width S. Thus, the set of B/B_p values only makes sense in combination with a specific gap width, which is S = 2.5 mm in the case of Fig.4.37.

On the other hand, the flux density distribution throughout the bulk of the volume enclosed by the lens casing with the exception of the space close to the pole-piece gap depends little on the actual pole-piece geometry. In order to obtain absolute values of the magnetic flux density, the B/B_p numbers have to be multiplied by a calculated value of $B_p = \mu_0 NI/S$, and this can be performed for any NI because S is the gap width used with the B/B_p set of numbers and must be known. This procedure is applicable everywhere except in the space close to the pole-piece gap.

So far as the magnetic flux in the partial volumes is concerned, if the total flux ω_{total} is not known by measurement or otherwise (e.g., by utilizing the methods described in Sects.4.3.2b,c), then it can be determined by first employing the procedure described in Sect.4.3.1a to calculate the magnetic flux passing through the space between the pole pieces. Dividing this by the corresponding value of ω/ω_{total}, ω_{total} is obtained immediately and so are the remaining absolute values of the magnetic flux in the partial volumes.

In the case of Fig.4.38, the partial volumes have been chosen in such a way that they carry equal amounts of magnetic flux, each conveying just 3% of the total. It was found experimentally that at NI = 5300 A, a total flux ω_{total} = 3.1 mV s = 3.1×10^5 G cm^2 is generated [4.73].

The tangential component of the magnetic field strength H is continuous at the wall of the central bore through the lens core. Therefore, the axial magnetic field that prevails throughout the lens core will also appear on the lens axis. It is common policy in strong electromagnet design to try and achieve constant magnetization and flux density throughout the lens core [4.80]. This requirement is still more essential in electron lens design because any marked change of the axial magnetic field within the ferromagnetic material of the lens core proper will be reflected in a corresponding gradient of the magnetic field in the bore and on the lens axis, thus possibly producing an unexpected spurious lens action and confusing the behaviour of the electron optical system of which the lens is part.

A bad example of this kind of disturbance is described in [4.73]. With a projector of a commercial electron microscope, a spurious reducing lens was generated in the bore of the lens core; this gave rise to smaller maximum effective magnification than the actual pole-piece system taken separately and shifted the lens strength for maximum magnification from about $NI/\sqrt{U} \approx 13$ $A/V^{1/2}$ to about $NI/\sqrt{U} \approx 10$ $A/V^{1/2}$.

In this respect the design reproduced in Fig.4.37 has proved to be quite satisfactory. The magnetic flux density was observed to change by less than about 10% throughout the whole length of the lens core, and its absolute value was found to be only a little larger than 20% of the nominal gap flux density $B_p = \mu_0 NI/S$ (with S = 2.5 mm, as explained above). With the projector lens core of Fig.4.38, the flux density is seen to vary somewhat more. This is due to the presence of the rather large conical bore, which was obligatory so that the projector could be used with large-bore pole pieces.

Finally, the choice of a suitable ferromagnetic material for the construction of the lens core can considerably affect the magnetic properties of the core and its electron optical fringe effects. The magnetic properties of some typical materials have been assembled in Fig.4.39. For pole-piece systems, it is customary and mostly advantageous to utilize rather expensive cobalt-iron alloys, which have especially high saturation magnetization, such as Vacoflux 48 or Permendur. Conversely, it would be a wrong conclusion to assume that cobalt-iron alloy would also be the best choice for the lens core. Quite the contrary! We are now concerned with a comparatively low magnetic field strength, say less than about 50 Oe (4A/mm), and the magnetic permeability μ of soft unalloyed iron is considerable larger in that range than the magnetic permeability of iron-cobalt alloys (Fig.4.39, where Hypern O is a a typical representative of an unalloyed soft iron). In other words, the same mag-

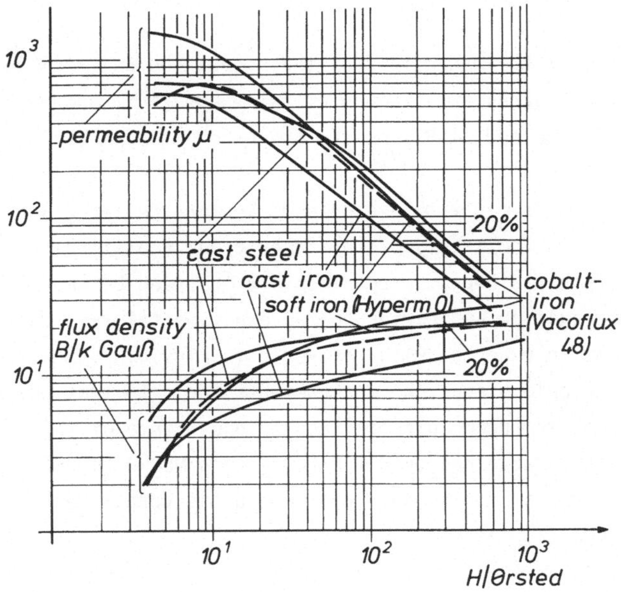

Fig.4.39. Magnetic properties of some typical commercial materials suitable for magnetic electron lens construction. Vacoflux 48 is supplied by Vacuumschmelze GmbH, Hanau (Germany), and Hyperm 0 by Krupp Widia Fabrik, Essen (Germany)

netic flux density B is now carried by a much lower magnetic field strength $H = B/\mu_r\mu_0$ in soft iron than in iron-cobalt alloy, and it is the axial magnetic field of the core that encroaches onto the lens axis region.

This leads us to the question, how much magnetic field strength should be permitted in the lens core material? It should be realized in this context that with lenses to be operated at about Φ = 100 kV, the total length L_m of the magnetic path in the pole pieces, the cores, and the shielding is about 0.3 m to 0.5 m, and that with lenses designed for megavolt-beam voltages it may be even more than an order of magnitude larger.

In order to generate a magnetic field of strength H along this path, a magnetic potential

$$(NI)_m = \frac{10}{4\pi} H L_m \approx 0.8 H L_m \tag{4.79}$$

is necessary. (Here, I_m is expressed in amperes, L_m in centimetres and H in Oersteds. If H is in A/m and L_m in m, we have $(NI)_m = HL_m$.) This magnetic potential has to be provided by the lens coil in addition to the main magnetic potential required for the gap. In the case of the 100 kV lenses, a field strength of $H \approx 10$ Oe ($\approx$ 1 A/mm), which already looks rather low, would require about $(NI)_m \approx 240$ to 400 A. This would claim roughly 5% of the nominal ampere-turns of the lens coil (about $NI \approx 5000$ to 8000 A for a lens to be employed at 100 kV). Thus, the field strength $H \approx 10$ Oe which corresponds to a flux density of 10 kG (= 1 T) in unalloyed soft iron should certainly not be exceeded in a well-designed lens core.

4.3.2b The HILDEBRANDT-SCHISKE Procedure for the Calculation of the Magnetic Flux Density in the Lens Core

In order to predict the distribution of the magnetic flux throughout the coil-filled space, two different mathematical approaches are available: fully numerical procedures based on Southwell's relaxation method [4.85] on the one hand and more analytical approaches on the other.

At first sight, the relaxation results might be expected to be more reliable and useful than the results obtained by the admittedly often rather crude analytical approach. Nevertheless, some still unexplained discrepancies are observed between flux distributions calculated by relaxation and corresponding experimental results obtained with actual lenses. Just as an example, with reference to Figs.4.37,38 we see that for the region far from the pole pieces the measurements indicate that the direction of the magnetic field tends to become parallel to the end flange of the lens casing, whereas relaxation results obtained by HESSE [4.71] and MUNRO [4.134, 135] predict more or less quarter-circular field lines centred on the corner between end flange and external cylindrical casing.

On the other hand, the predictions on the core shape obtained by utilizing the analytical approximation methods turned out to be surprisingly successful and thus quite satisfactory for practical applications. Two of these analytical methods will now be described. For a treatment of the relaxation methods the reader is referred to another part [4.85] of this book. We shall first describe the HILDEBRANDT-SCHISKE procedure [4.73].

This procedure is essentially an extension of the methods described in Sect. 4.3.1 for the calculation of the flux density traversing the space between the pole pieces, the principal notion also adopted here being the assumption of circular field lines which are centred on the centre point of the pole-piece system. This field shape suggests itself as a reasonable continuation of the field between the conical slopes of the pole-piece system into the coil-filled space. The type of lens casing considered is the symmetrical one (Fig.4.10a) and for the first step of the calculation the lens cores are assumed to be cylindrical. The diameter of the cores is equal to the diameter D_2 of the pole-piece bases, which are connected to the corresponding front faces of the cores.

A further simplification adopted for the first step of the calculation is to assume that instead of the two separate lens coils of the symmetrical design (Fig. 4.10a), one very long coil is employed which also bridges the gap between the conical slopes of the pole pieces. Despite this, the extension of the coils in the axial direction of the cores is not changed so that the position of their end faces is at the correct distance from the centre of the pole-piece system. For the external diameter of the coil, the actual external diameter planned for the twin coils is employed but only if all of the field lines considered still remain within the coil volume. If the twin coils are too slender for this, the external diameter of

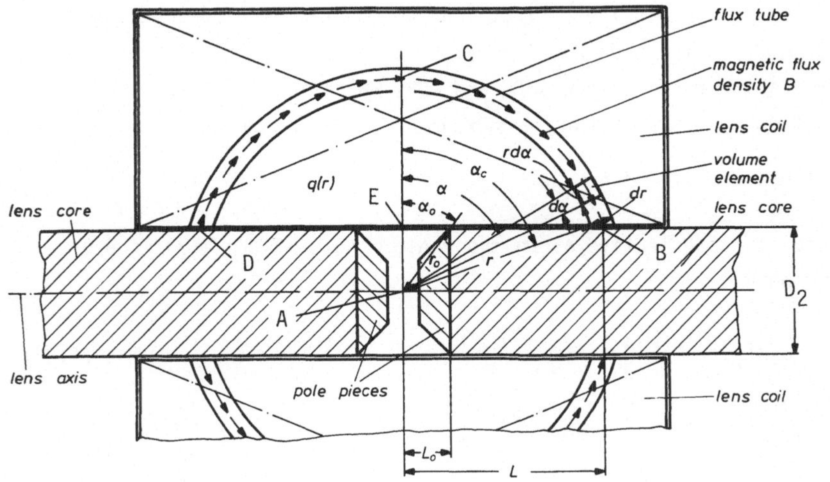

Fig.4.40. Representation of the relevant quantities for the determination of the magnetic flux density in the lens cores by means of the Hildebrandt-Schiske procedure

the substitute coil has to be increased sufficiently. It will become clear from the following that both these simplifications permit a somewhat less complicated mathematical treatment for the first step of the calculation where not much is as yet known about the actual technical shape of the cores and the coils.

The external magnetic short circuit, which is provided by the lens flanges and the outside tubular casing in the real construction, is assumed to be so far away that the arc-of-circle-shaped field distribution is not disturbed. This simplification is appropriate because only the contributions of the core flux up to the end faces of the coil are of interest, where in the actual lens the cores end, and where the flux is absorbed by flanges. Moreover, the flux contribution to the core decreases and is already comparatively small in the outermost parts of the lens coil.

The geometry on which the calculation is based is sketched in Fig.4.40. It is seen that the procedure deliberately disregards the fact that the angle of incidence of the field lines on the core should be close to 90° because of the high permeability in the core material.

In analogy with Fig.4.28 (but here with a = 0), the flux-carrying space in the coil volume is subdivided into flux tubes of nearly rectangular cross-section $r \cdot \cos\alpha \cdot d\phi \cdot dr$, which follow the path of the magnetic field lines and conduct a magnetic flux

$$d^2\omega_c = B(\alpha,r)\; r\; \cos\alpha d\phi dr \quad . \tag{4.80}$$

The magnetomotive force over the whole length of the flux tubes extending between the lens cores is equal to that fraction of the whole current NI of the coil which is situated outside the path of the corresponding flux tube. We now call Q the cross-sectional area of the whole lens coil and q(r) the area of that part of the coil cross-section which lies inside a circle having the radius r of the flux tube. In Fig.4.40, q(r) is the area $2\alpha_c r^2/2$ of the circular sector ABCD less the areas of the mirror symmetric triangles ABE and ADE, which are $D_2L/4$ in each case. With $\alpha_c = \arctan(2L/D_2)$ as half the angle subtended by the circular sector, we have

$$q(r) = \alpha_c r^2 - \frac{1}{2} D_2 L = \alpha_c r^2 - \frac{1}{4} D_2^2 \tan\alpha_c \quad . \tag{4.81}$$

The magnetomotive force effective along the whole length of the flux line of radius r is accordingly

$$\frac{r}{\mu_0} \int_{-\alpha_c}^{+\alpha_c} B(\alpha,r)\,d\alpha = NI\left[1 - \frac{q(r)}{Q}\right] \quad . \tag{4.82}$$

Substituting for $B(\alpha,r)$ from (4.80) gives

$$\frac{1}{\mu_0} \frac{d^2\omega_c}{d\phi dr} \int_{-\alpha_c}^{+\alpha_c} \frac{d\alpha}{\cos\alpha} = NI\left[1 - \frac{q(r)}{Q}\right] \quad . \tag{4.83}$$

The integral is elementary and can be solved immediately. After a second integration over the azimuth angle ϕ, we have

$$\frac{d\omega_c}{dr} = \frac{2\pi\mu_0 NI}{\ln\left(\frac{1+\sin\alpha_c}{1-\sin\alpha_c}\right)}\left[1 - \frac{q(r)}{Q}\right] \quad . \tag{4.84}$$

Here, $d\omega_c$ is the magnetic flux which passes between the cores over a "truncated" spherical shell of thickness dr. A further integration with respect to r yields the total magnetic flux $\omega_c(r)$ or $\omega_c(L)$ which passes between the lens cores up to an axial distance $L = \sqrt{r^2 - (D_2/2)^2} = r\sin\alpha_c$ from the centre of the pole-piece system. It should be kept in mind that because $D_2/2 = r\cdot\cos\alpha_c$, r and α_c are interdependent variables. Hence $dr = (D_2/2)\cdot(\tan\alpha_c/\cos\alpha_c)\cdot d\alpha_c$ and so the final integration may be expressed in two forms:

$$\omega_c(\alpha_c) = 2\pi\mu_0 NI \int_{\alpha_0}^{\alpha_c} \frac{1}{\ln\left(\frac{1+\sin\alpha_c}{1-\sin\alpha_c}\right)}\left[1 - \frac{D_2^2}{4Q}\left(\frac{\alpha_c}{\cos^2\alpha_c} - \tan\alpha_c\right)\right]\frac{\tan\alpha_c}{\cos\alpha_c}\,d\alpha_c \quad , \tag{4.85}$$

$$\omega_c(r) = 2\pi\mu_0 NI \int_{r_0}^{r} \frac{1 - \frac{r^2}{Q}\left[\arccos\frac{D_2}{2r} - \frac{D_2}{2r}\cdot\sqrt{1 - \left(\frac{D_2}{2r}\right)^2}\right]}{\ln(8r^2 - D_2^2 + 4r\sqrt{4r^2 - D_2^2}) - \ln D_2^2}\,dr \quad . \tag{4.86}$$

Recalling that $L = (D_2/2)\tan\alpha_c = \sqrt{r^2 - (D_2/2)^2}$, the results obtained from (4.85, 86) can easily be used to express the flux $\omega_c(L)$ passing through the walls of the cores as a function of the axial distance L from the pole-piece centre.

Obviously, the total flux $\omega_{c,total}$ can then be determined to a first approximation by adding the flux ω_c coming from the other core to the flux ω_p passing through the pole-piece base which is fitted to the core face:

$$\omega_{c,total} = \omega_c + \omega_p \quad . \tag{4.87}$$

The average magnetic flux density $B_c(L)$ at the distance L from the centre of the pole-piece system is accordingly given for the cylindrical core of diameter D_2 by

$$B_c(L) = \frac{\omega_{c,total}(L)}{\frac{\pi}{4}\, D_2^2} \quad . \tag{4.88}$$

If it is now stipulated, e.g., as the result of considerations described at the end of Sect.4.3.2a, that the average magnetic flux density should nowhere in the core exceed the critical value $B_{c,max}$, then the local core diameter must be increased accordingly to

$$D_c(L) = D_2 \sqrt{\frac{B_c(L)}{B_{c,max}}} = \sqrt{\frac{4\omega_{c,total}(L)}{\pi B_{c,max}}} \quad . \tag{4.89}$$

A preliminary design of the magnetic circuit of the lens can now be established, in which the technical shape of the core is based on the $D_c(L)$ values obtained from (4.89). The lens coil should have its proper size and shape and should be arranged in its prospective final position.

As the second step of the procedure and in obvious analogy with the first calculation just described, the amount of flux entering the revised core from the coil space is now obtained by integrating (4.84). Here, α_c, which is proportional to the length $2\alpha_c r$ of the flux tube, is still related to the axial coordinate L by $\alpha_c = \arctan[2L/D_c'(L)]$ where $D_c'(L) \gtrsim D_c(L)$ is the local diameter of the core of the preliminary design. $r = [L^2 + (D_c'(L)/2)^2]^{\frac{1}{2}}$ is the bending radius of the flux tubes which touch the cores at an axial distance L from the mid-plane of the pole-piece system.

Now, L is a more natural independent variable than r for describing the increase of the magnetic flux in the lens cores, and it is also a more convenient variable for integration, for both α_c and r can be expressed explicitly as functions of L while the converse can be achieved neither for α_c nor for r. After some elementary transformations, (4.84) can be written in integral form:

$$\omega_c(L) = 2\pi\mu_0 NI \int_{L_0}^{L} \frac{\left(2L + \frac{1}{2} D_c' \cdot \frac{dD_c'}{dL}\right)\left(1 - \frac{q(L)}{Q}\right) dL}{[\ln(8L^2 + D_c'^2 + 4L\sqrt{4L^2 + D_c'^2}) - \ln D_c'^2]\sqrt{4L^2 + D_c'^2}} \quad . \tag{4.90}$$

Here, q(L) is that part of the cross-sectional area of the lens coil which remains inside the path of the flux tube of radius $r(L) = \sqrt{L^2 + (D_c'(L)/2)^2}$. It should be realized that in a technical design the position and shape of the coil cannot in general be described in a mathematically simple form, and the same is even more true of q(L). In particular, (4.81) can be used only for the special case of a long continuous coil, and this coil shape has been selected for the first "run" of the procedure with just that intention. In order to obtain numerical values for q(L) with technical coil shapes, a graphical method which employs a cross-sectional drawing of the preliminary design can be a good approach with sufficient accuracy for practical purposes.

After numerical values have been obtained for the integral (4.90), they can be employed with (4.87-89) to obtain a second approximation to the technical core shape, and if necessary, corresponding changes are applied to the design. The procedure can be repeated until the core shape has stabilized.

In the above, the form of the HILDEBRANDT-SCHISKE procedure applicable to a symmetric lens design as in Fig.4.10a has been described. The procedure can also be applied to asymmetric lenses (Fig.4.10b), for here the lower surface of the upper lens flange can be regarded as a magnetic mirror. The roughly quarter-circular flux lines arrive at the surface of the flange nearly orthogonally because of the high permeability of the flange material, and thus the actual field distribution of the asymmetric lens is completed by its mirror image to create practically the same type of distribution as would arise with the corresponding symmetrical lens. Carrying through the calculations of the procedure with the substitute symmetrical lens model, the double amount 2NI of the actual ampere-turns NI of the asymmetric lens must be employed. The axial point of the mirror surface takes the place of the centre of the pole-piece system of the symmetrical lens as the centre of curvature of the quarter-circular flux tubes.

HILDEBRANDT and SCHISKE have performed numerical investigations on the cores of the lenses represented in Figs.4.37,38 and have checked their results experimentally (for details, see [4.73]). As a consequence of their findings, and if it is stipulated that the magnetic flux density in the core should remain approximately constant, the following rules emerge.

1) The cross-section of the lens core should increase with increasing distance from the core face.

2) The required rate of increase of the cross-section becomes smaller with increasing distance from the pole-piece system, and the first "run" calculation indicates that the core should almost look like a truncated parabola.

3) A design more suitable for fabrication in the workshop than the parabola of rotation is to employ a conical surface for that part of the core which is close to its face and to pass over to a cylindrical surface for those parts of the core that are more distant from the pole- piece system.

4) The angle of taper of the conical surface should be about 15°.

5) The conical surface should end and pass over into the cylindrical one after the cross-sectional area of the core has attained about three times the area

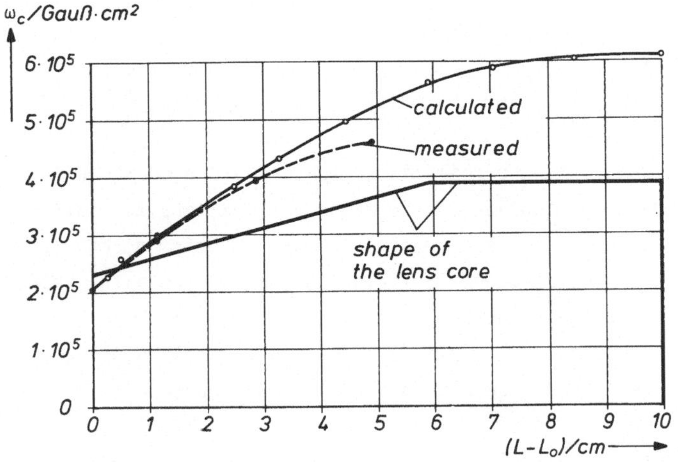

Fig.4.41. Shape of lens core for a symmetrical electron lens according to Fig. 4.37, determined by employing the set of rules resulting from the HILDEBRANDT-SCHISKE procedure. Calculated and measured values of the flux distribution in the core are also shown for comparison [4.73]

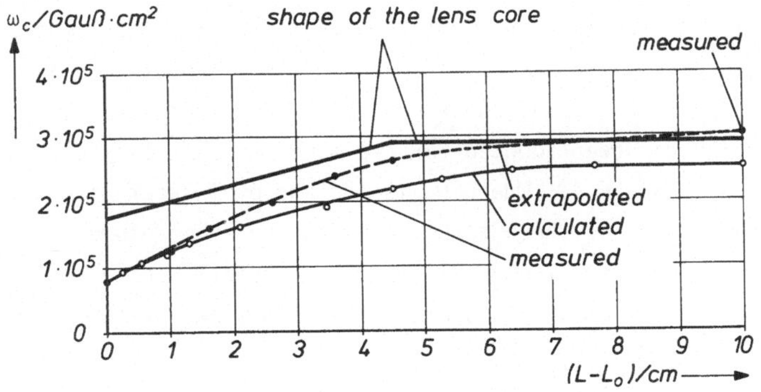

Fig.4.42. Shape of the lens core for an asymmetric electron lens according to Fig.4.38, determined by employing the set of rules resulting from the HILDEBRANDT-SCHISKE procedure. Calculated and measured values of the flux distribution in the core are also shown for comparison [4.73]

of the front face of the core. This rule is advantageous for keeping the internal diameter of the coil as small as possible and thus reducing electrical resistance and thermal losses within the lens coil.

For the lenses represented in Figs.4.37,38, HILDEBRANDT and SCHISKE have employed the second "run" procedure to investigate the magnetic properties of the actual technical shape of the cores with the coils in their proper position. The calculated values and corresponding results from measurements have been plotted in Figs. 4.41,42 and show satisfactory agreement. It can thus be expected that generally, the flux in the lens cores can be predicted well enough by this method for technical purposes.

4.3.2c A Perturbation Method for Obtaining the Magnetic Flux Distribution in the Core and Casing of a Magnetic Electron Lens

This method essentially continues into the coil-filled space the procedure described in Sect.4.3.1a for the determination of the magnetic flux density distribution in the space between the pole pieces, especially between their conical surfaces. It has been shown that there the magnetic field lines travel on paths which are nearly arcs of circles having their centres of curvature close to the center of the pole-piece system. The assumption of circular field lines was also utilized with satisfactory practical results for the HILDEBRANDT-SCHISKE procedure explained in Sect.4.3.2b.

Nevertheless, it is obvious that circular field lines cannot give a completely true picture of the flux distribution in the coil-filled space because some basic physical laws are violated: for example, concerning the angle of incidence of the field lines on the high permeability cores, as mentioned above. Another such violation will be discussed shortly. The following treatment aspires to achieve a more rigorous solution of the problem [4.168].

For this purpose, as a natural continuation of the double-cone pole-piece system model, a double-cone lens-core model suggests itself. In order to complete the iron circuit in a manner well suited to the use of spherical polar coordinates r, θ, ϕ, a spherical external lens casing has been adopted for the model lens. This yields the spherical lens model shown in Fig.4.43, where θ_0 is the angle of taper of the conical cores and r_a the inner radius of the external lens casing. The volume of the casing is assumed to be filled by the lens coil, with the exception of a small sphere of radius r_i around the centre of the lens where the pole-piece system is situated.

The field distribution in the coil-filled space is determined by MAXWELL's equations (4.77), which, if written in spherical polar coordinates, have the form

$$\mathrm{curl}_r\,\vec{H} = \frac{1}{r\cdot\sin\theta}\left[\frac{\partial}{\partial\theta}\,(\sin\theta\cdot H_\phi) - \frac{\partial H_\theta}{\partial\phi}\right] = 0 \quad , \tag{4.91}$$

$$\mathrm{curl}_\theta\vec{H} = \frac{1}{r}\left[\frac{1}{\sin\theta}\,\frac{\partial H_r}{\partial\phi} - \frac{\partial}{\partial r}\,(r\cdot H_\phi)\right] = 0 \quad , \tag{4.92}$$

$$\mathrm{curl}_\phi\vec{H} = \frac{1}{r}\left[\frac{\partial}{\partial r}\cdot(rH_\theta) - \frac{\partial H_r}{\partial\theta}\right] = j_\phi \quad . \tag{4.93}$$

Here, we have included the fact that in our case, only the azimuthal component j_ϕ of the current density $\vec{j}$ is different from zero: $j_r = j_\theta = 0$.

Now, the azimuthal component H_ϕ of the magnetic field strength must vanish identically as a consequence of AMPERE's law, i.e., $\oint H_\phi\cdot r\cdot d\phi = J_z$, because no axial

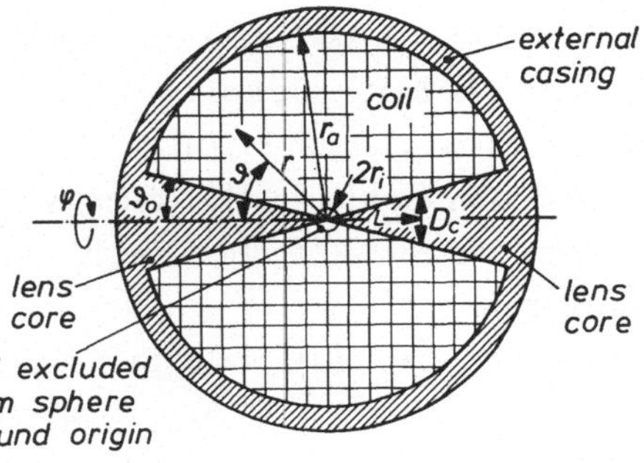

Fig.4.43. Spherical-shell lens model employed for calculating the magnetic flux and the flux density in lens cores and casing

component J_z of current exists, so $H_\phi = 0$. Moreover, because of the rotational symmetry of the lens around the polar axis, all the partial derivatives with respect to ϕ must vanish. Thus, only (4.93) remains to be solved.

If the circular field-line model were true, then the radial component H_r of the field strength would vanish throughout the coil space, i.e., we should have $H_r = 0$. With this assumption, (4.93) becomes an ordinary differential equation of first order in r which can be solved directly. The integration constant can be determined from the boundary condition $H_\theta(r_a) = 0$. Speaking in physical terms, this means that the tangential component $H_\theta(r_a)$ must vanish at the surface of the external spherical casing because of its high permeability. The result is

$$H_\theta = \frac{r}{2}\, j_\phi \cdot \left(1 - \frac{r_a^2}{r^2}\right) \quad . \tag{4.94}$$

On the other hand, for the coil-filled space where $\mu_r = 1$, the second MAXWELL equation (4.78) is identical with $\operatorname{div} \vec{H} = 0$ and can be written in spherical polar coordinates as

$$\operatorname{div} \vec{H} = \frac{1}{r^2}\frac{\partial}{\partial r}(r^2 H_r) + \frac{1}{r\cdot\sin\theta}\frac{\partial}{\partial\theta}(\sin\theta\cdot H_\theta) + \frac{1}{r\cdot\sin\theta}\frac{\partial H_\phi}{\partial\phi} = 0 \quad . \tag{4.95}$$

Putting as before $H_r \equiv 0$ and $H_\phi \equiv 0$, the solution is straightaway seen to be $\sin\theta\cdot H_\theta = \text{const}$. Adopting for boundary condition that on the surface of the core the product $\sin\theta_o \cdot H_{\theta_o}$ must be equal to the same constant, the following dependence of H_θ on θ results:

$$H_\theta = H_{\theta_o}\frac{\sin\theta_o}{\sin\theta} \quad . \tag{4.96}$$

This dependence on θ is not compatible with (4.93) if $H_r \equiv 0$. In particular, it is easily seen by substitution that

$$H_\theta = \frac{r}{2}\, j_\phi \cdot \left(1 - \frac{r_a^2}{r^2}\right)\frac{\sin\theta_o}{\sin\theta} \tag{4.97}$$

is not an acceptable solution.

It is thus again evident that circular field lines cannot represent the true field configuration in the coil-filled space, even in a special geometry where they would be particularly likely to arise. On the other hand, it was explained above that there are theoretical arguments and experimental evidence suggesting that with symmetrical iron circuits as in Fig.4.10a, the assumption of circular field lines cannot be completely wrong, at least in the region of higher flux density in the neighbourhood of the pole-piece system. Thus, the radial field strength H_r cannot vanish completely but will probably remain small.

It will now be explained that by adding a small perturbational field to a generally much stronger tangential field H_θ of the type described by (4.97), simultaneous solution of both MAXWELL equations (4.77,78) can be achieved. In this perturbation stage, a radial field is generated which is small in comparison with the tangential field throughout the bulk of the coil volume. Complementing this, a small diminution of the tangential field arises. The gist of the perturbation method will now be outlined briefly.

As the starting point, and employing the general form of (4.97), the tangential component of the magnetic field strength can be written in the form

$$H_\theta(r,\theta) = \frac{r}{2}\, j_\phi \cdot \left(1 - \frac{r_a^2}{r^2}\right) \cdot \frac{\sin\theta_0}{\sin\theta}\, K \quad . \tag{4.98}$$

Substituting this into (4.93), a first-order partial differential equation in θ for H_r results, which can be integrated at once:

$$H_r(r,\theta) = r j_\phi \cdot \left[\sin\theta_0 \cdot K \cdot \ln \frac{\tan(\theta/2)}{\tan(\theta_0/2)} - (\theta - \theta_0)\right] \quad . \tag{4.99}$$

In this equation the constant of integration has been chosen so that on the surface of the core the field component $H_r(\theta_0)$ parallel to this surface vanishes.

In (4.98,99), K is a calibration factor and can accordingly be determined by applying AMPERE's law:

$$\oint H ds = (1 - q)NI = (1 - q) j_\phi \cdot \left(\frac{\pi}{2} - \theta_0\right)(r_a^2 - r_i^2) \quad .$$

Here, the integration has to be performed along a closed path within the coil volume, and q is that fraction of the cross-sectional area of the coil which remains outside the path of integration. (In principle, the integration path is arbitrary, but the origin of the polar coordinate system must remain external to the area surrounded by the path of integration because there, H_θ has a singularity and becomes infinitely large.)

Thus, as equations describing a magnetic field distribution that obeys the MAXWELLian curl equations (4.91-93) and also AMPERE's law (with $r_i \ll r_a$ for simplicity), we obtain

$$H_\theta(r,\theta) = -j_\phi \cdot \frac{r}{2}\left(1 - \frac{r_a^2}{r^2}\right) \cdot \frac{\pi/2 - \theta_0}{\ln(\tan(\theta_0/2))} \cdot \frac{1}{\sin\theta} \qquad (4.100$$

$$H_r(r,\theta) = -\ j_\phi r\left[(\pi/2 - \theta_0)\frac{\ln(\tan(\theta/2))}{\ln(\tan(\theta_0/2))} - (\pi/2 - \theta)\right] \quad . \tag{4.101}$$

It is easy to verify that all the components of the magnetic field that are parallel to the surfaces of the high permeability ferromagnetic parts of the lens vanish, i.e., $H_r(\theta_0) = 0$, $H_r(\pi - \theta_0) = 0$, and $H_\theta(r_a) = 0$. Moreover, as is to be expected from the symmetry of the casing of the model lens, at constant radial distance r and for points having the same spacing from the central symmetry plane $\theta_s = \pi/2$, we obtain $H_\theta(r,\theta) = H_\theta(r,\pi - \theta)$ and $H_r(r,\theta) = -H_r(r,\pi - \theta)$: the path of the magnetic field lines is indeed symmetric with respect to the midplane $\theta_s = \pi/2$.

On the other hand, the field $\vec{H} = (H_r,H_\theta)$ described by (4.100,101) is not compatible with the second MAXWELL equation (4.78) or (4.95), for we obtain

$$\mathrm{div}\,\vec{B} = \mu_0\,\mathrm{div}\,\vec{H} = -3\mu_0 j_\phi\left[(\pi/2 - \theta_0)\frac{\ln(\tan(\theta/2))}{\ln(\tan(\theta_0/2))} - (\pi/2 - \theta)\right] = \hat{\rho} \quad .$$

This formally describes the presence of a magnetic space charge $\hat{\rho}$ which, of course, is a physical impossibility. In any case, div $\vec{B}$ not being zero is an indication that the field described by (4.100,101) is not a pure curl field.

But there is a remedy for this defect, which consists in formally introducing a compensating space charge $\hat{\rho}^* = -\hat{\rho}$. The field $\vec{H}^*$ generated by this charge is added to the field $\vec{H}$, so that the resulting field $\vec{H} + \vec{H}^*$ obeys the second MAXWELL equation $\mathrm{div}(\vec{H} + \vec{H}^*) = 0$. On the other hand, as the additional field $\vec{H}^*$ is assumed to be generated by a fictitious magnetic space charge, it can be described formally as the gradient $\vec{H}^* = -\mathrm{grad}\,U^*$ of a rotationally symmetric scalar potential field $U^*(r,\theta)$.

It is well known that a vector field which is the gradient of a scalar potential is a curl-free field: curl grad $U^* = -\mathrm{curl}\,\vec{H}^* = 0$. So, after adding the space charge field $\vec{H}^*$ to the original vector field $\vec{H}$ of (4.100,101), the first MAXWELL equation, (4.91-93), is still satisfied.

Moreover, U^* being a conservative potential field, the integral $\oint H^*\,ds$ taken over any closed path of integration vanishes. Therefore, when calibrating the factor K in (4.98,99) by employing AMPERE's law, the result of the calibration is not changed.

Finally, provided that the general properties of the field $\vec{H}^*$ of the compensating space charge $\hat{\rho}^*$ discussed above are respected, the actual distribution of the field $\vec{H}^*$ can be adapted to the physical requirements by stipulating suitable boundary conditions. In order to find these, we observe that the combined field $\vec{H} + \vec{H}^*$ must be orthogonal to the surfaces of the highly permeable ferromagnetic parts of the iron circuit. This is already a property of the original field $\vec{H}$ (4.100,101), and so the same is necessarily true of the compensating field $\vec{H}^*$. The required orthogonality can evidently be accomplished if, for the compensating potential U^*, the surfaces of the ferromagnetic parts are assumed to be equipotential surfaces; U^* and hence $\vec{H}^*$ can then be determined by conventional methods such as relaxation.

In this way, the distribution of the pure curl field $\vec{H} + \vec{H}^*$ can be obtained for the spherical-shell lens model. The field distribution $\vec{H} + \vec{H}^*$ is the rigorous solution to the problem of field determination, because it obeys both MAXWELL equations (4.77,78) and has for boundary condition the orthogonality of the field to all of the ferromagnetic surfaces. It is evident that this method can also be adaptec to other lens casing model configurations and other types of coordinate systems if a field distribution that fulfils the first MAXWELL equation (4.77) and the respective boundary conditions can be found as a starting point.

In the present case, for the practical purpose of predicting the magnetic flux in the different sections of the magnetic circuit and for determining the required cross-sectional areas of lens core and casing, considerable simplifications can be applied. For this, use is made of the fact that H_r, as given by (4.101), is smaller by about two orders of magnitude than the value which H_θ assumes throughout the bulk of the coil volume (4.100). This can easily be seen from

$$\frac{H_r}{H_\theta} = \frac{1}{1 - r_a^2/r^2} \cdot 2 \sin\theta \cdot \ln(\tan(\theta_0/2)) \cdot \left(\frac{\ln(\tan(\theta/2))}{\ln(\tan(\theta_0/2))} - \frac{\pi/2 - \theta}{\pi/2 - \theta_0}\right) ,$$

in which the contribution depending on the angles θ and θ_0 amounts to a few percent only for cone angles $\theta \geq \theta_0 \gtrsim 45^\circ$, and $\theta_0 \gtrsim 45^\circ$ is the range of core angles employed in actual lens core design. It is also evident that the values of the components H_r^* and H_θ^* of the compensating field must be smaller than or at most just about equal to the average value of H_r. There is thus an order of magnitude difference between the flux carried by H_θ on the one hand and H_r, H_r^* and H_θ^* on the other.

In a first approximation for a lens design, H_r^* and H_θ^* can be disregarded and the calculation of the cross-sectional areas of the iron circuit can be based on H_θ and H_r alone. As a "reward" for making a half-way correct guess about the field distribution H_θ, H_r employed in the first step of the perturbation calculation, it is not at all necessary in practice to determine the shape of the correction field U^*. This saves much effort and makes the application of the method comparatively simple and straightforward, as will become evident from the examples described in the following section.

Figure 4.44 shows an example of the type of field distributions H_θ, H_r described by (4.100,101). The rather low cone angle $\theta_0 = 15^\circ$ has been chosen in order to illustrate clearly the deviation of the field lines from the right circular shape. When employing larger cone angles $\theta_0 \gtrsim 45^\circ$ as required for actual design applications, the approximately circular path of the field lines prevails still farther out towards the spherical shell. When the field lines of Fig.4.44 are rotated around the lens axis, they subdivide the coil space into partial volumes. The particular field lines shown in Fig.4.44 have been chosen in such a way that each of the partial volumes carries about 5% of the magnetic flux passing through the symmetry plane $\theta = \pi/2$ beyond $r = 0.2\, r_a$.

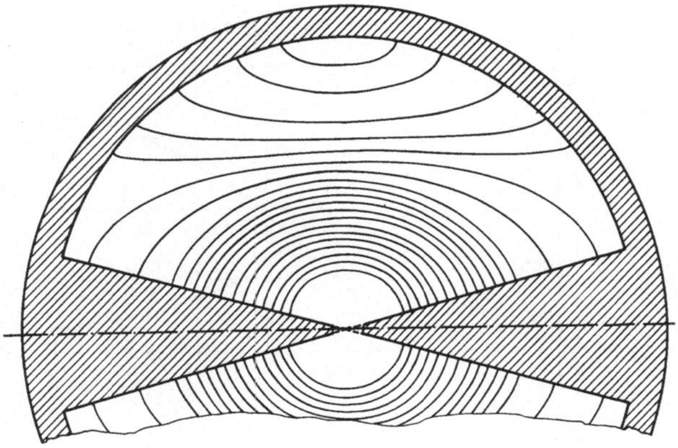

Fig.4.44. First-approximation field distribution in the spherical-shell model lens as described by (4.100,101) for a core half-angle $\theta_0 = 15°$

4.3.2d Determination of the Cross-Sectional Area of the Lens Core by Means of the Perturbation Method

It is immediately obvious that in the peripheral regions of the coil volume the magnetic flux density is comparatively small, only about 10% of the flux density of the innermost field lines shown. It can therefore be expected that on blowing the external casing up into the more usual can-shaped form, the flux distribution in the centre plane and the flux input into the cores will be essentially preserved. Predictions about these quantities obtained from (4.100,101) can therefore be expected to picture the actual physical situation with an accuracy satisfactory for practical applications.

Now, utilizing (4.100), the magnetic flux entering the conical cores from radius r_i outwards can be determined straightaway by integration:

$$\omega_c(r,\theta_0) = -\mu_0 \frac{\pi NI}{\ln(\tan(\theta_0/2))} \frac{r - r_i}{r_a^2 - r_i^2}\left[r_a^2 - \frac{1}{3}(r^2 + rr_i + r_i^2)\right] . \qquad (4.102)$$

The corresponding average flux density within the core is

$$B_c(r,\theta_0) = -\mu_0 \frac{NI}{\sin^2\theta_0 \cdot \ln(\tan(\theta_0/2))} \frac{r - r_i}{r^2(r_a^2 - r_i^2)}\left[r_a^2 - \frac{1}{3}(r^2 + rr_i + r_i^2)\right] . \qquad (4.103)$$

The expressions containing θ_0 are shown as curves in Fig.4.45; they determine how the magnetic flux and the flux density can be changed by varying the angle θ_0 of the taper of the cores. It is remarkable that the flux input ω_c changes rather slowly throughout the range of θ_0 values up to $\theta_0 \approx 50°$. As a rule of thumb, it can be said that in this range the flux increases by about 10% for an angular increment of $\Delta\theta_0 = 5°$. The average magnetic flux density B_c in the cores has its minimum at $\theta_0 = 49.6°$, but this minimum is rather broad, and a low and only slightly higher value of B_c is obtained throughout a wide θ_0 range, say between $\theta_0 \approx 40°$ and $\theta_0 \approx 60°$.

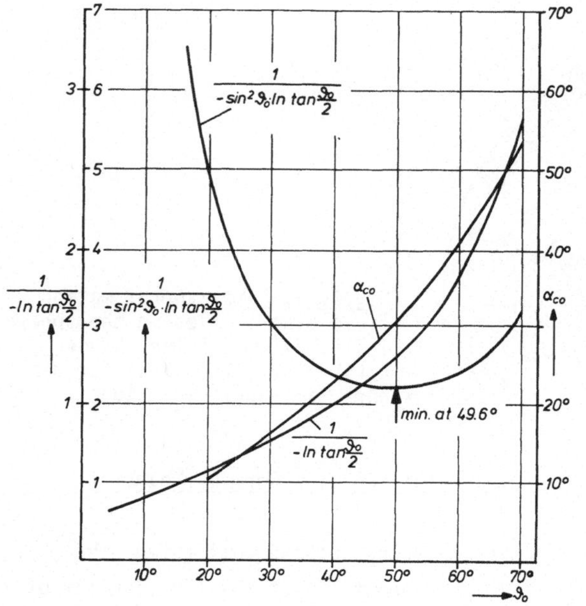

Fig.4.45. Dependence on the core half-angle θ_0 of the magnetic flux and the flux density in the cores of the spherical-shell model lens (4.102-105)

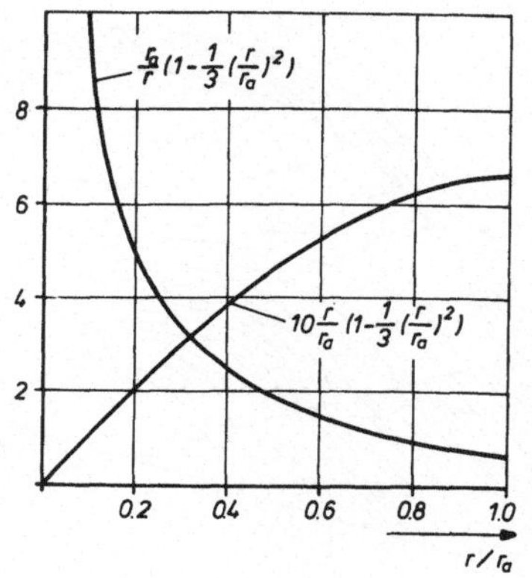

Fig.4.46. Radial dependence of the magnetic flux and the flux density in the cores of the spherical-shell model lens

In order to investigate the general trend of the radial variation of ω_c and B_c, it will now be assumed that the coil-free space is rather small so that $r_i \ll r_a$. Then (4.102,103) can be simplified to

$$\omega_c(r,\theta_0) \approx -\mu_0 \frac{\pi NI\, r_a}{\ln(\tan(\theta_0/2))} \cdot \frac{r}{r_a}\left[1 - \frac{1}{3}\left(\frac{r}{r_a}\right)^2\right] \quad . \tag{4.104}$$

$$B_c(r,\theta_0) \approx -\mu_0 \frac{NI}{r_a \sin^2\theta_0 \cdot \ln(\tan(\theta_0/2))} \frac{1 - \frac{1}{3}(r/r_a)^2}{\frac{r}{r_a}} \quad . \tag{4.105}$$

Here, the radial variations are expressed in a normalized form, depending only on r/r_a. These normalized functions are represented in Fig.4.46. It is seen that the magnetic flux density B_c decreases rapidly with increasing r/r_a, and this diminution only levels off rather close to the end of the core. On the other hand, the magnetic flux ω_c increases nearly in direct proportion to r/r_a at the lower r/r_a values and rather more slowly as the end of the core is approached. The cross-section of the conical core increases in proportion to $(r/r_a)^2$, and this evidently outweighs the slower flux intake and produces the rapid diminution of the average flux density in the core.

Clearly, employing a right conical core shape is not a particularly satisfactory design because the core will then have too large a cross-section and use up too much material at its base. Therefore, a reduction of the cross-section of the

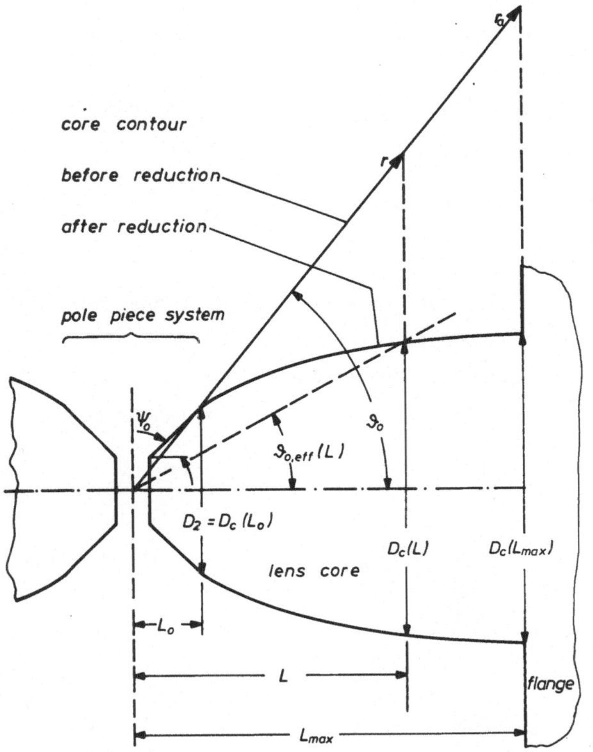

Fig.4.47. Schematic picture of quantities and definitions employed when calculating the lens core contour by applying the spherical-shell lens model in combination with the perturbation method

outermost parts of the core is called for, leading to a core shape more akin to a paraboloid of revolution.

The advantages of having constant (and low enough) flux density in the cores have been explained in Sect.4.3.2a. In order to achieve this aim, the above results concerning the spherical lens model can be employed. According to (4.102,104), the flux input to the core does not diminish drastically with a moderate reduction of the core taper angle θ_0. Evidently, there will be even less diminution of the flux input if only the more outlying parts of the core are reduced in their local diameter D_c and hence in effective core angle $\theta_{0,eff} = \arctan(D_c/2L)$ (Fig.4.47), while a larger angle of core taper is retained for the parts of the core close to the lens centre.

For practical design applications, it can therefore be stipulated that the magnetic flux in the cores is more or less independent of the angle $\theta_{0,eff}$ and – within rather wide limits – of the local diameter D_c of the core. We can thus determine the proper core shape required for constant flux density by reducing the cross-section of the core until the same flux density is obtained throughout the whole length of the core. Obviously, this flux density must also be equal to the average density $B_c(L_0)$ of the flux $\omega_p(L_0)$ prevailing in the base planes of the pole-piece system and thus also in the adjoining front faces of the cores, which are at a distance L_o from the centre of the lens.

The local diameter $D_c(L)$ of the constant flux density core shape is then determined by

$$\frac{\omega_{c,total}(r,\theta_o)}{\pi[D_c(L)/2]^2} = \frac{\omega_p(L_o)+\omega_c(r,\theta_o)}{\pi[D_c(L)/2]^2} = B_c(L_o) = \frac{\omega_p(L_o)}{\pi[D_c(L_o)/2]^2} \quad , \tag{4.106}$$

where $D_c(L_o)$ is the diameter of the front face of the core. $\omega_p(L_o)$ is the flux which originates in the pole-piece system and enters the core via its front face. $\omega_c(r,\theta_o)$ is the flux which passes into the core through its conical wall and up to a distance r from the lens centre. $\omega_{c,total}(r,\theta_o)$ is the flux taking both these contributions together.

From (4.106) it now follows that

$$D_c(L) = D_c(L_o)\sqrt{\frac{\omega_{c,total}(r,\theta_o)}{\omega_p(L_o)}} = D_c(L_o)\sqrt{1+\frac{\omega_c(r,\theta_o)}{\omega_p(L_o)}} \quad . \tag{4.107}$$

L can be determined from

$$L = r\cos\theta_o \quad , \tag{4.108}$$

an equation which corresponds to the right conical core shape before reduction. On the other hand, it can be seen in Fig.4.44 and also derived from (4.101) that in the space around the cores the path of the magnetic field lines is bent towards the lens axis. Therefore, the magnetic flux tube which impinges on the non-reduced surface at a distance r from the lens centre will now touch the surface of the reduced core at a somewhat smaller distance (Fig.4.47). This, in combination with the diminution of the polar angle $\theta_{o,eff} < \theta_o$ of the point of incidence on the reduced core, would actually give rise to a distance from the lens centre of the corresponding cross-sectional plane of the core which is a little larger than the distance L given by (4.108). To determine exactly how much would be a rather involved problem that could probably only be solved by numerical methods such as relaxation. This would go beyond the scope of the present treatment, which aims at providing easy-to-use formulae, suitable for actual design engineering. Moreover, the correction would remain small, and in any case, employing (4.106) means remaining on the safe side with the design. The model situation adopted here for the pole-piece region is outlined in Fig.4.47. The cone-taper angle θ_o of the core shape prior to reduction is the angle under which the radius $D_2/2$ of the pole-piece base is seen from the lens centre, i.e., $\theta_o = \arctan(D_2/2L_o)$. The flux in the lens core is composed of the sum of the flux ω_p coming from the pole-piece system and entering the core via its front face and the flux $\omega_c(r,\theta_o)$ traversing the coil space and entering the core via its conical wall.

In the model configuration, $r_0 = \sqrt{L_0^2 + (D_2/2)^2}$ is the distance of the circumference of the pole-piece base from the lens centre. We shall assume for the model that the lens coil begins at this radius, so that $r_i = r_0$, a situation similar to the geometry usually employed in actual lens design. The flux $\omega_c(r,\theta_0)$ which would pass through the conical surface of the original core can then be calculated by employing (4.102).

The flux $\omega_p(L_0)$ coming from the pole-piece base can be determined separately by utilizing the procedure described in Sect.4.3.1a, but a more simple and convenient calculation can be used here. For this, we go back to the double-cone pole-piece model with coincident cone tips of Sect.4.3.1a. Referring to (4.58), it is seen that the magnetic flux density on the conical surfaces $\theta = \theta_0$ is given by

$$B_\theta = -\frac{\mu_0}{r \sin\theta_0} \cdot \frac{NI}{2 \ln(\tan(\theta_0/2))} \quad . \tag{4.109}$$

From this, by straightforward integration, the magnetic flux in the conical pole piece is found to be

$$\omega_{cp}(r,\theta_0) = -\frac{\mu_0 \pi NI}{\ln(\tan(\theta_0/2))} \cdot r \quad . \tag{4.110}$$

Here, all of the flux that passes through the conical surface of the pole-piece up to a distance r from the centre of the pole-piece system has been taken into account

Now, with the actual shape of pole-piece systems and for the purpose of establishing the core contour, the double-cone model may be expected to be a good enough approximation for the pole pieces and for the determination of the magnetic flux in their end faces, which connect up to the cores. This appears rather plausible from Figs.4.48,49, in which the pole-piece and core contours are depicted for two lenses that have actually been constructed. In accordance with the notion of conical surfaces, the pole piece ends at the external rim of the cone and r_0 has to be measured accordingly; the sections PP' in Figs.4.48,49 have to be considered as the end faces of the respective pole-piece systems. It does not matter in this context that in actual practice the pole pieces are usually fabricated with a short cylindrical continuation serving as seat for fitting the pole piece into a nonmagnetic spacer. This is also the case with Figs.4.48,49, where the core end faces are in the planes QQ' and the cylindrical continuations between PP' and QQ' have to be considered as already being parts of the cores. To a sufficiently good approximation for the present purpose therefore, $\omega_{cp}(r_0,\theta_0)$ given by (4.110) can be employed for the magnetic flux in the effective end face of a pole piece. On the other hand, for a spherical coil-filled model lens with a right conical core, which has the same taper angle θ_0 and a coil extending nearly right up to the centre, the flux $\omega_c(r_0,\theta_0)$ accepted by the cores up to a radius r_0 can be calculated using (4.104). On comparing these equations, it is readily seen that they

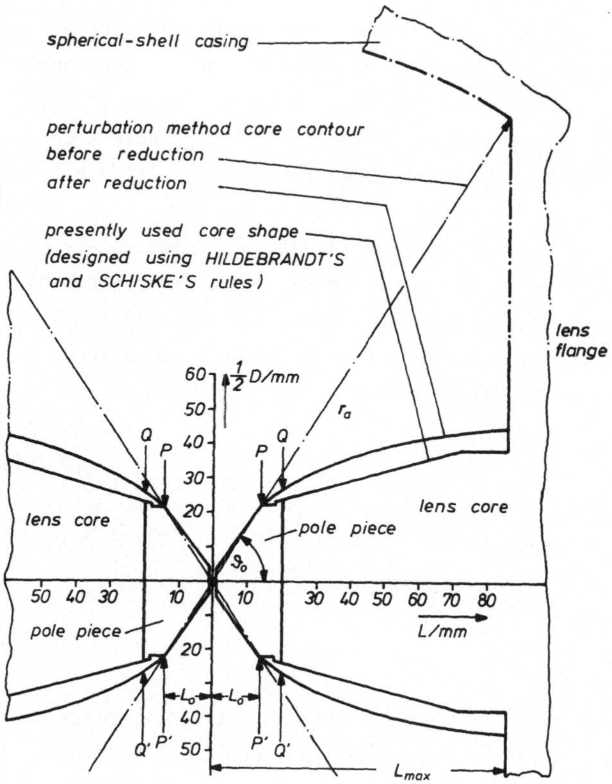

Fig.4.48. Illustration of the determination of the core contour for the symmetric lens shown in Fig.4.37

differ only by the factor.

$$\frac{\omega_c(r_0,\theta_o)}{\omega_{cp}(r_0,\theta_o)} = 1 - \frac{1}{3}\left(\frac{r_0}{r_a}\right)^2 , \tag{4.111}$$

which remains small and of the order of a few percent at most for all practical lens configurations.

The simplification can therefore be pushed a little further still by stipulating that the total flux $\omega_{c,total}(r,\theta_o)$ in the lens core is described well enough for the core shape determination by (4.104). Then, from (4.107,108), the reduced diameter $D_c(L)$ of the core is obtained for $r \geq r_0$ or $L \geq L_o$ from

$$D_c(L) = D_c(L_o)\sqrt{\frac{\omega_c(r,\theta_o)}{\omega_c(r_0,\theta_o)}}$$

$$= D_c(L_o)\sqrt{\frac{r\left[1 - \frac{1}{3}(r/r_a)^2\right]}{r_0\left[1 - \frac{1}{3}(r_0/r_a)^2\right]}}$$

$$= D_c(L_o)\sqrt{\frac{(L/L_{max})\left[1 - \frac{1}{3}(L/L_{max})^2\right]}{(L_o/L_{max})\left[1 - \frac{1}{3}(L_o/L_{max})^2\right]}} . \tag{4.112}$$

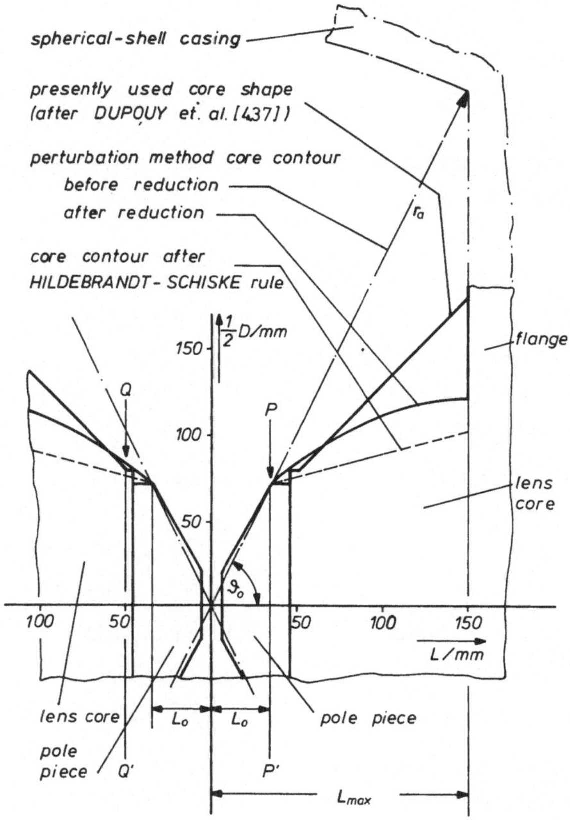

Fig.4.49. Illustration of the determination of the core contour for a symmetric objective lens employed in a 3 MeV electron microscope by DUPOUY et al. [4.37]

Here, $L_{max} = r_a \cos\theta_0$ is the distance of the end of the core from the centre of the lens (Figs.4.48,49). By differentiation of (4.112) with respect to L, we determine the local angle $\alpha_c(L)$ of inclination relative to the lens axis of the tangent drawn to the lens core at the axial position L, as follows:

$$\alpha_c(L) = \arctan\left[\frac{1}{2}\tan\theta_0 \cdot \frac{(L_0/L_{max})[1-(L/L_{max})^2]}{\sqrt{(L_0/L_{max})[1-\frac{1}{3}(L_0/L_{max})^2]}\ \sqrt{(L/L_{max})[1-\frac{1}{3}(L/L_{max})^2]}}\right] \tag{4.113}$$

From (4.112,113), some useful *rules* arise *for the design of a lens core for constant magnetic flux density*.

1) At the end of the conical surface of the pole piece, the effective pole-piece taper angle θ_0 drops abruptly to the incipient core angle $\alpha_c(L_0)$, such that

$$\alpha_c(L_0) = \arctan\left[\frac{1}{2}\tan\theta_0 \cdot \frac{L-(L_0/L_{max})^2}{1-\frac{1}{3}(L_0/L_{max})^2}\right] \;.. \tag{4.114}$$

2) For practical lens geometries,

$$\frac{1 - (L_o/L_{max})^2}{1 - \frac{1}{3}(L_o/L_{max})^2}$$

differs from 1 by a few percent at most; neglecting this rather small correction yields an approximate value (Fig.4.45):

$$\alpha_c(L_o) \approx \alpha_{co} = \arc\tan\left(\frac{1}{2}\tan\theta_o\right) \quad . \tag{4.115}$$

3) Towards the external end of the core ($L \rightarrow L_{max}$), the core surface becomes more and more parallel to the lens axis and finally orthogonal to the flange: $\alpha_c(L_{max}) = 0$. This reflects the corresponding diminution and final vanishing of the magnetomotive force between the corresponding parts of the cores.

4) The diameter of the core close to the flange ($L = L_{max}$) is given by

$$D_c(L_{max}) = D_c(L_o)\sqrt{\frac{2L_{max}/3L_o}{1 - \frac{1}{3}(L_o/L_{max})^2}} \quad . \tag{4.116}$$

This is a more precise formulation of the recommendation enunciated by HILDEBRANDT and SCHISKE, already reported above in Sect.4.3.2b.

5) The core contour can be determined by employing (4.112).

By way of example, core contours have been calculated according to these rules for two lenses actually constructed and are shown in Figs.4.48,49. Figure 4.48 shows a core suitable for use with a condenser-objective lens at a beam voltage of 100 to 150 kV [4.157]. The core shape presently employed is also shown in Fig.4.48 and has been designed according to HILDEBRANDT and SCHISKE's rules as discussed in Sect.4.3.2b. The magnetic flux along this core has been measured (Fig.4.41), and from this, the actual flux density distribution has been determined [4.73]. It turned out that for $L \lesssim 0.5\, L_{max}$, the flux density becomes about 10 to 20% larger than intended. Thus, a moderate increase of the cross-sectional area of the core appears to be advantageous, quite in agreement with the core contour resulting from the perturbation calculation.

The second example concerns the core of an objective lens suitable for 3 MeV electrons [4.37] and is illustrated in Fig.4.49. Here, in the present construction, pole pieces and lens cores have about the same taper angle and can be considered to a good approximation as forming the conical core of a spherical-shell lens. The average flux density in the cores can now be calculated from (4.103,105). It is found that with the present construction, the magnetic flux density in the core decreases rapidly, so that at the flange end of the core ($L = L_{max}$) it is less than one-third of the density at the interface between pole-piece and core ($L = L_o$). Therefore, a reduction of the cross-sectional area of the more outward parts of the core appears to be indicated, exactly as is found with a core shape predicted by (4.112).

For comparison, the core contour given by HILDEBRANDT and SCHISKE's rules is also represented in Fig.4.49. For this contour, the cross-sectional area of the core would increase less and a corresponding increment of the flux density would ensue. Clearly, the 15° core taper angle is too small here and the rules of Sect. 4.3.2b would have to be adapted appropriately by employing a larger angle, say of about 20° to 25°.

Evidently, the procedure described in the present section has a more universal character for design engineering and, in addition, requires little computational effort. Finally, it should be recalled that in the above, the ferromagnetic circuit of the lens has been assumed to be of the symmetric type of Fig.4.10a. But if the lens is of the asymmetric type (Fig.4.10b), the various considerations described in Sect.4.3.2b in connection with the HILDEBRANDT-SCHISKE procedure again apply, and the lower surface T'T" of the upper lens flange can be regarded as a magnetic mirror for the field distribution. The magnetic flux and the flux density in the core can then be calculated using the concept of the spherical-shell-casing model lens and (4.102-105). In these formulae the double amount 2NI of the actual ampere-turns NI of the asymmetric lens must be used for the magnetomotive force. The formulae (4.112-116) for the determination of the core contour also apply for the asymmetric lens because they have been derived in relative units.

4.3.3 Magnetic Design of the Lens Casing

Here again, symmetric lens casings (Fig.4.10a) and asymmetric lens casings (Fig.4.10b) must be distinguished. As in the case of flux determination in pole-piece systems with asymmetric lenses, the inner surface T'T" of the flange opposite the core can be regarded as a magnetic mirror which generates a (virtual) mirror image of core, casing, coil, and field distribution, so that the appropriate symmetric configuration arises as the sum of real and virtual elements. In the spherical-shell lens model the magnetic mirror surface of the flange has to be placed in the plane $\theta = \pi/2$ in order to produce the corresponding asymmetric configuration. Equations (4.100,101), which describe the magnetic field distribution, also apply in this case, and it is seen immediately that because $H_r(r,\pi/2) = 0$, the field is orthogonal to the mirror surface, as it should be.

Employing H_θ as given by (4.100), the magnetic flux entering the flange can be calculated immediately (again for $r_i \ll r_a$):

$$\omega_f(r) = + \frac{2\pi\mu_0\widehat{NI}}{\ln(\tan(\theta_0/2))}\, r\left[1 - \frac{1}{3}\,(r^2/r_a^2)\right] \quad . \tag{4.117}$$

Here, $\widehat{NI} = 0.5\, j_\phi\, r_a^2(\pi/2 - \theta_0)$ is the actual number of ampere-turns used in the coil of the asymmetric lens.

Within the flange, the flux lines extend in a radial direction, and the cross-sectional areas in the flange, which are orthogonal to the flux lines, are short cylinders concentric to the lens axis. If at a radial distance r the flange has a thickness h(r) (Fig.4.50), the corresponding effective cross-sectional area of the flange amounts to $Q(r) = 2\pi r \cdot h(r)$. For the average flux density $B_{r,f}(r)$ within the flange at the radial distance r we obtain

$$B_{r,f}(r) = \frac{\omega_f(r)}{Q(r)} = \frac{\mu_0 \widehat{NI}}{h(r)\ln(\tan(\theta_0/2))}\left[1 - \frac{1}{3}(r^2/r_a^2)\right] \quad . \tag{4.118}$$

As has been explained in Sect.4.3.2a, the magnetic field strength in the iron circuit of the lens should not exceed about 10 Oe (≈ 1 A/mm) in a well-designed lens casing. This corresponds to a maximum acceptable flux density B_0 between about 5000 to 10000 G (0.5 - 1 T) according to the ferromagnetic properties of the chosen flange material.

Substituting B_0 for $B_{r,f}$ into (4.118) and taking the absolute value for $\ln(\tan(\theta_0/2))$, the required flange thickness can be written as

$$h(r) = h_0\left[1 - \frac{1}{3}(r/r_a)^2\right] \quad . \tag{4.119}$$

Here

$$h_0 = h(r = 0) = \frac{\mu_0 \widehat{NI}}{B_0|\ln(\tan(\theta_0/2))|} \tag{4.120}$$

is the thickness of the flange at its centre.

It is seen from (4.119) that in principle, the flange could be made thinner with increasing off-axis distance r, down to $2h_0/3$ at its external periphery. In actual practice, a constant flange thickness will usually be preferred for the sake of easier fabrication. Moreover, this automatically mitigates the effect of neglecting the correction field B_θ^* of the perturbation in deriving (4.117-120) (Sect.4.3.2c), and it can easily be shown that this field would increase the flux in the flange by a few percent.

Finally, as is to be expected for a calculation based on the first-approximation field equations (4.100,101), comparison of (4.104,117) shows that the total magnetic flux $\omega_f(r_a)$ in the periphery of the flange is equal to the flux $\omega_c(L_{max})$ in the base of the core. (In making this comparison, it should be noted that $\widehat{NI}$ of the asymmetric lens corresponds to NI/2 of the symmetric lens.) This flux has to be carried by the flange to which the lens core is connected; clearly the largest thickness is required for the flange at a distance from the lens axis equal to the radius $D_c(L_{max})/2$ of the core at its end. The corresponding diameter $D_c(L_{max})$ is described by (4.116). Employing this and again writing B_0 for the upper limit of the acceptable flux density, the required thickness h' of the flange (Fig.4.50) is obtained as

$$h'\left(\frac{1}{2}D_c(L_{max})\right) = h_o \frac{1}{\sin\theta_o}\sqrt{\frac{2}{3}\frac{L_{max}}{L_o}\left[1 - \frac{1}{3}(L_o/L_{max})^2\right]} \quad . \tag{4.121}$$

Here, h_o is given by (4.120) for asymmetric lenses and by

$$h_o = -\frac{\mu_o NI}{2B_o \ln(\tan(\theta_o/2))} \tag{4.122}$$

for a symmetric design.

The flux coming from the core is practically the whole flux that has to be carried by the adjoining core flange. This is shown convincingly by Figs.4.37,38, which concern lenses actually measured and show that the direction of the flux lines in the close neighbourhood of the core flange is nearly parallel to the flange surface (so that it must tend to zero when closely approaching it). Also, with the spherical lens model and as demonstrated by the field distribution of Fig.4.44, the relative fractional amount of flux entering the spherical-shell casing is lowest in the region around the core end. Consequently, the thickness of the core flange could in principle be decreased as $h'(r) = h_o'\cdot D_c(L_{max})/2r$, in inverse proportion to the distance from the lens axis. This will rarely be made use of in actual practice however, because of the additional complications for the fabrication.

Finally, we have to consider the cylindrical part of the casing. This has to accommodate both the flux coming from the adjoining core flange and the additional flux passing through the space occupied by the lens coil. Once again employing the spherical lens model and the field distribution described by (4.100,101), the flux within the spherical-shell casing can be calculated straightforwardly. The radial field $H_r(r,\theta_o)$ of (4.101) provides all the flux coming from the coil. This flux can be written in the form

$$\omega_{s,r} = \omega_c(r_a,\theta_o)\cdot 2\ln(\tan(\theta_o/2))\left[\frac{1-\sin\theta_o}{\frac{\pi}{2}-\theta_o} - \frac{\ln(\sin\theta_o)}{\ln(\tan(\theta_o/2))}\right] \tag{4.123}$$

Here, $\omega_c(r_a,\theta_o)$ is the flux present in the lens core at its external end [cf. (4.102,104)] and the total flux in the spherical casing will be obtained by adding to this the flux coming from the core:

$$\begin{aligned}\omega_s &= \omega_{s,r} + \omega_c(r_a,\theta_o)\\ &= \omega_c(r_a,\theta_o)\left[1+2\ln(\tan(\theta_o/2))\left(\frac{1-\sin\theta_o}{\frac{\pi}{2}-\theta_o} - \frac{\ln(\sin\theta_o)}{\ln(\tan(\theta_o/2))}\right)\right]\end{aligned} \tag{4.124}$$

The angle θ_o-dependent term of the sum in square brackets is smaller than about 3% for $\theta_o \gtrsim 45^o$ and can therefore be neglected in practice.

Thus, the magnetic action of the lens casing can be summed up in first approximation as follows. The magnetomotive force generated by the coils gives rise to a magnetic flux which passes either between the cores and the pole pieces with a symmetrical design or between a flange and the opposing pole piece and core with an asymmetric lens. The flux then continues into the flange to which the core is connected and from this flange into the external cylindrical casing. By way of the casing, the flux is carried back to the opposite flange and into the other core with a symmetric design.

With the spherical lens model, the flux entering the spherical casing via the coil space is comparatively low relative to the flux passing over the core, a feature seen to hold true for the lens actually constructed shown in Fig.4.37.

The asymmetrical lens of Fig.4.38 has a more elongated casing and thus appears to be less similar to the corresponding model lens, and a considerable fraction of the flux entering the core originates from the external cylindrical casing. Nevertheless, in this case too, the total flux which passes over the coil space is concentrated at the base end of the core and also arises in the proximal end of the cylindrical casing. The flux in the casing gradually decreases with increasing distance from the core flange, and in principle, the wall thickness could be diminished accordingly. For the design of the casing, this would rarely be of advantage in actual practice.

Therefore, for both symmetrical and asymmetrical lenses, the thickness w of the wall of the casing can be determined from the flux $\omega_s \approx \omega_c(r_a,\theta_o)$ (4.124,104) by stipulating the maximum value of the acceptable flux density B_o. If D_{cc} is the inner diameter of the cylindrical casing and w the thickness of its wall (Fig. 4.50), the cross-sectional area of the casing taken at right angles to the lens axis is $Q_s = \pi(D_{cc}w + w^2)$, which can often be simplified to $Q_s \approx \pi D_{cc}w$, because in practice usually $w \ll D_{cc}$. Employing the definitions (4.120) and (4.122), the thickness required for the wall of the casing is

$$w = \frac{4}{3}\frac{L_{max}}{D_{cc}\cos\theta_o}h_o \quad . \tag{4.125}$$

This completes the specification for the design of the magnetic circuit of magnetic lenses.

In conclusion, for the determination and prediction

1) of a suitable core shape, we have (4.112,115,116),

2) of the necessary flange thickness, we have (4.119-122),

3) of the thickness of the cylindrical casing, we have (4.125).

It should be noted that the actual diameter D_{cc} of the cylindrical casing does not enter into the equations for core shape and flange thickness (but of course it does enter into (4.125) for the wall thickness). The design engineer therefore has ample freedom in selecting a suitable shape for the lens casing.

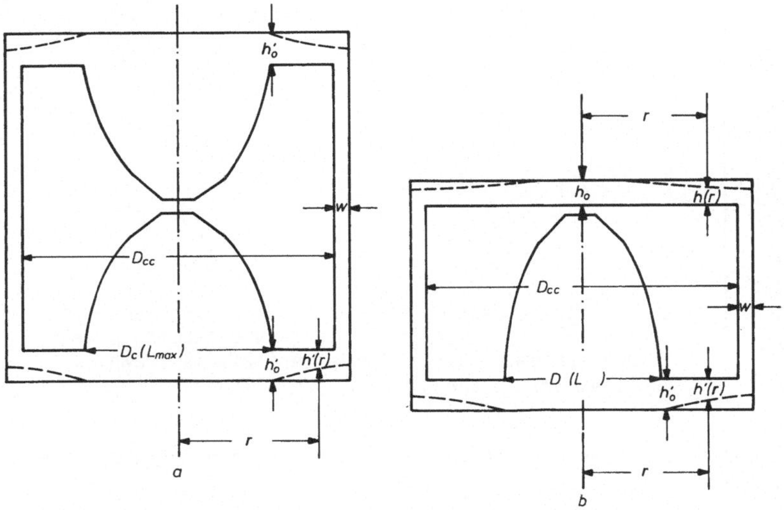

Fig.4.50a,b. Representation of quantities employed when calculating the thickness of the cylindrial lens casing and of the flanges

4.3.4 Comments on Lenses with Highly Saturated Pole-Piece Tips

Pole pieces operating at a flux density beyond the point of saturation were employed rather early in electron microscopy. Nevertheless, the expectations that the use of such highly saturated lenses would lead to a marked improvement in resolution unfortunately did not materialize.

All things considered, this is not particularly surprising, because for a given beam voltage, resolution is finally limited by spherical aberration and is proportional to the fourth root of its coefficient C_s (4.6). In order to improve resolving power appreciably, say by about a factor of two, a reduction of C_s by an order of magnitude would be necessary, and — at a given lens strength $NI/\sqrt{U}$ — this would require a reduction of the half-width of the lens field by the same amount. That this cannot be achieved by running the pole-piece tips into oversaturation can be concluded from the results of a number of experimental [4.73,94,176,183,217] and theoretical [4.8,25a-c,71,84,109,134,135] investigations. The possible improvements in the coefficient C_s of spherical aberration appear to remain at around a factor of two. Nevertheless, this may still be of interest for some special applications, for example, selected area microdiffraction [4.172].

Unfortunately, the investigations on the electron optical properties of highly saturated lenses such as those quoted above refer to a number of special pole-piece configurations only. A more general treatment comparable to the information published for unsaturated lenses [4.36,39,41,83,106,108] is not available to this day.

So far as the flux in the iron circuit is concerned, in principle only minor modifications to the treatment for non-saturated lenses are necessary. Retaining the

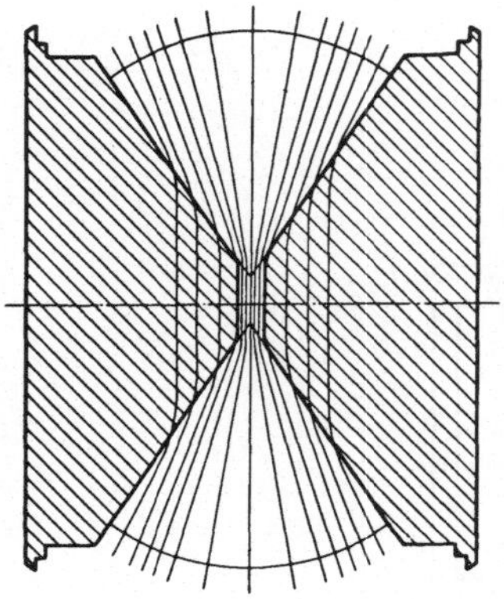

Fig.4.51. Schematic representation of the magnetic potential distribution in a highly saturated pole-piece system. A set of equipotentials is shown as well as two of the circular field lines

approximation of circular field lines between the conical pole-piece surfaces, the modification consists in taking into consideration a magnetic potential drop in the axial direction within the pole pieces. Consequently the conical surfaces are not equipotential surfaces any more (Fig.4.51).

In order to calculate the magnetic flux traversing the space between the conical pole-piece surfaces, this space is subdivided into flux tubes which follow the arc-of-circular path of the magnetic field lines. The magnetic flux passing through the flux tubes is obtained by multiplying the magnetic conductivity of each tube by the magnetomotive force between the ends of the tubes, i.e., between the points where the particular tube touches the pole surface of the pole pieces. The magnetomotive force is lowest for flux lines close to the pole-piece gap and increases gradually with increasing distance from the pole-piece centre.

Now, the flux in the pole piece is calculated by suitably summing up the contributions from the individual flux tubes. On dividing by the pole-piece cross-sectional area, the average flux density in the pole pieces is obtained. This may look rather simple, but in fact, at high magnetic saturation, the variation of the magnetomotive force on the pole-piece surfaces is itself a function of the magnetic flux density.

A theoretical investigation shows that the average magnetic flux density B(z) in the pole pieces can be determined from

$$\frac{dB(z)}{dz} = -\frac{4\tan\psi_0}{D_3(z)} + \frac{4M_0 + 8\int_0^z H(B(z))dz}{D_3(z)^2 \cdot G(z) \cdot \cos\psi_0} \quad , \tag{4.126}$$

which is essentially a second-order differential equation for B(z) after a first integration. Here, z is the distance from the pole face (Fig.4.52) and has the same meaning as l in Fig.4.31a, D_3 is the diameter of the pole piece at that distance, and ψ_0 the taper angle of the conical surface (Fig.4.31a). H(B(z)) is the magnetic field strength present in the pole-piece material at a flux density B(z) and is hence essentially the hysteresis curve, commonly shown in the form B(H(z)).

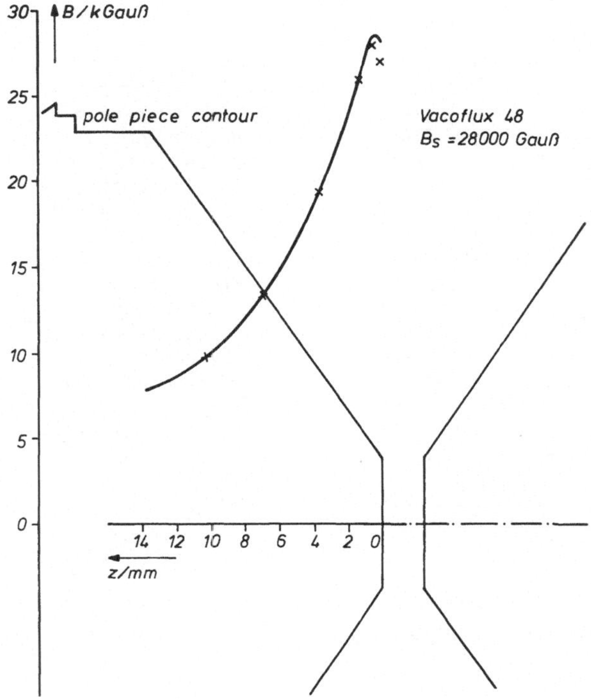

Fig.4.52. Variation of the average magnetic flux density in one
of the pole pieces of a highly
saturated pole-piece system.
Calculated values based on
(4.126,127) are shown as a continuous curve, measured values
as (x). The pole pieces were
fabricated from Vacoflux 48
[4.217]

$M_0 = SB_s/\mu_0$ is the magnetomotive force at the flux density B_s across the pole-piece
gap of width S. Finally in the denominator

$$G(z) = \frac{1}{\sqrt{1 - \left(\frac{2a}{D_3(z)}\right)^2}} \ln \left| \frac{\left(1 - \frac{2a}{D_3(z)}\right)\tan\left(\frac{\pi}{4} - \frac{\psi_0}{2}\right) + \sqrt{1 - \left(\frac{2a}{D_3(z)}\right)^2}}{\left(1 - \frac{2a}{D_3(z)}\right)\tan\left(\frac{\pi}{4} - \frac{\psi_0}{2}\right) - \sqrt{1 - \left(\frac{2a}{D_3(z)}\right)^2}} \right| , \quad (4.127)$$

in which

$$a = \frac{D_1}{2} - \frac{S}{2}\tan\psi_0$$

and

$$r = \frac{2z + S}{2\cos\psi_0}$$

have the meaning indicated in Fig.4.28, so that r is the radius of the flux line
which touches the conical pole-piece surfaces at an axial distance z from the pole
face. The numerical solution of (4.126) has been described by ZEMLIN and satisfactory agreement between calculated and measured results has been found [4.217] as
shown by way of example in Fig.4.52.

It should be realized that for the calculation, the flux density B_s in the gap
is stipulated and not NI. The total magnetomotive force required is $NI = M_0 +$
$+ 2\int_0^{z_0} H(B(z))dz$ and is therefore a result of the calculation. The value of z_0

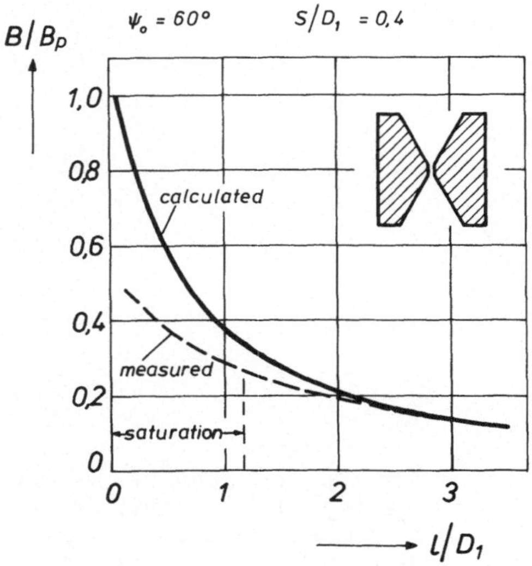

Fig.4.53. Comparison of measured and calculated values of the average magnetic flux density in one of the pole pieces of a highly oversaturated pole-piece system. For the calculation, the methods described in Sect.4.3.1a and originally designed for the treatment of non-saturated pole pieces have been employed. The calculated values tend to approach the experimental ones for the pole-piece cross-sections farther away from the gap. They can thus be safely employed as a convenient and sufficiently accurate approximation for predicting the flux density at the junction surface to the lens core. Again, $B_p = \mu_0 NI/S$ is the hypothetical flux density for the gap width S

has to be chosen large enough to carry the upper limit of integration into a region of unsaturated pole-piece material where the magnetic field strength H and the additional contributions to the integral tend to vanish.

Finally, as a simple first-approximation procedure, the magnetic flux and the flux density in the pole pieces and their diameter can be formally calculated using the methods described in Sect.4.3.1a. This is equivalent to assuming that no saturation occurs and that the relative magnetic permeability is always very high. The actual values of flux and flux density in the pole-piece system determined in this way will always be smaller than the calculated values, and the two will tend to converge with increasing diameter of the pole pieces. This is obvious from Fig.4.53, where calculated and measured values are compared by way of example.

4.4 Other Aspects of Magnetic Lens Design

By employing the methods described in Sects.4.2-3, the general layout required for the lens can be reliably predicted. The importance of carefully designing the magnetic circuit can hardly be overestimated because, with a poor magnetic design, the lens will not perform according to expectations and a later amelioration of the iron circuit is usually not feasible.

Besides the magnetic design, other construction details have to be considered, such as lens coils, stigmators, apertures, special specimen stage features essential for a satisfactory electron optical performance of the lens, and protective measures to guard against environmental disturbances. Unfortunately, there is often little detailed information available about these aspects, or at least not as much as the design engineer would like to have in order to avoid pitfalls. The following discussion consequently remains rather sketchy in some places.

4.4.1 The Lens Coil and Its Cooling System

In electron lens construction, the design of the lens coil is nearly always chosen along quite conventional lines. The coil proper consists of a couple of thousand turns of round enamelled copper wire wound on a spool made of brass or other non-magnetic material. This spool (Fig.4.19) consists of a cylindrical tube which fits around the ferromagnetic lens core and is joined at its ends to the two flat end flanges. The flanges confine the coil volume proper and by taking up the winding pressure prevent the coil windings from spilling out over the spool ends.

The surfaces of the spool tube and flanges which face the coil proper are covered by several layers of insulation paper (typically 0.05 to 0.1 mm thick). At least several millimetres of overlap have to be provided between adjoining sheets of paper in order to create sufficient electrical isolation of the windings from the grounded metal body of the spool. In the windings, DC voltages, usually of several hundred volts relative to ground, are generated by the lens current power supplies, and the same potential difference arises between the different parts of the coil. In order to suppress leakage currents that could arise from failure of the wire insulation, it is advantageous to insert an additional sheet of insulating paper after every couple of layers of the winding, so that the path length of electrical leakage is increased rather strongly. These intermediate sheets of paper are also useful for stabilizing the layered construction of the coil (Fig.4.54a).

The winding and the remainder of the electrical circuit of the lens have also to be protected against a high transient overvoltage, which can be generated if the current in the coil is interrupted suddenly, be it by switching off the power supply too abruptly or by breaking the circuit by accident. The self-induction of an iron-shrouded lens coil easily amounts to several hundred henries, and so, if the lens current is switched off within a time of a few milliseconds, a fairly high voltage is induced across the now disconnected lens coil by the collapse of the magnetic flux in the lens core and casing. By way of example, for typical objective or projector lenses employed at around 100 kV, the unprotected transient switch-off overvoltage might easily attain 30 kV or more, evidently a serious danger to the insulation of the coil and the current leads connecting it to the switch. Cases have been known where unprotected coils have been damaged or destroyed in this way.

As a means of protecting the coil, the generation of too-high transient voltages can be averted by connecting a capacitance of a few tens of μF across the switch contacts. When the contact is broken, the current in the coil is continued for a short time by the exponentially decreasing current now charging the capacitor, and the interruption of the current is stretched over a long enough time (Fig.4.54b).

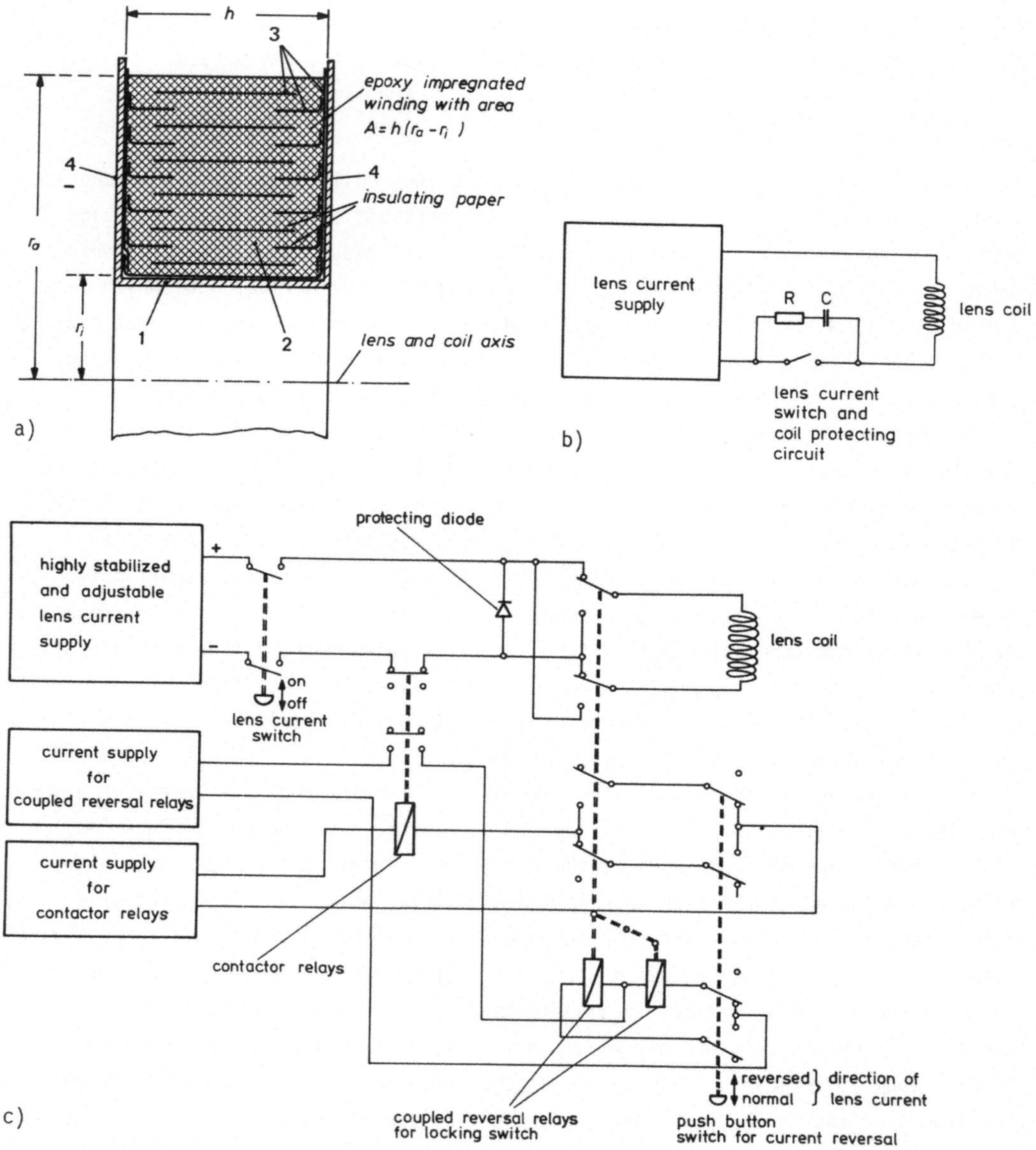

Fig.4.54. (a) Schematic representation of the construction of a typical lens coil: Spool 1, winding space 2, paper insulation 3, external flange surfaces 4. Hints for the fabrication: The paper insulation indicated here would be suitable for a winding to be vacuum-impregnated with epoxy resin. In this case, in the volume of the winding, the consecutive coil-stabilizing sheets of paper insulation should not extend uninterruptedly between the flanges. The evacuation of the coil volume and the influx of the resin is then not obstructed during impregnation. It must not be forgotten that while curing, the epoxy resin contracts rather strongly. This may lead to a noticeable dishwise deformation of the end flanges. Therefore, if satisfactory thermal contact between the coil and a separate cooling chamber has to be maintained (as in the design of the lenses represented in Figs.4.37,38), the external flange surfaces 4 should be machined flat after the curing of the resin as a final step of the fabrication. (b) RC-circuit for protecting the lens coil from transient overvol-

tages during interruption of the lens current. A capacitor C of 10 to 50 μF is usually employed, and the resistor R of some 100 Ω limits the discharging current of the capacitor to values that do not damage the contacts of the switch when it is closed. (c) Solid-state circuit for protecting the lens coil from overvoltages during interruption of the lens current and current reversal. The rest positions of the switches actuated by the push-button and the relays correspond to the 'normal' lens current direction. These rest positions are indicated in the circuit diagram. In order to understand the function of the circuit, it is essential to realize that the movement of the switch actuated by the coupled relays is backed up by a bent spring in such a way that after disconnecting one of the switch positions the opposite position of the switch is forcibly assumed by action of the spring. A particular merit of this specific circuit design is that after the required switching actions have been performed the currents in the relays are set to zero and remain so during the actual operation of the electron lens. Thus, any disturbance of the electron beam path, which might be caused by the magnetic stray field of the relay coils, is avoided

A still more elegant concept for the protection of the lens coil against cut-off overvoltages consists in connecting the end leads of the lens coil by a (solid state) diode in such a manner that for the operational polarity of the coil, the diode is non-conducting and can withstand the normal voltage difference present between the coil ends. If now the lens current is interrupted suddenly, the voltage induced in the lens coil by the collapse of the magnetic field is in the opposite sense to normal operational polarity. Therefore, for the induced transient voltage, the diode acts as a short circuit between the coil ends and the voltage differences in the volume of the coil cannot rise to dangerous levels.

Care has to be taken over the protection of the lens coil if a means for fast operational lens current reversal is to be provided. (This is a powerful and nearly indispensable method of determining the alignment state of the lens with respect to the electron optical system of which it is part [4.170,171].) Damage can be prevented by employing a system of mutually delayed switches which ensure that the lens current has been interrupted before the protecting diode is disconnected (Fig.4.54c). For the magnetic lenses employed in MeV electron microscopes, overvoltage transients could be particularly dangerous and must be carefully avoided. Thus, in the 3-MeV electron microscope at Toulouse [4.37] the lens current is gradually diminished to zero before the current leads are interchanged for reversal.

In any case, in order to safeguard a long operational life of the coil and improve the insulation, vacuum impregnation employing epoxy resin to fill the whole volume of the winding is advantageous. This not only considerably improves the electrical insulation but also prevents damage to the coil in the case of leakage in the water-cooling system and removes the danger of water condensing on the unprotected windings if they are cooled but not energized (and thus heated electrically) for longer periods.

It has been explained in Sect.4.2, that for electron lens construction the number of ampere-turns NI required is a given quantity completely determined by the beam voltage and the mode of operation for which the lens has been designed. Thus, a cross-section of the coil has to be determined which will accommodate the required number of ampere-turns without becoming too hot for safe operation.

A reasonably exact calculation of the dependence of the temperature rise ΔT in the coil on the ampere-turns NI applied would be extremely complex, due to the many parameters involved and the difficulty of expressing accurately the physical properties of the actual coil construction in mathematical terms. On the other hand, by introducing some plausible simplifications, formulae can be derived relating the average current density j_w and the ampere-turns NI that can be accepted for a maximum temperature rise ΔT in a coil of a given cross-sectional area A.

For this, the basic assumption is that at thermal equilibrium the power P dissipated by the coil surface is proportional both to the surface area S of the coil and to the temperature rise ΔT. This power, of course, has to be provided by the ohmic loss of the lens current I at the coil resistance R_c:

$$P = \frac{1}{2} C_T \cdot S \cdot \Delta T = I^2 R_c \quad . \tag{4.128}$$

Here, C_T is a constant and the factor 1/2 has been included in order to give a more simple form to the actual design equations, (4.129,130) below.

An equation of the general form of (4.128) has already been applied by TEUCHERT [4.202] to the treatment of the thermal properties of transformer and relay coils. The constant $C_T/2$ is determined by the mechanism responsible for the removal of the power dissipated by the coil to its environment, e.g., by conduction to air or water-cooled surfaces or by radiation.

Thus C_T is determined by the general construction and shape of the system but not by its actual size, which vanishes because the diffused power is assumed to be proportional to the surface area S of the coil. It should be realized that the assumed proportionality of P to S tacitly implies that the temperature does not vary too much throughout the bulk of the winding and so is elevated on the average by ΔT, the temperature drop to the environmental temperature taking place in the immediate neighbourhood of the coil surface. How such a temperature distribution arises can be readily understood in terms of the difference between the thermal conductivity of the copper wire and the thermal conductivity of the insulating materials; the latter is smaller by more than three orders of magnitude.

Equation (4.128) can now be applied rather easily to actual design problems. By way of example, we shall consider coils of rectangular cross-section which are the most important in actual practice. For these, starting from (4.128), the average current density j_w that can be accepted in a winding of cross-sectional area A with an average temperature rise ΔT can be derived:

$$j_w = \sqrt{C_T \sigma q \frac{\Delta T}{h_0}} \cdot \sqrt{\eta + \frac{1}{\eta}} \quad . \tag{4.129}$$

The corresponding number of ampere-turns is

$$NI = \sqrt{C_T h_0^3 \sigma q \Delta T} \cdot \sqrt{\eta + \frac{1}{\eta}} \quad . \tag{4.130}$$

Here, C_T (in W m^{-2} $^oC^{-1}$) is the cooling efficiency of the coil as defined above; σ is the specific electrical conductivity of the wire material; q=Na/A is the space factor of the coil winding, where a is the actual cross-sectional area of the copper wire employed; h_0 represents a characteristic length for the size of the coil cross-section and is defined by $h_0^2 = A$; and $\eta = h/h_0$ specifies the shape of the coil cross-section with h as the extension of the coil in the direction of the lens axis (Fig.4.54a). If b is the radial extension of the cross-section of the (rectangular) coil, its 'aspect ratio' h/b can be easily seen to be connected with η by $h/b = \eta^2$. The value $\eta = 1$ indicates a square cross-section, and it can easily be shown that this also corresponds to a minimum of $\sqrt{\eta + 1/\eta}$. So far as the thermal properties are concerned, therefore, a square cross-section is the least appropriate for the coil. On the other hand, $\sqrt{\eta + 1/\eta}$ rises rather slowly around $\eta = 1$ and only by about 10% up to $\eta = 2$ or down to $\eta = 0.5$. It should also be noted that coils having aspect ratios that are the reciprocals of each other have the same thermal limits on the number of ampere-turns and the average current density that can be accepted.

Therefore, when choosing the coil shape for a given cross-sectional area, the design engineer can do little to improve the thermal limitations on the ampere-turns, except when going to extremes with the coil cross-section aspect ratios η^2. By way of example, for a 50% increase in the number of ampere-turns, a very long and comparatively thin cylindrical coil would be required with a cross-section aspect ratio $h/b = \eta^2 \approx 20$, or a rather extended and thin pancake-shaped coil with $h/b = \eta^2 = 1/20$. Both of these coil types would be rather awkward in their consequences for the magnetic circuit of the lens casing, and the same objective (the 50% increase in NI) could be achieved more appropriately, according to (4.130), by a 30% increase of the linear dimensions b and h of the coil which are both proportional to h_0.

It should also be noted that the thermal limits on the average current density j_W in the winding and on the ampere-turns of the coil are the same, regardless of the average diameter of the winding. Also, the space factor q of the winding changes little throughout the range of wire diameters suitable for lens design. In practice, $q \approx 0.65$ can be used with sufficient accuracy for carefully constructed compact windings. Thus, the cooling efficiency can serve as the quality factor for the thermal properties of the coil design: a higher value of C_T indicates a thermally better coil.

A particularly efficient construction results if the water-cooling chamber is an integral part of the spool, so that one wall of the cooling chamber acts as a spool flange. Such a design has been employed with the condenser lens represented in Fig.4.22 and with the condenser-objective lens shown (rather sketchily) in Fig.4.37 (for a technical drawing see [4.157]). With these lenses, the winding has been split up into two halves which are separated by the integral water-cooling chamber. A supplementary cooling action can be provided as with the condenser-objective by pressing separate additional cooling chambers against the flat external surfaces of the spool. $C_T \approx 85$ W m^{-2} $^{\circ}C^{-1}$ has been achieved in this manner, corresponding to an average current density in the winding of $j_w \approx 2.5$ A/mm^2 for a temperature rise of $\Delta T \approx 70^{\circ}C$.

If the lens coil unit does not include its own 'integrated' cooling chamber and is not in immediate contact with an extended water-cooled surface, then the cooling efficiency is much lower and drops to about $C_T \approx 40$ to 50 W m^{-2} $^{\circ}C^{-1}$. These conditions prevail if the water-cooling chamber is incorporated in the external cylindrical lens casing as for example with RUSKA's magnetic objective lens of 1934 (Fig.4.1) or with the early commercial electron microscopes (e.g. [4.19a,178]), where average current densities around $j_w \approx 1.8$ A/mm^2 could be employed for a temperature rise of about $\Delta T \approx 70^{\circ}C$. With this construction, the lens core enclosed by the coil tends to become quite hot. However, this may even be beneficial from the point of view of vacuum technology because it induces a partial degassing of the vacuum wall.

If the lens is to be operated without any water cooling at all, the whole lens tends to become rather hot. The temperature rise should be limited to about $\Delta T \approx 40^{\circ}C$, or else the lens casing could not be touched any more with the unprotected hand, and damage to soft-soldered connections might even occur.

Several efforts have been made to improve the thermal quality of the lens coil, but although some of these designs look quite promising, the "classical" construction described above is still the only one generally employed for magnetic lenses with perhaps a few laboratory exceptions.

Thus in a comparatively "classical" coil design employed for the lenses of a 500-keV electron microscope [4.99] the spool has been shaped in the form of a single layer bifilar coil of soldered 3/16" O.D. copper tube through which the cooling water runs. In contrast to the coils discussed further below, the winding (18-gauge poly-thermalize insulated copper wire) is still fairly conventional, the only uncommon feature being the use of a 60% by volume dispersion of a beryllium oxide filler in the insulating epoxy resin, which considerably improves its thermal conductivity without degrading the electrical insulation. However, some fabricational problems have arisen due to the possible toxicity of beryllium. An average current density j_w of 4.25 A/mm^2 was attained for a temperature rise of $\Delta T \approx 120^{\circ}C$ and about 3 A/mm^2 for $\Delta T = 70^{\circ}C$ [4.114], corresponding to $C_T \approx 120$ W m^{-2} $^{\circ}C^{-1}$. The improvement nevertheless remains rather moderate if compared to a well-designed conventional coil ($C_T \approx 80$ W m^{-2} $^{\circ}C^{-1}$).

With some other experimental coils, still more efficient removal of the heat generated is achieved by bringing the current-carrying conductors into direct contact with the coolant. This is accomplished by fabricating the winding from fibreglass-insulated hollow copper tube which carries both the current and the cooling water [4.47,195]. The winding contains between 100 to 200 turns of 6.3 mm O.D., 4.4 mm I.D. tube and carries an average current density of 3.1 A/mm^2 for a temperature rise of only $\Delta T \approx 18^{\circ}$C and 4.6 A/mm^2 for $\Delta T \approx 30^{\circ}$C. This type of coil was actually employed in practice with a 600-kV scanning electron microscope [4.27]. In order to avoid overheating the winding, a fairly large inner diameter of the tube is required to allow a sufficient throughput of cooling water. An actual lens coil fabricated from 2.9 mm I.D. tube turned out to be inadequate in this respect and did not perform any better than the more conventional coil [4.114] discussed just above. A disadvantage of this coil construction is that rather large currents up to some 100 A and powers up to several kilowatts are necessary, so that the current stabilization and the means for current reversal may present some additional problems.

Another type of coil having particularly high cooling efficiency is the ribbon or tape-wound coil. It consists of a long tape of copper foil, some 10 μm thick and of the order of one to several centimetres wide, and an equivalent tape of insulating material (e.g., Mylar, some micrometres thick). Both are wound into a reel of rectangular cross-section where successive copper layers are separated by the insulation and represent the lens-current carrying conductors [4.70,128]. Due to this particular construction, the thermal conductivity in the axial direction of the coil is high and in the radial direction low. Therefore, the temperature does not change in the radial direction. The temperature rise between the mid-plane of the ribbon coil and its end faces can be calculated from

$$\Delta T = \frac{\rho}{8\lambda q}\left(\frac{NI}{Q}\,h\right)^2 \quad , \tag{4.131}$$

where λ is the average specific thermal conductivity in the axial direction and ρ the electric resistivity. The cooling water flows freely across the end faces of the coil and hence over the bare edges of the tape. In order to avoid local overheating generated by trapping of water between protruding pieces of insulation, the tape reel should first be vacuum impregnated by epoxy resin and then the end faces machined flat and down to the copper tape.

It can be seen by employing (4.131) that even with a reel cross-section Q as small as about 1 cm^2 and for a number of ampere-turns of the order of 10^4 A, the temperature rise T will amount to only a few degrees. On the other hand, for a winding with a space factor q, the electrical power

$$P = \frac{\rho L_m}{qQ}\,(NI)^2 \tag{4.132}$$

dissipated in a ribbon coil having an average length L_m for one turn of the tape can easily run up to several kilowatts under these conditions. The ensuing difficulties for high-stability current regulation and for a sufficient throughput of coolant set a limit on the practical application of these coils. An average current density j_w up to 30 A/mm^2 has been obtained experimentally.

A comparable performance can also be obtained by utilizing wire-wound coils of similar size which are brought into immediate contact with the coolant [4.129]. Coils of this type have been operated successfully with experimental 100-kV [4.132] and high-voltage [4.131] electron microscopes (for detailed discussion, see Chap.5).

4.4.2 Stigmators

The short stretch of strong magnetic field that represents the magnetic electron lens proper would be exactly rotationally symmetric with an ideal lens. In actual practice, the field distribution is disturbed by small amounts. The most frequent causes for this are [4.2]:

- the pole-piece bores are not exactly round;
- the pole faces are not exactly plane, they are 'corrugated';
- the magnetic properties of the pole-piece material are such that it is either not isotropic or not homogeneous, or may even have both defects simultaneously;
- the axis of the pole-piece bores are not exactly coincident, they may be parallel displaced or tilted with respect to each other, or they may even have both defects simultaneously, i.e., they are "skew".

Although these disturbances are only small they give rise to optical effects which are quite noticeable, the main ones being axial astigmatism and transverse displacement of the image. Whereas a displacement does not impair the resolution, axial astigmatism may become quite serious if the full resolving power of the lens is to be utilized, and so some kind of correction is required for a high-quality lens.

Obviously, it is always good policy to fabricate the pole pieces as precisely as possible. Throughout the past decades of electron microscope development, a considerable effort has been invested in trying to improve standard workshop methods up to the accuracy required. After utilizing a high-precision lathe for shaping the pole-piece contours and the bore, a final improvement of bore roundness and pole-face flatness is achieved by employing honing and lapping methods, or by spark erosion. Between successive steps of machining, the pole pieces should be heat treated in vacuo or in protective gas in order to release mechanical constraints which might give rise to magnetic anisotropy.

Good quality rotationally symmetric bores can be obtained by employing a method suggested by F. STÖCKLEIN (unpublished). Here, advantage is taken of the fact that very accurately round, hardened steel balls are commercially available and normally used for precision ball bearings. The ball must have the diameter D required for the bore which, in turn, is machined on the lathe to a diameter D - ε, a little smaller than the ball by $\varepsilon \approx$ 3-5 µm. Finally, the ball is pressed through the bore by force, starting at its gap end. The surplus material is thus pushed towards the core side and a well-rounded bore is obtained.

Taking all these techniques together, with a well-fabricated pole piece of a few millimetres bore diameter, the deviation of this diameter as measured in different azimuthal directions should not exceed 0.5 to 1 µm. This gives rise to astigmatic differences of the focal length of about the same magnitude which have to be removed by employing a stigmator. The basic stigmator types may be classified as follows:

1) stigmators operating by deformation of the magnetic field of the lens;

2) stigmators working independently of the lens field as refracting electron optical elements:
   - magnetic quadrupoles and hexapoles,
   - electric quadrupoles.

Historically speaking, the field-deforming stigmators were the first to appear on the scene, the original idea probably being to pull the defective magnetic field back into its proper shape, but this would have been asking too much in practice. Thus, the stigmator distortion of the field is chosen in such a manner that its optical effect in the image counteracts the original aberration. This principle also applies to the independent stigmators, giving considerable latitude in the position and design of the multipole systems.

The first stigmator ever designed to compensate second-order astigmatism in an electron objective lens is due to HILLIER [4.75] and was one of the pioneering steps forward in corpuscular optics. It consisted of eight little iron screws, which were positioned in the free space between the slopes of the pole-piece system and hence right in the region of strong magnetic field (Fig.4.55). Their axes were arranged like the spokes of a wheel, having an angle of 45° between each pair of adjacent axes. The distortion of the magnetic lens field was adjusted by screwing the little screws further into the lens field or retracting them, and as this could only be done after removing the pole-piece system from the objective lens, a rather tedious trial and error process resulted [4.77]. This principle was quickly abandoned with the advent of "dynamic" stigmators, which could be adjusted during operation of the electron microscope. Moreover, a correction does not hold long: The astigmatism tends to change a little from day to day according to magnetic history and charging-up disturbances. So, resetting the stimator is often called for.

Even to-day, however, the HILLIER stigmator may be useful for "saving" a strongly disturbed pole-piece system by removing the bulk of the astigmatism and leaving only a comparatively minor amount for dynamic correction. To provide a strong enough astigmatic compensation could be a difficult task for most of the dynamic stigmators,

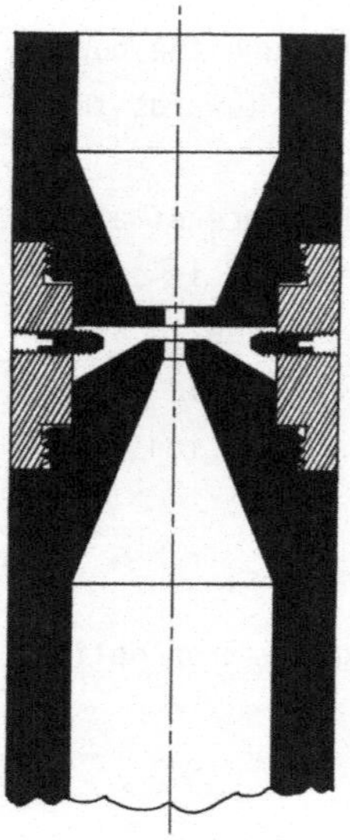

Fig.4.55. Cross-section of a pole-piece system equipped with an eight-screw stigmator for compensating the second-order axial astigmatism [4.77]. The stigmator consists of eight little soft iron screws arranged in the pole-piece system gap like the spokes of a wheel. They can be adjusted individually in the radial direction. Other configurations with six or twelve screws can also be used, in analogy to the design possibilities for electromagnetic stigmators with coils or electrostatic stigmators with electrodes. To achieve the proper compensation by radial adjustment of the screws, the pole-piec system has to be removed from the electron objective lens for every step of the procedure: a rather tedious process. Nevertheless, this type of stigmator has been employed successfully in the RCA EMU electron microscopes

whereas the HILLIER stigmator can easily be made as strong as required simply by pushing the screws far enough into the region of high magnetic field strength. Moreover, it is quite inexpensive.

The first dynamic field deformation stigmator was developed for a magnetic electron objective lens by LEISEGANG [4.101]. It consisted of a short piece of thin-walled non-magnetic brass tube which carried two minute specks of ferrite cemented onto the tube at points radially opposite each other. The tube was screwed into a fine-pitch thread located on the inside of a corresponding slightly larger brass tube press-fitted into the bore of the image-side pole piece of the objective lens. The little tube can be moved in the axial direction and with it the ferrite specks, from a position at the lower fringe of the lens field up into a region of progressively increasing field strength. The magnetization of the little specks is likewise increased and hence so is their electron optical effect. As the tube is advanced by means of a screw, the specks are simultaneously rotated, together with their correcting field, around the lens axis. In this manner, both strength and azimuthal orientation can be adjusted by actuating one and the same 'screwdriver', which is connected via a gear and a vacuum seal to a control knob. This has a certain disadvantage, namely, that for a given azimuthal orientation the correction strength can only be set in small steps and not continuously. These steps correspond to half turns of the corrector tube and to a focal difference of about $\Delta f \approx 0.1$ μm [4.101], which is just the limit at which the presence of astigmatism can still be detected by observing FRESNEL fringes [4.204]. Therefore, for very high resolution work where still smaller astigmatic focal differences are required, the LEISEGANG stigmator is not accurate enough.

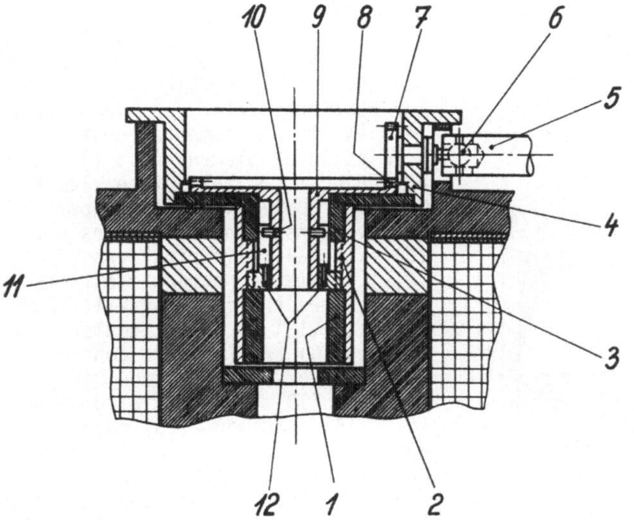

Fig.4.56. Cross-section of the pole-piece region of a magnetic intermediate-projector lens equipped with a LEISEGANG stigmator [4.149]. The two ferrite specks (12) of the stigmator are carried by a bronze ring which can be screwed up or down in a corresponding thread in the non-magnetic spacer (2), which connects the pole pieces (1) and (3). The ring is provided with two slits (11), which accept corresponding non-magnetic pins (10). The pins are fixed to a bronze tube (9), which is connected to a disc provided with a gear rim (8). Thus, axial position and azimuthal orientation of the stigmator ring can be varied during operation of the instrument by rotating the gear wheel (7), which engages with the rim (8). The rotation is actuated from outside the lens casing by turning the rod (5), which connects to the gear wheel (7) by means of a sliding universal joint (6). A second universal joint (not shown) connects the rod (5) to its rotary control knob. Thus, a 'yielding' connection between the control knob and the gear wheel (7) results, which permits the pole-piece system to be aligned transversely (Fig.4.72) without affecting the setting of the stigmator. The brass collar (4) carries both the pole-piece system and the bearing of the gear wheel (7)

As another disadvantage, there is little space available in the objective lens pole-piece bore and its adjoining lens core, and so the design is rather delicate there. The little corrector tube and the screwdriver, which is also a hollow tube to allow the electron beam to pass, both have rather narrow bores and there is a danger of charging up. With larger bore pole-pieces, these restrictions do not arise so that for condenser and for intermediate-projector lenses, this type of dynamic stigmator can be utilized successfully (Fig.4.56, [4.149]).

A less crowded design for a dynamic stigmator results if the field-distorting ferromagnetic chunks are positioned outside the pole-piece system but still within the magnetic stray field emanating from its conical parts. The chunks are some millimetres in size and so comparatively large. They distort the stray field, and this distortion propagates into the field region close to the axis. This design was employed with the objective lens of the well-known ELMISKOP 1 electron microscope (Fig.4.57,[4.182]). Here, two pairs of chunks were used, each of the

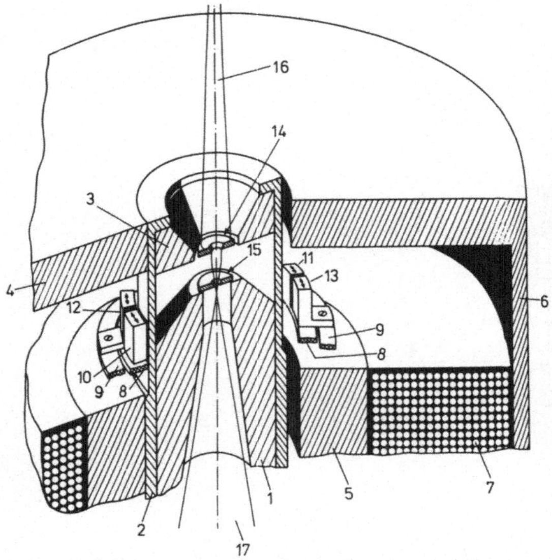

Fig.4.57. Schematic representation of the stigmator developed by RUSKA and WOLFF for the objective lens of the ELMISKOP 1 electron microscopes (cf. [4.182]; documentation by kind permission of Siemens AG). (1) lower pole piece. (2) pole-piece sleeve. (3) upper pole piece. (4) upper objective lens flange. (5) lens core. (6) external cylindrical casing. (8,9) rotatable nonmagnetic rings which carry the ferromagnetic chunks (10,11,12,13) of the stigmator; the chunks (10) and (11) are connected to the inner ring (8), the pieces (12) and (13) to the outward ring (9). (14) specimen. (15) objective aperture. (16) illuminating electron beam. (17) imaging electron beam. The pole-piece system represented in the drawing is a so-called 'Hülsenpolschuh' (an approximate translation would be: tubular-shell-confined pole-piece system). Here, the pole pieces are introduced into a tightly fitting ferromagnetic sleeve which aligns them accurately on a common axis. At a low value of the ampere-turns in the lens coil (7), the sleeve acts as a magnetic short circuit across the gap, but with increasing ampere-turns it rapidly oversaturates and becomes magnetically ineffective. The advantages of this design are that it is possible to exchange defective pole pieces for more perfect ones and to reduce the basic astigmatism of the complete pole-piece system by rotating the two pole pieces relative to each other until the mutual azimuthal orientation of minimum astigmatism is attained. (For simplicity, the bronze spacer that separates the pole pieces and determines the gap width has been omitted)

pairs being connected to its own gear ring, with the two chunks on a common diameter on opposite sides of the lens axis. Pinion drives were provided for rotating the rings from outside the lens. The maximum corrective action is obtained if adjoining chunks of the different pairs just touch. If, starting from this position, the two rings are rotated in opposite senses but through the same angles, the correcting action is slowly decreased because the second-order distortion gradually transforms into a fourfold distortion. Complete four-fold distortion is reached if all four chunks are at 90° to each other in azimuthal orientation, and the respective aberration can be neglected because it is of the third order and small in comparison with spherical aberration. If, starting from any of the end or intermediate position the gear rings are rotated in the same sense, the respective angles between the chunks and the correction strength is maintained whereas the azimuthal orientation of the correcting field can be varied (Fig.4.57). With this type of field-distortion stigmator, both amplitude and azimuth of the correction can be adjusted continuously On the other hand, the maximum strength of correction is limited, for the ELMISKOP 1 stigmator to about $\Delta f \approx 1$ µm, whereas with the LEISEGANG stigmator $\Delta f \approx 3$ µm and even more can be attained [4.100].

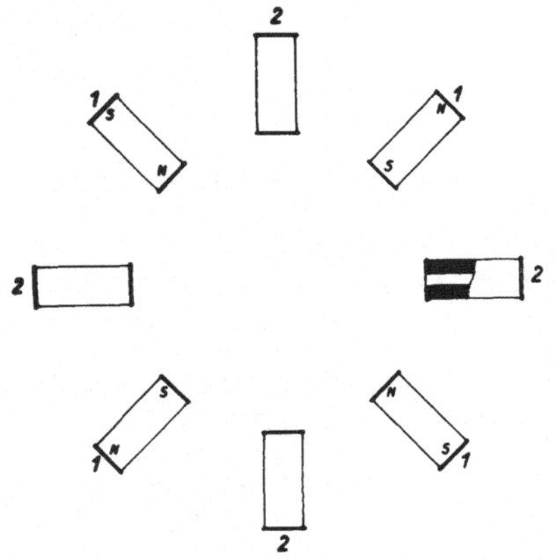

Fig.4.58. Schematic representation of an eight-coil magnetic stigmator for the correction of second-order astigmatism. The axis of the magnetic lens and the electron beam would pass through the point of intersection of the coil axes and be orthogonal to the plane of the drawing. In actual engineering practice, the off-axis ends of the coils are connected by a ferromagnetic ring, which absorbs and short circuits the external magnetic stray field

The other alternative for providing the lens with a stigmator is to employ an optically independent multipole unit [4.10,11], either of the magnetic or of the electric type.

A magnetic stigmator suitable for the correction of second-order astigmatism is shown schematically in Fig.4.58. It consists of two sets of four coils each, numbered 1 and 2 respectively. The sense of the currents in the coils is arranged so that with one opposite pair of coils of each set, virtual magnetic north poles are generated in the coil faces looking towards the electron beam and magnetic south poles with the other respective pairs. (This is shown schematically in Fig. 4.58 for set No. 1.) In order to generate quadrupole lenses, all the virtual poles of each set should have the same absolute value of the magnetic pole strength, and this is obtained automatically with coils having the same shape and number of turns and connected in series.

The optic action of the magnetic quadrupole can be vizualized quite easily. Let us look at the quadrupole coils in the direction of flight of the electron beam and let us assume that the virtual magnetic poles of the coils have been rotated clockwise by 45° with the electron beam as the axis of rotation. If we now furthermore assume that the virtual magnetic poles have been replaced by electric poles of the same respective polarity, then the attractive and repulsive forces which would be exerted by these hypothetical charges on the beam electrons give a qualitative picture of the real forces exerted on the same electrons by the magnetic field of the quadrupole. Thus, the directions of converging and diverging optical action become immediately obvious. By suitably balancing the directions and the absolute values of the currents in the two sets of coils, i.e., by superposing the two quadrupoles 1 and 2 which are rotated by 45° relative to each other, a quadrupole of the required strength can be generated in any desired direction.

An eight-coil stigmator with elongated coils constructed according to the schematic representation in Fig.4.58 requires much space. Usually, this space will be

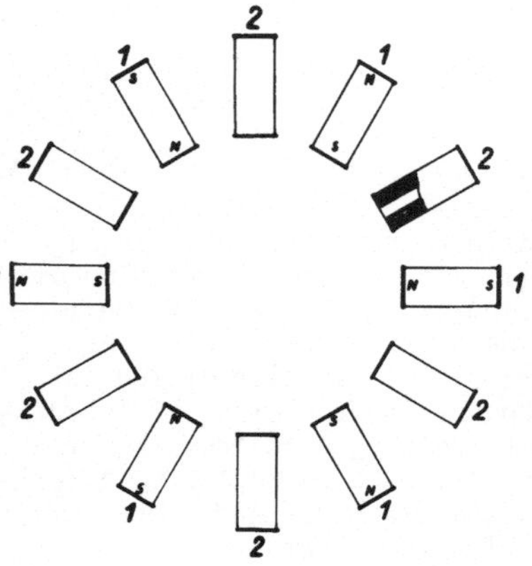

Fig.4.59. Schematic representation of a twelve-coil isoplanator for the correction of third-order astigmatism. The two hexapoles numbered 1 and 2 are rotated relative to one other by 30°. In practice, a ferromagnetically short-circuiting ring is employed to remove the external stray field. Experience has shown that for multipole systems of comparable size, the ampere-turns required for the isoplanator are larger by nearly an order of magnitude than the ampere-turns required for a stigmator of the type shown in Fig.4.58

available with long focal length condenser lenses, which have comparatively large bore diameters and gap widths. Experience has shown that with long-focal-length lenses, third-order astigmatism must be corrected too [4.150]. This aberration can be compensated by employing a magnetic hexapole and in order to permit any required orientation in azimuth, a total of twelve coils are provided, divided up into two sets of six (Fig.4.59). Each set represents one hexapole, and by way of example, the signs of the virtual magnetic poles of the set No.1 have been indicated in Fig.4.59. Again, in analogy to what has been described above, the optic action of the hexapole can be vizualized more easily by an imagined exchange of the actual magnetic poles for electric charges, but now the clockwise rotation has to be performed by the angle of 30° only.

The twelve-coil corrector of Fig.4.59, which is also known as an "isoplanator", has been employed in combination with an eight-coil stigmator as in Fig.4.58 for the condenser lens represented in Fig.4.22 [4.150]. There, the external ends of the coils nearly touch the cylindrical lens casing, which short circuits and absorbs their outward magnetic stray field.

Turning now to quadrupole stigmators suitable for objective lenses, the problems connected with having very little space available become of prime importance. Here, it helps that it is not essential to place the stigmator within the strong magnetic lens field itself. Putting it 1 to 2 cm beyond the gap will cause some image distortion, but only a few tenths of a percent and therefore so little that it does not matter in practice.

The electron optics of magnetic quadrupole lenses has been treated rather thoroughly (e.g. [4.61,64]), including the possibility of reducing the quadrupole aberrations by carefully shaping the poles and hence the magnetic field distribution. For the design of stigmators, detailed considerations of this kind need not be taken into account because the refractive power of stigmators is very low in comparison with the refractive power of the lens to be corrected and only their

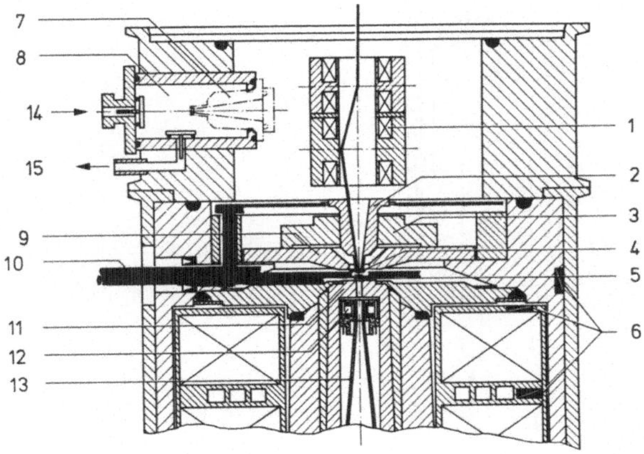

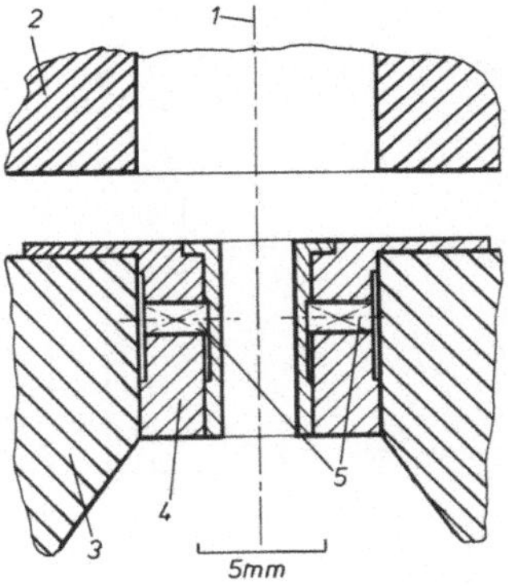

← Fig.4.60. ↑ Fig.4.61.

Fig.4.60. Cross-section of a magnetic objective lens equipped with an electromagnetic eight-coil stigmator and of its adjacent specimen-stage section [4.4]. (1: alignment coils, 2: specimen cartridge, 3: specimen stage, 4: specimen, 5: objective aperture, 6: thermistors, which serve as temperature sensors for minimizing the thermal image-drift by regulation of the cooling-water temperature, 7: position of specimen cartridge in airlock 8, 9: upper objective lens flange, 10: cooling rod connected to anticontamination surfaces, 11: lower objective pole piece, 12: eight-coil electromagnetic stigmator, 13: protection tube, 14: flooding valve, and 15: evacuation line for airlock 8)

Fig.4.61. Particularly small eight-coil stigmator positioned within the bore of the image-side pole piece of a magnetic objective lens [4.206]. (1: lens axis, 2: specimen-side pole piece, 3: image-side pole piece, 4: non-ferromagnetic carrier for stigmator coils 5)

first-order optical properties count. It can thus be understood that stigmators of widely different construction may be employed successfully, and some possibilities will be briefly described below.

A stigmator that still corresponds rather closely to the general scheme of Fig. 4.58 has been incorporated in the lower objective lens pole piece with the ELMISKOP 101 and 102 electron microscopes [4.4,212]. Here, the lower part of the image-side objective pole-piece has been bored out to almost 2 cm in diameter, and into this cavity a set of eight coils, each about 5 mm in size, has been introduced, nearly 2 cm below the objective lens gap (Fig.4.60). A shielding tube separates the coils from the beam and prevents charging-up disturbances in the image, which might arise if stray electrons could reach the insulation of the stigmator coils. A related design is shown in Fig.4.61, but here the eight little coils are carried in eight corresponding bores which have been drilled radially into a cylindrical brass spacer that fits into the bore of the image-side pole piece [4.206].

A quite different construction for a magnetic quadrupole stigmator is represented schematically in Fig.4.62. With this design [4.86] (cf. also [4.200]), the magnetic field is generated by the current flowing in four conductors which are extended parallel to and at equal distances from the lens axis, and which have a mutual angular spacing in azimuth of 90°. The current direction is the same for conductors

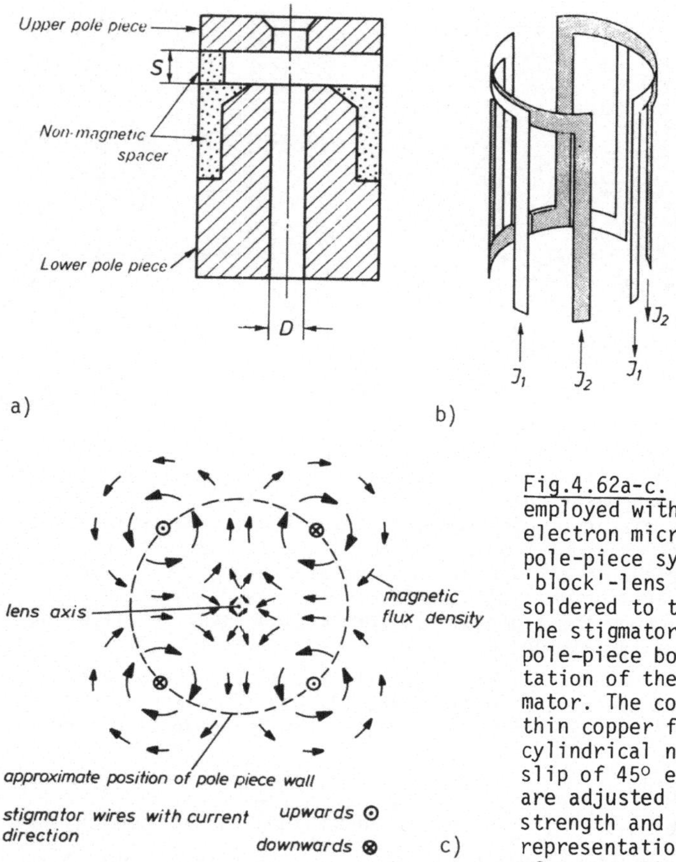

Fig.4.62a-c. Electromagnetic stigmator employed with objective lenses of HITACHI electron microscopes. (a) Objective lens pole-piece system, constructed as a 'block'-lens where the pole pieces are soldered to the non-magnetic spacer. The stigmator is positioned in the lower pole-piece bore. (b) Schematic representation of the construction of the stigmator. The conductors are fabricated from thin copper foil and are wound around a cylindrical non-magnetic carrier with a slip of 45° each. The currents J_1 and J_2 are adjusted independently in both strength and direction. (c) Schematic representation of the field distribution of a magnetic quadrupole lens consisting of four axially extended current-carrying conductors. The essentials of this field distribution remain unchanged by the addition of a pole piece with its bore walls close to the conductors. In particular, the twofold field periodicity around the axis is maintained, as is required for the quadrupole stigmator. (a,b) after [4.200]

on opposite sides of the lens axis and reversed for the other two conductors, so that a quadrupole type field distribution is generated (Fig.4.62c). In actual practice, the conductors are made from copper foil and fixed onto a cylindrical carrier. The complete stigmator is composed of two of these four-conductor quadrupoles which are rotated relative to each other by 45° in azimuthal direction (Fig.4.62b). It can be fabricated with an external diameter that is so small that it fits into the pole-piece bore of the objective lens represented in Fig.4.62a. Here, for each quadrupole, the four long conductors extending parallel to the lens axis are interconnected by shorter circular conductors. The corresponding "pieces" of current contribute a magnetic field which in the beam region is predominately a weak axial field and produces a focusing action similar to a round lens of just one turn. This is only a minor disturbance.

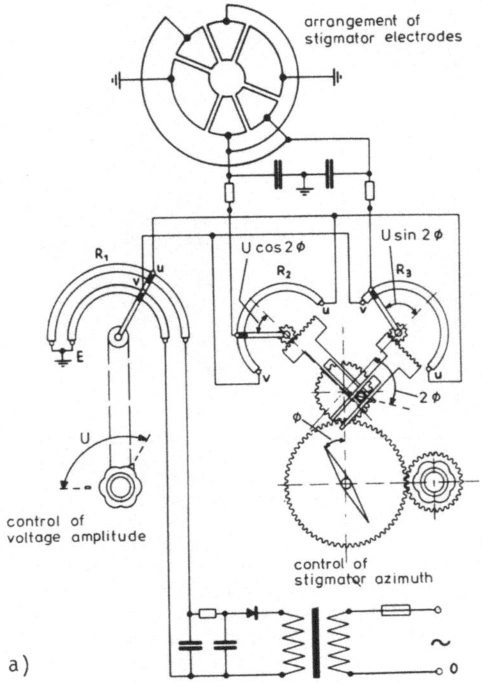

Fig.4.63a,b. Electrostatic stigmator after RANG (adapted from [4.143]). (a) Schematic representation of the electrode arrangement and of the corresponding electrical supply. It should be noted that with this circuit there is no way of electrically aligning the stigmator quadrupole field relative to the axis of the lens field to be corrected. The mechanical alignment device shown in (b) was provided for that purpose. (b) Cross-sectional drawing of a stigmator design actually built. (1) stigmator electrodes, (2) micrometer knob for the alignment of the stigmator with respect to the lens axis, (3) micrometer knob for the alignment of the aperture (4) which simultaneously serves as the actual aperture of the associated objective lens and as a protection against stray electrons for the stigmator. (Return springs are not shown.) A related stigmator design is described in [4.143a]

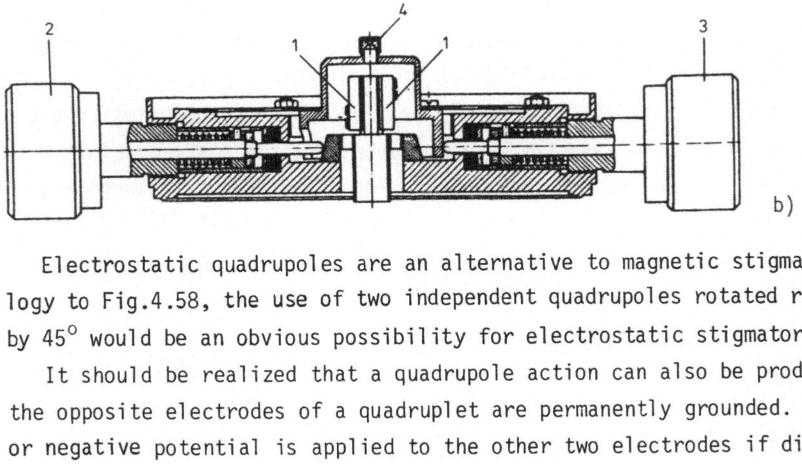

Electrostatic quadrupoles are an alternative to magnetic stigmators, and in analogy to Fig.4.58, the use of two independent quadrupoles rotated relative each other by 45° would be an obvious possibility for electrostatic stigmators too.

It should be realized that a quadrupole action can also be produced if two of the opposite electrodes of a quadruplet are permanently grounded. Then, positive or negative potential is applied to the other two electrodes if diverging or converging quadrupole action, respectively, is required in their direction. With this potential distribution, the potential on the lens axis is also slightly different from ground, and a weak round-lens action results; in practice this usually does not cause much disturbance.

A stigmator designed according to this principle is represented in Fig.4.63. There, the grounded electrodes of the two individual quadrupoles have been condensed in pairs into two large electrodes of double the size of the smaller electrodes, to which voltages different from ground are applied. In azimuth, there is a 45°

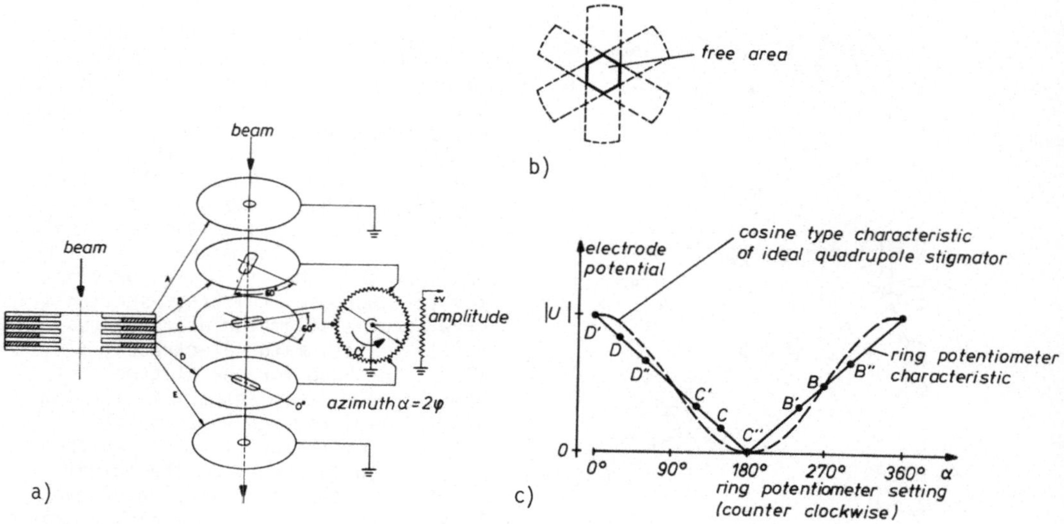

Fig.4.64a-c. Schematic representation of the electrostatic slotted-electrode stigmator (Fig.4.64a adapted from [4.146]). (a) Exploded view of the stigmator and representation of its associated electrical supply circuit. (b) Orientation of the slots as seen in the direction of the beam and resulting beam-transparent free area. (c) Comparison of the voltage characteristic of the three-terminal ring potentiometer with the ideal quadrupole potential distribution at a fixed radial distance. By way of example, combinations of voltages assumed by the slotted electrodes are indicated by the point triplets (B, C, D), (B', C', D') and (B", C", D")

angular spacing between adjacent small electrodes and also between the small electrodes and their respective proximal half-areas of the large electrodes.

It can be shown that for this particular 45° angular spacing, the potentials to be applied to the pairs of small electrodes are proportional to $\cos 2\phi$ or $\sin 2\phi$, where ϕ is the azimuth of the required resulting quadrupole lens as measured with respect to the cos-electrode pair. The angle is to be taken as positive for the sense of rotation that would transfer the cos-electrode into the nearest sin-electrode. A sine-cosine gear system performing according to this specification is also shown schematically in Fig.4.63 [4.143].

Another interesting possibility for a stigmator design is represented in Fig. 4.64. Here, the stigmator is composed of a set of three slotted electrodes which are insulated electrically from each other and from ground [4.146]. In azimuth, the electrodes are oriented in such a way that the long sides of the slots include angles of 60°. Thus, when projected in the beam direction, the common unobstructed free area represents a regular hexagon, each of its opposite sides being formed by the opposite sides of the slot in the corresponding electrode. The whole stigmator is constructed as a "sandwich" of alternate layers of electrode and insulation, with round-bore grounded electrodes terminating the stack on both sides.

The potential is supplied to the inner electrodes by connecting them to a three-terminal ring potentiometer (Fig.4.64). It is seen that at any arbitrary setting of the potentiometer, all the inner electrodes have a common voltage polarity, which is again equivalent to employing a (real or virtual) grounded electrode. Then, with a pure-quadrupole stigmator, the ideal potential distribution at a fixed distance r from the lens axis would have to obey the equation $U(r,\phi) = U_0(r)\cdot(1 + \cos 2\phi)/2$, where ϕ is the angular distance from the orientation of the quadrupole and $U_0(r)$ its potential. It can also be seen in Fig.4.64 that this cosine type characteristic is modelled reasonably well by a straight line characteristic of the ring potentiometer.

The main advantage of the slotted electrode stigmator appears to be its smallness so that it is fully compatible even with a rather cramped objective lens design. An actual stigmator has been constructed which was only 0.75 mm high in the direction of the beam, had a free bore of the same size and a correction sensivity of 0.1 μm/V. The total range of correction was of the order of at least 10 μm [4.146].

As an alternative to the electric rotation of a quadrupole having a sufficient number of electrodes (six separate off-axis electrodes or three slots at the minimum), mechanical rotation of a single quadrupole can also be applied [4.59].

Finally, a set of equations will be listed and explained for predicting the electron optical performance of a stigmator design [4.167]. For this, we shall assume an objective lens of focal length f and image distance b. This objective lens is to be equipped with a stigmator capable of compensating an astigmatic focal length difference Δf_a which corresponds to a mutual spacing

$$\Delta b_a = \Delta f_a \cdot \frac{b^2}{f^2} \approx \Delta f_a \cdot (M + 1)^2 \qquad (4.133)$$

of the two astigmatic image planes. Here, M is the magnification for the imaging process. Furthermore, we shall assume that a quadrupole stigmator is placed at a distance d beyond the objective lens as seen in the direction of the beam.

The stigmator can be of either the magnetic or the electric type, with an effective axial length L of the refracting field (in terms of quadrupole theory this means employing the rectangular field model). In actual stigmator design practice, rather crude approximate values for L serve their purpose adequately. Thus, for the magnetic stigmator the diameter of the coils may be employed for L, and for the electrostatic stigmator the axial length of the quadrupole electrodes.

Now it can be shown that for the correction of Δf_a with the quadrupole stigmator, a specific lens strength q^2 is required which is given by

$$q^2 = \frac{1}{2}\,\frac{\Delta f_a}{f}\cdot\frac{1}{Lf(1 - d/b)^2} \quad . \qquad (4.134)$$

The quadrupole lens stigmator will then have the following focal lengths:

- in the converging plane:

$$f_{s+} = \frac{1}{q\cdot\sin(qL)} \approx \frac{1}{Lq^2} \tag{4.135}$$

- in the diverging plane:

$$f_{s-} = -\frac{1}{q\cdot\sinh(qL)} \approx -\frac{1}{Lq^2} \quad . \tag{4.136}$$

(The approximate formulae are the first terms of the series expansions of the circular and hyperbolic functions. They suffice here because in actual stigmator design practice $qL \lesssim 0.1$.)

Typically, q^2 is of the order of $q^2 \lesssim 10^{-5}$ mm^{-2} and so $q \lesssim 3 \cdot 10^{-2}$ mm^{-1}. It is important to note that the specific quadrupole lens strength q^2 does not depend on the axial extension of the field. The stigmator focal lengths are of the order of $f_{s+} \approx |f_{s-}| \gtrsim 10$ m.

With the magnetic stigmator, the magnetomotive force (ampere-turns) N_sI_s required for each of the coils is then obtained from

$$N_sI_s = \frac{1}{\mu_0}\sqrt{\frac{m_0}{2e}}\, r_s^2 \sqrt{U}\cdot q^2$$

$$= 1.34 \sqrt{U/V}\cdot r_s^2 q^2 \text{ A} \tag{4.137}$$

Here, N_s is the number of turns required for each of the stigmator coils; I_s is the field-generating current, and r_s is the radius of the ferromagnetic short circuiting ring which surrounds the coils. It can be seen right away that with the objective lens stigmators described above only rather weak coils are required, e.g., of about $N_sI_s \lesssim 0.1$ A at a beam voltage around $\Phi = 100$ kV, so that even with currents I_s of a small fraction of an ampere, only a few turns are necessary. It can also be seen from (4.134) that the specific lens strength required for the quadrupole is not affected much by the actual position d of the stigmator with respect to the objective lens.

If electrostatic quadrupoles are employed which have no permanently grounded electrodes, the voltage U_s required at the electrodes of the stigmator for a beam of voltage Φ is

$$U_s = \pm\, r_s^2 q^2 \Phi \quad . \tag{4.138}$$

Here, r_s is the distance of the electrodes from the lens axis. Obviously, for quadrupoles having a pair of permanently grounded electrodes, double the voltage given by (4.138) is necessary. In actual practice, U_s is of the order of 10 to 100 V.

It has been mentioned that, as regards the correction strength of the stigmator, its distance d from the objective lens is not particularly critical. On the other hand, if this distance d becomes large in comparison with the objective focal length f, a noticeable amount of elliptic distortion may arise which is given by

$$\frac{\Delta M}{M} = 2\,\frac{M_+ - M_-}{M_+ + M_-} \approx \frac{d}{f} \cdot \frac{\Delta f_a}{f} \quad . \tag{4.139}$$

Here, M_+ is the magnification in the direction of divergence of the stigmator, and M_- the magnification in the direction of convergence. Thus, if the astigmatic focal difference to be expected for the pole-piece system is known, an upper limit for d can be established from the specified limit of distortion by employing (4.139). In practice $\Delta f_a/f \lesssim 10^{-3}$, and if the distortion should remain within $\Delta M/M < 1\%$, (4.139) gives $d \lesssim 10f$. This provides an idea of the range available for d.

Finally, in an actual lens, it cannot be expected that with an electrostatic quadrupole stigmator all the electrodes will be at exactly the same radial distance from the magnetic field axis of the pole-piece system. A similar misalignment must likewise be expected with a magnetic stigmator for the coils and perhaps the ferromagnetic short-circuiting ring. Thus, the axis of the quadrupole stigmator could be transversely displaced with respect to the lens axis, and on energizing the stigmator, an image shift proportional to the correction strength would result [4.206]. This may become rather inconvenient in practice for the binocular observation of the progress of correction, and some remedy is called for.

Clearly, a mechanical means of alignment would require a comparatively complicated and expensive construction, electrical 'balancing' being much cheaper. The idea behind the balancing is to adapt the exact number of the ampere-turns of each of the coils of the magnetic stigmator or the exact voltage at each electrode of the electrostatic stigmator according to (4.137,138), so that the individual values of r_s for each coil or electrode are employed. Shunting the coils by suitable resistors and drawing the electrode potentials from voltage dividers will easily serve this purpose. In practice, potentiometers would be used in both cases so that the proper balancing can be carried through in a controlled manner while observing the image. A rather large range of variation must be provided for the balancing elements because both ampere-turns and electrode potentials depend on the square of r_s and hence change appreciably even with small variations of r_s.

4.4.3 Apertures and Their Alignment

No electron lens can perform to satisfaction unless its imaging beam angles are limited by suitable apertures. Additional apertures are also required with objective lenses if the specimen has to be surrounded by liquid-nitrogen-cooled surfaces to suppress specimen contamination by freezing out the hydrocarbon gas molecules and thus drastically reducing their partial pressure. Narrow apertures then have to be provided in order to permit the electron beam to enter and leave the cold chamber without admitting too large an inflow of contaminating gas molecules from the surrounding vacuum space. Combining both types of apertures together with a

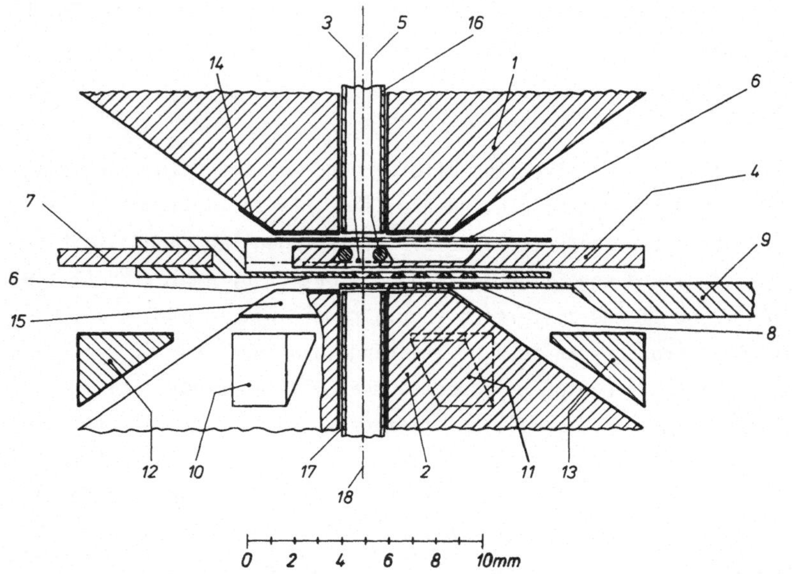

Fig.4.65. Schematic representation of the different types of apertures typically employed with electron objective lenses and of their position within the lens gap. The configuration shown here corresponds closely to the design employed with the condenser-objective lens of the electron microscope described in [4.157].
(1) illumination-side pole-piece, (2) image-side pole-piece, (3) specimen clamped to specimen carrier (4) by an expanding ring-shaped spring (5) [4.160a]. With the condenser-objective lens, in addition to the specimen stage movement orthogonal to the lens axis, provision must be made for traversing the specimen in the axial direction (Fig.4.70). (6) multiple-bore aperture blades cooled to liquid-nitrogen temperature for the prevention of specimen contamination. (7) thermally insulating plastic strip for connecting the cooled blades (6) to their alignment device (Fig.4.69). (8) multiple-bore objective aperture blade and (9) rod connecting it to its alignment device. (10), (11), (12), (13) ferromagnetic chunks of a RUSKA-type magnetic stigmator similar to the one represented in Fig.4.57. (14), (15) molybdenum caps for protecting the pole-piece surface against mechanical damage. (16), (17) molybdenum protection tubes which can be removed for cleaning. (18) lens axis

specimen cartridge (preferably of the side-entry type) within the narrow gap of an objective lens pole-piece system may lead to a rather crowded design (Fig.4.65).

The bore diameter of the apertures usually amounts to 50 μm or even less. It would therefore be a formidable task for the workshop to keep the fabrication tolerances narrow enough to permit the use of a rigidly fixed aperture having its bore firmly centred on the lens axis. Moreover, an aperture exchange device containing a supply of several apertures is most useful. This permits replacement of the contaminated ones with new clean apertures without breaking the vacuum and interrupting the operation of the electron optical apparatus (Fig.4.65). Thus, some kind of alignment device is called for.

The most straightforward system suitable for this purpose comprises two micrometer screws mounted at right angles to each other (Fig.4.66). This arrangement

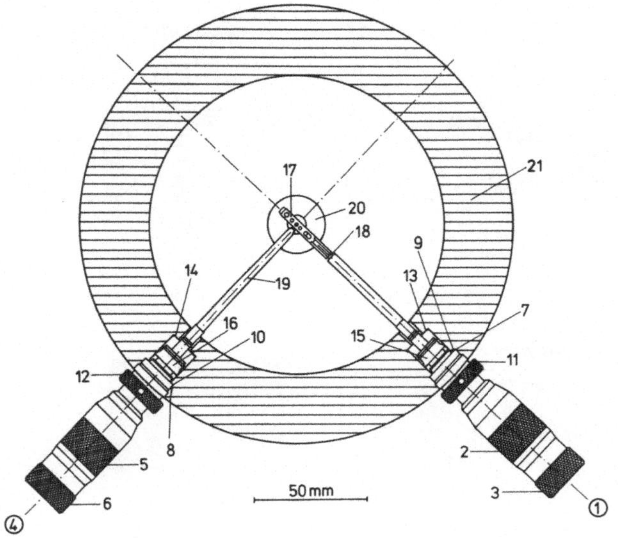

Fig.4.66. Cross-section of an exchange and alignment device for lens apertures, employing two micrometer controls arranged at right angles to each other and to the electron beam. (1) micrometer control for exchange and fine adjustment of apertures (for details see Fig. 4.67a). (2) control ring for exchanging apertures. (3) control knob for fine adjustment of aperture position. (4) micrometer control for transverse adjustment of aperture position. (5) control ring for preliminary coarse adjustment of the aperture position. (6) control knob for fine adjustment of the aperture position. The micrometers are locked to the lens casing (21) by means of the connecting faces (7), (8), which are pressed against the shoulders (9) (10) by tightening the screws (11), (12). The shoulders (13), (14) prevent the vacuum gaskets (15), (16) from being inadvertently pushed into the space inside the casing. (17) aperture carrier, soldered to the end of resilient rod (18). (19) push rod. (20) pole piece. (21) lens casing and vacuum wall of the instrument. (The particular design represented here has been developed for the magnetic objective lens of the electron optical bench described in [4.181])

has been employed in many commercial electron microscopes (e.g. [4.120,175a]). Micrometer screws suitable for this purpose are represented in cross-section in Fig.4.67.

With an aperture system containing several apertures such as the one illustrated in Fig.4.65, the provision of two alignment rods for each aperture would result in a rather crowded and complicated construction. A more advantageous design is depicted schematically in Fig.4.68. Here, the objective apertures and the two aperture plates for the liquid-nitrogen-cooled anti-contamination chamber are each carried and traversed by a single rod. The double-screw alignment systems for driving the rods are shown in cross-section in Fig.4.69 [4.161].

4.4.4 Special Aspects of Specimen Stages Relevant for the Optical Performance of the Magnetic Lens

Electron microscope specimen stage design is intimately related to the construction of objective lenses and so to magnetic lens design. A comprehensive discussion of specimen stage design and the whole gamut of associated cartridges for diverse specimen treatments would, however, go far beyond the scope of this book. Therefore, only a few points will be briefly mentioned here which are relevant for the operation or the optical performance of the associated magnetic lens.

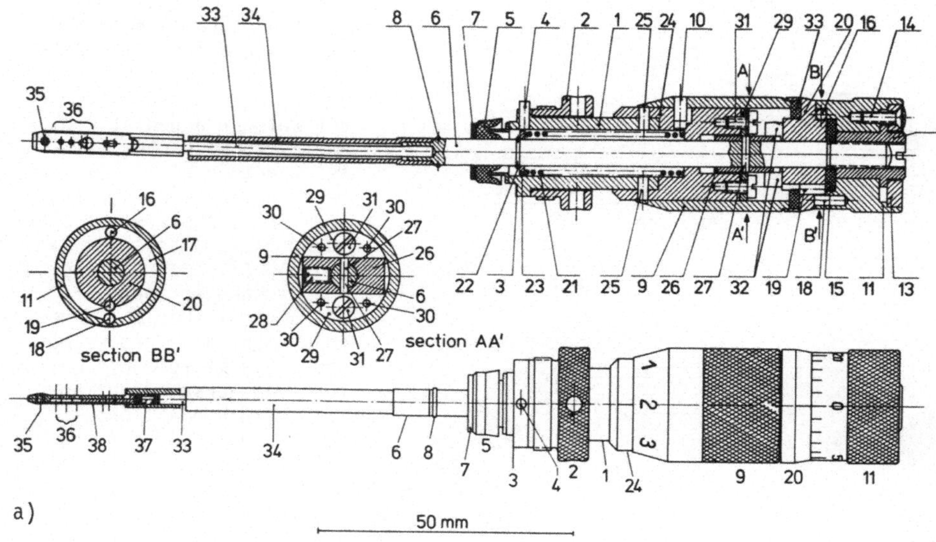

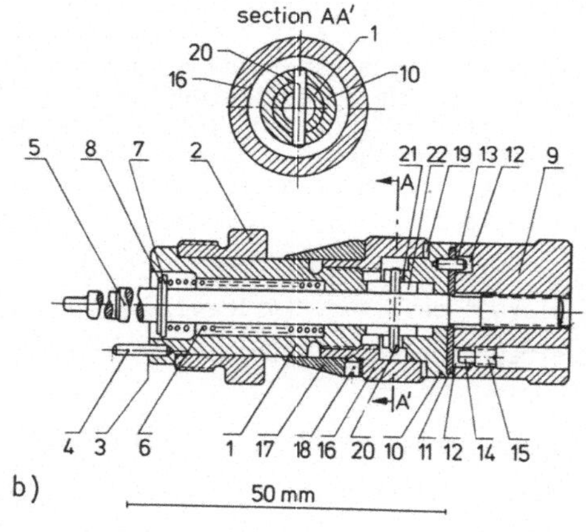

Fig.4.67. (a) Cross-section of micrometer control for exchange and alignment of four apertures. The main body 1 of the micrometer is locked to the lens by tightening the ring screw 2 and pressing the contact surface 3 against a corresponding surface provided in the lens casing (Fig.4.66). The orientation of the micrometer with respect to the lens is secured by guiding the pin 4 into a corresponding recess in the lens casing. As a vacuum seal, the gasket 5 is slipped over the push rod 6 and is secured by a plastic washer 7 and the circlip 8. Thus, the gasket 5 is always retained on the rod 6 and cannot stick to its seat in the lens casing. The mutual exchange of the four apertures is effected by rotating the knurled collar 9 whereby the pin 10 is slipped into related recesses of a coulisse guide. The resulting axial stepping of the collar 9 generates a corresponding axial movement of the control knob 11 and the push rod 6. The end of the the push rod 6 is threaded and screwed into a cylindrical nut 12. This nut is locked to the control knob 11 by means of a clamping ring 13, which is tightened by three screws 14 (only one is shown). Then, on rotating the control knob 11, the push-rod 6 is moved back and forth. If a fine enough thread is employed and a hard plastic washer 15 introduced for reducing friction, a smooth and accurate fine alignment of the aperture results. The range of adjustment is limited to a distance corresponding to a little less than two complete turns of the knob 11 by a design indicated in the transverse section BB'. Here, a ball 16 rolls freely in groove 17 and is chased by the moving pin 18 fixed to the knob 11. A second pin 19 connected to the non-rotating block 20 obstructs the movement of the ball and thus limits the knob 11 to a little less than two turns. With respect to the lens axis, the position of the range of alignment can be adjusted after slackening the clamping screws 14 by engaging a screwdriver in the slots of the cylindrical nut 12 and rotating it without altering the angular position of the knob 11. The alignment mechanism is tightened and any slackness removed by the action of the spring 21 which extends between the circlip 22 retained ring 23 and the coulisse block 24, connected to the main body 1 by

pins 25. In order to ensure correct functioning of the micrometer, any possible rotation of the rod 6 is prevented by mounting it in the guide block 26 by means of the pin 27 and the grub screw 28. The guide block 26 glides on the edges of the plates 29, which are locked into their exact position against the coulisse block 24 by pins 30 and screws 31. A hard plastic washer 39 can be useful to remove friction between the block 20 and the aperture exchange control collar 9. The aperture holder proper is connected to the adjustable rod 6 by means of a resilient bronze spring-rod 33, which is pre-stressed in the direction against the second micrometer of the pair. If the second micrometer has not been introduced into the lens, the resilient rod 33 is halted by the protection tube 34, so that mechanical damage to the aperture holder is safely prevented. In addition to a large diameter bore 35, three small bore apertures 36 can be inserted simultaneously in the form of the widely used platinum or molybdenum discs of 2 mm external diameter. The apertures are introduced into a box-shaped channel and advanced into position by a little spring 37 actuated slide bar 38. Single or multiple bore apertures with bores ranging from about 5 μm to some 100 μm diameter are commercially available from Günther Frey Comp., Berlin, for example, or industrial companies manufacturing electron microscopes. (The micrometer has been designed by F. Stöcklein (unpublished). Drawing by kind permission of the Fritz-Haber-Institut der Max-Planck-Gesellschaft, Berlin-Dahlem. Some minor modifications were made for the present purpose.) (b) Cross-section of a control micrometer for the transverse alignment of the aperture, suitable for use with an aperture exchange and alignment device, e.g., as represented in Fig.4.67a (cf. Fig.4.66). The main body 1 of the micrometer is rigidly connected to the magnetic lens by tightening the locking screw 2 in order to press the connecting surface 3 against a corresponding shoulder in the lens casing. The proper azimuthal orientation of the micrometer is determined by the pin 4. The push rod 5 is advanced against the lens axis by the helical spring 6, which extends between a seat in the main body 1 and a ring 7 retained on the rod 5 by a circlip 8. A fine thread cut on the off-axis end of the rod 6 is gripped by the correspondingly threaded fine control knob 9. This knurled knob 9 abuts on block 10, and a plastic disc 11 is interposed between the two to reduce friction. By turning the control knob 9, a fine displacement of the push rod 6 is produced. The rotation of knob 9 is restricted to a little less than one turn by a system of two pins extending into the circular grove 12. One pin 13 is firmset in block 10, and the other 'pin' is the cylindrical end 14 of the grub screw 15 located in knob 9. Thus, pin 13 blocks the path of rotation of the cylinder 14 and limits the angle of turn available for knob 9 in the practical operation of the lens. When selecting the range of fine alignment, the grub screw 15 is retracted and thus obstruction to the rotation of knob 9 removed. By rotating the knurled ring nut 16 which has a thread of coarser pitch, a rapid axial translation of the block 10 and thus a coarse alignment of the apertures can be effected. A scale-ring 17 can be fastened in a chosen angular orientation by employing the grub screw 18 so that the relocation of the aperture in a specific position is facilitated. Friction between nut 16 and block 10 can be reduced by employing a plastic ring 19. In order to ensure proper operation of the fine and coarse alignment controls, possible rotation of the push rod 5 and the block 10 must be prevented. This is achieved by pin 20 which protrudes from rod 5 and travels in corresponding slots 21 and 22 machined into block 10 and the main body 1

Axial displacement of the specimen is a most useful stage feature. It is nearly indispensable for a condenser-objective lens, because there, the specimen has to be positioned in axial direction to an accuracy of a few micrometres at most. This is necessary so that the demagnified image in the specimen plane of the field-limiting aperture, which serves as a sub-micron electron probe, and also the magnified image of the specimen and the probe on the final screen are reasonably well in focus at the same time (cf. [4.164] and Fig.4.9a). In practice, the specimen is usually a little bent and thus it extends in axial direction by more than these

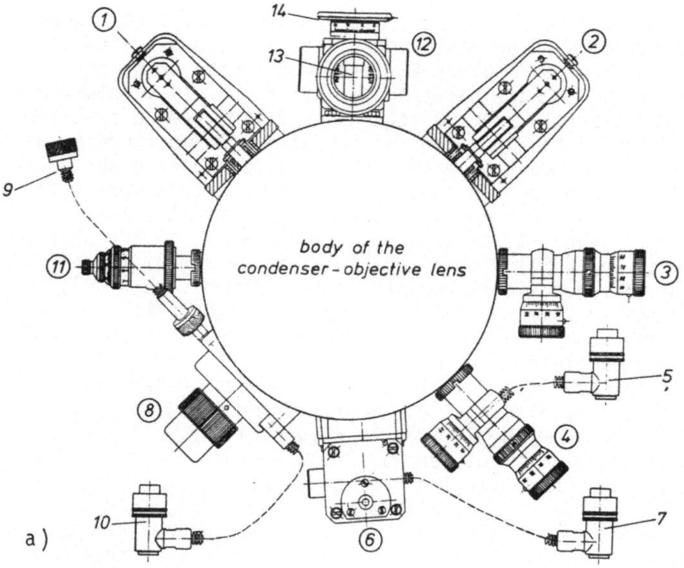

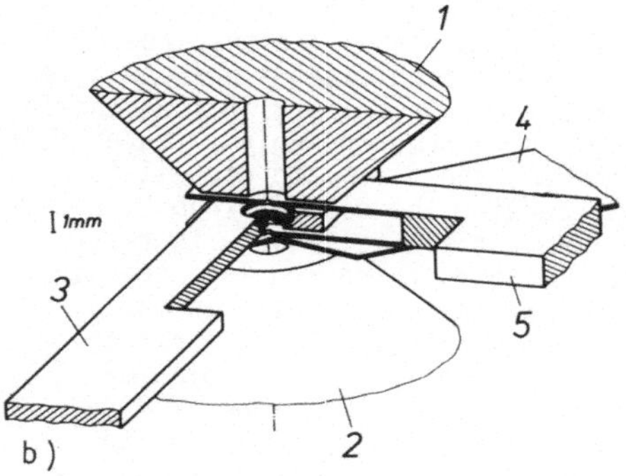

Fig.4.68a,b. From this illustration the potential advantages of a more crowded design, if a single-rod aperture-alignment system is employed, become clear. Here, the apertures are introduced into a holder, rigidly connected to the end of the alignment rod, which is in turn actuated by two micrometer screws at right angles to each other. The space-saving single-rod construction is particularly useful for objective lenses because not only the actual lens aperture but also the specimen and the entrance and exit apertures of the liquid-nitrogen cooled anticontamination chamber have to be precisely aligned about the lens axis. As an example, the special design situation that arises in the case of a particular condenser-objective lens is illustrated; the lens in question is schematically represented in a condensed functional cross-section in Fig.4.65 and employed with the electron microscope described in [4.157]. a) Axial view from above of the arrangement of the actuating micrometers, which are connected to the casing of the condenser-objective lens at the level of the pole-piece system. (1) (2) micrometers and stepping-down bent levers for specimen traverse (the return springs are contained in the pole-piece unit and so are not shown here). (3) single-rod objective-aperture micrometer control. (4) single-rod anticontamination-chamber apertures control (shown in cross-section in Fig.4.69). 5 plug with leads to thermocouple used for measuring the temperature of the wall of the anticontamination chamber. (6) small liquid-nitrogen dewar connected to and cooling the anticontamination chamber. 7 plug connecting to thermocouple used for measuring the temperature of the dewar. (8) specimen airlock (for a detailed description cf. [4.193]). 9 pumping line for evacuation of specimen airlock. 10 plug of safety circuit, which disconnects the beam voltage while inserting or withdrawing the specimen. (11) adjustment controls for a RUSKA-type magnetomechanical stigmator (cf. Fig.4.65). (12) pumping line connection for the evacuation of the pole-piece space, containing the "screwdriver"-type control-rod 13 for the axial translation of the specimen (cf. Fig.4.70), which is actuated by the toothed wheel 14 and a corresponding gear wheel (not shown). b) Schematic perspective drawing of the pole-piece space, illustrating the mutual orientation of the single-rod aperture micrometer controls. Owing to the smallness

of the gap space available, the single-rod aperture-control design will be distinctly preferable to a two-rod design as in Fig.4.66. 1 illumination-side pole piece. 2 image-side pole piece. 3 specimen carrier and specimen. 4 objective aperture plate carried by its associated control-rod. The part of the rod that extends towards the lens axis is shaped as a flat resilient strip tension-biased against the image-side pole piece. Thus, the aperture plate is always in contact with the pole-piece face and possible vibrations are removed by friction damping (cf. Fig. 4.69). 5 beam entrance and exit aperture-plates of the liquid-nitrogen-cooled anticontamination chamber. The two plates are jointly carried by a single control-rod and their aperture bores are prealigned with respect to each other. By guiding the thin aperture plates in corresponding slideways provided in the thermally insulated anticontamination chamber, thermal contact of the plates with either pole piece or specimen carrier is safely avoided. (For a description of the anticontamination chamber see [4.161].)

few micrometres. Therefore, the required positioning accuracy cannot be attained without a means for precision axial displacement of the specimen during operation of the electron microscope.

A means of effecting such a displacement, which proved to work highly satisfactorily in practice, is represented in Fig.4.70. Here, the specimen 3 is introduced into a block 2, which is connected to the body 6 of the specimen stage by a plate spring 4. On turning the screw 10 by actuating the loosely catching screwdriver 11, the little lever 9 pushes the block 2 and hence the specimen down, or permits it to be pulled back up by the action of the spring 8. The movement of the block 2 is made possible through the bending of the plate spring 4 which acts as a "virtual" hinge with its turning joint about half way along the spring length. Thus, because the specimen never loses its rigid connection to the stage body 6, the axial displacement becomes extremely smooth and free from jitter, and it is not accompanied by any noticeable transverse-to-axis specimen movement.

Quite generally, care should be taken with the construction of those parts of specimen stages that are introduced into the region of strong magnetic field close to the lens axis (Fig.4.68b). Evidently, non-ferromagnetic material must be employed here so that the rotationally symmetric distribution of the magnetic lens field is not disturbed. In order to be on the safe side, for the magnetic permeability μ_r the condition $\mu_r - 1 < 10^{-3}$ should be observed. Unfortunately, materials such as brass or bronze which are considered to be non-magnetic, sometimes do not live up to this expectation, and commercial materials specifications usually just do not consider this point. Materials that can be relied upon to be sufficiently free from magnetism are beryllium-copper alloys.

In cases of doubt, it is always good policy to check the magnetic properties by actual measurement (for methods see e.g. [4.5]), even after the parts have been finished in the workshop. There is always a danger of their becoming magnetic in the fabrication process by picking up minute morsels of the tools. For instance, in the fabrication of the little spring 8 employed in the specimen stage represented in Fig.4.70, even if the spring was made from non-magnetic wire, winding of the wire

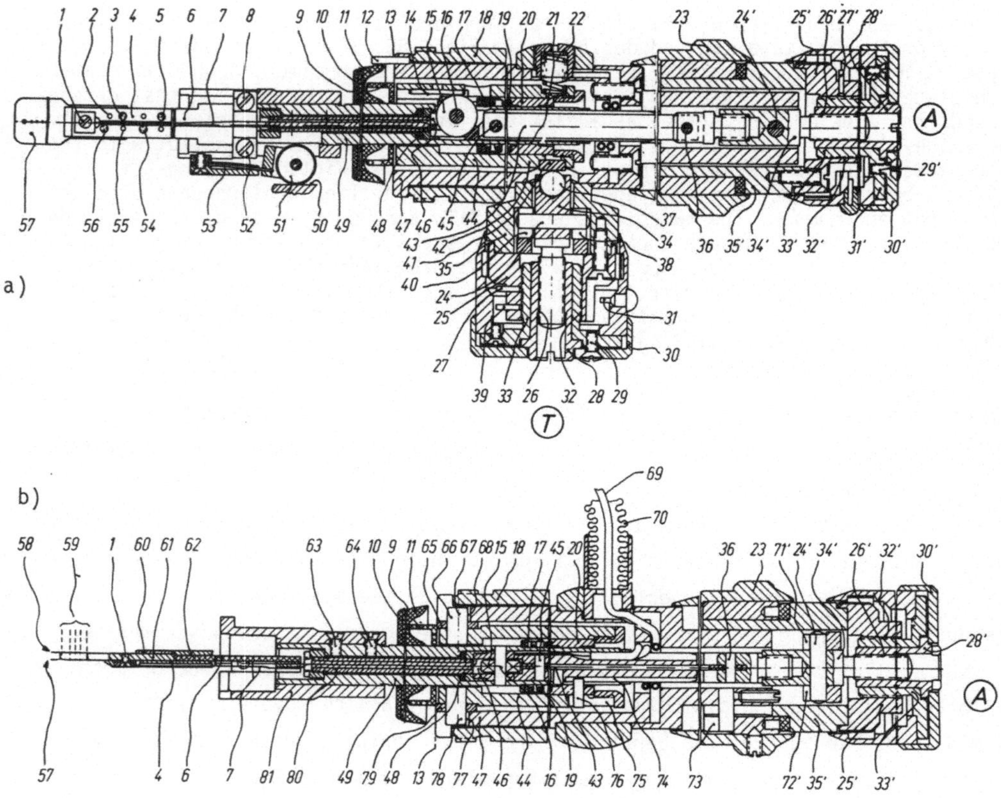

Fig.4.69. Cross-sections of a single-rod aperture alignment and exchange device (adapted from [4.161]). Here, (a) shows the horizontal and (b) the vertical section, and both sections contain the rod axis. The particular device represented here by way of example has been designed to permit manipulation of the entrance and exit apertures of a liquid-nitrogen-cooled anticontamination chamber for a condenser-objective lens and is indicated by ④ in Fig.4.68a. The objective aperture exchange and alignment device designated by ③ in Fig.4.68a is of essentially identical construction, but in this case only a single aperture blade is employed and neither thermal insulation nor a thermocouple is required (see below).

The main body 47 of the micrometer is locked to the lens body by tightening the ring screw 18 and so pressing the contact surface 13 against a corresponding shoulder machined into the cylindrical casing of the lens. The proper orientation of the micrometer is secured by a pin 12, which has to be inserted into a corresponding bore in the casing. The cylindrical pins 66, 78 with washers 65, 79 serve as pivots for the tubular shell 44, which contains the actual aperture-carrying rod 49. The pins 66, 78 rest in seats 68, 77 machined into the micrometer body 47 with a common axis 67. Thus, shell 44 and rod 49 can be tilted together through small angles around the axis 67; this results in a transverse movement of the aperture blades in a direction at right angles to the axis of rod 49.

This movement is effected by employing the micrometer screw Ⓣ. The latter has a push-rod 34, which abuts against the tubular shell 44, and a helical spring 22 provides the backing-up force. In order to achieve a well-defined point contact, rod 34 is tipped by a hardened steel ball 37 (from a ball-bearing), which abuts against a hardened and polished steel platelet 41 fixed to the end 42 of tube 44. The tubular shell 44 with the enclosed aperture-carrying rod 49 forms a lever, having the axis 67 of the cylinders 66, 78 as pivots. With the present ratio of the lever arms, the displacement of the micrometer screw Ⓣ is transformed into a 2.5 times magnified transverse displacement of the aperture plates. Therefore, a particularly sensitive advance is required for the push rod 34. This is obtained by means of a differential screw system: The threaded end 26 of rod 34 is gripped by the tubular nut 28, 33 which in turn is threaded on its outside 32 and screwed into a further nut 25. Nut 25 is mounted on the body 35 of the transverse micrometer screw (T). The two threads 26, 32 are fabricated with slightly different pitch so that the movement of rod 34 is governed by the difference between the pitches. This difference can easily be made equal to a comparatively small fraction of the actual pitches of threads 26 and 32. (With the micrometer shown here, the difference pitch amounts to 0.1 mm.) An arbitrary rotation of rod 34 is prevented by its carrying a transverse pin 24, which travels in corresponding slots 38, 40. The tubular nut 28, 33 is clamped to the control knob 30 by tightening screws 29 (only one of a total of three is shown). The number of turns available for knob 30 and nut 28, 33 is limited by two fixed pins 27, 39 and a moving pin 31 which is inserted into the control knob 30 and is intercepted by the fixed pins at the limits of the required range of travel (five turns are available with the design represented here). When setting the micrometer during coarse prealignment, the screws 29 are slackened so that the inner tubular nut 28 can be rotated individually with a screwdriver while the outward screw-part 33 is held fast with knob 30.

The micrometer (A) provided for traversing the aperture-carrying rod 49 in the direction of its axis is seen to be of analogous design. Thus, corresponding parts have been indicated by the same reference numbers, but marked by primes for micrometer (A). Rod 49 is flexibly connected to rod 34' of micrometer (A) by means of a flat metal bar 43, suspended between the pins 36, 45, which serve as pivots. It is observed from a) that tilting the aperture-carrying rod 49 produces a transverse motion of its end, which carries the pin 45. The bar yields to this motion by tilting around its pivot 36. Any possible play in the mechanical chain consisting of rod 49, bar 43, and rod 34' is removed by the tension of a helical spring 19 extending between a collar 17 connected to rod 49 and the tubular ring 20 mounted within the tube 44. A second purpose of the spring 19 is to force the tubular shell 44 backwards and the pivot pins 66, 78 into their seats 68, 77 in the micrometer body 47. Ring 20 carries a pin 76, which in connection with slots 74, 75 determines the proper angular orientation of rod 49 with respect to its axis.

Any possible play of rod 49 within its tubular shell 44 is removed by means of the spring 53 cushioned wheel 51 which abuts against a corresponding flat surface 50 machined into the lens casing. A second wheel 15 is inserted into a recess 46 in rod 49 and the axis 16 of wheel 15 is fitted into its walls. Due to the transverse force exerted by the spring 53, rod 49 is pressed in a lever-like manner against the internal wall of the shell 44 at two points: At the point of contact of the wheel 15 with the bearing surface 14 provided within shell 44, and at the point of contact of the opposite end 21 of the external wall of rod 49 and the tubular ring 20.

The coarse traverse required for moving different aperture bores onto the lens axis is effected by turning the knurled sleeve ring 23, which acts on a coulisse guide 73 and shifts the whole micrometer (A) back and forth.

The aperture blades 57, 58 (five aperture bores 59 are indicated) are fabricated from 0.1 mm thick molybdenum sheet and clamped against a copper block 61, by two thin copper plates 4, 60. The plates are tightened together by copper screws 3, 56; pins 2, 55 are provided to safeguard the proper mutual orientation of the aperture bores. Thus, the aperture blades can be exchanged easily if they are damaged or if other bore diameters are required. The whole aperture-blades assembly is clamped to a slightly resilient blade 6 (made from plastic for thermal insulation) by a copper clamping block 62 with screws 5, 54.

The plastic blade 6 is screwed (8, 52) against a bronze cage 81 which also carries the spring 53 and wheel 51 assembly and is screwed (63, 64) to the rod 49.

A thermocouple is clamped to the copper block 61 by means of screw 1, so that the cooling efficiency of the apertures can be monitored. Its leads 7 are threaded through two bores in a thin ceramic tube 80, which fits into a corresponding bore in rod 49. For vacuum sealing, the bores in the ceramic tube were filled with

(continued see page 290)

epoxy resin, and a gasket 48 is compressed into the space between the ceramic tube and a shoulder in rod 49. The thermocouple leads are brought out via a flexible, bellows 70 protected line 69. As vacuum seal to the vacuum space of the lens, a gasket 11 and associated plastic support rings 9, 10 have been placed over rod 49, as close as possible to its pivot axis 67

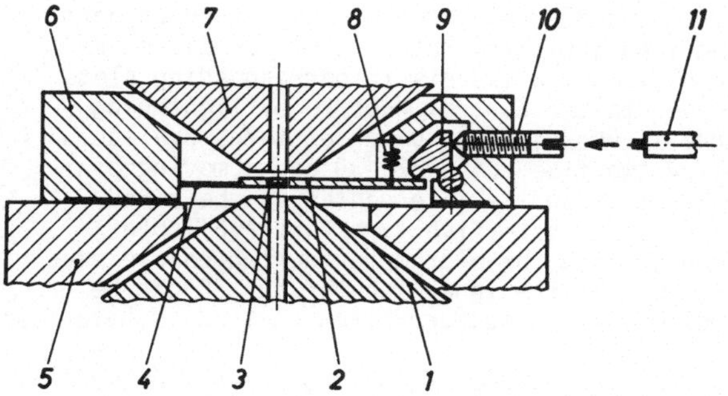

Fig.4.70. Schematic representation of a device permitting high-precision in situ axial displacement of the specimen [4.164]. An actual stage constructed after this principle has been employed with the condenser-objective lens of the electron microscope described in [4.157], permitting precisely controlled axial specimen adjustment down to a fraction of a μm without any disturbing transverse specimen movement. 1 image-side pole piece and 7 illumination-side pole piece. 5 cage for RUSKA-type magnetomechanical stigmator (Fig.4.65), having a flat polished surface on which the specimen stage 6 can glide smoothly and without sticking. The specimen 3 is clamped to a flat specimen carrier (not separately shown and inserted into a corresponding mounting machined into a flat bronze tongue 2. This tongue is connected to the stage 6 via a thinned resilient portion 4 which can be bent like a flat spring and acts as a loose-free hinge. The other end of tongue 2 is pulled against a nose of the lever 9 by a small helical spring 8. The axial position of the nose of lever 9 and thus of the specimen can be adjusted by means of the screw 10 which is threaded into the stage 6 and actuated by the screwdriver-type control 11. The edge of the screwdriver 11 can slide in the slit of screw 10, and moreover, the control rod 11 (13 in Fig.4.68a) is flexibly connected to its adjustment gear wheel (14 in Fig.4.68a). Therefore, the axial position of the specimen is unaltered if the specimen stage 6 is traversed. The high precision of the adjustment movement of the specimen benefits from the fact that the axial movement of the speçimen is less than the movement of the nose of lever 9 by a factor in direct proportion to the respective distances of lever nose and specimen from the centre of the 'virtual hinge' portion 4

around a steel mandrel was sufficient to render the finished spring so magnetic that a noticeable deterioration of the alignment of the lens resulted. Such a disturbance could be avoided by employing a brass mandrel. For machining, it is advisable to use carbide-tipped cutting tools.

Much useful information about specimen stages is contained in [4.67,205].

4.4.5 Mechanical Lens Alignment Devices

When pole-piece systems are actually fabricated, the axes of the two bores never completely coincide due to unavoidable machining tolerances, the main error being a mutually parallel displacement of the axes. If, for objective lenses, this parallel displacement exceeds a few micrometres, a quite noticeable deterioration of the resolving power results [4.2]. Keeping the machining tolerances within limits as narrow as this is quite a problem, even if, with a block-type lens (Fig. 4.62a), both bores are made during a single clamping on the lathe. Therefore, some device for aligning the pole pieces with respect to each other is of considerable advantage for objective lenses.

For this, two different approaches are available.

1) Alignment of the pole-piece system after withdrawal from the objective lens.

2) In-operation alignment while observing an image obtained with an alignment ray path.

Although at first sight 2) seems to be the method of choice, 1) has for its own particular advantage the fact that the μm-precision alignment can be performed without subjecting the pole pieces to the operational magnetic forces. These forces can well amount to the order of 10^3N, attracting the pole pieces against the lens cores and causing so much friction that the alignment movement becomes rather jerky and difficult to control. Alignment procedures are available for both approaches [4.165,170].

A pole-piece system of type 1), suitable for alignment after withdrawal from the magnetic lens body, is represented in some detail in Fig.4.71 [4.160]. The pole pieces 1 and 15 are connected by the non-magnetic bronze spacer 3, but this connection is completely rigid only for the upper pole piece 1, which fits tightly into the spacer 3 and is pressed against it by the tapered ring 2. The latter is pulled against the spacer 3 by little screws, which are not shown here. The lower pole piece 15 is joined to the non-magnetic ring 13 by press fit and has some clearance with respect to the other parts of the pole-piece system, so that it can be traversed a little at right angles to the lens axis 30 after the screws 12 have been slackened. No additional screws have been provided for moving the pole piece 15 by the prescribed amount. Experience has shown that a quite satisfactory alignment is obtained by placing the pole-piece system upside down under a light microscope (i.e., with the pole piece 15 upwards), traversing the pole piece 15 by tapping the ring 13 slightly at the obvious places with the blunt end of a pencil and controlling the proper amount of movement at high magnification under the light microscope. After the alignment, the pole piece 15 is fixed in position with respect to the spacer 3 and the other pole piece 1 by tightening the axial screws 12. Teflon washers between the screw heads and the ring 13 prevent any noticeable transverse forces from arising when tightening the screws 13; such forces might deteriorate the alignment just reached.

As an example of a pole-piece alignment device that can be employed during operation of the lens, reference is made to Fig.4.60, where the upper pole plate 9 of the objective lens represented here can be transversely aligned by a system of two driving rods and one counter spring (none of these is shown in Fig.4.60, because unfortunately, no technical details have been published). A similar system is incorporated in the lenses of the PHILIPS EM300 and 301 electron microscopes.

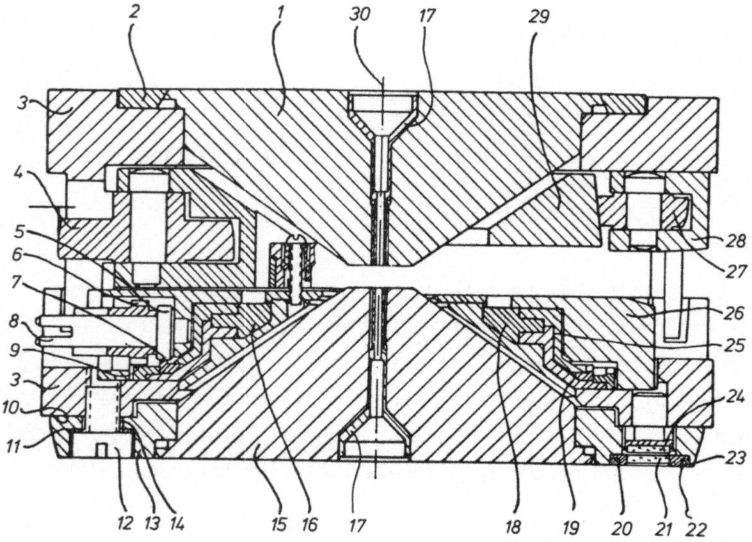

Fig.4.71. Condensed cross-section of the transversely alignable pole-piece system employed with the condenser-objective lens of the electron microscope described in [4.157]. 1 illumination-side pole-piece system. 2 tapered clamping ring. 3 bronze spacer which connects both pole pieces into a single unit. 4 roller with axis connected to the specimen stage; the roller is traversed by the push rod of the stepping-down lever represented as (2) in Fig.4.68a. 5, 6 gear wheels for adjusting a magnetomechanical stigmator after the design principle illustrated in Fig. 4.57. 7 toothed ring catching with gear wheel 5. 8 slot for insertion of the edge of a screwdriver control provided for rotating the gear wheel 5 (the screwdriver control is indicated by (11) in Fig.4.68a). 9 toothed ring catching with gear wheel 6. 10 polished surface supporting the ring 13, which carries the image-side pole piece 15 and can be traversed smoothly after screws 12 (only one of a total of five is shown) have been slackened. 11 teflon washer. 12 screws for tightening ring 13 and the lower pole piece 15 into position. 14 clearance gaps for allowing a transverse displacement of ring 13. 16, 18 ferromagnetic chunks of magnetomechanical stigmator, mounted onto the conical ring 19, which bears the toothed rim 9. 17 protection tubes, made from molybdenum for ease of cleaning.
21 indicates a glass plate with crosshairs which is cemented to a frame 22 and can be secured into its position on ring 13 by tightening the clamping ring 23 (by means of screws located just before and behind the plane of drawing and thus not visible here). Suitable gaps 20 are provided to permit the glass plate to be transversely displaced by some 0.1 mm with respect to the ring 13. A second glass plate 24 with crosshairs is cemented rigidly to a pin, which is press-fitted into the bronze spacer 3. Another pair of glass plates is arranged on the same face of the pole-piece system, approximately opposite the first one with respect to the lens axis 30. In both cases, the crosshairs are located on the surface facing each other, and as these surfaces are brought quite close to contact by carefully fashioning the frames 22, the crossover points can be adjusted to coincidence during observation with a high magnification light microscope. In this manner, after the pole pieces have been mutually aligned, the corresponding location of ring 13 can be recorded and thus the lower pole piece repositioned, after removal for cleaning purposes for example.
25 carrier ring for the second pair of ferromagnetic chunks (not shown) of the magnetomechanical stigmator. 26 cage for stigmator mechanism. 27 indicates a roller which has its axis located in the block 28 and is forced against the specimen stage 29 by action of the stage return springs (not shown). The taper of the rollers 4 and 27 generates a downward component of the return spring force, which presses the specimen stage against its surface of travel, namely, the lid of the stigmator cage 26

Another type of mechanical adjustment concerns the alignment of the electron optical system. For this, traversing the pole-piece system with respect to the remainder of the lens body looks like a simple method and has been employed with a projector lens having a pole-piece revolver of the type represented in Fig.4.17 [4.182]. Here, the rotation of the pole-piece turret by small amounts provides one of the translations and moving the turret axis at right angles to this the other one. Unfortunately, the actual alignment translations of the pole-piece systems proved to be rather jerky, due to the considerable friction arising between the pole pieces on the one hand, and the lens core or the upper lens flange on the other, which is caused by the large magnetic attraction forces between those parts. Moreover, the surfaces in frictional contact are both of the same material and, still worse, of iron or soft steel, a notoriously unpleasant combination in respect to sliding properties. Lubricants cannot be applied there because of the nearness of the beam and the danger of disturbances due to charging up.

It is much better if an immediate contact between the pole pieces and the connecting surfaces of the iron circuits can be avoided. How this can be done is shown by way of example with the intermediate projector lens [4.149] represented in Fig.4.72, where a gap of about 0.1 mm width is interspersed between the lower pole piece 6 and the adjacent face plate 5 of the lens core 4. The gap between the upper pole piece 8 and the lens flange 2 is a few tenths of a mm wide and thus a little larger (it would be difficult to adjust both connecting gaps to about 0.1 mm or less simultaneously), but the large collar-like extension of the pole piece 8 provides for a low enough value of magnetic resistance across to the flange 2 [4.148]. The pole pieces are joined together by the bronze spacer 7, and the whole assembly is carried by the brass collar 9 which slides on a cylindrical extension 19 of the lens flange 2. At the sliding interface, the pairing of a soft and a hard metal (brass and steel) generates noticeably less friction than in the above case with the iron and steel combination. Moreover, lubricants can safely be applied here because the gliding surfaces are far from the beam and completely shielded by the brass collar 9.

The pole-piece traverse is actuated by two couples of push rods 10, 13 and their corresponding return springs 14 (only the couple visible in the plane of drawing is shown, the other couple is orthogonal to this plane). In order to separate the two linear motions from each other, a cross-slit disc 20 (made from brass) is provided, with the slits lying parallel to the respective directions of motion of the push rods and catching two pairs of spigots 15, 21 and 17. The spigots 17 are fastened to the fixed flange 18 (brass), whereas the spigots 15, 21 are connected to the collar 9.

An important point in the design is the presence of a rather large gap between the cylindrical pole pieces and the internal wall of the adjoining parts of the magnetic circuit, a gap that is larger by about an order of magnitude than the total range of traverse possible for the pole-piece system. There is then no danger that, on displacing the pole piece, the magnetic lens field distribution in the gap will noticeably change and lose its rotational symmetry, thus causing axial astigmatism.

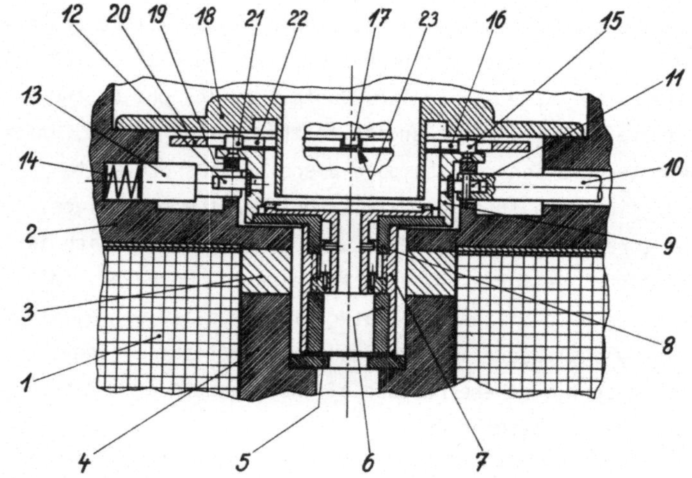

Fig.4.72. Cross-section of the pole-piece region of an intermediate-projector lens designed for alignment during operation of the electron-optical system. 1 lens coil. 2 upper lens flange. 3 bronze spacer. 4 lens core. 5 ferromagnetic face plate, retouched in thickness during the actual construction of the lens so that a sufficiently narrow gap of about 0.1 mm is obtained between plate 5 and lower pole piece 6. 7 brass spacer. 8 upper pole piece. 9 traversable brass collar, which carries the pole-piece system. 10 micrometer-operated push rod. 11 ball-bearing butting against a hardened and polished platelet. 12 ball-bearing and associated platelet. 13 return push rod. 14 return spring. (A second pair of actuating and returning push rods is at right angles to the drawing and so not visible here.) 15, 21 spigots and 16, 22 corresponding slots for guiding the pole-piece system traverse parallel to the plane of drawing. 20 cross-slit disc. 17 spigot and 23 corresponding slot for guiding the pole-piece traverse orthogonal to the plane of the drawing. (The second spigot and slot required for this is above the plane of drawing and so not visible here.) 19 cylindrical extension of the lens flange. The lens is equipped with a magnetomechanical stigmator as represented in Fig.4.56

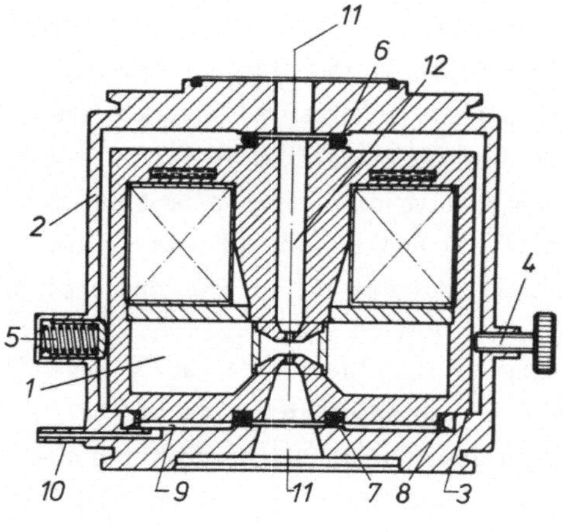

Fig.4.73. Simplified cross-section of a transversely alignable electron projector lens equipped with a pneumatic locking device. 1 electron projector lens proper. 2 external instrument-fixed casing. 3 supporting shoulder. 4 lens traverse micrometer. 5 return spring. 6, 7, 8 vacuum gaskets. 9 pneumatic chamber. 10 pumping duct. 11 axis of instrument-fixed casing. 12 axis of the electron lens proper, usually parallel displaced with respect to axis 11

Another particularly advantageous design for the mechanical lens alignment is shown schematically in Fig.4.73. Here, the whole magnetic lens 1 is traversed within an additional external can-shaped casing 2 which connects it to the other units of the electron optical column and carries their weight. The lens 1 slides on the shoulder 3 and can be traversed by means of two couples of push rods 4 and return springs 5, which are at right angles to each other. In the actual construction, the lens 1 is suspended between cup-spring-cushioned ball bearing balls (three to each flange) which run on highly polished surfaces. (These cushioned balls are not shown in Fig.4.73 but are similar to those of the design represented in Fig.4.22.) By carefully balancing the forces exerted on the lens via the balls by the cup springs, the contact pressure for the lens can be chosen virtually at will and reduced to a value (of around 1 kp $\approx$ 10 N) which enables a smooth translation with μm accuracy.

Since lens 1 bears only very lightly on the shoulder 3, it has to be locked in place with respect to the external casing 2, after the alignment has been completed. For this, a pneumatic locking device [4.162] can be employed, essentially consisting of a thin ring-shaped vacuum chamber 9, which extends between the vacuum gaskets 7 and 8. If the chamber 9 is evacuated (a few torrs, i.e., a few hundred pascals is sufficient) via the pumping duct 10, then a part of the atmospheric pressure surrounding lens 1 is removed, and the latter is pulled downwards against the shoulder 3 with considerable pneumatic force. Typically, with the area enclosed on the surface of lens 1 by the gaskets 7 and 8 being of the order of 100 cm^2, the lens is pulled against the shoulder 3 by forces of the order of 100 kp (1 kN) and thus safely locked in place by corresponding frictional forces. Of course, for alignment traverse, the vacuum chamber 9 has to be brought up to atmospheric pressure. It should be noted that with the pneumatic chamber 9 evacuated, the alignment remains stable even if the instrument column has to be flooded, for removal and cleaning of apertures, for example. (A pneumatic locking device for specimen stages has been proposed in [4.174].)

4.5 Measures for Reducing the Sensitivity of the Magnetic Electron Lens to Environmental Disturbances

The performance of an electron lens can be strongly impaired by environmental disturbances, especially if it is to be employed as the objective lens of an electron microscope.

Low frequency mechanical vibrations and magnetic disturbances are particularly dangerous in this respect, and possible design countermeasures for these will be discussed below. Further disturbances can be caused by audio frequency sound waves, which may give rise to resonant vibrations of parts within the lenses. This problem arises rather rarely, and each case must be treated individually: a general recipe for avoiding it in the design of the lens can hardly be given. Moreover, it is not difficult to remove the acoustic resonance by applying suitable damping.

Other disturbances may be caused by environmental temperature changes which unbalance the influx of thermal energy to the electron optical column and deform the optical elements with varying thermal expansion. To some degree these effects can be counteracted by providing the lenses with thermal sensors (Fig.4.60) and regulating the throughput of the lens-coil cooling water accordingly [4.4]. This also takes care of variations in the input of electrical energy (changes of the lens current), of the gradual heating up of the lens coil to working temperature, and of temperature fluctuations in the city water supply.

4.5.1 Design Considerations for Protecting the Lens Against Low Frequency Environmental Vibrations

Low frequency vibrations in the range from a few hertz up to about 20 Hz with amplitudes having 0.1 μm as typical order of magnitude are nearly universal and always present in laboratory buildings. The vibrations are then propagated onto the electron optical instrument and its column, which oscillates with a similar amplitude. Thus, forces of inertia are generated which tend to bend the column and periodically deform the lenses (cf. Fig.4.74); this in turn results in a corresponding oscillation of the image [4.113,169].

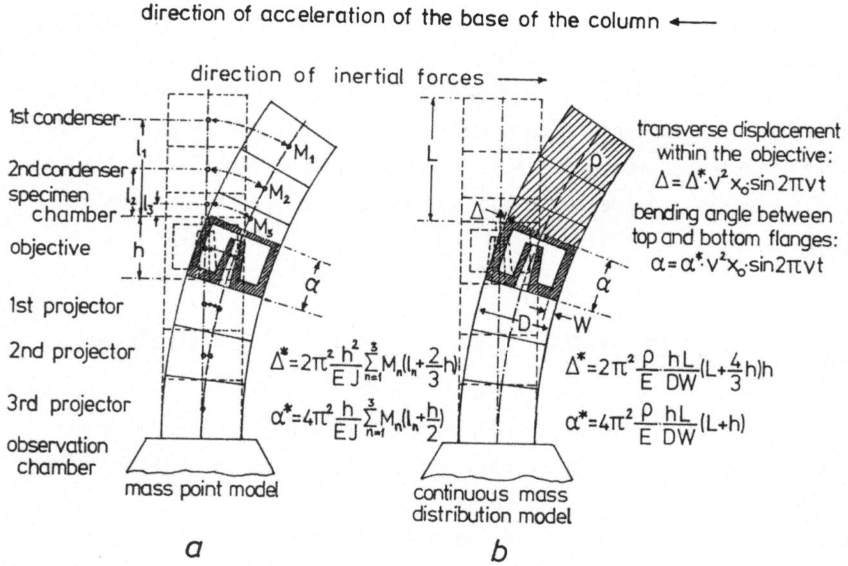

Fig.4.74a,b. Illustration of the deformation of the column of an electron optical instrument caused by environmental vibrations. The deformation of the objective lens is treated here by way of example.
The column is mounted on the observation chamber, which is connected to the foundation of the instrument, and this foundation is subjected to forced vibrations

transferred from the ground on which it stands. (Experience has shown that quite universally the predominating disturbances of electron optical systems are caused by vibrations with amplitudes between 1 μm and 0.1 μm in order of magnitude and are carried by frequencies around 10 Hz.)
The forced vibrations of the column give rise to a bending deformation both of the column as a whole and of the individual lenses. This is shown in the drawing for the objective lens of an electron microscope. The deformations Δ and α can be established either (a) by using the mass point model whereby the masses M_n of the lenses above the objective lens are assumed to be condensed into their respective centres of gravity or (b) by employing a continuous averaged mass distribution for the column part above the objective lens. In the formulae printed in the drawing, x_0 and ν are the amplitude and frequency of forced vibration of the observation chamber, h and W are the height and thickness of the casing of the objective lens and E the modulus of elasticity of the material used for the casing.
For details of the resulting deformations of the pole-piece system and other electron optical disturbances caused by the casing deformations Δ and α, see [4.113,169]

Quantitatively, it is found that with magnetic objective lenses and at a resolving power below about 5 Å = 0.5 nm, the performance of the lens may be adversely affected by the usual level of environmental vibrations, if no special precautions are taken, a result well confirmed in actual practice. Efforts have been made to reduce the harmful effects of the environmental vibrations by installing electron microscopes on spring or rubber or air-bag cushioned foundations [4.26,52,184,185] with very low natural frequency. It seems however, that there is a fundamental limit to the improvement that can be expected from such a system, because the necessary damping of the cushioned platform cannot be effected without transferring a certain amount of vibrational energy to it from the environment. Therefore, reducing the vibration sensitivity of the magnetic lens by an appropriate design is called for, especially for objective lenses.

For this purpose, one can in principle try to make the lens casing stiff enough to withstand the torque exerted on its top flange relative to its bottom flange by the oscillating inertial forces. In order to increase its rigidity sufficiently, the external cylindrical casing has then to be constructed with a wall thickness of at least a few centimetres and thus much thicker than is usually necessary for the magnetic circuit. The appropriate value of the wall thickness depends on the weight and configuration of the parts of the column stacked on top of the objective lens and can be determined by employing formulae reported in [4.169].

A quite different approach to the vibration-proof construction is to surround the magnetic lens proper by a second external can-shaped casing; this takes up the inertial bending forces and the weight of the column above the 'protected' lens, which thus does not experience any external torque and deformation. Some possibilities for combining a magnetic objective lens with its external vibration-protection casing are shown schematically in Fig.4.75; all three arrangements have actually been incorporated in electron microscopes. The design indicated in Fig.4.75a with the objective lens joined to the bottom flange of the external casing has been adopted for the Mark II version of the condenser-objective lens of the electron

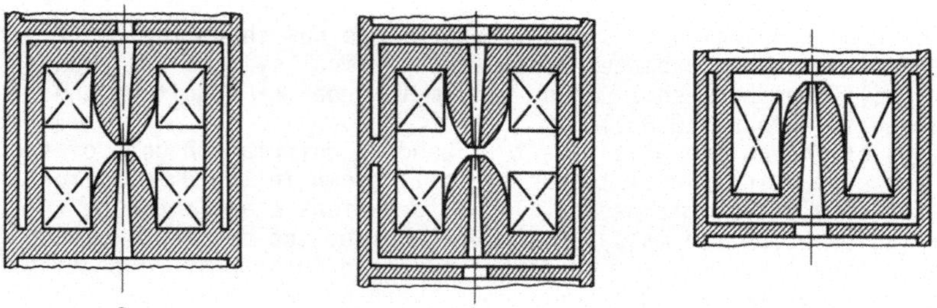

Fig.4.75a-c. Schematic representation of design principles suitable for protecting electron lenses from deformations which could be caused by environment-generated forced vibrations of the foundations of the electron optical instrument

microscope described in [4.157]. The form represented in Fig.4.75b, where the inner lens is carried by a 'belt' connecting it to the middle of the external casing, is currently employed with the PHILIPS EM400 commercial electron microscope [4.46]. Figure 4.75c illustrates another commercial design utilized with the SIEMENS 101 [4.4] and 102 [4.125] electron microscopes where the objective lens is suspended from the top flange of the external casing.

It should be clearly realized that although the environmental vibration-induced deformation of the objective lens and the corresponding disturbance can be eliminated in this way, the vibrational tilting of the objective lens axis remains unaffected. Fortunately, the image movement due to the tilting of the objective lens as a whole is smaller by an order of magnitude than the image movement that would have been induced by objective lens deformation in a design without an external protective casing [4.169]. With regard to making the tilt disturbance small too, the design represented in Fig.4.75a is advantageous because there the tilt angle of the objective axis remains lowest.

4.5.2 Protective Measures in the Lens Design Against Stray Magnetic Field Disturbances

In laboratory buildings, mains-generated magnetic stray fields with frequencies of 50 or 60 Hz and corresponding higher harmonics, with magnetic flux densities of the order of 1 to 10 mGauß = 0.1 to 1 μT are a nearly ever-present source of danger threatening the performance of electron beam equipment. With an electron microscope, if a small but still noticeable fraction of the transverse stray field penetrates onto the instrument axis, it generates an oscillatory deflection of the beam and correspondingly of the image, thus causing a loss of resolution. Typically, with an electron microscope operating at a beam voltage Φ = 100 kV, the magnetic AC stray field should be less than about 10^{-6} to 10^{-7} Gauß = 100...10 pT if the limit of resolution is to be attained [4.119].

For the construction of the magnetic shielding, even a moderately accurate calculation of the properties of an actual technical form would be extremely complex and so cannot be used for design engineering. In practice, however, the design may be safely based on some simple approximate formulae and guide rules:

1) The low frequency transverse magnetic screening efficiency of the cylindrical lens casing can be estimated from [4.116]:

$$\frac{B_{ext}}{B_{int}} = \bar{Q}_{scr}\,\frac{\mu_r w}{2r_w} \tag{4.140}$$

where w is the wall thickness of the cylindrical casing of radius $r_w >> w$; $\mu_r = dB/\mu_0 dH$ is the relative differential magnetic permeability; B_{ext} is the magnetic flux density that would be present without the column and B_{int} is the flux density in the interior space of the tube. Equation (4.140) is valid for stray fields having a frequency of the order of magnitude 1 Hz or below and thus also applies to the slow variations of the magnetic field of the earth which can be of the order of 1 mGauß ≈ 0.1 μT. With a lens casing designed according to the rules outlined in Sect.4.3.3, we have $Q_{scr} \approx 100$ in order of magnitude.

2) More strictly speaking, (4.140) applies to a very long cylindrical tube and to those parts of it that are at a distance from the tube ends much greater than its diameter $2r_w$. At the open end itself the transverse field is reduced to about $B_0 \approx 0.35\ B_{ext}$ of its value in the absence of the ferromagnetic tube. It penetrates into the open tube end and decreases with the distance z from the end as

$$B(z) = B_0 \exp(-3.83\ z/r_w) \quad . \tag{4.141}$$

(Cf. [4.115]. The constant 3.83 is the first zero of the first-order Bessel function.)

3) Therefore, open ends of cylindrical ferromagnetic parts having a comparatively large inner diameter should be closed by a lid with a small central bore to allow the electron beam to pass [4.139].

4) The screening efficiency of a long ferromagnetic tube against medium and high frequency magnetic fields has been treated by KADEN [4.82]. The formula obtained is rather complex, but fortunately, for the conditions of interest in electron lens design, it can be simplified considerably. The screening efficiency can be written in the form

$$Q_{scr} = \bar{Q}_{scr}\,\frac{\exp(w/\delta)}{2w/\delta} \tag{4.142}$$

and this equation is valid for $w \gtrsim 3$. Here, $\bar{Q}_{scr}$ is the screening efficiency for low frequency fields as in (4.140) and δ the thickness of the equivalent conducting layer at frequency ν:

$$\delta = \sqrt{\frac{\rho}{\pi\nu\mu_r\mu_0}}$$

$$= 503 \sqrt{\frac{\rho[\Omega\ mm^2/m]}{\nu[Hz]\cdot\mu_r}}\ mm \quad .$$

ρ is the specific resistivity of the ferromagnetic material. For soft steel, at $\nu = 50$ Hz, δ is about 1 mm in order of magnitude and differs somewhat for the different materials employed in lens design.

5) Magnetic screening strongly improves with decreasing thickness of the conducting layer, i.e., with increasing frequency. This can be seen from the accompanying table, which shows the frequency-dependent correction factor employed in (4.142). It is evident that the physical mechanism responsible for screening the electron beam in the magnetic lens depends to a large extent on skin effects and skin current phenomena.

w/δ	$\exp(w/\delta)/(2w/\delta)$
3	3.35
4	6.82
5	14.84
6	33.62
7	78.33
8	186.3
9	450.2
10	1101

6) Thus, for medium and high frequency magnetic fields, the penetration of the field into the interior of the casing is prevented by skin effect currents, which generate a compensating magnetic field in opposition to the original one (cf. Fig. 4.76a). If the flow of the skin currents is impaired, if the casing is divided by a thin slit into two sections for example, then the magnetic stray field can seep into the casing in the slit region. Worse still, the skin currents adjacent to the slit region give rise to a magnetic field parallel and not in opposition to the original one, so that the stray field is "sucked" into the casing at this spot. It should be realized that even if the two sections are connected together directly, the resistance of the slit interface will still be comparatively high. For the low voltages induced, the situation is thus not very different from that in which a slit is actually present.

7) Therefore, if the casing has to be subdivided into two sections, it is advisable to slide the two parts into each other with a sufficiently long 'telescopic' overlap. This is shown schematically in Fig.4.77 and can also be seen to have been employed in the condenser lens represented in Fig.4.22.

8) For $w < 3\delta$, the conducting layer gradually fills the whole thickness of the wall as the relative wall-thickness w/δ is decreased. The screening efficiency Q_{scr} gradually decreases from the lowest value $3\,\overline{Q}_{scr}$ described by (4.142) to the low-

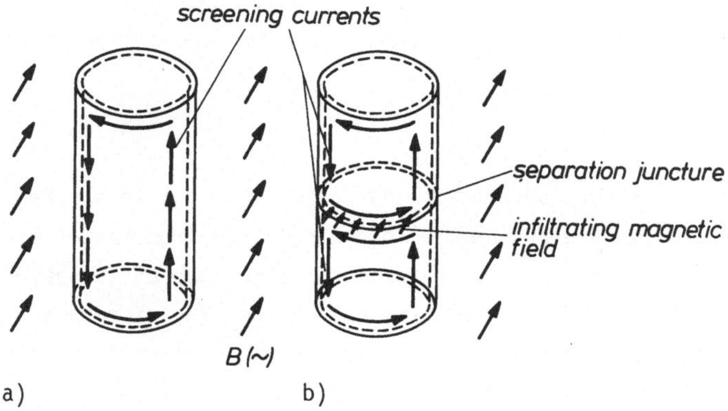

Fig.4.76a,b. Schematic representation of the distribution of the screening skin-currents in a cylindrical lens casing, exposed to an alternating transverse magnetic field (adapted from [4.139]). (a) Screening current distribution in a long individual cylindrical casing. (b) Screening current distribution in a divided (and not overlapping) cylindrical casing. Here, at the separation of the two cylinders a screening 'window' is opened through which the transverse magnetic field can penetrate into the interior of the casing

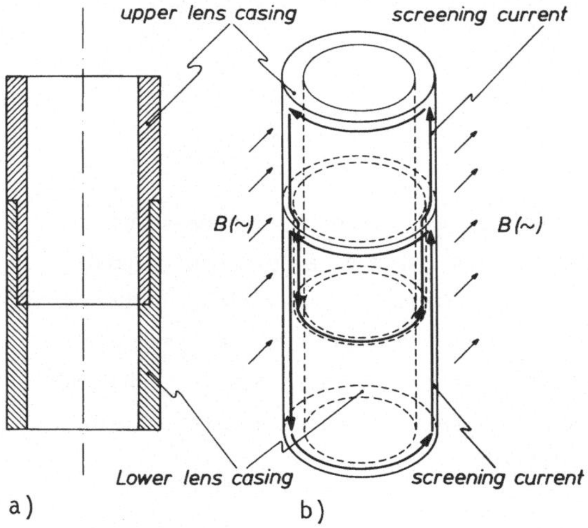

Fig.4.77a,b. Schematic representation of the improvement in the screening efficiency with a divided casing if a telescopic overlap of the two casing sections is applied. (a) Cross-section of the cylindrical casing. (b) Schematic representation of the distribution of the screening currents and of their overlap

frequency screening efficiency $\overline{Q}_{scr}$ and remains nearly constant below $w = 0.5\delta$. This decrease is *not* described by (4.142), and if the transition range values are required, the more rigorous treatment of [4.82] has to be applied.

9) With a narrow-bore pole-piece lens, the lens core gives rise to additional magnetic screening. Here, the low-frequency screening efficiency can be written [4.115] in the form

$$\bar{Q}_{scr} \approx \mu_r \frac{D_c^2 - D_b^2}{4D_c^2} \qquad (4.143)$$

with D_c as external diameter of the core. D_b is the diameter of the central bore, which is much smaller than D_c in most practical cases. Thus, for the core we have approximately

$$\bar{Q}_{scr} \approx \frac{1}{4}\mu_r \quad . \tag{4.144}$$

For practical lens design, it can be concluded that with lenses having a lens core, the region within the core is doubly shielded and so completely safe from magnetic stray field effects. There is some stray field penetration into the gap of the pole-piece system, which is only singly shielded by the external casing. The disturbed path length is comparatively short however and the disturbance will generally remain small enough unless the stray fields are extremely strong (10 mGauß or more, cf. [4.53]). With rather 'open' condenser lenses, which have a comparatively large gap-width, it is especially important to employ a thick enough wall for the casing to provide sufficient shielding against medium (50 to 60 Hz) frequency disturbances (cf. 5). A value of $w \gtrsim 5\delta$ (i.e., 7-10 mm) would be a reasonable choice, even if such a thickness is not required for the lens-field-generating magnetic circuit.

In extremely adverse cases, the external magnetic stray field can be reduced up to two orders of magnitude by compensation coils and associated electronic devices [4.9,117].

4.6 Permanent-Magnet Lenses

In the introductory Sect.4.1, it has already been explained that, because of AMPERE's Law, a one-gap magnetic electron lens energized by a permanent magnet must necessarily give rise to a strong (constant) magnetic stray field on the instrument axis and is therefore not satisfactory for applications. For permanent-magnet lenses, combinations of two or more gaps have to be arranged in a shielded design, which prevents the magnets from giving rise to an external stray field (Fig.4.6).

4.6.1 The General Layout of Electron-Optical Lens Systems with Permanent-Magnet Lenses

The simplest possible permanent-magnet lens arrangement is obviously the two-gap lens represented schematically in Fig.4.6. It was actually used in some of the so-called 'simplified' electron microscopes, in which one magnet energized objective lens and projector in parallel. Both axial [4.23,24,88,144,145,179] and radial [4.122-124] magnets have been applied.

By adding a second magnet, electron optical systems with three lens gaps can be realized (Fig.4.78). Here, the two magnets can be arranged magnetically in series so that each of the magnets independently energizes one gap and the magnetic potential at the third gap is generated as the sum of the potentials of the single

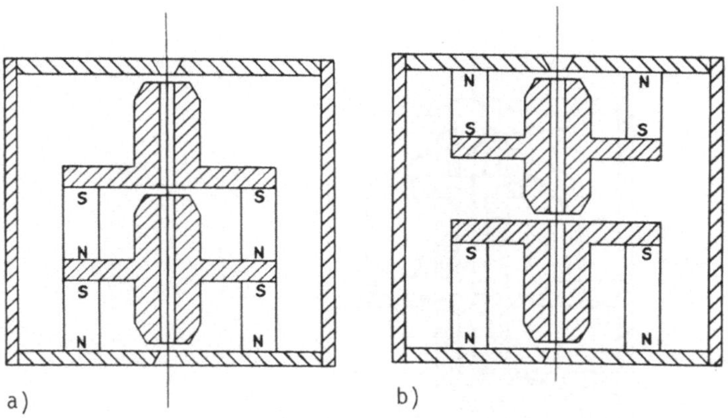

Fig.4.78a,b. Basic forms of three-lens permanent-magnet electron optical systems (after [4.121]). (a) Three-lens system with the magnets arranged in series, whereby the sum of the magnetomotive forces of the two magnets is applied to the upper lens gap. (b) Three-lens system with the magnets arranged in opposition so that the difference between their magnetomotive forces becomes effective at the central lens gap. Here, therefore, in order to establish genuine three-lens operation, the magnetomotive forces of the two magnets must be chosen quite different from each other, in contrast to a) where two magnets of the same power can be employed

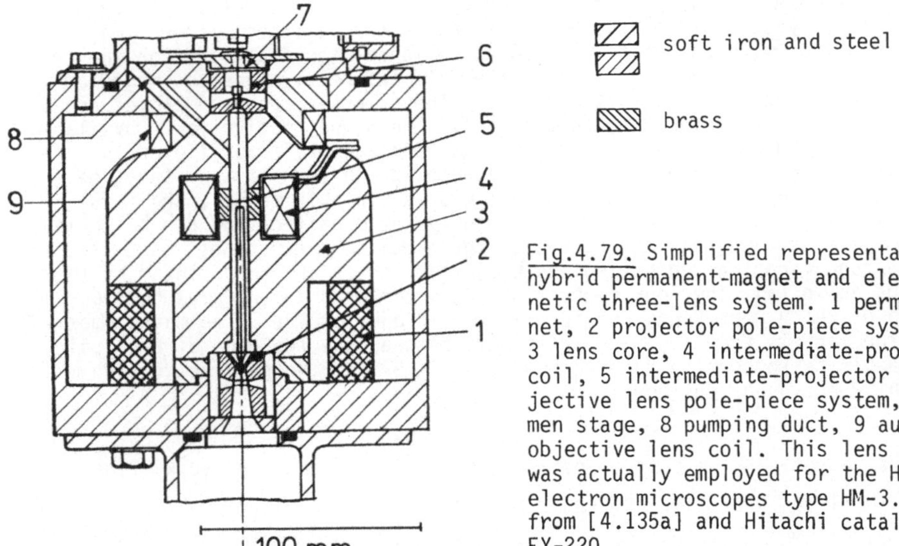

Fig.4.79. Simplified representation of a hybrid permanent-magnet and electromagnetic three-lens system. 1 permanent magnet, 2 projector pole-piece system, 3 lens core, 4 intermediate-projector coil, 5 intermediate-projector gap, 6 objective lens pole-piece system, 7 specimen stage, 8 pumping duct, 9 auxiliary objective lens coil. This lens system was actually employed for the Hitachi electron microscopes type HM-3. Adapted from [4.135a] and Hitachi catalogue EX-220

magnets (cf. Fig.4.78a). If, on the other hand, the two magnets are employed in opposition, then the magnetomotive force at the third gap is the difference between their magnetic potentials (Fig.4.78b). Evidently, in the difference arrangement of Fig.4.78b, the magnetomotive forces of the two magnets must be different from each other, whereas for the sum arrangement of Fig.4.78a, two magnets of the same type can be used.

An interesting hybrid design with three lens gaps was used for the HITACHI HM-3 electron microscope [4.89]. Here, only a single permanent magnet was employed for energizing the objective lens and the projector in parallel. For the intermediate

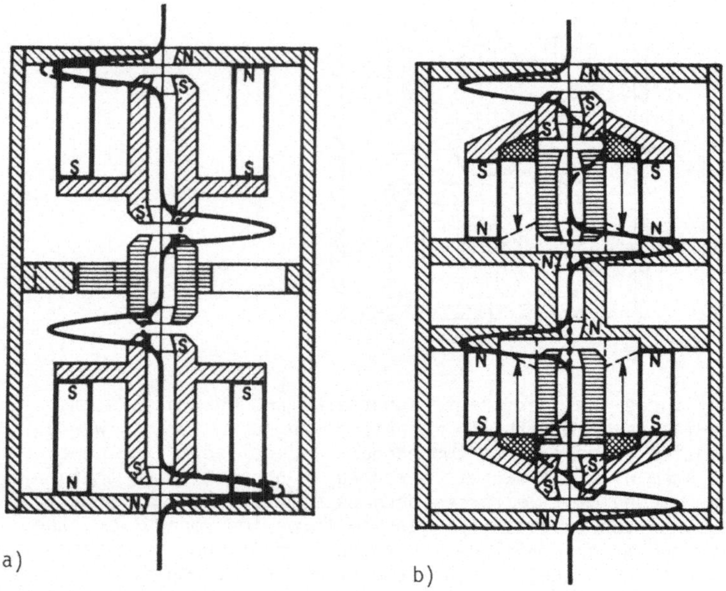

a) b)

Fig.4.80a,b. Examples of basic forms of four-lens and six-lens permanent-magnet electron optical systems (after [4.180]). (a) Four-lens permanent-magnet imaging system for electron microscopes to be used at variable magnification. For the two outward lenses positioned at the flanges, the full and nearly constant magnetomotive force of the magnets is always available. These lenses can be employed as objective and final projector lens respectively. The magnetomotive force available at the two central gaps, which generate the intermediate-projector fields, can be adjusted by applying a variable series resistance in the form of the star-type control-yoke configuration described in Sect.4.6.2a. In this manner, the focal lengths of the intermediate projectors and the system magnification can be varied. (b) Six-gap permanent-magnet imaging system for electron microscopes to be used at variable magnification. Objective and projector lenses are represented by the outward gaps as in (a). The four central gaps are employed in two consecutive pairs, each of which represents one intermediate-projector lens. The magnetomotive forces of the magnets are variably partitioned between the two gaps of each pair by displaceable ferromagnetic tapping rings, which bridge the gaps betwen the magnets and their enclosed lens cores

projector, a conventional electromagnetic lens was fitted into the bulky lens core connecting the lower objective and upper projector pole pieces (Fig.4.79). A second coil of small volume encircles the objective lens gap and contributes a fraction of the magnetomotive force of the objective lens; it can thus be used for focusing by correspondingly varying the coil current.

Finally and obviously, a four-lens electron optical system can be realized by stacking two of the two-gap one-magnet systems on top of each other (cf. [4.180] and Fig.4.80). If two magnets of the same type are used, or more generally two magnets having equal magnetomotive forces, and if the magnetic moments of the magnets are arranged in opposition, then not only is the complete four-lens system free from image rotation but so too is the two-lens system composed of the two intermediate lens gaps, a property that is useful for minimizing aberrations.

As an illustration of the technical form of permanent-magnet lenses, cross-sections of a two-lens system [4.180] and a three-lens system [4.121] are presented in Figs.4.81 and 4.82.

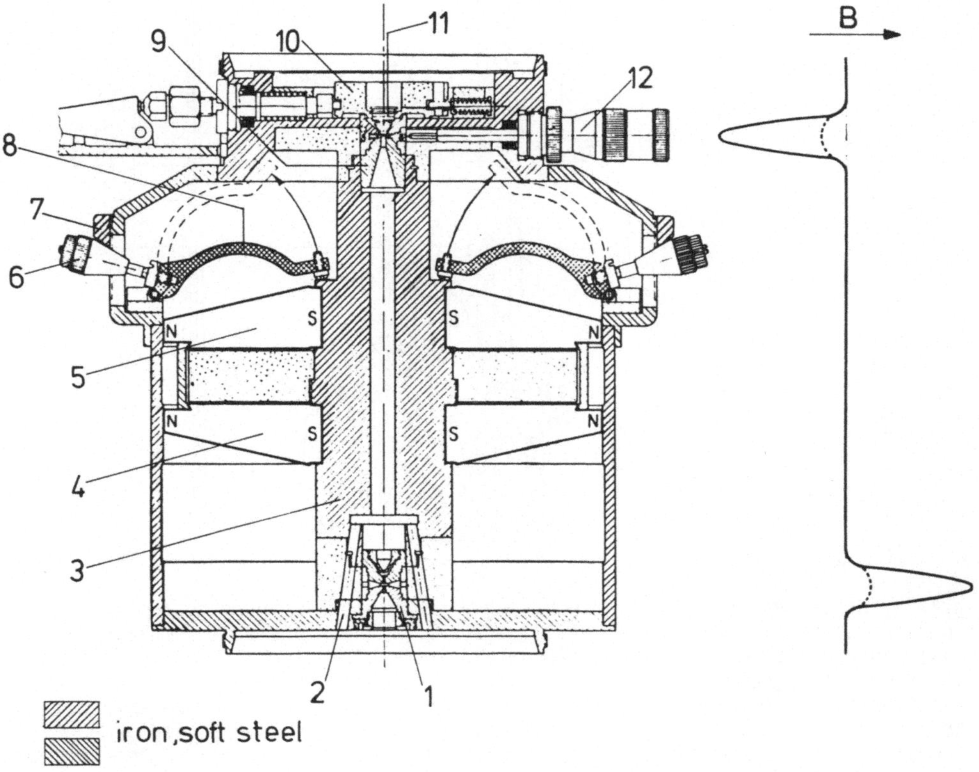

Fig.4.81. Permanent-magnet two-lens system with ferromagnetic flaps for varying the focal length of the lenses. 1 Projector pole-piece system; 2 pumping duct; 3 lens core; 4, 5 permanent magnets; 6 rotary knob for fine adjustment of flap position; 7 rotary ring-nut for the simultaneous coarse adjustment of the positions of the set of flaps; 8 soft-iron flaps for the variable shunting of the permanent magnets; 9 objective lens pole-piece system; 10 specimen stage; 11 specimen cartridge; 12 objective lens aperture control [4.180]

(Fig.4.82 see next page)

4.6.2 Special Design Aspects for Permanent-Magnet Lenses

Two particular problems arise in the construction of permanent-magnet lenses: The variation of the magnetic field strength for focusing and magnification change, and the selection of a suitable permanent magnet.

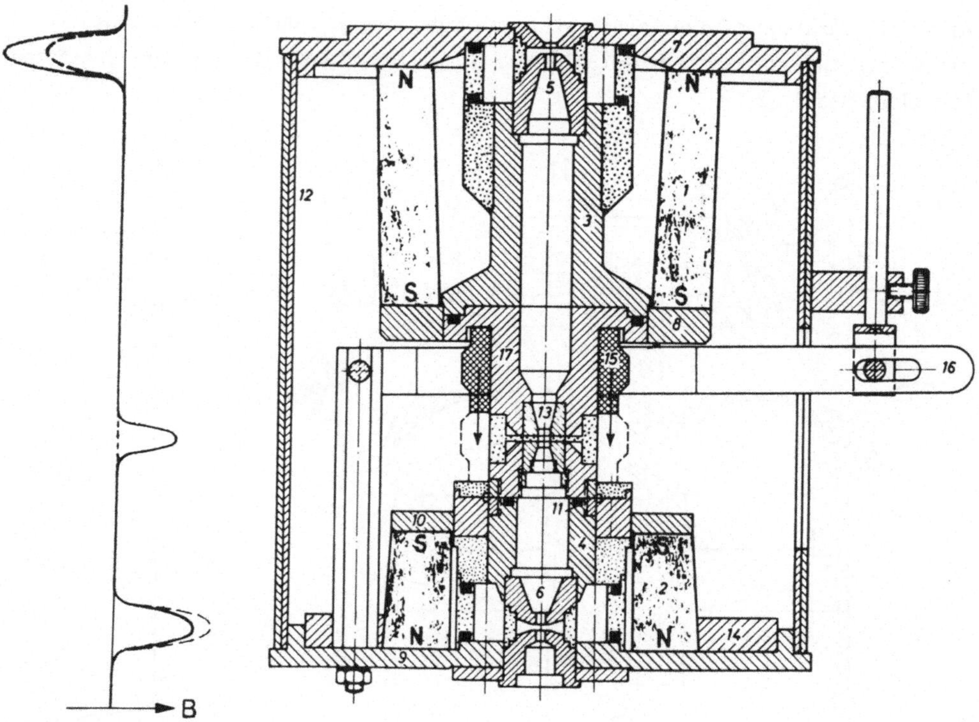

Fig.4.82. Cross-section of a permanent-magnet three-lens imaging system for electron microscopes (condensed from [4.121,179,180]). The magnets are employed in opposition and according to the difference principle represented in Fig.4.78b. 1, 2 permanent magnets. 3, 4, 17 lens cores. 5 objective pole-piece system. 6 final projector pole-piece system. 7, 9 outward lens flanges. 8, 10 internal flange rings. 11 vacuum gasket. 12 lens casing. 13 intermediate-projector pole-piece system. 14 ferromagnetic short-circuiting ring, which decreases the magnetomotive force of the smaller magnet 2 and thus – according to the difference principle – increases the magnetomotive force available for the intermediate projector. This gave an increase in total magnification of the system over its performance without ring 14. 15 soft-iron ring employed for controlling the refractive power of the intermediate-projector lens (Sect.4.6.2a and Fig.4.83c). 16 lever for sliding ring 15 in axial direction, i.e., for changing the magnification

4.6.2a Means for Changing the Focal Length of Permanent-Magnet Lenses

If the magnetic field of the lens is left untouched, a small focal length change can be achieved by correspondingly changing the beam voltage so that the required value of the lens strength is reached [cf. (4.13) and (4.15)]. This method can be employed for small focal length changes, such as those required for the precise focusing of the specimen with the objective lens [4.122-124], but not for a large variation of the focal length because this would also throw all the other lenses off balance simultaneously. Another method for providing a modest change of focal length has been described above, namely, the addition of a small coil providing a few ampere-turns magnetically in series with the permanent magnet (Fig.4.79).

A much stronger variation of the focal length can be achieved by suitable mechanical manipulation of the magnetic circuit. Two basic methods are available:

1) Changing the magnetomotive force of the magnet by varying its magnetic flux load.

2) Transferring a part of the magnetomotive force of the magnet to a gap which is gradually opened in the magnetic circuit, far from the axis, so that no electron optical disturbance results.

In actual practice, both techniques are often employed in combination.

From the typical demagnetization curve for a permanent magnet (Fig.4.86), it can be seen right away that by increasing the magnetic flux load, i.e., increasing the flux density in the magnet, the demagnetizing field strength H_m in the magnet can be made smaller and thus so can its magnetomotive force $H_m \cdot l_m$ (with l_m as the effective length of the magnet). Conversely, decreasing the magnetic flux load produces a corresponding increase in the demagnetizing field H_m and in the magnetomotive force $H_m \cdot l_m$.[2]

Some methods that permit the focal length to be varied over a larger range will now be discussed, and their application to two-gap permanent-magnet lens systems considered.

In Fig.4.83a, a radial magnet system is represented schematically; it is equipped with a set of soft-iron flaps that can be adjusted to different angles with respect to the magnet surface [4.179]. In this manner, the load on the magnets can be adjusted and the focal length of both lenses varied in synchronism. The wide-open position of the flaps provides the smallest load, the maximum field in the lens gaps and the minimum focal length. An actual permanent-magnet lens system designed to use this regulation mechanism is shown in Fig.4.81.

Another method of varying the load applied to the magnet is represented schematically in Figs.4.83b,c. Here, a cylindrical soft-iron ring is slipped over the lower lens core and can be traversed in the axial direction. When the ring is moved downwards the magnetic resistance between the core-and-ring combination on the one hand and the lower flange of the casing on the other is decreased so that the magnetic flux and the load increase. The field strength in the gaps and the refractive power of the lenses then diminish. With this design however there is a danger of the region of the core close to the magnet becoming oversaturated if the core cross-section is not large enough. Thus, on sliding the ring down, a spurious mag-

---

2 *Permeability and permeance*: It should be noticed that the magnetic flux density and the magnetic field strength in a permanent magnet point in opposite directions. This is quite different from the situation in soft magnetic materials, where the magnetic flux density and the field strength have the same direction. In order to underline this difference, the term permeance is employed for the ratio of the absolute values of magnetic flux density and field strength in permanent magnets; permeability is reserved for the quantity $\mu_r = \vec{B}/\mu_0 \vec{H}$ or $\vec{B}^*/\vec{H}^*$ used with soft magnetic materials. With permanent magnets, therefore, the permeance is defined as B/H in rational units or as B^*/H^* in the Gaussian system (B, B^*, H and H^* denote numerical values and not vectors as $\vec{B}$, $\vec{B}^*$, $\vec{H}$ and $\vec{H}^*$).

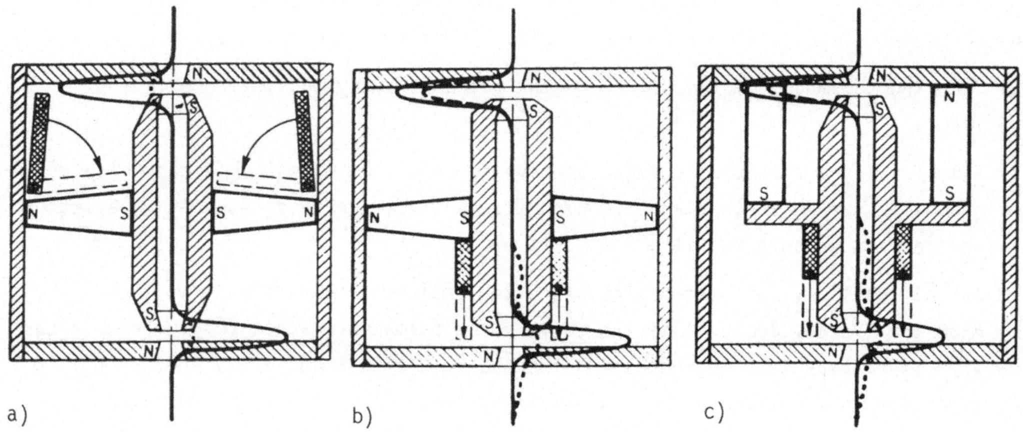

Fig.4.83a-c. Schematic representation of devices suitable for changing the focal length of permanent-magnet electron lenses by varying the magnetic flux load on the permanent magnets [4.179,180]. (a) Two-gap permanent-magnet lens system equipped with flux-shunting flaps for varying the focal lengths of both lenses simultaneously. The amount of flux shunting at the magnets is determined by the inclination of the flaps and affects both lenses to the same degree. (b) Two-gap permanent-magnet lens system employing radial magnets. The device is equipped with an axially traversable ferromagnetic sleeve ring that slides on the lower lens core and changes the stray flux of the magnet as a function of the ring position. As long as no magnetic saturation arises in the iron circuit, the magnetomotive forces at the gaps remain equal and are changed in synchronism. If it is necessary to render one of the lenses much weaker than the other, magnetic oversaturation in part of the lens core and a corresponding parasitic stray field lens have to be accepted. (c) Two-gap permanent-magnet lens system equipped with a focusing ring as shown in (b), but here employing an axial magnet instead of the radial magnets of the other system

netic field (indicated as a broken line in Figs.4.83b,c) may appear on the instrument axis far from the gaps, giving rise to a parasitic lens which lives at the expense of the refractive power of the lower gap. This device can be applied to both radial (Fig.4.83b) and axial (Fig.4.83c) magnet systems.

As an alternative to the design represented in Fig.4.83c but basically following the same principle, the load of the magnet can be varied by traversing within a fixed ferromagnetic sleeve tube a pole-piece system, which has a sliding fit inside the tube bore. The sleeve tube is interrupted by a gap coincident in axial position with the pole-piece gap at one end of the traverse range [4.23,24].

It should be clearly realized that according to AMPERE's Law, $\int H dz$ must vanish if the integration is performed along the whole length of the two-gap system axis. Therefore, as a consequence of basic physical laws, if it is necessary to decrease the refractive power of one of the gaps more quickly than that of the other, a parasitic lens will inevitably be generated, regardless of the specific magnetic configuration in question. The task of shaping the magnetic circuit in such a way that the spurious field will do the smallest possible optical damage is left to the design engineer.

Similar considerations concerning the oversaturation of the core and the appearance of a parasitic lens field also apply to the systems represented in Figs.4.84a, b. Here, the load on the magnet can be varied by traversing in the axial direction a soft-iron annular ring, which is mounted between the axial magnet and the upper lens core [4.179] and virtually produces a magnetic short circuit over a fraction of the magnet length. Here again, the core may become oversaturated magnetically, whereupon part of the magnetomotive force will be drawn from the upper gap to the oversaturated core region.

On moving the annular disc upwards, the field in the lower gap of the one-magnet system of Fig.4.84a is decreased too. If the lower lens is to remain unaffected, a second axial magnet parallel to the first one so far as the sense of magnetization of the gaps is concerned can be employed. The second magnet substantially maintains the magnetomotive force at the lower gap at its original value [4.179], which means that as seen from it the magnetic load represented by the first magnet and its iron circuit does not change much on moving the annular disc.

A ferromagnetic annular ring can also be employed to bridge the gap between the permanent magnet and the external cylindrical casing magnetically (Fig.4.84c,d). This again generates a magnetic short circuit over part of the length of the permanent magnet. Now, however, the region of magnetic oversaturation arises in the external casing.

If only one permanent magnet is employed (Fig.4.84c), its full magnetomotive force always appears at the upper lens gap. In the other magnetic path, the magnetomotive force of the magnet is subdivided between the lower lens gap and the oversaturated region of the casing. The magnetomotive force extended across the lower gap is therefore always smaller than the force across the upper gap: the position of the annular ring determines both the instantaneous magnetomotive force of the magnet and the weakening of the magnetic field in the lower lens gap. The magnetomotive force of the magnet becomes smaller the more the annular ring is moved downwards while at the same time the difference between the magnetic fields generated in the gaps becomes increasingly more pronounced.

The magnetically asymmetric design of Fig.4.84c can be made symmetric by adding a second magnet of the same type, as represented in Fig.4.84d. The annular ring now acts as a ferromagnetically conducting bridge for simultaneously determining the shunting loads of both permanent magnets. From Fig.4.84d, it is immediately obvious that the instantaneous magnetomotive force of one of the magnets is available for each of the lens gaps; this magnetomotive force is determined by the instantaneous shunting loads and hence by the position of the annular ring. The path of the two magnetic shunting circuits follows the respective parts of the external cylindrical lens casing determined by the ring position. It is important to realize that the same magnetomotive force is available to each of the two shunting circuits. For the shunt relating to the upper magnet, the effective magnetomotive force is tapped off

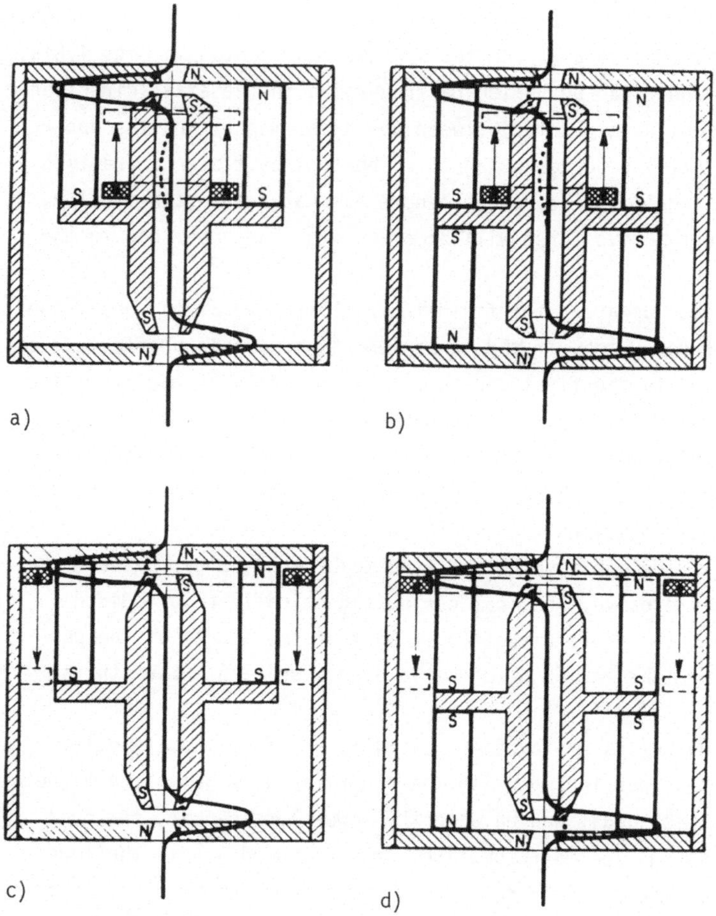

Fig.4.84a-d. Schematic representation of devices suitable for changing the focal length with permanent-magnet lens systems by varying the flux load on the magnets [4.179,180]. (a) Two-gap permanent-magnet lens system employing a single cylindrical axial magnet and an 'internal' axially traversable ferromagnetic regulating ring. The physical effect of the ring can be expressed by saying either that the magnetomotive force of the magnet is decreased by increasing its stray flux or that part of the magnet is short-circuited over the highly permeable path formed by the regulator ring and part of the central lens core. Below saturation of the magnetic circuit, both lens fields are weakened in synchronism whereas an asymmetric weakening of the lens field in the gap surrounded by the magnet can only be obtained at the cost of oversaturating the adjacent region of the lens core and thereby generating an extended parasitic lens. (b) Two-gap permanent-magnet lens system similar to a) but with two axial magnets employed in opposition. The stray flux from the additional (in the drawing lower) magnet is practically unaffected by the position of the shunting ring, so that the magnetomotive force at the associated lower gap is stabilized. (c) Two-gap permanent-magnet lens system similar to (a) but with an external shunting ring instead of the internal one. If this shunting ring is moved downwards (in the direction indicated by the arrow) then the magnetomotive force provided by the magnet gradually decreases. As long as the cross-section of the magnetic circuit remains large enough to avoid magnetic saturation, the magnetomotive force applied to both gaps is the same and is changed in synchronism.
On the other hand, the magnetomotive force at the lower lens gap can be weakened more than that at the upper gap if the cylindrical casing is made thin enough. Then,

if the shunting ring is moved downwards, saturation gradually sets in and increases in the part of the casing between the shunting ring and its proximal lens flange. Whereas the (somewhat reduced) magnetomotive force of the full length of the magnet always applies across the upper lens gap, a series resistance is generated with respect to the lower lens gap over the saturated part of the casing, which consumes part of the magnetomotive force of the magnet. The magnetomotive force remaining available for the lower gap just corresponds to that length of the magnet which is tapped off by the shunting ring. Nevertheless, it should be realized that with this regulating principle a substantial magnetic stray field is generated externally to the lens which may give rise to disturbances due to spurious lens fields or deflections of the beam (Fig.4.5). (d) Two-gap permanent-magnet lens system similar to (c) but with two axial magnets employed in opposition. Here, the stray flux and thus the magnetomotive force of the additional (in the drawing lower) magnet is practically unaffected by a change of position of the shunting ring. In this manner, the focal length of the lower gap can be stabilized

as a more or less extended cylindrical part of the upper permanent magnet. The remainder of the upper magnet is opposed to the lower magnet. Therefore, if both magnets are of the same type, the magnetomotive force available for the second shunt is just equal to the force available for the first shunt.

The magnetic resistances of the two shunt circuits are not very different, and nor therefore are the shunt loads and the magnetomotive forces generated over the magnets.

It is important to note that with the one-magnet design of Fig.4.84c, where one of the fields is weakened more strongly than the other, the generation of magnetic stray fields cannot be avoided by any kind of clever design: this intention would would run counter to AMPERE's Law, a basic physical principle which, in the present case, requires $\int Hdz = 0$ with the integration performed along the instrument axis over the length of the two-lens system and beyond it. On weakening the lens fields in an 'unbalanced' manner, $\int Hdz$ taken over the lens system proper becomes different from zero, and this difference must manifest itself as a magnetic stray field beyond the confines of the permanent-magnet two-lens system proper.

Evidently, this disadvantage does not necessarily arise with the 'symmetric' two-magnet system of Fig.4.84d, where the integral $\int Hdz$ always assumes equal absolute values for each lens gap. Here the integrals have different signs so that within the confines of the two-lens system proper, $\int Hdz$ vanishes and an external stray field does not necessarily arise. With a well balanced design the cross-section of the ferromagnetically conducting parts of the shunting circuits should be made large enough to avoid saturation and the corresponding generation of stray fields.

Finally, we discuss a design in which the focal length of a permanent-magnet lens is adjusted by providing additional gaps of variable size between the lens gap and the magnet. In the design represented schematically in Fig.4.85, one of the lens gaps always remains at the full magnetomotive potential of the permanent magnet. The other gap is connected to the magnet via three ferromagnetic yokes, arranged like the spokes of a wheel and cut through along planes inclined to the radial direction. The outward lying parts of the yokes are connected rigidly to the external

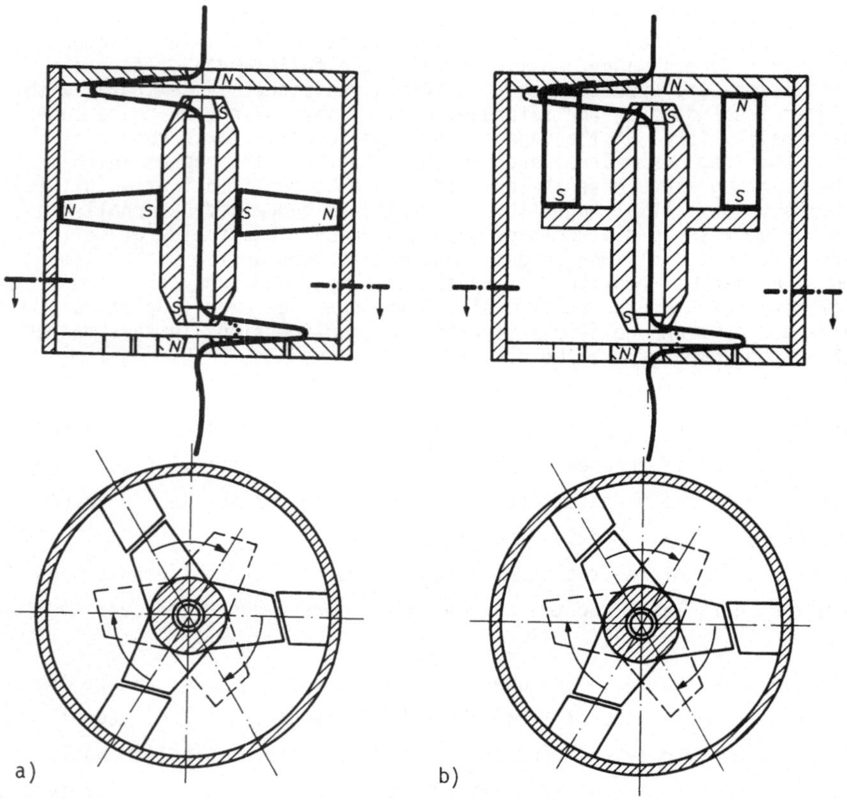

Fig.4.85a,b. Two-gap permanent-magnet lens system with variable series-resistor focal-length control [4.179,180]. One of the lens flanges is replaced by a spokes-of-a-wheel type regulator yoke. (a) Lens system employing radial magnets. (b) Lens system employing an axial cylindrical magnet. If the regulator is opened, the resistance of the magnetic circuit branch belonging to the lower part becomes larger and the magnetic flux traversing the magnet is reduced. This gives rise to an increase in the magnetomotive force of the magnet and is reflected in a corresponding increase in the magnetic field in the upper gap

casing while the 'star' of the central yoke stubs can be rotated to some extent around the lens axis. Thus, the cuts in the yokes can be gradually opened into more or less wide gaps which use up a corresponding fraction of the total magnetomotive force provided by the magnet for the circuit of the corresponding lens gap.

4.6.2b The Permanent Magnet and Its Design

The characteristics of some typical permanent-magnet materials are shown in Fig. 4.86. If the magnet has been mounted into its circuit and then magnetized, a point on the characteristic will indicate the working point of the magnet, i.e., a specific combination of magnetic flux density B_m and 'coercive' field strength H_m prevailing throughout its volume. We shall presently see that the flux density B_m is proportional to the coercive field strength H_m, inversely proportional to the cross-

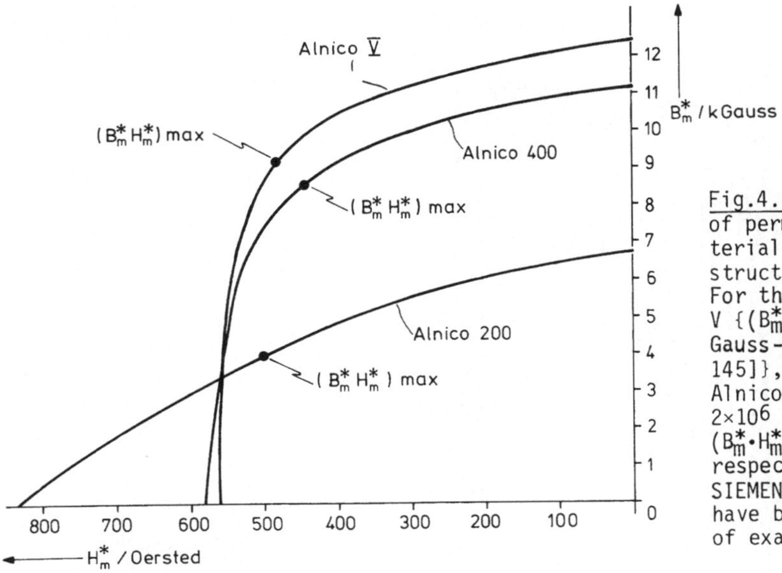

Fig.4.86. Remanence curves of permanent magnet materials suitable for constructing electron lenses. For this drawing, Alnico V $\{(B_m^* H_m^*)_{max} \approx 4.5 \times 10^6$ Gauss-Oe, data from [4.144, 145]}, Alnico 200 and Alnico 400 $[(B_m^* \cdot H_m^*)_{max} \approx 2 \times 10^6$ Gauss-Oe and $(B_m^* \cdot H_m^*)_{max} \approx 4 \times 10^6$ Gauss-Oe respectively, data from SIEMENS AG documentation] have been selected by way of example

sectional area Q_m and proportional to the length h_m of the magnet. Therefore, B_m depends heavily on the amount of the magnetic stray field carried by the ferromagnetic lens casing and the ferromagnetically 'empty' volume surrounded by it.

The design engineer can determine the performance of the magnet by suitably choosing the cross-section Q_m and length h_m of the magnet and by shaping the lens casing accordingly. The essentials of this process will be discussed in the following paragraphs as applied to a two-gap permanent-magnet lens represented schematically in Fig.4.6. Two modes of operation have to be distinguished for the permanent magnet.

1) The static operation mode where the magnetic circuit remains essentially untouched during operation and throughout the whole life of the lens [4.23,24,89,122-124, 144,145].

2) The dynamic operation mode where magnetic shunts are applied to vary the focal length of the lenses by means of corresponding changes of the flux load of the magnet and hence of its magnetomotive potential [4.121,179,180].

The design of the magnetic circuit of the lens and the prediction of the magnetomotive force to be expected at the gaps can be based on two fundamental physical laws:

- AMPÈRE's Law, which has to be written down here for two magnetic paths, each path crossing one of the gaps (the different quantities are distinguished by the subscripts 1 and 2). For the present purpose, the integral $\int H ds = 0$ can be simplified to:

$$H_1 S_1 = H_m h_m - H_{Fe} L_1 - L_f H_f = \eta_1 H_m h_m \qquad (4.145)$$

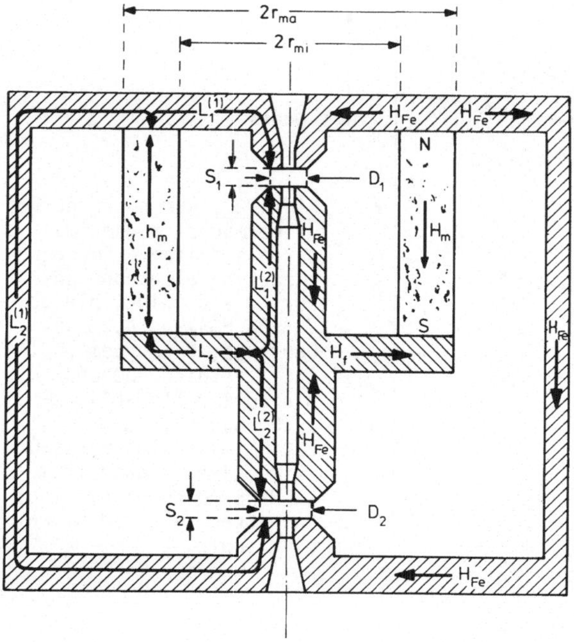

Fig.4.87. Schematic representation of the magnetic circuit of a two-gap permanent-magnet electron lens system

$$H_2S_2 = H_mh_m - H_{Fe}L_2 - L_fH_f = \eta_2H_mh_m \quad . \tag{4.146}$$

- The law of preservation of magnetic flux, which can be written in the form:

$$\mu_oH_1 \cdot \frac{\pi}{4} D_1^2 + \omega_1 + \mu_oH_2 \frac{\pi}{4} D_2^2 + \omega_2 = B_mQ_m \, . \tag{4.147}$$

In the above equations, H_1 and H_2 indicate the magnetic field strength in the pole-piece gaps 1 and 2, which are assumed to have gap widths S_1 and S_2 respectively and pole face diameters D_1 and D_2 (Fig.4.87). H_m is the coercive field strength and B_m the corresponding flux density in the permanent magnet of length h_m and cross-sectional area Q_m. [With an axial magnet of straight cylindrical form, we have $Q_m = \pi(r_{ma}^2 - r_{mi}^2)$.] With the magnetic path assumed in the ferromagnetic casing for (4.145) and (4.146), L_1 and L_2 indicate the lengths of those segments of the paths of integration that belong to only one of the paths through the gaps 1 and 2 whereas L_f is common to both of them. H_{Fe} and H_f are the corresponding field strengths, which have been written in such a way that H_{Fe} applies to both L_1 and L_2 and is thus the same for both circuits.

For the present purpose of lens design it is essential that the magnetic potential difference over the gaps should be much larger than the potential difference over the lens casing. In analogy to the requirements discussed above for satisfactory construction of an electromagnetic lens (cf. Sect.4.3.2), at most about 5% of the total magnetomotive force H_mh_m provided by the magnet should be lost along the casing with a well-designed permanent-magnet lens system. Therefore, h_m,L_1,L_2 and L_f being of comparable magnitude, the field strengths H_{Fe} and H_f in the casing must be smaller than the coercive field H_m by at least two orders of magnitude, and within this limitation neither H_{Fe} nor H_f need necessarily be designed to be constant along the paths L_1, L_2 and L_f of integration. Therefore, the essence of (4.145) and (4.146) can be expressed in terms of quality factors η_1 and η_2, which describe the fraction of the magnetomotive force of the magnet available at the gaps. For a satisfactory design, $1-\eta_1 \lesssim 0.05$ and $1-\eta_2 \lesssim 0.05$. It should be realized that strictly speaking η_1 and η_2 are subject to slight variations during large variations of H_m and B_m because, as a consequence of complex demagnetization effects in the magnet and of the hysteresis of the pole-piece material, neither H_{Fe} nor H_f is rigorously proportional to H_m. As before, however, we shall see that a small variation of η_1 or η_2 does not matter in practice, so that in the following we can use $\eta_1 \approx \eta_2 \approx \eta$ as a simplification.

Finally, ω_1 and ω_2 are the stray magnetic fields associated with the magnetic circuits 1 and 2 respectively and the two remaining terms on the left-hand side of (4.147) describe the useful flux contributions running across the parallel gaps of the pole-piece systems (Fig.4.87). Thus, the 'useful' fraction σ of the total magnetic flux can be written as

$$\sigma = \frac{\mu_0 \pi}{4B_m Q_m}(H_1 D_1^2 + H_2 D_2^2) = \frac{B_m Q_m - \omega_1 - \omega_2}{B_m Q_m} \quad . \tag{4.148}$$

In analogy to the magnetic flux distribution prevailing in electromagnetic lenses (cf. Sect.4.3, especially Fig.4.37), the flux traversing the parallel gap is again a small part of the total flux with permanent-magnet lenses, and σ is thus a comparatively small fraction.

The actual magnetic flux in the gaps is slightly different from $\mu_0 H_1 D_2^2 \pi/4$ and $\mu_0 H_2 D_2^2 \pi/4$, due to the presence of the bores and to the 'fringing out' of the off-axis parts of the field. Therefore, the use of these expressions in (4.148) should be understood as the introduction of a convenient magnetic flux unit, only slightly different from the flux in the gaps.

After some elementary transformations of (4.145-148), a set of equations is obtained that can serve as the basis for the design of the permanent magnet. For this, it is convenient to introduce $H_1 S_1 = H_2 S_2 = (NI)_{equ}$, the equivalent ampere-turns of the permanent-magnet lens, and $S_1/D_1 = A_1$ and $S_2/D_2 = A_2$, the aspect ratios of the two gap spaces. In many practical cases, both pole-piece systems will be constructed with the same aspect ratio $A \approx A_1 \approx A_2$, whereupon a particularly simple form results for the design equations.

In general then, we obtain:

- for the energy product of the magnet

$$B_m H_m = \mu_0 \cdot \frac{\pi}{4\sigma\eta} \cdot \frac{(NI)^2_{equ}}{A^2 Q_m} \cdot \frac{S_1 + S_2}{h_m} \quad , \tag{4.149}$$

- for the permeance of the magnet

$$\frac{B_m}{H_m} = \mu_o \frac{\pi}{4} \cdot \frac{\eta}{\sigma} \frac{(S_1 + S_2)h_m}{A^2 Q_m} , \tag{4.150}$$

and as a consequence of (4.149) and (4.150)

- for the magnetomotive force available at the gaps

$$(NI)_{equ} = \eta h_m \sqrt{\frac{B_m H_m}{B_m/H_m}} = \eta h_m H_m \quad . \tag{4.151}$$

In the above equations, rational (SI) units have been employed: the flux density B_m is expressed in teslas and the field strength H_m in A/m. Nearly all the available data on permanent magnets are given in Gaussian units however, with the flux density B_m^* measured in Gauss and the field strength H_m^* in Oersteds; we therefore list the corresponding forms of (4.149-151):

- for the energy product

$$B_m^* H_m^* = \frac{4\pi^3}{10^2 \sigma\eta} \cdot \frac{(NI)_{equ}^2}{A^2 Q_m} \cdot \frac{S_1 + S_2}{h_m} , \tag{4.152}$$

- for the permeance

$$\frac{B^*}{H^*} = \frac{4\pi^3}{10^2} \frac{\eta}{\sigma} \frac{(S_1 + S_2)h_m}{A^2 Q_m} , \tag{4.153}$$

- for the magnetomotive force at the gaps

$$(NI)_{equ} = \frac{10}{4\pi} \eta h_m \sqrt{\frac{B_m^* H_m^*}{B_m^*/H_m^*}} = \frac{10}{4\pi} \eta h_m H_m^* \quad . \tag{4.154}$$

In (4.154), $(NI)_{equ}$ is measured in amperes.

In the literature and in data sheets concerned with permanent magnets and employing Gaussian units, the magnetomotive force is often expressed in Gilberts[3]. Instead of (4.154) we then write, with h_m, S_1, S_2 in cm:

- for the magnetomotive force at the gaps (in Gilberts)

$$F_S = H_1 S_1 = H_2 S_2 = \eta h_m \sqrt{\frac{B_m^* H_m^*}{B_m^*/H_m^*}} = \eta h_m H_m^* \quad . \tag{4.155}$$

There is yet another property of magnetic devices energized by permanent magnets, a physical property associated with the energy content of the magnetic fields:

---

3 The Gilbert is the unit of magnetomotive force in the e.m.u. system: 1 ampere-turn = $4\pi/10$ Gilberts = $4\pi/10$ Oe·cm.

Quite generally, the magnetic energy contained in the magnetic field external to the permanent magnet is just equal to the energy product $\frac{1}{2} B_m H_m$ integrated over the volume of the magnet [4.118,187]. As will be shown below, this energy relation can be quite helpful during the process of optimization.

It will now be explained how equations (4.148) to (4.155) and the energy relation can be applied to the design of the permanent magnet required for a specified lens system.

The *static operation mode* will be considered first. Here, the lens system and its magnet can be optimized so that a high value of the magnetomotive force $(NI)_{equ}$ or HS across the gaps is obtained in the most efficient way.

From (4.149) and (4.152), it is immediately obvious that for a given design the highest magnetomotive force $(NI)_{equ}$ is obtained if the magnet is operated at the maximum value $(B_m H_m)_o$ of the energy product $B_m H_m$. Numerical values of $(B_m H_m)_o$ or $(B_m^* H_m^*)_o$ for commercially available permanent-magnet materials are to be found in the literature, mostly expressed in units of MGauss·Oe (e.g. [4.118,187]) or W·s/dm^3 (e.g. [4.51]). These units are related as follows:

$$1\text{MGauss}\cdot\text{Oe} = \frac{10^2}{4\pi}\ \text{W}\cdot\text{s/dm}^3 = \frac{10^2}{4\pi}\ \frac{\text{kW}\cdot\text{s}}{\text{m}^3} \quad .$$

The maximum value $(B_m H_m)_o$ occurs at a specific point of the B_m - H_m curve (sometimes also called the remanence curve or hysteresis curve) of the material (see Fig.4.86, or the more schematic representation of Fig.4.88). It is important to realize that the permanent magnet can be operated at the $(B_m H_m)_o$ working point only if it is magnetized in its mounted position within the magnetic circuit, and that this is a necessary but by no means sufficient requirement for attaining $(B_m H_m)_o$.

From Fig.4.88, it becomes obvious that working at the $(B_m H_m)_o$ point requires a corresponding particular value $(B_m/H_m)_o$ of the magnetic permeance of the lens system, which can be calculated from (4.150) or (4.153). The permeance does not depend on the specific magnetic properties of the permanent-magnet material, at least as long as the flux distribution in the device (and hence σ) is not altered because of magnetic saturation in a casing with poor magnetic design. It should also be noted that the permeance of a lens system is not changed if the design is scaled up or down, i.e., if S_1, S_2 and h_m are multiplied by the same factor and Q_m by its square.

Hence, the following procedure can be adopted for the design of the permanent magnet to be employed with a permanent-magnet lens system:

1) Specify S_1, S_2, A and $(NI)_{equ}$, to satisfy the electron optical requirements, and η as a measure of quality of the magnetic circuit design to be achieved.

2) Select a suitable permanent-magnet material, preferably with a particularly high value of the maximum energy product $(B_m H_m)_o$. Alnico V (4.144,145),

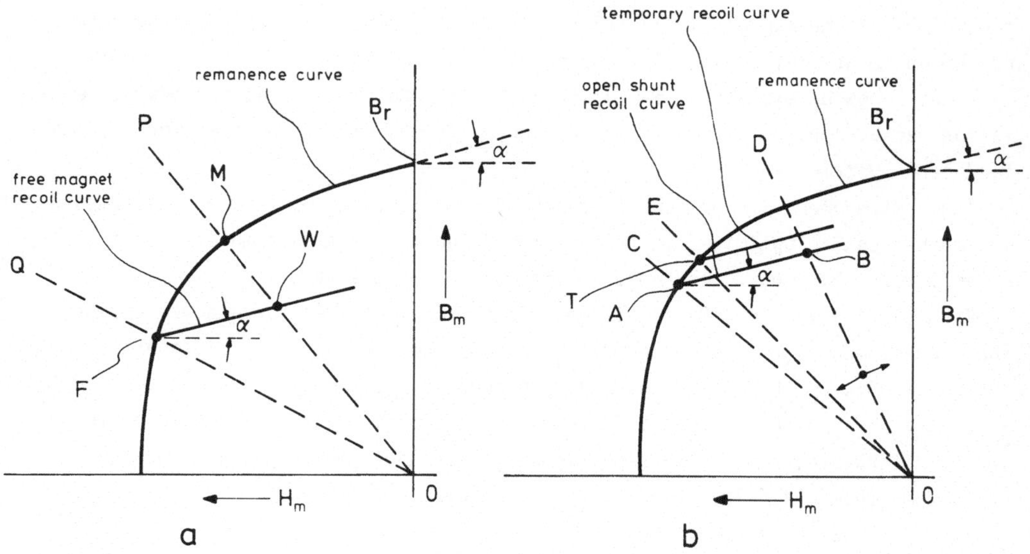

Fig.4.88a,b. Determination of the working points of permanent-magnet electron lenses. (a) Here, the conditions are shown for permanent magnets operated at fixed magnetomotive force. OP is the permeance line for the magnet mounted into its position in the lens casing. If the magnet is magnetized in this position, the working point is the point of intersection M of the permeance line OP and the remanence curve of the magnet. On the other hand, if the magnet is magnetized before assembly of the lens, it first assumes the working point F, the point of intersection of the remanence curve with the permeance curve OQ for the magnet suspended in free space. On mounting the magnet into its position within the lens, the magnetic resistance in the space around the magnet becomes smaller so that its permeance increases and exhibits the behaviour indicated by the line OP. The working point W now assumed is the point of intersection of the recoil curve extending from point F and the permeance line OP. With a given design and with a view to obtaining a large magnetomotive force, the advantage to be gained by magnetizing the magnet in its final mounted state is obvious. (b) Here, the case of magnets equipped with variable shunts for regulating their magnetic loads and hence their magnetomotive forces is illustrated. Different positions of the shunts give rise to changes in the magnetic resistance externally loading the magnet and hence to corresponding variations of the gradient of the permeance lines OD. The permeance line OC of smallest slope corresponds to the fully opened shunt condition and determines the working point A if the magnet is magnetized ready mounted in its circuit. Then, on actuating the shunt, the working point travels along the recoil line extending from point A and is determined as the point B of intersection of the recoil curve and the corresponding permeance line OD. It should be realized that the recoil curve extending from A is always taken to be the locus of the working points as soon as the shunt has been fully opened once: if the magnet is magnetized with the shunt in a partially closed position, e.g., corresponding to the permeance line OE, the working point T is first assumed. On further closing the shunt the working point travels along the 'temporary' recoil curve extending from T. If the shunt is now fully opened, the working point first comes back to T and then slides down the remanence curve to point A. From then on, regular performance is resumed and the working point will remain on the 'open shunt' recoil curve extending from A

Alnico 400 [4.121], and Alnico 200 [4.123] should be named here as examples of permanent-magnet materials successfully used for lenses actually constructed. The Ferroxdure type materials [4.211] could be another promising possibility.

3) For the working point of maximum energy product, determine numerical values of $(B_m^*H_m^*)_o$, $(B_m^*/H_m^*)_o$ and H_{mo}^* or H_{mo}. By way of example, for the Alnico magnets, $(B_m^*H_m^*)_o$ can be expected to amount to several MGauss·Oe (up to around 5), $(B_m^*/H_m^*)_o$ to about 10 to 20, and H_{mo}^* to values in the range of 500 to 800 Oe.

4) Employing $(NI)_{equ}$ and η from step 1), H_{mo}^* from step 3), and (4.151) and (4.155) the length h_m of the magnet can be determined.

5) The 'useful' fraction σ of the total magnetic flux defined by (4.148) has to be determined, a point discussed more thoroughly below.

6) Finally, all the necessary data are now available for the determination of the cross-sectional area Q_m of the magnet. For this, either (4.149) and (4.152) or (4.150) and (4.153) can be employed.

The only 'problem child' of this procedure is the determination of the fraction σ of the flux traversing the pole-piece gaps. Unfortunately, for permanent-magnet lenses, the distribution of magnetic flux within the space surrounded by the lens casing has been very little investigated. No universally applicable formulae for the flux determination of the kind reported in Sect.4.3, so convenient for the design of electromagnetic lenses, are available. Of course, the field distribution can be determined by a numerical procedure, employing the relaxation method. This has been actually carried through for one special lens and is described in [4.23, 24].

On the other hand, a sufficiently true 'network' representation of the actual shape of the pole pieces (even neglecting the bores) and of the bulk of the iron-free space enclosed by the lens casing might well require a major computational effort. In these cases, the design engineer may well resort to approximate values for σ, obtained by making a plausible estimate of the flux distribution. For this, some of the results of Sect.4.3 can be adapted to the present case or even used unchanged, the formulae for the flux between the pole pieces for example, where there will evidently be little difference between the flux distributions of permanent-magnet lenses on the one hand and electromagnetic lenses on the other.

It should be realized that, in cases of doubt, a trend to lower estimated σ values is more on the safe side, because at worst a cross-sectional area Q_m of the magnet will be calculated that is somewhat too large. The actual permeance of the circuit will correspondingly become lower than $(B_m/H_m)_o$, leading to a working point with $B_m < B_{mo}$ and $H_m > H_{mo}$. Often, below the $(B_mH_m)_o$ point of optimum energy product, the remanence curve approaches the H_m-axis of the B_m-H_m diagram (cf. Fig.4.86) rather steeply. Decreasing the actual permeance below $(B_m/H_m)_o$ will thus have comparatively little effect on H_m and, from (4.151) and (4.154), on the equivalent ampere-turns $(NI)_{equ}$ available at the gaps, and $(NI)_{equ}$ is what counts electron optically. Even if H_m and $(NI)_{equ}$ are appreciably larger than is required by the specification, the specified ampere-turns can be subsequently obtained if need be without any large modification of the design by employing a new magnet having a correspondingly reduced cross-section or by introducing an additional shunt to increase the magnetic flux external to the magnet, thereby decreasing σ. The progress of this correction as well as the proper magnetic performance of the permanent-magnet lens system has to be checked by measuring the actual magnetomotive force across the gap. Ballistic and vibrating coil measuring devices have been employed for this task and are described in the literature (e.g. [4.73,121,179,180]). If the actual coercive field strength $H_m^{(m)}$ in the magnet has been established by employing the measured value of the equivalent ampere-turns $(NI)_{equ}^{(m)}$ across the gaps together with (4.151) and (4.154), the field strength $H_m^{(s)}$ corresponding to the specified equivalent ampere-turns $(NI)_{equ}^{(s)}$ is given by

$$H_m^{(s)} = H_m^{(m)} \frac{(NI)_{equ}^{(s)}}{(NI)_{equ}^{(m)}} \quad , \tag{4.156}$$

as a consequence of (4.151) and (4.154). After extracting the corresponding values $(B_mH_m)^{(s)}$ and $(B_mH_m)^{(m)}$ from the remanence curve of the magnet material, and making use of the energy relation discussed above, the cross-sectional area $Q_m^{(s)}$ of the magnet corresponding to the specification can be predicted as

$$Q_m^{(s)} = Q_m^{(m)} \cdot \frac{(B_mH_m)^{(m)}}{(B_mH_m)^{(s)}} \cdot \frac{[(NI)_{equ}^{(s)}]^2}{[(NI)_{equ}^{(m)}]^2} \quad . \tag{4.157}$$

Evidently, with a properly designed magnetic circuit that remains free from saturation, we should obtain $H_m^{(s)} \approx H_{mo}$ and a lens system working at the maximum energy product point if H_{mo} has been defined appropriately (as in 3).

It should be noted that (4.149) and (4.152), in the form

$$\sigma = \frac{\mu_o}{(B_mH_m)^{(m)}} \cdot \frac{\pi}{4\eta} \frac{[(NI)_{equ}^{(m)}]^2}{A^2Q_m^{(m)}} \cdot \frac{S_1 + S_2}{h_m} \tag{4.158}$$

$$= \frac{4\pi^3}{10^2\eta} \cdot \frac{[(NI)_{equ}^{(m)}]^2}{(B_m^*H_m^*)^{(m)}A^2Q_m^{(m)}} \cdot \frac{S_1 + S_2}{h_m} \quad ,$$

yield an 'experimental' method for the determination of the 'useful' fraction of the magnetic flux enclosed in the casing.

Clearly, this procedure for establishing σ is not restricted to the working point corresponding to the maximum energy product $(B_mH_m)_o$, but can be carried through for any measured value of the equivalent ampere-turns $(NI)_{equ}$ if the corresponding energy product B_mH_m can also be determined, and this does not present any particular problems for working points on the remanence curve.

As mentioned above, working points on the remanence curve are obtained if the permanent magnet is magnetized with the magnet mounted in its final position within the magnetic circuit of the lens. The magnetic field externally applied to the magnet in the magnetization process should be high enough to drive the magnetic flux density within the magnet completely into saturation. For this, as a rule of thumb, a field strength about five times the maximum coercive field strength that can be attained by the magnet is required [4.45]. The magnetizing field can be generated by means of a strong electromagnet or by discharging a large bank of capacitors through a coil of a few turns, wound around the casing [4.23,24,144,145]. Equipment suitable for magnetization can be obtained commercially (cf. e.g. [4.44]).

The pole pieces, the lens cores and the lens casing are completely oversaturated while the magnetizing field is applied, and the direction of the magnetizing field must be oriented so that it tallies with the direction of the permanent magnetic moment to be impressed onto the magnet. Then, if the magnetizing field is cut off, the transient working point of the magnet slides down along the remanence curve until the static working point is reached. The latter is the point of intersection of the remanence curve with the straight line drawn through the origin of the $B_m - H_m$ coordinate system and defined by (4.150) or (4.153), which represents the constant permeance of the design and hence all possible combinations of B_m and H_m that can be assumed by the magnetic circuit configuration of the lens (Fig.4.88a).

If, on the other hand, the permanent magnet is magnetized in its unmounted state, the permeance now effective corresponds to the free magnet and is considerably lower than that of the 'mounted' geometry; the 'free' magnet working point is correspondingly shifted to lower B_m values (cf. Fig.4.88a). On then introducing the permanent magnet into its position within the magnetic circuit of the lens system, its working point moves to the right and a little upwards along the so-called recoil curve. For the purposes of practical lens design, the recoil curve can be approximated by a straight line which has the same slope as the remanence curve at its remanent magnetization point $H_m = 0$, $B_m = B_r$ [4.40,92,138]. The static working point of the permanent-magnet lens system designed for 'external' magnetization can then be determined as the point of intersection of its permeance line with the recoil line extended from the free magnet working point. It is evident from Fig.4.88a that, quite generally, 'external' magnetization must lead to a considerably lower magnetomotive force at the pole-piece gaps than mounted magnetization.

Finally, the design of the permanent magnet and its associated magnetic circuit will be discussed for lenses equipped with variable shunts for changing the magnetomotive forces across the pole-piece gaps (cf. Sect.4.6.2a). By varying the shunts, the permeance of the magnetic circuit is changed: decreasing the magnetic resistance of the circuit energized by the magnet increases the permeance, and vice versa.

After magnetization with the shunt in the 'open' position, the working point assumed is the point A at which the remanence curve intersects the permeance line corresponding to the currently effective shunt resistance, which should be the highest resistance of the available range (Fig.4.88b). On varying the shunt and hence its corresponding resistance, the working point travels along the recoil line extending from the working point on the remanence curve. Again, the effective working point B is the point of intersection of the recoil line with the actual permeance line OD. The permanent-magnet lens system is then operating in the *dynamic working mode*. It will be advantageous to employ a magnetic circuit design that places the working point A at or close to the working point $(B_m H_m)_o$ of maximum energy product.

Finally, so far as the magnetization is concerned, it is evident from the above that magnetizing an axial magnet system mounted in its circuit does not present any major problems. On the other hand, with a radial magnet system where the magnets are fabricated in the form of ring-shaped discs, only 'external' magnetization is possible. For this, rather intricate procedures may become necessary in order to ensure a sufficiently homogeneous magnetization (cf. e.g. [4.123], where this has been achieved by magnetizing the discs in a succession of ring-shaped sections of comparatively narrow radial extension).

In another design [4.144,145], the radial magnet system was composed of a set of four bar magnets, arranged like the spokes of a wheel. Here, the magnets could be magnetized ready mounted in the magnetic circuit by winding a few turns of wire around the individual magnets and flashing the resulting magnetizing coils with a strong current.

4.7 Superconducting Electron Lenses

The superconducting electron lens is a comparatively recent development. Originally, the extremely high stability of the lens current that could be attained in the so-called persistent current mode seemed to be a feature of great value. Nevertheless, a constant lens current was not sufficient to stabilize the focal length and other electron optical properties of the lens to a comparable degree because the fluctuations of the beam voltage remained at their full value.

It was also hoped that by using superconductors it would be possible to generate extremely strong lens fields and thereby considerably reduce the geometric optical aberrations. However, this would make sense only if associated with a correspondingly drastic reduction of the half-width of the lens field so that the lens strength k^2 remained at values that safeguarded a reasonable performance of the field as an electron optical device. Unfortunately, there has been little progress in this respect and there is not much hope of a breakthrough in the near future. It should be realized that for the actual construction of magnetic objective lenses, fields having peak flux densities far above the just saturated, two-tesla region could scarcely be used for the conventional range of electron energies around 100 keV: the field half-width would amount to a few tenths of a millimetre only and this would not leave enough space for a mechanically stable and trouble-free specimen stage and objective aperture system. In this respect, fields with a half-width of about one to a few millimetres would be more satisfactory, and such fields would be suitable for megavolt electron microscopes. Nevertheless, there seems to be little actual progress in this direction.

At present, then, the principal advantages to be derived from the use of cryogenic superconductor technology for electron-lens design appear to arise from 'fringe' benefits. By way of example, the liquid-helium-cooled walls of the vacuum space in superconducting objective lenses produce a vigorous pumping action and improve the already high vacuum to extreme UHV-values, thereby protecting the specimen from hydrocarbon contamination and etching by residual-gas ions [4.18,32]. Liquid-helium-cooled superconducting or low resistance instrument walls are most effective shields against AC magnetic stray fields [4.30]. The resistance of the specimen to radiation damage is markedly increased at liquid-helium temperatures in comparison with ambient temperature microscopy [4.34]. Thermal specimen drift is strongly reduced at low temperatures since the thermal conductivity of the relevant specimen stage materials is higher by some orders of magnitude [4.33].

The following presentation of the design of superconducting lenses should be read with these points in mind. Unfortunately, the information published about the actual technological aspects of superconducting lens design often remains rather sketchy, and it is to be feared that this must inevitably slow down general progress in the field. A rather large number of designs have been built but regrettably, the particular advantages and disadvantages of the individual solutions are rarely clear from the published accounts.

The following treatment cannot aim to fill these gaps, and it will concentrate on the iron-shrouded superconducting lens and the superconducting shielding lens. During recent years these types have become generally accepted as 'practical' and applied superconducting lenses. For a treatment of the historical development of the field, a useful specialized monograph [4.32] should be consulted, which also contains a survey of various aspects of superconducting electron microscope systems; [4.65] contains a very full bibliography.

4.7.1 Superconducting Electron Lenses with Ferromagnetic Pole Pieces

The first superconducting lenses had been constructed as iron-free coils without any pole pieces [4.48,50]. As expected, the magnetic fields turned out to be extremely stable and — when employing the persistent current mode — could be so maintained for several hours [4.49]. Nevertheless, as electron lenses, these iron-free superconducting coils did not compare favourably with conventional ambient-a)

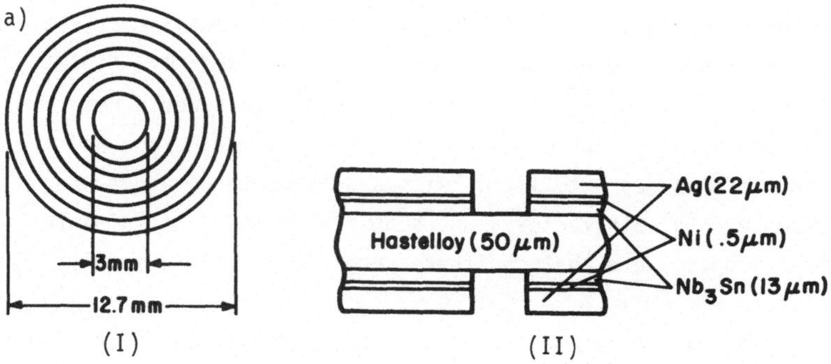

Fig.4.89a. Simplified illustration of the construction of superconducting Nb_3Sn ring discs composed of numerous annular conductors [4.72]. (I) Top view of a disc as seen in the direction of the lens axis, showing the annular subdivision of the superconducting Nb_3Sn layers. (II) Magnified schematic cross-section of a disc displaying the region around the gap between two of the individual zones of the Nb_3Sn layer that form the annular structure

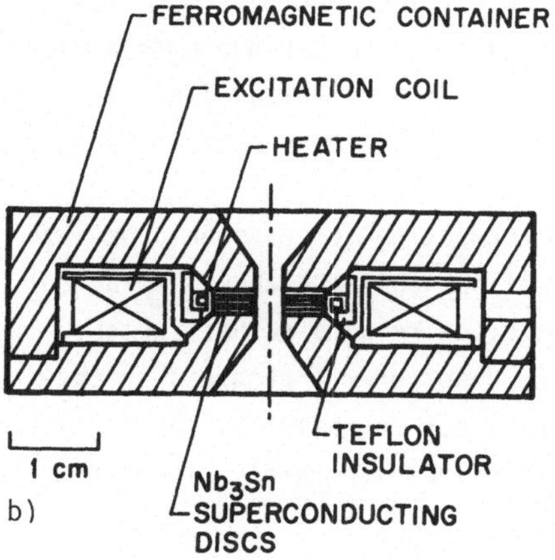

b)

Fig.4.89b. Simplified cross-section of a superconducting electron objective lens equipped with a stack of Nb_3Sn ring discs (Fig.4.89a) for generating the lens field [4.72]. An essentially conventional ferromagnetic circuit and pole-piece design is used and the stacks are composed of between twenty and a hundred discs. The complete lens structure is immersed in a liquid-helium bath-cryostat (Fig.4.90)

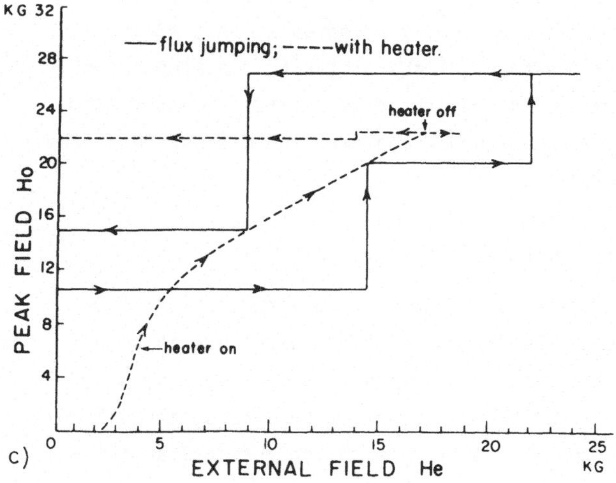

c)

Fig.4.89c. Trapping of the magnetic lens field by starting a persistent current in a superconducting ring disc stack of an electron objective lens as shown in Fig.4.89b. The trapping process is shown here as the dependence of the peak field on the lens axis on the magnitude and temporal history of the external field generated by the excitation coil. The continuous curve describes flux trapping without using the heater, by overrunning the critical field strength of the discs. It is seen that with increasing external field the magnetic field penetrates into the axial region of the stack in rather big jumps, and that on reducing the external field to zero, part of the magnetic flux is lost in a flux jump of comparable magnitude. The actual size of the flux jumps and hence of the remaining field can neither be reproduced nor predicted accurately. A method better suited to actual practice consists in using the heater indicated in Fig.4.89b to raise the temperature of the stack above its critical value. The stack is now normally conducting and does not obstruct the continuous penetration of the external field onto the lens axis where it is frozen in by switching off the heater current. This procedure is indicated by the interrupted line and is seen to generate a higher field strength than the flux-jump charging method [4.72]

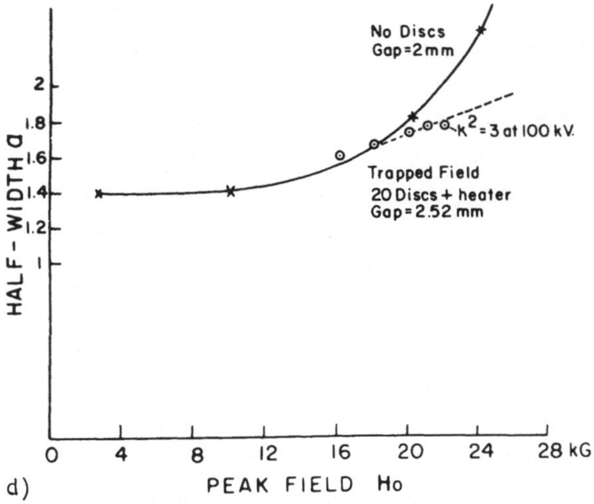

d)

Fig.4.89d. Representation of the half-width (in mm) of the axial magnetic field distribution in a superconducting electron lens of the type shown in Fig.4.89b. The continuous curve describes the field half-width obtained in a lens where the superconducting discs have been removed, so that the field is generated by the current in the excitation coil and shaped by an essentially classical iron circuit. The broken line indicates the field half-width obtained in a lens equipped with twenty superconducting annular discs, energized by using the heater procedure illustrated in Fig.4.89c. Comparison of the two curves shows that as soon as magnetic saturation of the pole-piece tips sets in, better compression of the field is obtained with the current concentration provided by the discs, which is highly compact and close to the axis [4.72]

temperature iron-shrouded lenses because it was not possible to generate strong, highly peaked magnetic fields with half-widths of the order of millimetres. It is well known that with an iron-free coil, the half-width of the field is about the same size as the inner diameter or the axial length of the coil. Therefore, in order to produce a magnetic field with a half-width comparable to or even shorter than conventional pole-piece-concentrated lens fields at a comparable or even higher magnetic flux density, extremely small coils would have to be employed with current densities reaching 10^6 A/cm^2, a formidable or nearly impossible task, even for modern high-field, high-current superconducting wire materials (cf. e.g. [4.32]). Moreover, the current density throughout the cross-section of a small wire-wound coil is not homogeneous enough to generate a magnetic lens field of sufficient rotational symmetry for objective lens requirements.

Better rotational symmetry can be achieved by employing a small superconducting ring in place of the coil and operating the ring in the persistent current mode [4.12]. Here again, the current density distribution may deviate from complete rotational symmetry, due to defects of the ring material, for example. The deviation of the current paths in the ring from the required circular pattern can be suppressed by subdividing the massive toric superconductor into a number of tiny flat annular rings [4.72,190]. Such rings have been fabricated in the following manner: Discs with an external diameter of about 1/2" and a 3 mm bore were punched out of thin platinum foil or Hastelloy ribbon, which had previously been coated on both sides with a thin vapour-deposited layer of niobium-tin Nb_3Sn (cf. Fig.4.89a; Hastelloy is an alloy consisting mainly of Ni with smaller amounts of Fe, Mo, Cu and Al). By subsequent etching, employing photo-lithographic techniques, a set of concentric rings of superconducting material (cf. Fig.4.89a) was produced [4.60,140]. Twenty to a hundred discs were assembled into a stack a few millimetres thick, with a space factor ratio 1:3 of superconducting to non-superconducting material. The stacks are then clamped between the pole pieces of an essentially conventional iron circuit (Fig.4.89b). With this structure too, a rather large basic astigmatism may arise ($\Delta f \approx 3$ µm was observed [4.72]), and a stigmator must be provided to remove this aberration. A resolving power of better than about 1 nm can then be expected for this type of ring lens [4.72], for which vibrations caused by the boiling of helium may be the limitation and not the electron optical properties of the magnetic field.

For the actual operation of the ring lens, an additional superconducting coil (typically fabricated from Nb-Ti wire) must be provided to generate the magnetic field that is to be trapped by the persistent current in the ring structure. In principle, with the ring structure cooled down to the superconducting state, the coil field can be pushed along the axis of the ring lens if it is first made stronger than the critical field strength H_c of the ring material. The coil current is then decreased slowly and the persistent current is started as soon as the coil field-strength falls below H_c. In another method for pushing the field onto the axis of the ring lens, the coil field is increased rapidly so that large flux jumps in the stacks of discs are generated [4.72].

Unfortunately, with both methods, the amount of magnetic flux pushed onto the axis of the ring lens cannot be controlled very well. It would therefore be most difficult to adjust the field to a particular value determined by the electron optical requirements. In a more advanced construction, which avoids these restrictions, the stack is surrounded by a small heater, consisting of a few turns of manganin wire for example (cf. Fig.4.89b). Then, in order to introduce the magnetic flux into the ring lens, the stack is heated to a temperature at which the ring conducts normally and loses its property as a superconducting shield. The magnetic field of the coil can now advance unobstructed into the region close to the axis, surrounded by the ring, where it is trapped after cutting out the heater current and coming back to the superconducting state [4.72]. In this way, the lens field can be adjusted to the required value in a controlled manner (cf. Fig.4.89c). In the practical design, a thermal shield should be interpersed between the heater and field-

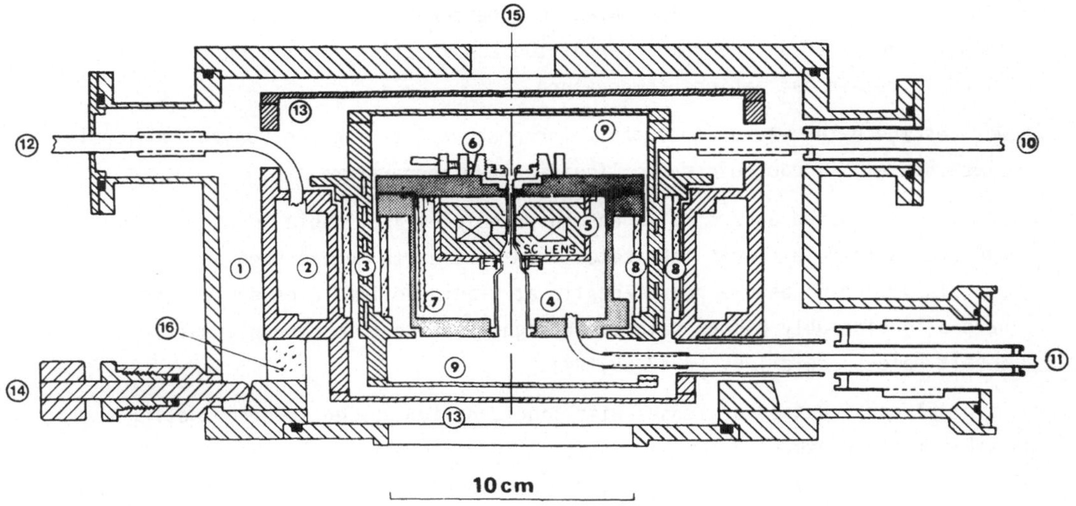

Fig.4.90. Simplified cross-section of the annular-disc-type superconducting lens of Fig.4.89b, shown ready mounted in its associated liquid-helium cryostat (adapted from [4.72]). 1 vacuum space. 2 liquid-nitrogen bath. 3 evaporated-helium gas-cooled radiation shield. 4 liquid-helium bath. 5 superconducting annular disc lens. 6 specimen stage. 7 resistance thermometer for monitoring the liquid-helium level. 8 thermal insulators fabricated from cylindrical glass tubes. 9 boiled-off-helium gas-cooled radiation shields. 10 boiled-off-helium gas-exhaust. 11 helium supply duct. 12 liquid nitrogen supply duct. 13 liquid-nitrogen-cooled radiation shield. 14 micrometer control for transverse alignment of lens and cryostat. 15 lens axis. 16 epoxy thermal insulator

generating coil, a teflon insulator as shown in Fig.4.89b, for example. The heater can then be energized without danger of heating the coil up to normal conductivity.

A cross-section of a complete stack-type ring-lens is represented in Fig.4.90, which also shows the cryostat. It is an interesting feature of stack-type ring-lenses that the axial distribution of the trapped field alone is very close to a Gaussian distribution $B(z) = B_0 \exp[-(z/a)^2]$, whereas a distribution approaching GLASER's bell-shaped field $B(z) = B_0/[1 + (z/h)^2]$ is obtained if the coil current is not disconnected after trapping the flux close to the axis [4.72]. The half-width is practically the same with both field distributions, and it is nearly the same as the half-width of the corresponding pole-piece field without a stack, at least until the magnetic flux density approaches saturation. If the field is increased beyond saturation, the iron-shrouded ring-stack lens becomes superior to the corresponding pole-piece lens, which has a noticeably larger half-width (Fig.4.89d).

The majority of the iron-shrouded superconducting lenses employed today do not utilize field-trapping ring structures. Essentially, they are conventional iron circuit lenses equipped with superconducting coils of comparatively small cross-sectional area. Here, when establishing the design principles, three cases must be distinguished.

1) Lenses where only the coil is cooled to superconductivity at liquid-helium temperature, whereas the remainder of the construction, including the iron circuit, remains at ambient temperature.

2) Devices where the whole lens: coil, ferromagnetic casing and conventional pole-piece system, is cooled down to liquid-helium temperature.

3) Lenses constructed according to design principles applicable to group 2) but equipped with pole pieces fabricated from rare earth metals such as dysprosium or holmium, which become ferromagnetic at liquid-helium temperatures where they have a considerably higher saturation magnetization than conventional iron-cobalt pole-piece alloys [4.42,69,194].

The typical construction of lenses with liquid-helium cooled superconducting coils and ambient temperature magnetic circuit is shown schematically in Fig.4.91. The coils with their cryostatic environment are suspended freely within the ferromagnetic casing and supported by the laterally extending feeding tubes for the liquid-helium supply. Depending on the position of the coil and the pole-piece gap with respect to the casing, a symmetric (Fig.4.91a) or asymmetric (Fig.4.91b) construction can be adopted. So far as the design of the pole-piece system and the lens casing is concerned, the methods described in Sect.4.3 can clearly be applied here too.

The symmetrical design (Fig.4.91a) is not so well adapted for lenses that require adjustable apertures or specimen stages for their operation, because access to the gap is effectively blocked by the cryostat. This is therefore the typical design for condenser or projector lenses [4.95,189]. It should be observed that there is a considerable electromagnetic attraction between the coil winding and the casing. Therefore, only if the lens construction is completely mirror symmetric throughout and hence the lens coil symmetric about the mid-plane is the coil in a position of (unstable) equilibrium. If the coil is displaced from the equilibrium position, either upwards or downwards, forces attracting the coil towards the nearer lens flange arise. These forces increase in direct proportion to the displacement and in proportion to the square of the ampere-turns in the coil [4.96,189]. Typically, for a lens as represented in Fig.4.91a and operated at $NI \approx 10^4$ A, these forces amount to 6 kp or 60 N per mm displacement.

In the design of a superconducting lens of this type, placing the coil in the position of unstable equilibrium around the mid-plane of the casing would create a danger of mechanical vibration of the coil, initiated by bubbling during the boiling-off of the liquid helium for example. It has been reported that a vibrational amplitude of 0.1 mm can produce an electron optical effect equivalent to a relative lens current fluctuation of $\Delta I/I \approx 5 \times 10^{-5}$ [4.6] (also reported in [4.18]), a dangerous disturbance, at least for objective lenses. A more stable construction results if the lens coil is placed a little off-centre (e.g., by 0.25 mm in the lens design represented in Fig.4.91a).

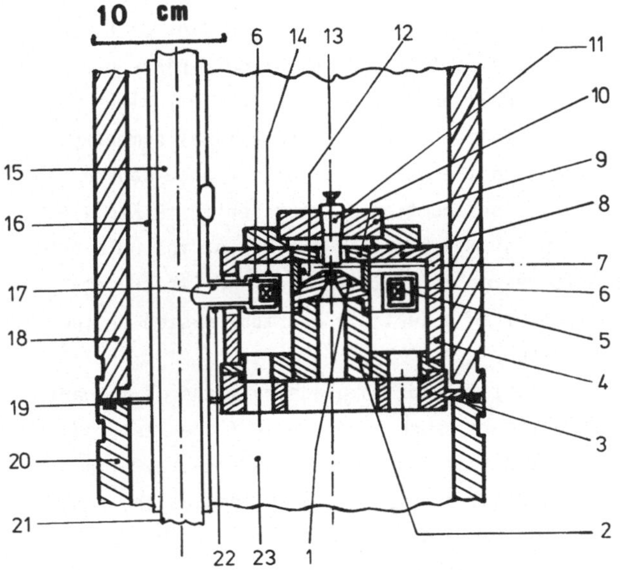

b)

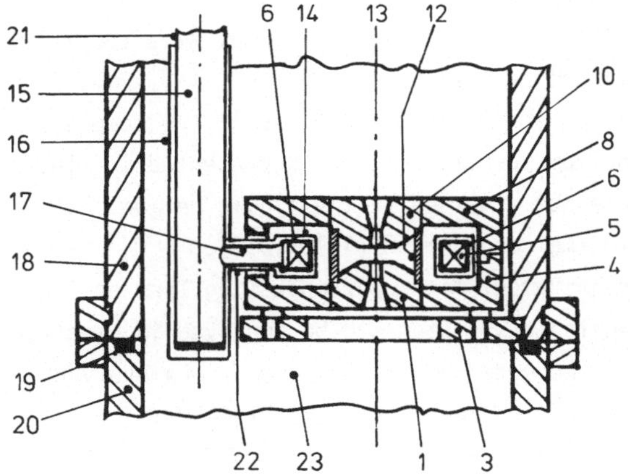

a)

Fig.4.91a,b. Simplified cross-section of two superconducting electron lenses which employ liquid-helium cooled coils but pole pieces and lens casings at room temperature (condensed from [4.7, 95,189]). (a) represents a final-projector lens (an intermediate-projector lens or a short-focal-length condenser would be of quite similar design), and (b) an objective lens where the superconducting coil has been so far displaced from the position of symmetry with respect to the pole-piece system that the insertion of apertures into the gap is not obstructed. The superconducting coils 5 are enclosed within toroidal containers 6 (made from brass, using indium-tin solder), which are connected to the main liquid-helium reservoir 21 by short tubes 17. This liquid-helium reservoir 21 consists of a long stainless-steel tube running parallel to the axis 13 of the electron optical column and carries the toroidal containers and coils of all its lenses. Thus, the toroidal containers are filled with liquid helium in which the coils are immersed, if the helium level in the volume 15 of the reservoir is maintained high enough. The liquid-helium reservoir 21, the toroidal containers 6 and the connecting tubes 17 are surrounded by liquid-nitrogen-cooled radiation shields 16, 22, 14. For cooling the radiation shields, two copper tubes are provided, which have been soldered to the outside surface of the tubular shield 16 and are filled with liquid nitrogen. (These copper tubes lie above and below the plane of the drawing and so are not visible here.) Other parts indicated in the drawing: 1 lower pole piece, 2 lens core, 3 lens support plate, 4 cylindrical lens casing. 7 plane of insertion of objective aperture. 8 upper lens flange. 9 specimen stage. 10 upper pole piece. 11 specimen cartridge. 12 non-magnetic spacer. 13 lens axis and optic axis of the column. 18, 20 external wall of the column. 19 vacuum gasket. 23 vacuum space

Another disadvantage of the completely symmetrical lens construction of Fig. 4.91a results from the comparative closeness of the coil to the pole-piece gap. Since the number of turns in the coil is small and the windings carry a high current, the winding structure and any deviation from exact rotational symmetry is reflected in the field distribution close to the axis. This disturbance can be reduced somewhat by employing an asymmetric construction as represented in Fig.4.91b where the lens coil is placed well below the mid-plane of the gap of the pole-piece system [4.7]. Another advantage of this design is that the coil and its cryostat no longer obstruct the lateral introduction of an adjustable aperture system or the utilization of a side-entry specimen stage.

With the asymmetric design, the coil must be kept in place by means of mechanical supports having a low thermal conductivity, such as epoxy or fibreglass rods [4.96]. In this context, it should be noted that the thermal conductivity of a rod-shaped support remains the same regardless of its length as long as the ratio of the cross-sectional area of the rod to its length is kept constant. On the other hand, so far as the elastic properties of the rod are concerned and with the same assumption of constant cross-section to length ratio, the rigidity of the rod is improved if it is made shorter. Therefore, optimizing the supporting insulators usually means making them rather short, the practical limit being set at a few millimetres diameter, below which plastic deformations set in [4.12].

Two examples of superconducting lenses belonging to group 2), with the complete magnetic circuit of the lens at liquid-helium temperature, are represented in Figs. 4.92 and 4.93. Here, the actual lenses are not immersed in the liquid helium, but encased in copper blocks and cooled by thermal conduction [4.191,215].

Mechanical rigidity is an important requirement for superconducting lens design, especially for objective lenses or for whole lens systems to be employed in electron microscopes. So far, however, no generally accepted design philosophy has evolved, even for rather basic aspects. Thus in the design represented in Fig.4.92, for example, liquid nitrogen is employed as coolant for the radiation shields. On the other hand, avoiding liquid-nitrogen cooling was claimed to be important for attaining high mechanical stability in the construction shown in Fig.4.93 [4.214].

A liquid-nitrogen intermediate temperature level can be useful for storing a supply of specimens to permit comparatively rapid specimen exchange at the liquid helium temperature level [4.192a]. These specimens can be cooled down from liquid-nitrogen to liquid-helium temperature within a few minutes.

For objective lenses and lens systems, a thermally 'integrated' design is important, as for example in the design represented in Fig.4.92, where the objective lens and two intermediate projectors are combined into a single liquid-helium-cooled unit [4.192]. A thermally integrated design of the pole-piece system with its associated specimen stage is especially important for superconducting objective lenses to be employed in high-resolution electron microscopes. Any large temperature difference

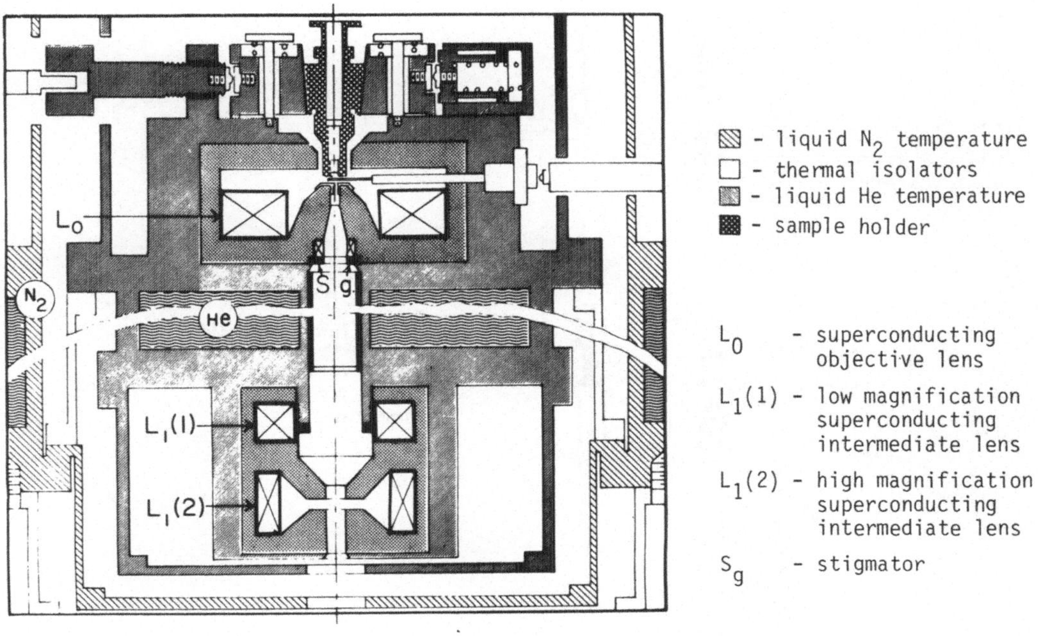

Fig.4.92. Simplified cross-section of a superconducting electron lens system composed of an objective lens and two intermediate projectors [4.192]. Here, the lenses are not actually immersed in liquid helium. They are clamped in a copper block containing the liquid-helium bath and both the lenses and the specimen stage are cooled by thermal conduction. Thus, mechanical and thermal integrity of the design is achieved. The liquid-helium-cooled block is surrounded by a liquid-nitrogen-cooled radiation shield. The magnetic construction of all three lenses is essentially classical: a field-generating coil is used and no superconducting discs are provided

between stage and pole pieces could cause the specimen to drift with respect to the lens axis, a problem known to be serious with liquid-helium-cooled specimen stages incorporated into conventional electron objective lenses [4.43,68,79].

In principle, the thermally integrated design can be accomplished for both groups 1) and 2) of superconducting lenses. With group 1) lenses, specimen stage and pole pieces remain at room temperature, so that the wealth of experience acquired on the construction of stable specimen stages can be utilized. With the group 2) lenses, both the specimen stage and the complete lens proper are cooled down to liquid-helium temperature, and the advantages of this design originate in the fact that the thermal expansion coefficients of the bulk materials suitable for stage and lens construction are smaller by about three orders of magnitude at liquid-helium temperatures than the values of the same coefficients at room temperature [4.32]. A specimen drift originating in temperature differences in the stage and pole-piece region would therefore be reduced in the same manner. Evidently, any alignment push rods coupling low temperature components to room temperature micrometer drives would always be a source of possible drift trouble owing to the

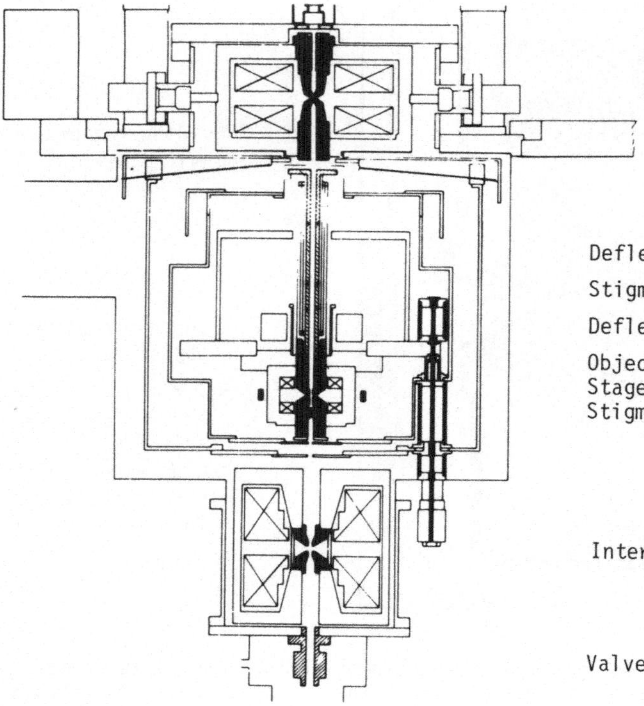

Fig.4.93a.

large temperature differences involved. It is therefore desirable to arrange that the push rods can be decoupled from the low temperature parts after the required position has been attained. This design principle is also asserted to be rather important in the literature and it is unfortunate that no details have been published on actual constructions and on experimental experience therewith (decoupling principles for room-temperature specimen stages have been discussed in [4.67]).

Quite generally, changes of the temperature distribution due to variations in the level of the liquid helium should be minimized by fabricating the liquid-helium chamber from high thermal conductivity material, copper for example. As in conventional magnetic lenses, rotational symmetry should be maintained as far as feasible in the design of superconducting lenses, so that drift and change of alignment are largely avoided in the cooling-down operation.

Severe misalignment can also arise if the magnetic properties of materials employed in the construction of the superconducting lens change markedly between room and liquid-helium temperatures. By way of example, some varieties of stainless steel and German silver fall into this category. They are practically non-magnetic at room temperature and are known to become ferromagnetic in the low temperature range [4.32]. So far as very high permeability materials such as mumetal, which are extensively used for shielding stray AC magnetic fields at room temperature, are concerned, some may well lose an order of magnitude in permeability if they are

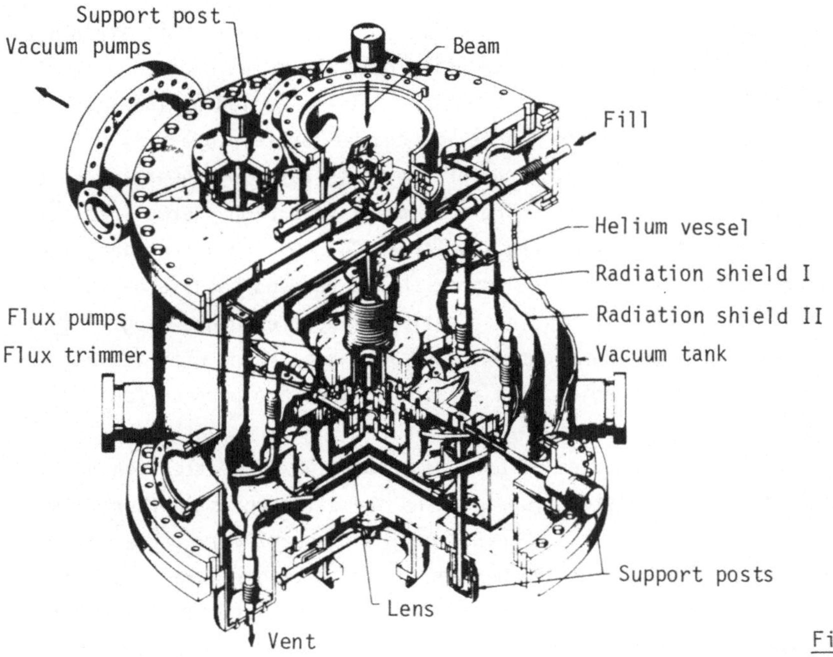

Fig.4.93b.

Fig.4.93a,b. Superconducting electron objective lens designed for condenser-objective or second-zone operation (Sect.4.2.1d) at a beam voltage of 150 kV. A symmetrical iron circuit design is utilized, of the general shape represented in Fig. 4.10a. (a) Simplified cross-section of the lens and its cryostat, also showing the associated condenser and intermediate lenses, which are mounted into the vacuum space of the cryostat but are room-temperature lenses. For the sake of mechanical stability, the entire cryostat assembly is rigidly supported on four posts (only one is shown in the drawing). Each post is assembled as a set of epoxy-glass cylinders, which provide the required spacing of the radiation shields, both mutually and from the cryostat proper. The set of glass-epoxy cylinders is tightened into a rigid post structure by employing a 3 mm stainless-steel rod located on the axis of the cylinders [4.214]. (b) Perspective drawing of the superconducting objective lens and its cryostat [4.215]. A flux pump is employed for generating the magnetic lens field which is then maintained in the persistent-current mode, see [4.98] and Fig.4.98

cooled down to low temperatures. Others, such as Cryoperm, have been developed specially for cryogenic applications and essentially retain their magnetic properties throughout the whole range from room to liquid-helium temperatures [4.32].

It should be mentioned here that materials that are superconducting at liquid-helium temperature can be employed there as perfect diamagnetic stray-field shields. For this, not only binary compound and alloy high-field superconductors can be employed, but also superconducting pure elements such as lead [4.192].

Some commercially available stainless steels are also sufficiently free from ferromagnetism at low temperatures to be employed for fabricating flexible bellows and thin-walled tubes, which can serve as ducts and vents for liquid or gaseous helium. The A/S steels 303 and 304 may be mentioned here as examples of this type of material [4.32]. So far as the thermal conductivity at low temperatures is concerned, it is smaller, by about three orders of magnitude for the stainless steels, than that of copper (the difference being only about two orders of magnitude at room temperature). Therefore, if radiation shields cooled by the still cold helium gas produced by evaporation in the helium chamber are employed, it will be advantageous to use stainless steel tubing for the vents carrying the gas from the chamber to the shields, thereby maintaining a comparatively high thermal resistance between low and more elevated temperature levels [4.214]. On the other hand, for the cooling pipes soldered onto the shields, copper should be used because of its high thermal conductivity and the correspondingly efficient heat exchange (e.g., Fig. 4.93). An actual heat exchange chamber fabricated from copper and with its walls shaped and serving as radiation shield can accomplish the same purpose (see Fig. 4.90).

Another point about the properties of bulk materials to which attention must be paid in superconducting lens construction concerns the differences in thermal expansion between the various materials. Because of the comparatively large temperature difference of about 300 K occurring between the room temperature of assembly and the liquid-helium operating temperature of the lens, the bulk fabrication materials used and the actual design have to be carefully combined in such a way that a thermally harmonized construction results. In the present context, the term harmonized means that no dangerous constraints should arise and no parts become loose, particularly during low temperature operation. Suitable springs and cup washers can be introduced for this purpose at critical points and flexible bellows can be inserted into the helium ducts and vents to prevent them from pulling the low temperature parts against the external lens casing at room temperature. Similarly, electrical connections should not be tightly stretched but have enough 'loopiness' to accommodate any contraction in their length during cooling down. In this context, it should not be forgotten that brass has a larger coefficient of thermal expansion than copper, normal and stainless steel and most of the other bulk materials of interest for superconducting lens design (notable exceptions are lead and titanium) [4.208]. Therefore, employing brass screws usually results in improved clamping of the parts at low temperatures [4.105].

Turning to the cryostat, the bath cryostat is now quite generally adopted for superconducting lens design because it is expected to cause less vibrational disturbance than evaporator cryostats [4.32]. A cryostat designed for the simultaneous cooling down of all five superconducting lens coils of an electron microscope has been described in [4.95]. Useful hints about relevant aspects of low temperature technology may be obtained from this paper and from [4.105].

Attention has been paid to the low temperature properties of special materials that could be employed with advantage for magnetic electron lens design; the pronounced ferromagnetism attained by some metals of the rare-earth group at liquid helium temperatures appears to be particularly promising [4.15]. Thus, dysprosium

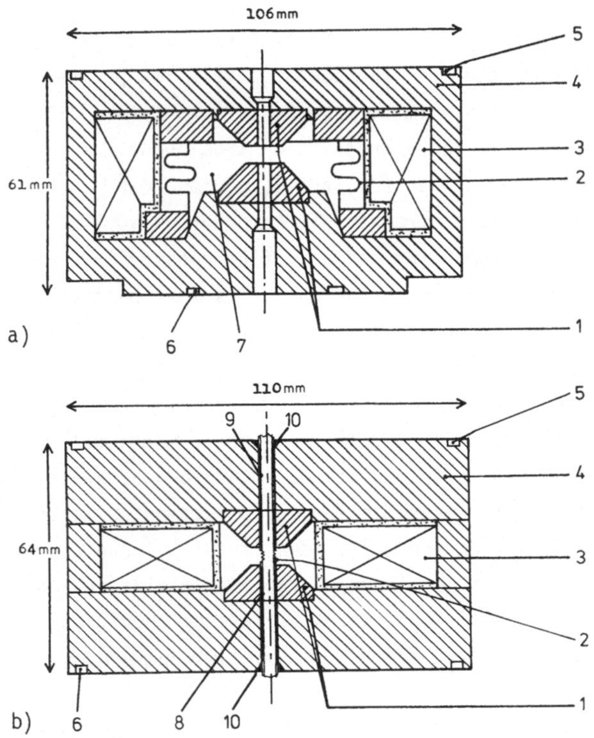

Fig.4.94a,b. Simplified cross-sections of superconducting electron lenses equipped with pole pieces fabricated from rare-earth metals [4.16]. Design (a), with the pole-piece system asymmetrically displaced against one of the flanges, could be employed as the objective lens of an electron microscope, the symmetrical design (b) as the projector and condenser lenses. With (a), the space enclosed by the bellows and extending between the pole pieces is directly connected to the vacuum space of the electron optical column. To allow for the passage of the electron beam with the design (b), a thin-walled stainless-steel tube is slipped into the lens bores and soft soldered to the lens flanges. In the actual construction, the tube is composed of two parts connected by stainless-steel bellows, so that a certain flexibility is provided and constraints between the tubes and other parts of the lens cannot arise. The lenses are immersed in a liquid-helium bath, the helium penetrating into the coil space by way of bores drilled through the wall of the cylindrical casing. Details indicated in the drawings: 1 pole pieces fabricated from holmium or dysprosium. 2 stainless-steel bellows. 3 superconducting coils (about 2000 to 3500 turns of superconducting wire on plexiglass spools). 4 lens casing. 5, 6 indium gaskets to provide vacuum seals to the adjacent parts of the electron optical column. 7 vacuum space. 8, 9 stainless-steel tubes. 10 soft-solder vacuum seal

and holmium single crystals exhibit a magnetic saturation approaching 4 T (3.7 T for Dy and 3.75 T for Ho) [4.16]. Unfortunately, due to their hexagonal close-packed structure, Dy and Ho single crystals are magnetically strongly anisotropic, with the direction of easy magnetization parallel to the hexagonal lattice planes whereas the hexagonal axis is the direction of hard magnetization [4.16,42,69,147,194]. Therefore, pole pieces fabricated from this single-crystalline material would generate fairly complex magnetic lens fields, deviating rather strongly from rotational symmetry, which would be quite unsuitable for use as electron lenses.

On the other hand, the polycrystalline materials obtained from the melt are magnetically isotropic, and can therefore be used for pole pieces. The magnetic saturation of the polycrystalline materials is about 3.1 T for Dy and 3.4 T for Ho, about 50% higher than the magnetic saturation of iron and its high permeability Co-alloys [4.188]. These materials are the basis for the construction of the group 3) superconducting magnetic electron lenses, which have a nearly 'classical' magnetic circuit but with polycrystalline Dy or Ho pole pieces and a superconducting coil (cf. Fig.4.94).

These lenses have their own particular problems in spite of their rather 'classical' appearance, problems which arise mainly from the comparatively low permeability of bulk polycrystalline dysprosium and holmium (see Fig.4.95 and [4.14]). It is seen that in the range of particularly high magnetization from about 2 T upwards, the permeability quickly drops from about 7(Ho) and 4(Dy) to around 2.5. Consequently, a rather high magnetic field strength of the order of 10 kOe ($\approx$ 1000 A/mm) is required to drive the corresponding magnetic flux density through the pole pieces. In particular, the field strength in the pole tips must be much higher than the field strength prevailing in conventional not yet saturated pole-piece systems. This situation will be closely akin to the physical situation arising at the tips of highly oversaturated conventional soft iron or cobalt-iron pole-piece systems (cf. Sect.4.3.4). Quantitatively, in comparison with conventional cobalt-iron pole-piece systems operated at high oversaturation, a tangible advantage of employing rare-earth pole-piece systems for magnetic lens construction seems to become apparent only if magnetic lens fields having a flux density of 4 T and above are required. For discussion and prediction of the shape of the lens fields and the electron optical properties of rare-earth pole-piece lenses, reference should be made to results based on measured field distributions [4.16], or on a numerical procedure [4.17].

Finally, some particular points should be mentioned, which have to be carefully observed during the construction of lens coils and their associated circuits intended for superconducting lenses running in the persistent current mode. Quite generally, the coils are fabricated from niobium-titanium wire [4.13], typically with a 0.25 mm diameter superconducting NbTi core protected by a stabilizing copper sleeve and formvar insulation, resulting in a total wire thickness of about 0.35 mm. These coils are capable of carrying an effective current density around 3×10^4 A/cm^2 Single filament wires as well as multifilament cables can be used for fabricating the coils, the latter type permitting a smaller winding radius, down to a few millimetres only. As the wires are destined to carry a comparatively heavy current, it is advisable to cement the turns into place to avoid degradation of the coil due to displacement of the wires by the mutual current attraction.

The performance of the superconducting coil circuit in the persistent current mode (Fig.4.96a) is largely determined by the quality of the superconducting connection of the two free ends of the coil. This connection is fabricated by welding or soldering the two coil ends to a short piece of thicker superconducting wire or to a superconducting metal strip. In this procedure, the superconducting cores of the wire ends have to be laid bare by removing the insulation and the stabilizing copper sleeve, employing suitable organic solvents and acids. For solder, lead-indium alloy can be used. The soldering should be carried out by melting the solder under vacuum in order to maintain good contact between the different materials [4.189].

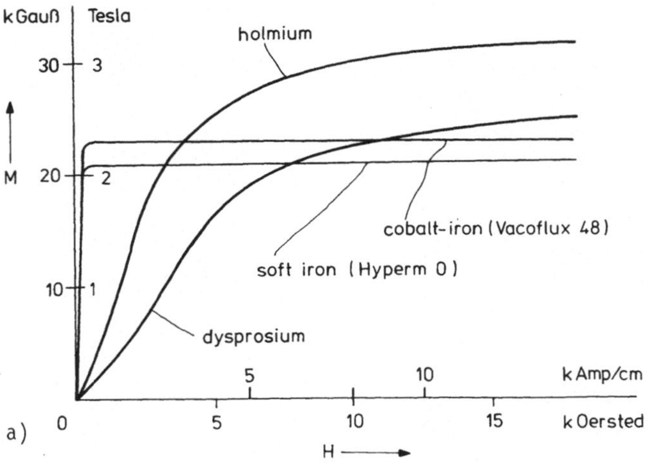

Fig.4.95. Illustration of the ferromagnetic properties of the rare-earth metals holmium and dysprosium relevant for electron lens design. The properties represented here refer to the materials in a polycrystalline state and at temperatures well below the critical temperatures: T_C = 20 K for holmium and T_C = 85 K for dysprosium. (a) Comparison of the magnetization M of polycrystalline holmium and dysprosium with that of the conventional pole-piece materials cobalt-iron and soft iron. The magnetization is shown as a function of the applied magnetic field strength H. At low to moderate field strength, the conventional materials perform much better than the rare-earth metals (data on Ho and Dy after [4.16]; cf. also Fig.4.39 where it should be observed that the total range of the horizontal scale of Fig.4.39 corresponds to a few mm only of the horizontal axis of the present drawing)

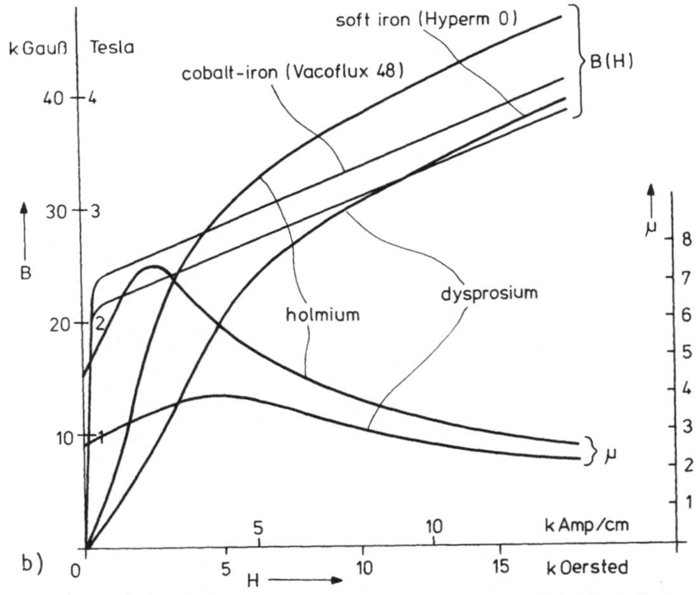

Fig.4.95b. Comparison of the magnetic flux density B carried by polycrystalline holmium and dysprosium with the flux density that can be obtained using conventional pole-piece materials, both shown as functions of the magnetic field strength H applied ($B = \mu_0 H + M$). For the rare-earth metals, the relative magnetic permeability $\mu_r = B/\mu_0 H$ is also shown. Comparison with the values reproduced in Fig. 4.39 shows that, with regard to permeability and in their usual range of application, the conventional pole-piece materials are superior to the rare-earth metals by orders of magnitude

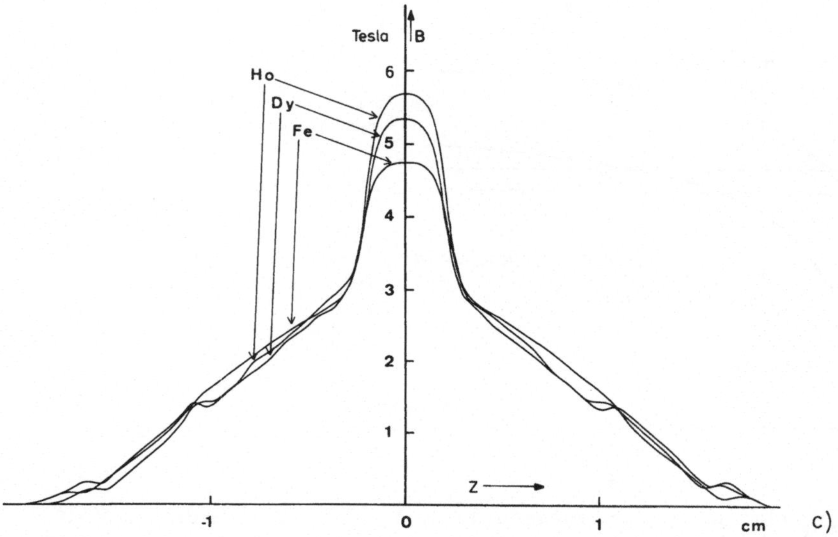

Fig.4.95c. This set of curves shows the change in the magnitude and distribution of the magnetic flux density on the lens axis that was observed experimentally when employing the pole-piece materials holmium, dysprosium and soft iron (after [4.16]). A gap width S = 5 mm and bore diameter D = 2 mm were used in all three cases at a magnetomotive force NI = 55 kA. It should be realized that a cobalt-iron pole-piece system having this geometry would already run into magnetic saturation at a little below NI = 10 kA, so that the oversaturation at NI = 55 kA is very high. It is seen that rare-earth metals can be expected to be somewhat superior to iron under these extreme operating conditions, but not as much as might have been hoped

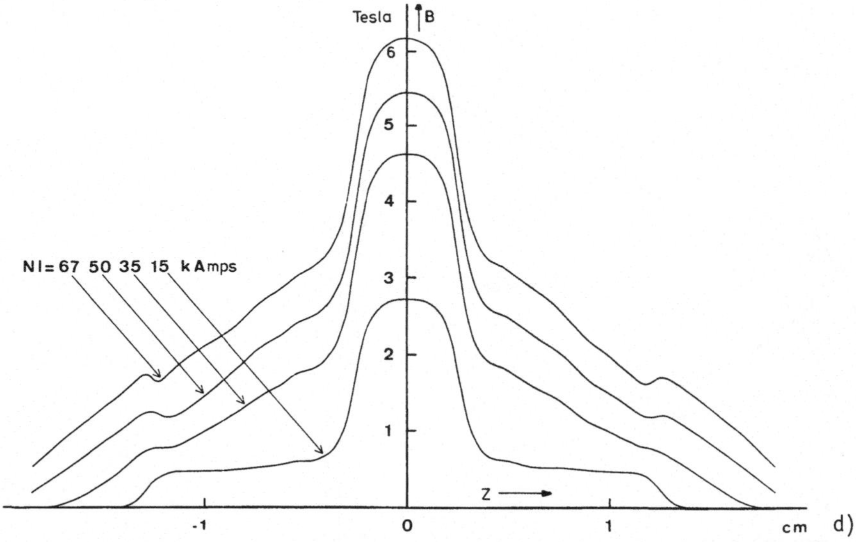

Fig.4.95d. This set of curves shows the magnitude and distribution of the magnetic flux density obtained experimentally with a pole-piece system fabricated from holmium (again with S = 5 mm, D = 2 mm) and used at a set of different values of the magnetomotive force NI [4.16]. Even at the comparatively low magnetomotive force of NI = 15 kA, a flat 'foot' of slowly varying flux density is observed to spread out from the central bell-shaped peak over a considerable distance along the lens axis. This is a peculiarity of the rare-earth metal pole-pieces, which require a comparatively high internal magnetic field strength in order to carry even a moderate amount of magnetic flux. This field will also appear in the bore because of the physical law that the tangential component of the magnetic field must be continuous

It is easily seen from Fig.4.96a,b that, in the persistent current mode the degradation in time t of the current J can be described by the differential equation

$$L_c \frac{dJ}{dt} + RJ = 0 \quad . \tag{4.159}$$

which has for solution

$$J = J_0 \exp[-(t - t_0)/\tau] \quad , \tag{4.160}$$

where L_c denotes the self-inductance of the coil, R the total ohmic resistance within the circuit and J_0 the current at time t_0. τ is the time constant, given by

$$\tau = \frac{L_c}{R} \quad . \tag{4.161}$$

The resistance R is mainly determined by the quality of the superconducting connection and should be of the order of 10^{-11} Ω or less, so that for a typical self-inductance L ≈ 1 H of the superconducting coil the current degradation will remain within some ppm per day.

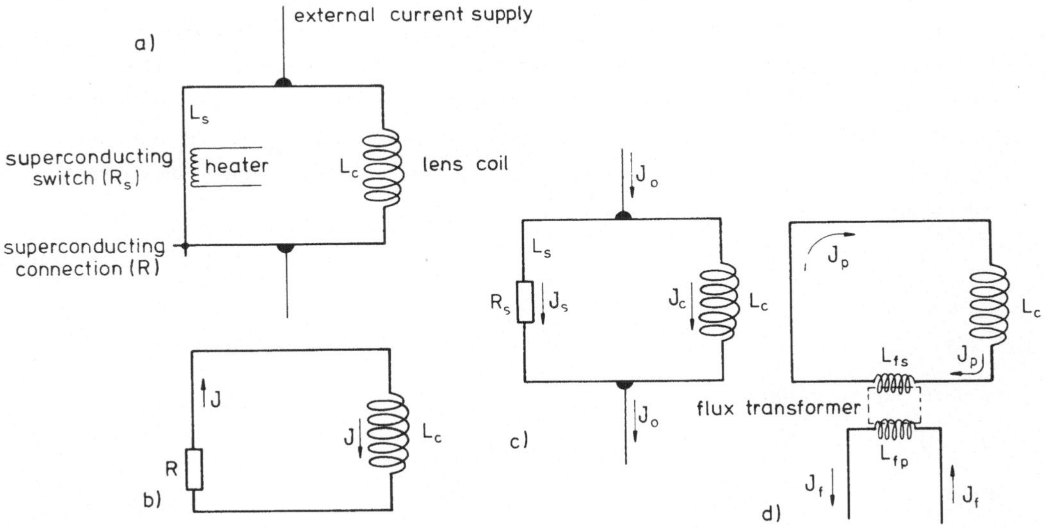

Fig.4.96a-d. Schematic representation of the layout and the operation of a circuit suitable for running a superconducting lens in the persistent current mode [4.56]. (a) General layout of the circuit: The superconducting lens coil has a self-inductance L_C and negligibly small ohmic resistance. L_S is the self-induction of the circuit branch containing the superconducting switch, which, if actuated, has a resistance R_S. R is the resistance of the superconducting connection. (b) Equivalent electric circuit for operating the lens in the persistent current mode. The residual resistance R of the superconducting connection is responsible for and determines the slow decay of the lens current J. (c) Equivalent electric circuit representing the 'charging' of the lens coil with the current to be trapped in the persistent current mode. Here, the resistance R ≈ 10^{-11} Ω of the superconducting connection can be neglected in comparison with the resistance R_S ≈ 1 Ω of the superconducting switch. (d) Equivalent electric circuit when employing a flux transformer to provide a fine adjustment of the lens current if a superconducting lens is operated in the persistent current mode. (The residual resistance R of the superconducting connection has been omitted here as not being relevant in this context)

The 'charging' of the lens coil with the current to be 'caught' in the persistent current mode can be effected either by employing an external current source (cf. Fig.4.96c and [4.56,189]) or by directly transferring the magnetic flux into the coil using a flux pump integrated into the low temperature part of the lens construction (cf. Fig.4.98 and [4.98,214]). For both circuits, superconductivity-interrupting switches have to be provided. There, as thermal interrupter switches, short pieces of superconducting wire can be heated above the critical temperature so that they become normally conducting. In actual practice, carbon resistors having a room temperature resistance of about 10 Ω are employed as heating elements. The insulation and the stabilizing copper sleeve are removed from the corresponding length of the wire, and its bare superconducting core is brought into immediate thermal contact with the resistor surface and cemented to it with araldite. Typically, a power of the order of 20 mW is dissipated in the carbon resistor and a normal conductivity resistance R_s of about 0.5 Ω is generated in the wire [4.189].

The value R_s of the normal-conduction switch-resistance together with the self-inductance L_c of the coil essentially determine the time constant τ_c controlling the current charging of the coil. If (Fig.4.96c) J_c and J_s are the currents passing through the coil and the interrupter switch respectively and J_o is the constant (but adjustable) current supplied by the external current generator, then the current charging process is governed by the equations

$$J_o = J_c + J_s \quad , \tag{4.162}$$

$$R_s J_s = L_c \frac{dJ_c}{dt} \quad . \tag{4.163}$$

They have for solution

$$J_c = J_{co} + (J_o - J_{co}) \cdot \{1 - \exp[-(t - t_o)/\tau_c]\} \tag{4.164}$$

with

$$\tau_c = \frac{L_c}{R_s} \tag{4.165}$$

as the time constant for current charging. J_{co} is the current flowing through the coil at the moment t_o when the superconductivity-interrupting switch is opened. Writing $\Delta J_c = J_o - J_c$, the deviation of the actual coil current J_c from its rated value J_o, and stipulating that after a charging time $t - t_o$ the relative deviation $\Delta J_c/J_o$ of the coil current shall become smaller than a certain specified value (say e.g., $\Delta J_c/J_o < 10^{-5}$), then using (4.164) and (4.165), the circuit requirements can be expressed in the following form (with lg as the common logarithm)

$$\frac{1}{\tau_c} = \frac{R_s}{L_c} > \frac{2.3026}{t - t_o}\left(\lg \frac{1}{\Delta J_c/J_o} + \lg \frac{J_o - J_{co}}{J_o}\right) \quad . \tag{4.166}$$

It can thus be predicted when designing the lens whether a specific pair of R_s and L_c values would be acceptable from the point of view of a quick enough response of the lens current to its control. By way of example, the typical values given above of $R_s \approx 0.5\ \Omega$ and $L_c \approx 1$ H would be quite satisfactory for obtaining $\Delta J_c/J_o < 10^{-5}$ within a time $t - t_o \approx 30$ s.

If the superconductivity-interrupting switch is now closed by cutting off the heater current, the lens circuit passes into the persistent current mode. For the design of the circuit, it is important to realize that the original current J_c prevailing at the time of switch closure is not exactly 'caught'. Instead, the magnetic flux $J_c \cdot \lambda_c L_c - J_s \cdot \lambda_s L_s$ surrounded by the now completely superconducting closed circuit at the moment of switch closure is caught. Here, $J_c \lambda_c L_c$ is the flux contributed by the branch of the circuit containing the coil and $J_s \lambda_s L_s$ the flux generated by the circuit branch containing the switch; L_s denotes the self-inductance of this branch. $|\lambda_c|$ and $|\lambda_s|$ (both <1) indicate the fractions of the total magnetic flux that are generated by the two circuit branches and actually accepted by the closed superconducting circuit. It should be noted that the currents in the two branches are directed in such a way that the corresponding magnetic flux directions are opposed to each other; the flux contributions $J_c \lambda_c L_c$ and $J_s \lambda_s L_s$ have thus to be subtracted in order to determine the flux ω_p that is actually 'caught' by the persistent current J_p (cf. Fig.4.96c):

$$\begin{aligned} \omega_p &= J_c \lambda_c L_c - J_s \lambda_s L_s \\ &= J_p(\lambda_c L_c + \lambda_s L_s) \quad . \end{aligned} \tag{4.167}$$

In practice, $L_s \ll L_c$ and nearly always $J_s \ll J_c$. Equation (4.167) can then be simplified to

$$J_p \approx J_c\left(1 - \frac{\lambda_s}{\lambda_c} \frac{L_s}{L_c}\right) \quad . \tag{4.168}$$

Thus, on switching into the persistent current mode, a defocusing

$$\frac{\Delta J}{J_c} = \frac{J_c - J_p}{J_c} \approx \frac{\lambda_s}{\lambda_c} \cdot \frac{L_s}{L_c} \quad . \tag{4.169}$$

takes place.

In order to avoid this handicap for the fine focusing of a superconducting objective lens, the superconductivity-interrupting circuit branch should be so designed that its self-inductance L_s is smaller by at least six orders of magnitude than the self-inductance L_c of the lens coil (in practice neither λ_s nor λ_c is very different from 1). Thus, when constructing the superconductivity-interrupting switch it would be quite wrong to wind the wire in the form of a coil around the carbon heater resistor. Employing a bifilar wiring technique, preferably up to the points of connection with the leads to the current source would be the method of choice. In actual constructions, switch circuit branch self-inductances down to about 10^{-7} H have been attained [4.56].

The persistent current J_p given by (4.168) is the actual coil current if the original current fed in from the source has been cut off. If, on the other hand, current continues to be supplied to the persistent mode circuit, this current $J_c' + J_s'$ must split up into the two circuit branches in such a way that the additional magnetic flux enclosed by the persistent mode circuit vanishes:

$$J_c'\lambda_c L_c - J_s'\lambda_s L_s = 0 \quad . \tag{4.170}$$

Thus, an additional current appears at the coil, given by

$$J_c' = J_s' \frac{\lambda_s L_s}{\lambda_c L_c} \quad . \tag{4.171}$$

This circuit must not cause any tangible defocus of the lens, which gives us yet another argument for designing the circuit with L_s very much smaller than L_c. These considerations are of special relevance for the construction of a lens which is at times operated in the persistent current mode with a strong external current as well. This mode of operation comes into play if several persistent-current-mode lens circuits are wired in series and 'charged' by one and the same current source [4.55].

For the fine focusing of an objective lens, the method explained above for trapping a persistent current has a particular disadvantage: it is difficult to pinpoint the J_p current value with high accuracy around $\Delta J_c/J_c \lesssim 10^{-5}$ while the lens current is still floating—with the comparatively long time constant of the circuit—against the current value supplied by the source, which must itself be regulated at least to the required focusing accuracy.

A more elegant procedure is to permit some error margin when trapping the proper value of the lens current and to perform the fine adjustment by means of a flux transformer (cf. Fig.4.96d). The principle is to provide a second coil, which represents the secondary winding of the flux transformer, in the persistent current loop in series with the lens coil. The flux encompassed by the secondary winding can be adjusted by feeding the primary winding with an appropriately chosen current. In the persistent current mode, the sum of the magnetic flux of lens coil and flux transformer remains constant. If the flux in the transformer is increased, the flux encircled by the lens coil decreases in compensation by the same amount and a corresponding diminution of the persistent current J_p results.

If now J_{po} is the persistent current with zero current in the primary winding, the performance of the flux transformer can be predicted from

$$J_p = J_{po} - J_f \frac{L_{fp}}{L_{fs} + \lambda_c L_c} \quad . \tag{4.172}$$

Here, L_{fp} and L_{fs} are the self-inductances of the primary and secondary winding of the flux transformer and J_f the current in its primary. The above equation can also be cast into a form expressing the relative lens current variations:

$$\frac{\Delta J_p}{J_{po}} = \frac{J_p - J_{po}}{J_{po}} = - \frac{J_f}{J_{po}} \cdot \frac{L_{fp}}{L_{fs} + \lambda_c L_c} \quad . \tag{4.173}$$

A particularly advantageous design for a flux transformer is represented in Fig. 4.97. The transformer is rotationally symmetric and essentially shaped like a very small electron lens without a bore and with two independent windings. Thus, for calculating the magnetic flux in the transformer and its coefficients of self-inductance, L_{fp} and L_{fs} the methods described in Sect.4.3 can be applied. Just like the casing of the magnetic electron lens, the ferromagnetic casing of the transformer prevents the generation of magnetic stray fields. The function of the gap is to effect a shearing of the hysteresis curve so that a practically linear response of the transformer flux to the primary current J_f is generated [a tacit assumption with (4.172) and (4.173)]. (For details, see [4.56].)

The flux transformer is closely akin to the flux pump, which can be employed as a second alternative method for charging the superconducting lens coil with magnetic flux. The principle is represented schematically in Fig.4.98. The flux pump consists of a primary winding on a toroidal iron core and a one-turn secondary winding, which can be opened at two points by superconductivity-interrupting switches S-1 and S-2. The operation is cyclic:

1) Switch S-1 is closed, the lens coil is in the persistent current mode.

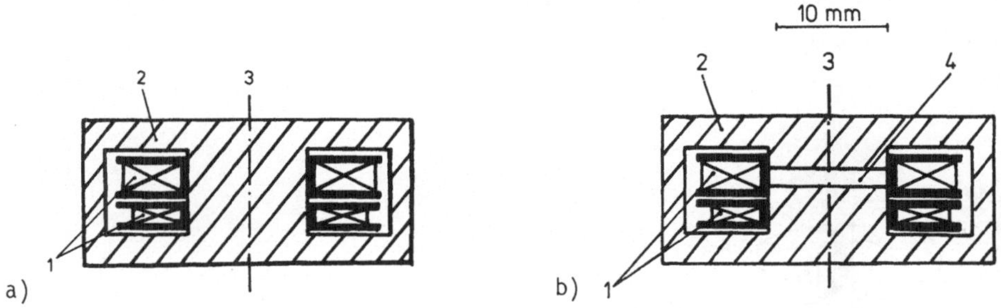

Fig.4.97a,b. Simplified cross-section of iron-shrouded flux transformers, by means of which the fine focus of superconducting lenses operated in the persistent current mode may be adjusted (adapted from [4.54]). The primary coils can be made of copper wire whereas the secondary coils, which are in series with the lens coils (cf. Fig.4.96d), must evidently consist of superconducting wire. The rotationally symmetric ferromagnetic casing is fabricated from soft iron. Transformers without (a) and with (b) a gap in the iron circuit can be employed; the gap causes a rather strong shearing of the hysteresis curve of the casing material so that a more linear response of the flux change relative to the primary current is generated. The number of turns in the secondary coil and the persistent lens current determine the working point of the transformer (defined for zero primary current), and about 100 turns (a) to 40 turns (b) are typical. A range of flux change of several mV·s can be provided. The flux transformers can be optimized to keep their mass and volume as small as possible, in order to avoid any unnecessary burden on the cryogenic system (for details cf. [4.56]). Parts indicated in the drawings: 1 coils. 2 ferromagnetic casing. 3 axis of rotation. 4 shearing gap

2) Switch S-2 is opened and the primary winding energized by the current J_{fp} so that the secondary loop (still open) encloses the magnetic flux $J_{fp}\,L_{fp}$.

3) First, switch S-2 is closed and then the primary current J_{fp} reduced to zero. The flux $J_{fs}\,L_{fs}$ is now caught by the persistent mode current $J_{fs} = J_{fp}\,L_{fp}/L_{fs}$ induced in the one-turn secondary winding.

4) Switch S-1 is now opened and the flux $J_{fp}\,L_{fp}$ redistributes itself between the secondary loop and the lens coil. After this, the cycle can be repeated by starting at 1) with the closing of switch S-1.

It can be shown that after many cycles the limiting current J_{pso} that can be pumped into the lens coil is just equal to the current J_{fs} induced in the secondary loop [4.98]. It does *not* depend on the number of turns in the lens coil, so that quite high fields may be obtained by pumping if the number of turns in the lens coil is large enough. It is important to generate a high magnetic flux in the transformer because the lens current will increase in direct proportion to it, and it would be completely wrong policy to employ more than one turn for its secondary winding.

Finally, some kind of protection should be provided for superconducting lens coils to safeguard them against accidental electrical damage. Sudden release of the electrical energy stored in the magnetic field, amounting to about 100 W·s in order of magnitude, might seriously damage or even destroy the windings. This can happen if part of the circuit accidentally becomes normally conducting, with a comparatively high resistance, a dangerous point in this respect being the superconducting soldered connections and the superconductivity-interrupting switches. As a suitable protective measure, a passive non-linear dipole can be wired across the particularly ticklish parts of the circuit, the dipole consisting of two diodes

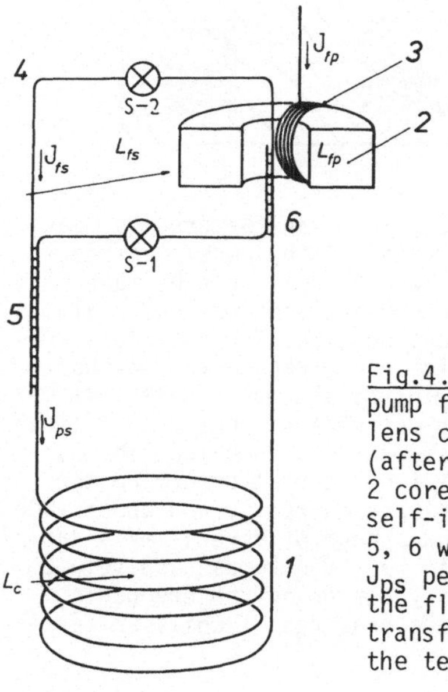

Fig.4.98. Schematic representation of an electrical flux pump for 'charging' the current into a superconducting lens coil to be operated in the persistent current mode (after [4.98]). 1 lens coil with self-induction L_c. 2 core and 3 primary coil of flux transformer, having self-induction L_{fp}. S-1, S-2 superconducting switches. 5, 6 welded or soldered superconducting connections. J_{ps} persistent lens current. J_{fs} secondary current of the flux transformer. J_{fp} primary current of the flux transformer. (The operation of the pump is described in the text)

connected head-to-tail and capable of carrying a current at least as large as the maximum current arising in the lens coil. Thus, a voltage limitation can be created of the order of one volt (see [4.189]).

4.7.2 The Superconducting Shielding Lens

The most recent child of the superconducting lens family is the superconducting shielding lens, which has turned out to be a very successful design. Fundamental for the shielding lens is the fact that superconducting materials act as perfect diamagnetic shields.

If a limited length of the lens axis is surrounded by a narrow bore superconducting tube, the magnetic field is effectively excluded from this region. Superconducting tubes of this kind can be incorporated into the design of the superconducting lenses described above, placed over the sloping 'tails' of the lens field; the latter are thus suppressed so that the actual lens field is compressed into a shorter axial extension [4.54]. This principle of compression can be applied to the construction of both iron-free 'air'-coil lenses and to iron-shrouded pole-piece lenses. It should be realized that the shielding tubes must already have been cooled down to the superconducting state before the lens field proper is applied.

In practice, this principle can also be applied to lenses which employ superconducting rings to create the lens field. Here, after starting the current in the rings and thus generating the magnetic lens field, a noticeable compression of the

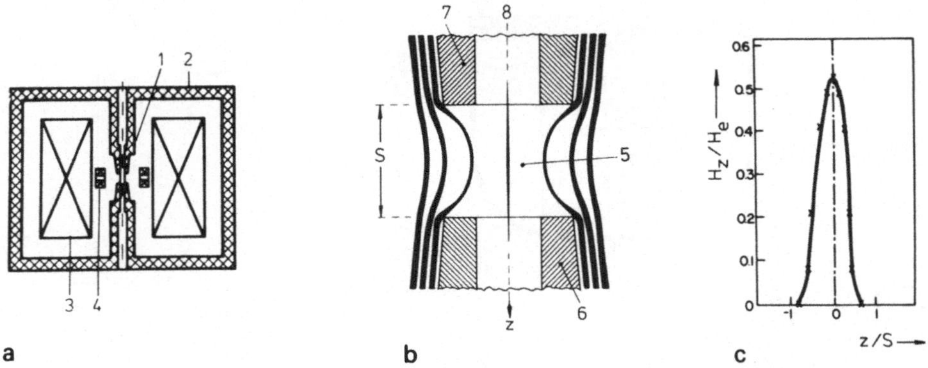

Fig.4.99a-c. Schematic representation of the basic construction and of the main physical properties of the superconducting shielding lens (adapted from [4.31a]). (a) Basic construction of the lens: 1 superconducting lens cores with field-shaping snouts. 2 superconducting lens casing. 3 superconducting lens coil. 4 field-deformation-correcting coils. (b) Schematic representation of the penetration of the external magnetic field, which is generated by the coil within the space surrounded by the casing, into the gap space 5 between the snouts 6, 7 and onto the lens axis 8. (c) Typical distribution of the magnetic field strength H_z on the lens axis z, shown relative to the coil-generated external field H_e which is nearly parallel to the lens axis over the length of the shielding cores. z = 0 indicates the centre of the gap and the axial coordinate z is normalized with respect to the gap width S

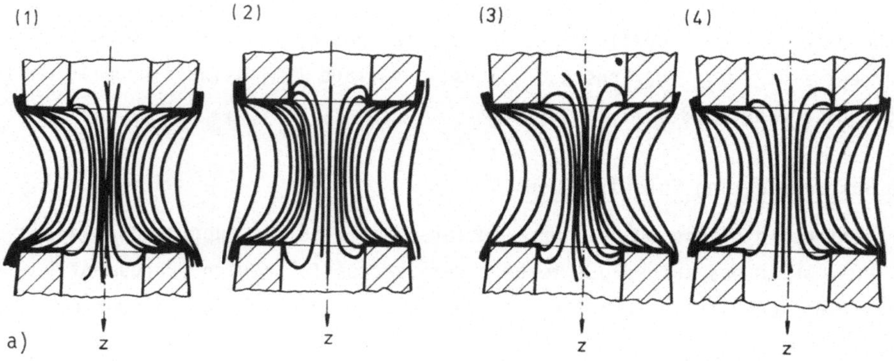

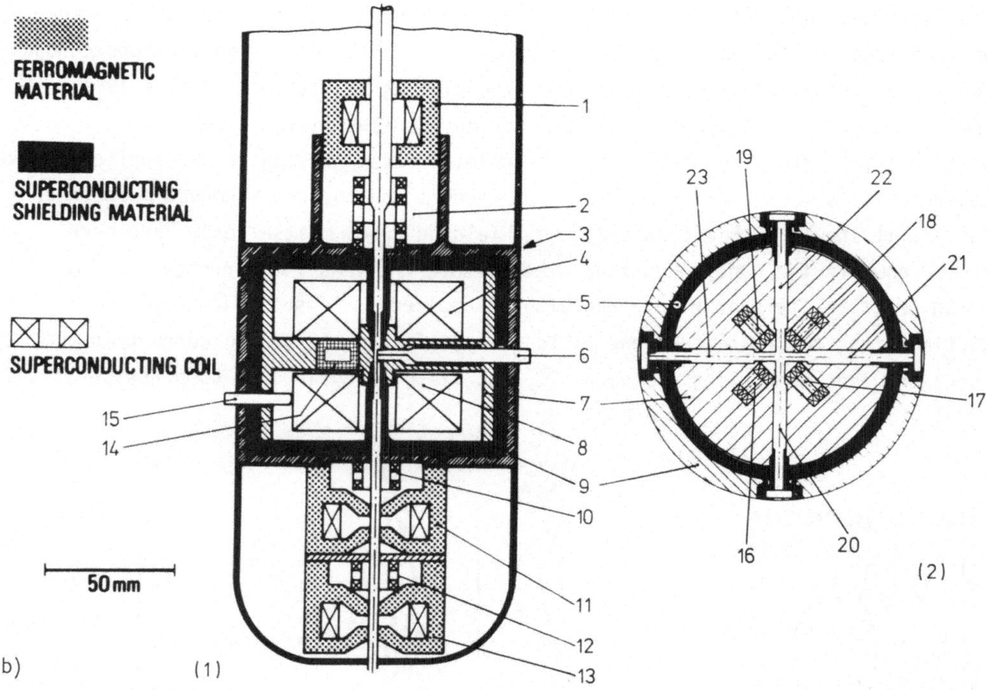

Fig.4.100a,b. Disturbance of the rotational symmetry of the lens field of the shielding lens by mutual misalignment of the shielding snouts, and first-order compensation of this defect by means of a superposed dipole field (adapted from [4.31]). The compensating dipole fields are magnetic deflection fields produced either by employing small coils positioned around the lens gap (Fig.4.99a) or by tilting the lens coil so that a radial field component arises. If a lens coil subdivided into two separate halves is used (i.e., with a symmetrical design analogous to that of Fig.4.10a, cf. also Fig.4.101), a radial field component can also be generated by traversing one of the half-coils with respect to the other. (a) (1) Lens field disturbed by mutual parallel displacement of the bore axes of the shielding snouts. (2) Lens field as in (1), its disturbance compensated by superposition of a transverse dipole field. (3) Lens field disturbed by mutual tilt of the bore axes of the shielding snouts. (4) Lens field as in (3), its disturbance compensated by superposition of a transverse dipole field.
(b) Schematic illustration of ways of compensating field disturbances caused by mutual misalignment of the shielding snouts, taking the example of a shielding objective lens employed with a four-lens superconducting lens system. (1) simpli-

fied cross-section of the system: 1 condenser lens. 2 two-stage deflection system. 3 shielding objective lens. 4 fixed upper lens coil. 5 superconducting lens casing. 6 specimen carrier rod. 7 stainless-steel coil support. 8 traversable lower lens coil. 9 stainless-steel lens casing. 10, 12 single-stage deflection systems. 11, 13 intermediate projectors. 14 a dipole coil which can be used to compensate the field disturbance. 15 control rod for the compensating traverse of the lower lens coil. (2) orthogonal-to-axis cross-section through centre of shielding lens: 16, 18, and 17, 19 pairs of dipole coils for the compensation of field disturbances. (The coils have their planes of winding parallel to meridional planes through the lens axis, so that the compensating action is effected by their lateral stray fields. In the actual design, each dipole coil is sandwiched between two stigmator coils.) 20 to 23 ducts for insertion of specimen and aperture carrier rods and for push-rods effecting their transverse alignment [4.33]

field distribution can subsequently be attained if mechanical means of shifting axis-centred superconducting shielding tubes or ring stacks (Fig.4.89) relative to the centre of the lens field are available [4.190]. Whereas such schemes can be understood as hybrid forms of 'conventional' superconducting lenses with additional shielding, the shielding lens developed by DIETRICH and coworkers during the last two decades embodies a drastically different construction [4.30]. Its principle can be understood from the schematic drawing of Fig.4.99. The lens field is generated by means of a superconducting lens coil which has a cross-sectional area similar to that of conventional normally conducting coils operated at room temperature. The external magnetic stray field of the current running in the coil cannot escape through the superconducting casing, which surrounds the coil and acts as a perfect diamagnetic shield. Narrow bore superconducting shielding tubes also surround the axial region over nearly the whole length of the casing with the exception of a central gap a few millimetres wide. There, the coil field can penetrate onto the axis and thus generate the lens field proper.

Both the can-shaped casing and the shielding tubes are fabricated from sintered niobium stannide (Nb_3Sn), the only material at present available possessing the required shielding and current-carrying properties up to magnetic flux densities of several teslas. There is considerable latitude in both the range and the number of parameters that determine the properties of this material (for details of a thorough investigation, see [4.31]). Typically, as raw material, niobium and tin powder having a grain size of about 60 μm is carefully mixed in the ratio 3:1. The mixture is then isostatically pressed at 1500 atm, sintered at 1000°C for 1 h and finally precision ground to the required dimensions [4.28,31].

It is important to realize that the number of ampere-turns $(NI)_c$ at which the superconducting coil is operated is larger by more than an order of magnitude than the electron optically effective number of ampere-turns $(NI)_{eo}$ required in the lens field proper and extending over the gap region of length S. If L_s is the total length of the two shielding tubes taken together, the actual ampere-turns required for the lens coil can be estimated from $(NI)_c = (NI)_{eo} L_s/S$. The magnetic flux density on the lens axis assumes its maximum in the centre of the gap where its value can be expected to amount to about 75% of the field that would be produced

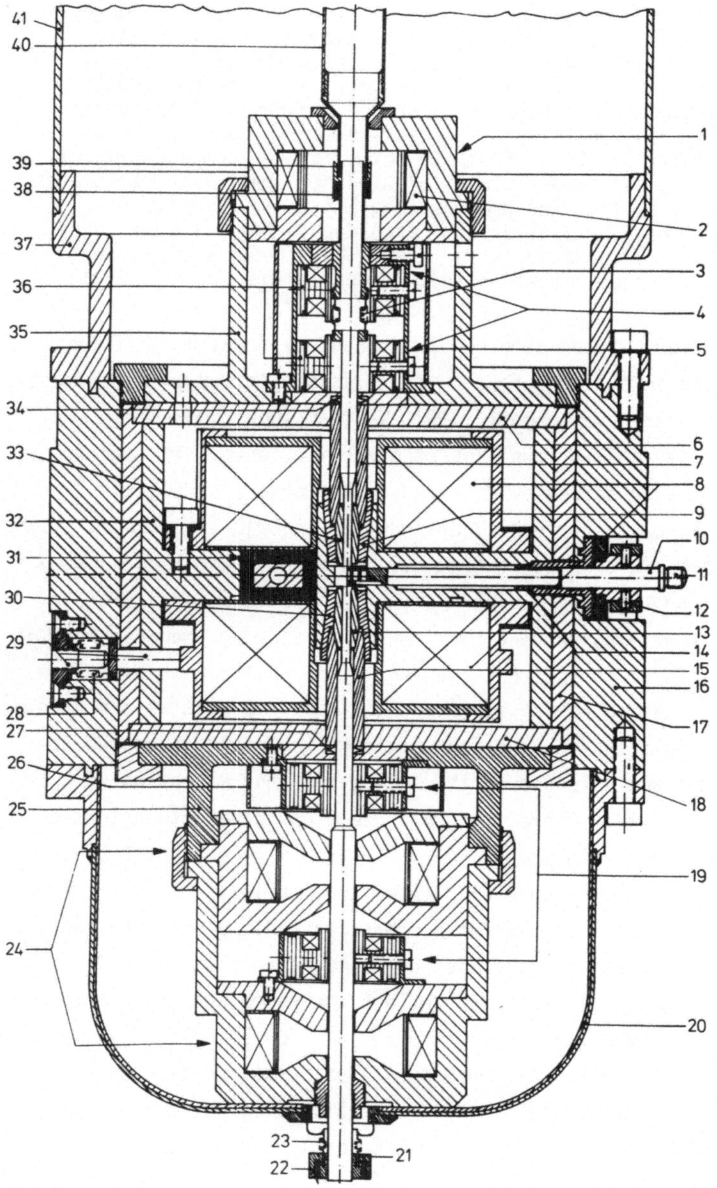

Fig.4.101. Cross-section of a superconducting four-lens system for electron microscopes, employing as objective lens a superconducting shielding lens with a symmetrical (Fig.4.10a) coil arrangement. The shielding lens is operated in the condenser-objective mode and is complemented by the addition of one condenser and two intermediate-projector lenses, all three of which have superconducting coils and conventional ferromagnetic casings (adapted from [4.31]).
1 condenser lens with conventional ferromagnetic casing; 2 superconducting coil of condenser lens 1; 3, 23, 28 flexible bellows for preventing constraints from arising on the vacuum tube during cooling down; 4 two-stage superconducting deflection system; 5, 26 mumetal screening tubes; 6, 18 superconducting Nb_3Sn flanges of shielding lens; 7, 15 Nb_3Sn lens cores of shielding lens; 8 lens coils (Nb-Ti multifilament

wire; 13, 33 Nb_3Sn screening snouts; 10 specimen carrier rod (a similar rod for transverse alignment of the specimen is arranged at right angles to the plane of the drawing); 11 thread for connecting rod 10 to a demountable manipulator rod used for inserting or removing the specimen; 12 pins and lever arm employed for traversing the specimen in the radial direction (in the actual construction, a similar design is provided for introducting and aligning apertures; cf. Fig.4.100b and [4.31] for details); 14 indium ring gasket for sealing the vacuum space from the liquid-helium bath; 16 stainless-steel casing for shielding objective lens, also serving as part of the wall for the liquid-helium bath; 17 cylindrical Nb_3Sn casing of shielding lens; 19 single-stage deflection systems; 20 two-layered wall of liquid-helium bath; the inner layer consists of lead which becomes a diamagnetic shield at liquid-helium temperatures; 21, 39 dip soldering baths serving as demountable vacuum-tight connections to the room-temperature parts of the vacuum wall around the electron beam path; 22, 38 heaters for dip soldering baths; 24 intermediate projector lenses with conventional ferromagnetic casing; 25 support tube rigidly connecting the intermediate projector lenses to the objective lens; 27, 34 spring washers forcing the cores 7, 15 and the screening snouts 9, 13 towards the lens centre (the magnetic interaction between the screening currents and the lens coil currents tries to force them apart against the flanges 6, 18); 29 push-rod system for alignment traverse of lower lens coil; 9, 30 conical adaptors aligning the screening snouts; 31 stigmator and deflection coils; 32 coil housing fabricated from stainless steel; 35 support rigidly connecting the condenser lens 1 to the objective lens; 36 deflection coils fabricated from single-strand NbTi wire and wound on epoxy spools; 37, 41 wall of the liquid-helium bath; 40 vacuum wall tube surrounding the electron beam region and leading to electron gun; the walls of the liquid-helium bath are formed by the parts indicated by 16, 20, 37, 41. Thus, if the bath is filled, all the coils are flooded by liquid helium while the electron beam path is protected by tube 40 and its continuation (fabricated from platinum-iridium alloy for the section in the lens)

by the lens coil alone without the shielding casing and tubes [4.28] (for the calculation of the magnetic field generated by unshielded iron-free coils, see [4.38]). Unfortunately, there is at present not much information available about the dependence of the axial field distribution on the gap width to bore diameter aspect ratio of the shielding tubes. For design work, use can be made of measured field distributions of shielding lenses actually constructed [4.28,31,34]. As might be expected, the superconducting shielding lens has its own particular design problems. One of these is a consequence of the fact that the actual lens field is produced as the difference between two comparatively large fields: the field generated by the currents flowing in the lens coil (an 'air coil' type field) and the field created by the shielding currents running in the casing. This compensation works quite well as long as both fields are perfectly rotationally symmetric about one and the same axis. In practice, however, the coil field will deviate a little from the ideal because of the winding 'microstructure' of the coil. Moreover, the bore axes of the two shielding tubes may be parallel displaced and/or tilted with respect to each other, with the shielding currents displaced correspondingly. A relative field distortion then arises, which has the same character as the field distortion caused by an analogous mutual displacement of the ferromagnetic pole pieces of conventional magnetic electron lenses [4.165].

These deviations from the ideal rotationally symmetric field shape are particularly serious for the combination of fields in the shielding lens. For here, as already mentioned, with two mutually compensating fields, the number of ampere-turns is larger by more than an order of magnitude than the number applied to the pole-piece system of the conventional lens, and the corresponding difference in magnitude also applies to the field components disturbing the rotational symmetry. Therefore, whereas the rotationally symmetric main fields largely compensate each other, leading effectively to a reduced number of ampere-turns for the actual lens field, this compensation does not apply to the randomly oriented disturbing fields.

For a given magnitude of the actual lens field therefore, the electron optical deterioration due to the disturbances can be expected to be rather more serious than with conventional lenses having a corresponding mechanical misalignment, the most conspicuous being a strong transverse displacement of the image. In order to straighten the lens field and bend the optic axis back again, a compensating transverse deflection field can be applied in the gap region (cf. Fig.4.100a). With two electrically independent pairs of coils producing fields at right angles to each other (cf. Fig.4.100b), the compensating field can be tuned to the disturbance actually present, guided by the electron optical imaging [4.29,31].

Another unique design consideration which must be carefully observed with the shielding lens results from the fact that the sense of the shielding currents in the tube and in the casing is opposite to that of the current in the coil. Repulsive forces between the coil currents and the shielding currents thus arise, which are equivalent to an internal pressure that threatens to explode the can-shaped casing. The pressure exerted from within onto the cylindrical casing can be calculated from

$$p \approx \frac{rH_W^2}{8\pi w} \frac{kp}{cm^2} \quad \text{or} \quad \text{atm.} \tag{4.174}$$

where 2r is the diameter and w the thickness of the wall of the cylindrical casing and H_W the magnetic field strength in kilo-Oersteds at the inner surface of the casing [4.28]. For example, with a casing of about 2r = 80 mm diameter and w = 5 mm wall thickness, which is subjected to a field strength of about H_W = 20 kOe (2000 A/mm), an internal pressure of about p ≈ 100 atm (≈ 10 MPa) is to be expected.

The fracture stress of sintered Nb_3Sn at liquid-helium temperatures has been found to amount to about 10^3 atm, but in casings actually constructed, the fracture strength may be considerably less due to voids, notches, and microcracks in the material. With an experimental lens, the shielding container exploded at about 150 atm internal pressure. It is therefore advisable to reinforce the shielding casing by arming it, with a copper tube for example. Likewise, the forces exerted on the flanges of the container may amount to some 100 kp (≈ 1 kN), and suitable supports have to be provided.

The cross-section of a shielding lens actually constructed is represented in Fig.4.101.

Acknowledgments. The present article could not have been written without extensive use of figures prepared and published by well-known experts in the field. The author is most obliged to his colleagues and to the publishers of the various scientific journals and books for permission to use this material. The sources of these figures are given in the captions, and the author has taken the liberty of making some minor modifications in order to adapt the illustrations to the present purpose.

References

4.1 K. Akashi, M. Mashimo, H. Tochigi, H. Uchida, S. Shirai: In *Proc. 7ième Cong. Int. Microscopie Electronique*, Grenoble 1970, ed. by P. Favard (Société Française de Microscopie Electronique, Paris 1970) Vol.1, pp.143-144

4.2 G.D. Archard: J. Sci. Instrum. *30*, 352-358 (1953)

4.3 M. von Ardenne: Kolloid Z. *108*, 195-208 (1944)

4.4 A. Asmus, K.-H. Herrmann, O. Wolff: Siemens Z. *42*, 609-619 (1968)

4.5 ASTM Technical Standard A 342-364 (1976)

4.6 J.L. Balladore: "Propriétés et applications des lentilles électroniques à enroulement supraconducteur"; Thèse, Toulouse (1972)

4.7 G. Balossier, A. Laberrigue, D. Génotel, J.C. Homo: In *Proc. 6th European Cong. Electron Microscopy,* Jerusalem 1976, ed. by D.G. Brandon (Tal International, Jerusalem 1976) Vol.1, pp.338-339

4.8 M. Bauer: J. Sci. Instrum. (J. Physics E) Ser.2, *1*, 1081-1089 (1968)

4.9 H.J. Beier: Z. Angew. Phys. *23*, 344-349 (1967)

4.10 F. Bertein: C.R. Acad. Sci. Paris *225*, 801-805 (1947)

4.11 F. Bertein: Ann. Radioélectr. *3*, 49-62 (1948)

4.12 H. Boersch, O. Bostanjoglo, B. Lischke: Optik *24*, 460-465 (1966/1967)

4.13 G. Bogner, C. Albrecht, R. Maier, C.P. Parsch: *Proc. 2nd Int. Cryogenic Engineering Conf. (ICEC2)*, Brighton 1968 (Iliffe Science and Technology Publications, Guildford, 1968) pp.175-178

4.14 P. Bonjour, A. Septier: C.R. Acad. Sci. Paris, B *264*, 747-750 (1967)

4.15 P. Bonjour, A. Septier: In *Proc. 4th European Regional Conf. Electron Microscopy*, Rome, 1968, ed. by D.S. Bocciarelli (Tipografia Poliglotta Vaticana, Rome 1968) Vol.1, pp.189-190

4.16 P. Bonjour: "Un nouveau type de lentille magnétique supraconductrice à pôles d'holmium pour très haute tension"; Thèse, Orsay (1973)

4.17 P. Bonjour: In *Proc. 8th Int. Cong. Electron Microscopy,* Canberra, 1974, ed. by J.V. Sanders, D.J. Goodchild (Australian Academy of Science, Canberra 1974) Vol.1, 148-149

4.18 P. Bonjour: In *Proc. 6th European Cong. Electron Microscopy,* Jerusalem, 1976, ed. by D.G. Brandon (Tal International, Jerusalem 1976) Vol.1, pp.73-78

4.19 B. von Borries, E. Ruska, J. Krumm, H.O. Müller: Naturwissenschaften *28*, 350-351 (1940)

4.19a B. von Borries, E. Ruska: Siemens Z. *20*, 217-227 (1940); cf. also M. von Ardenne: *Tabellen der Elektronenphysik, Ionenphysik und Übermikroskopie* (Deutscher Verlag der Wissenschaften, Berlin 1956) esp. Vol.1, p.407

4.20 B. von Borries: German Patents DRP S 153 465 (1942) and DBP 869 995 (19.10.1944)

4.21 B. von Borries, E. Ruska: DRP (German Patent) Nr. 923 616 (11.10.1944)

4.22 B. von Borries: *Die Übermikroskopie* (Verlag Dr. Werner Saenger, Berlin 1949) pp.204-206

4.23 B. von Borries: Z. Wiss. Mikrosk. *60*, 329-358 (1951)

4.24 B. von Borries, I. Johann, J. Hupperts, G. Langner, F. Lenz, W. Scheffels: "Die Entwicklung regelbarer permanentmagnetischer Elektronenlinsen hoher Brechkraft und eines mit ihnen ausgerüsteten Elektronenmikroskopes neuer Bauart"; Forschungsber. Wirtschafts- und Verkehrsministerium Nordrhein-Westfalen Nr. 156 (Westdeutscher Verlag, Köln, Opladen 1956)

4.25 H. Busch: Arch. Elektrotech. *18*, 583-594 (1927); Ann. Phys. (Leipzig) 4, *81*, 974-993 (1926)

4.25a J.R.A. Cleaver: Optik *49*, 413-431 (1978)
4.25b J.R.A. Cleaver: Optik *57*, 9-34 (1980)
4.25c J.R.A. Cleaver: Optik *58*, 409-432 (1981)
4.26 V.E. Cosslett: *Proc. 5th Int. Cong. High Voltage Electron Microscopy*, Kyoto 1977, ed. by T. Imura, H. Hashimoto (Japanese Soc. of Electron Microscopy, Tokyo 1977) pp.87-90
4.27 J.M. Cowley, A. Strojnik: In *Proc. 27th Annual EMSA Meeting*, St. Paul, Minn., 1969, ed. by C.J. Arceneaux (Claitor, Baton Rouge, Louisiana 1969) pp.106-107
4.28 I. Dietrich, R.G. Maier, R. Weyl, H. Zerbst: In *Proc. 3rd Int. Cryogenics Engineering Conf.* (ICEC III), Berlin 1968 (Iliffe Science and Technology Publications, Guildford 1970) pp.422-425
4.29 I. Dietrich, R. Weyl, H. Zerbst: In *Proc. 7ième Cong. Int. Microscopie Electronique*, Grenoble, 1970, ed. by P. Favard (Société Française de Microscopie Electronique, Paris 1970) Vol.II, pp.101-102
4.30 I. Dietrich, G. Lefranc, R. Weyl, H. Zerbst: Optik *38*, 449-453 (1973)
4.31 I. Dietrich, F. Fox, E. Knapek, G. Lefranc, K. Nachtrieb, R. Weyl, H. Zerbst: "Supraleitende Linsen für Höchstspannungsmikroskopie"; Forschungsber. T 75-44 of the Bundesministerium für Forschung und Technologie, Bonn (1975)
4.31a I. Dietrich, E. Knapek, R. Weyl, H. Zerbst: Cryogenics *15*, 691-699 (1975)
4.32 I. Dietrich: *Superconducting Electron-Optic Devices* (Plenum, New York, London 1976)
4.33 I. Dietrich, F. Fox, E. Knapek, G. Lefranc, K. Nachtrieb, R. Weyl, H. Zerbst: Ultramicroscopy *2*, 241-249 (1977)
4.34 I. Dietrich: In *Proc. 9th Int. Cong. Electron Microscopy*, Toronto, 1978, ed. by J.M. Sturgess (Microscopical Society of Canada, Toronto 1978) Vol.I, pp.173-184
4.35 A.C. van Dorsten, J.B. Le Poole: Philips Tech. Rev. *17*, 47-59 (1955)
4.36 J. Dugas, P. Durandeau, Ch. Fert: Rev. Opt. *40*, 277-305 (1961)
4.37 G. Dupouy, F. Perrier, L. Durrieu: J. Microsc. (Paris) *9*, 575-592 (1970)
4.38 E. Durand: *Magnétostatique* (Masson, Paris 1968)
4.39 P. Durandeau, Ch. Fert: Rev. Opt. *36*, 205-234 (1957)
4.40 A. Edwards: "Magnet design and selection of materials". In *Permanent Magnets and Magnetism*, ed. by D. Hadfield (Iliffe Books, London 1962) pp.191-296
4.41 A.B. El-Kareh, J.C.J. El-Kareh: *Electron Beams, Lenses and Optics*, 2 Vols. (Academic, New York, London 1970)
4.42 J.F. Elliott, S. Legvold, F.H. Spedding: Phys. Rev. *94*, 1143-1145 (1954)
4.43 T. Etoh, S. Suzuki, H. Watanabe, T. Yanaka: In *Proc. 8th Int. Cong. Electron Microscopy*, Canberra, 1974, ed. by J.V. Sanders, D.J. Goodchild (Australian Academy of Science, Canberra 1974) Vol.1, pp.174-175
4.44 H. Fahlenbrach, W. Baran: *Dauermagnete und ihre Anwendung in Betrieben* (Carl Hauser Verlag, München 1965)
4.45 H. Fahlenbrach: Fortschr. Ber. VDI-Z, Reihe 9, *6* (VDI-Verlag, Düsseldorf 1968)
4.46 J.S. Fahy, F.H. Plomp, C.J. Rakels, M.N. Thompson: Philips Electron Optics Bulletin EM 110-1977/3, pp.6-13
4.47 J.L. Farrant: In *Proc. 7ième Cong. Int. Microscopie Electronique*, Grenoble, 1970, ed. by P. Favard (Société Française de Microscopie Electronique, Paris 1970) Vol.1, pp.141-142
4.48 H. Fernández-Morán: Proc. Natl. Acad. Sci. USA *53*, 445-451 (1965)
4.49 H. Fernández-Morán: In *Proc. 6th Int. Cong. Electron Microscopy*, Kyoto, 1966, ed. by R. Uyeda (Maruzen, Tokyo 1966) Vol.I, pp.147-148
4.50 H. Fernández-Morán: In *Proc. AMU-ANL Workshop on High Voltage Electron Microscopy*, Argonne National Laboratory 1966 (Clearinghouse for Federal Scientific and Technical Information, NBS, Springfield, VA 1966) pp.51-58
4.51 J. Fischer: *Abriss der Dauermagnetkunde* (Springer, Berlin, Heidelberg 1949)
4.52 M. Fotino: In *Proc. 31st Annual EMSA Meeting*, New Orleans, 1973, ed. by C.J. Arceneaux (Claitor, Baton Rouge, Louisiana 1973) pp.8-9
4.53 A. Gemperle, J. Novák, J. Kaczer: J. Phys. E *7*, 518-520 (1974)
4.54 D. Génotel, C. Séverin, A. Laberrigue: J. Microsc. (Paris) *6*, 933-944 (1967)
4.55 D. Génotel, G. Balossier, C. Séverin, M. Girard, J.-C. Homo, A. Laberrigue: C.R. Acad. Sci. Paris, B *272*, 1961-1964 (1971)

4.56 D. Génotel, G. Balossier: Rev. Phys. Appl. *10*, 443-451 (1975)
4.57 W. Glaser: *Grundlagen der Elektronenoptik* (Springer, Wien 1952)
4.58 M.E. Haine, R.S. Page, R.G. Garfitt: J. Appl. Phys. *21*, 173-182 (1950)
4.59 M.E. Haine, T. Mulvey: J. Sci. Instrum. *31*, 326-332 (1954)
4.60 J.J. Hanak: RCA Rev. *25*, 551-569 (1964)
4.61 P.W. Hawkes: *Quadrupole Optics*, Erg. Exakt. Naturwiss., Vol.42 (Springer, Berlin, Heidelberg, New York 1966)
4.62 P.W. Hawkes: J. Microsc. (Paris) *9*, 435-454 (1970)
4.63 P.W. Hawkes: In *Proc. 7ième Cong. Int. Microscopie Electronique*, Grenoble, 1970, ed. by P. Favard (Société Française de Microscopie Electronique, Paris 1970) Vol.2, pp.17-18
4.64 P.W. Hawkes: *Quadrupoles in Electron Lens Design* (Academic, New York, London 1970)
4.65 P.W. Hawkes, U. Valdrè: J. Phys. E *10*, 309-328 (1977)
4.66 P.W. Hawkes: This volume, pp.1-56
4.67 H.G. Heide: Mikroskopie *24*, 179-185 (1969)
4.68 H.G. Heide, K. Urban: J. Phys. E *5*, 803-808 (1972)
4.69 W.E. Henry: J. Phys. Rad. *20*, 192-194 (1954)
4.70 K.-H. Herrmann, K. Ihmann, D. Krahl: In *Proc. 8th Int. Cong. Electron Microscopy*, Canberra, 1974, ed. by J.V. Sanders, D.J. Goodchild (Australian Academy of Science, Canberra 1974) Vol.1, pp.132-133
4.71 M.B. Hesse: Proc. Phys. Soc. (London) B *63*, 386-401 (1950)
4.72 M. Hibino, D. Hardy, D.F. Plomp, F.H. Kawakatsu, B.M. Siegel: J. Appl. Phys. *44*, 4743-4748 (1973)
4.73 H.-J. Hildebrandt: "Zur Dimensionierung des Eisenkreises elektromagnetischer Elektronenlinsen"; Diploma-Thesis, Free University of Berlin (1954)
4.74 H.-J. Hildebrandt, W.D. Riecke: Z. Angew. Phys. *20*, 336-342 (1966)
4.75 J. Hillier: J. Appl. Phys. *17*, 307-309 (1946)
4.76 J. Hillier: J. Appl. Phys. *17*, 411-419 (1946)
4.77 J. Hillier, E.G. Ramberg: J. Appl. Phys. *18*, 48-71 (1947)
4.78 J. Hopkinson, E. Hopkinson: Phil. Trans. Roy. Soc. London *177*, 331-358 (1886)
4.79 E.M. Hörl: In *Proc. 8th Int. Cong. Electron Microscopy*, Canberra, 1974, ed. by J.V. Sanders, D.J. Goodchild (Australian Academy of Science, Canberra 1974) Vol.1, pp.170-171
4.80 Y. Ishikawa, S. Chikazumi: Jpn. J. Appl. Phys. *1*, 155-173 (1962)
4.81 J.D. Jackson: *Classical Electrodynamics* (Wiley, New York 1975) pp.54-55
4.82 H. Kaden: *Wirbelströme und Schirmung in der elektrischen Nachrichtentechnik*. Technische Physik in Einzeldarstellungen, Band 10 (Springer, Berlin, Heidelberg, New York 1959)
4.83 W. Kamminga, J.L. Verster, J.C. Francken: Optik *28*, 442-461 (1968/1969)
4.84 W. Kamminga:"The numerical investigation of some electron optical designs". Doctoral Thesis, Groningen (1976); Optik *45*, 39-54 (1976)
4.85 E. Kasper: This volume, pp.57-118
4.86 S. Katagiri: 19th Meeting of the Japanese Electron Microscopy Soc., A-3 (1963
4.87 H.-J. Kempin, K.-H. Müller, M.v. Rauch, N. Schäfer, K. Vollborn: Beitr. Elektronenmikroskop. Direktabb. Oberfl. *8*, 339-346 (1975)
4.88 H. Kimura, T. Fujioka: Hitachi Hyoron *36*, 1517-1526 (1954)
4.89 H. Kimura, Y. Kikuchi: J. Electron Microsc. (Japan) *7*, 1-4 (1959)
4.90 E. Kinder, A. Pendzich: Jahrb. AEG-Forsch. *7*, 23-26 (1940)
4.91 E. Kinder: Z. Phys. *122*, 192-208 (1944)
4.92 J. Koch, K. Ruschmeyer: *Permanentmagnete I, Grundlagen* (Hrs. Valvo, Hamburg 1967)
4.93 M. Kubozoe, T. Kimura, S. Ozasa, S. Kasai: In *Proc. 7ième Congrès Int. Microscopie Electronique*, Grenoble, 1970, ed. by P. Favard (Société Française de Microscopie Electronique, Paris 1970) Vol.II, pp.47-48
4.94 W. Kunath, W.D. Riecke, E. Ruska: In *Proc. 6th Int. Cong. Electron Microscopy*, Kyoto, 1966, ed. by R. Uyeda (Maruzen, Tokyo 1966) Vol.1, pp.139-140
4.95 A. Laberrigue, P. Levinson, J.-C. Homo: Rev. Phys. Appl. *6*, 453-458 (1971)
4.96 A. Laberrigue, D. Génotel, M. Girard, C. Séverin, G. Balossier, J.-C. Homo: In *Proc. 3rd Int. Conf. High Voltage Electron Microscopy*, Oxford, 1973, ed. by P.R. Swann, C.J. Humphreys, M.J. Goringe (Academic, London, New York 1974) pp.108-113

4.97 A. Laberrigue, G. Berjot, P. Bonhomme, D. Gênotel, M. Girard, J.-C. Homo, C. Sêverin: In *Proc. 8th Int. Cong. Electron Microscopy*, Canberra, 1974, ed. by J.V. Sanders, D.J. Goodchild (Australian Academy of Science, Canberra 1974) Vol.1, pp.144-145

4.98 H.L. Laquer: Cryogenics *3*, 27-30 (1963)

4.99 K.R. Lawless, R.H. Geiss: In *Proc. 27th Annual EMSA Meeting*, St. Paul, Minn., 1969, ed. by C.J. Arceneaux (Claitor, Baton Rouge, Louisiana 1969) pp.94-95

4.100 S. Leisegang: Optik *10*, 5-14 (1953)

4.101 S. Leisegang: Optik *11*, 49-60 (1954)

4.102 F. Lenz: This volume, pp.119-161

4.103 J.B. Le Poole: Philips Tech. Rundsch. *9*, 33-46 (1947)

4.104 J.B. Le Poole: In *Proc. 3rd European Reg. Conf. Electron Microscopy*, Prague, 1964, ed. by M. Titlbach (Czechoslovak Academy of Sciences, Prague 1964) Appendix, p.6

4.105 P. Levinson, A. Laberrigue, O. Testard: Cryogenics *5*, 344-346 (1965)

4.106 G. Liebmann, E.M. Grad: Proc. Phys. Soc. London B*64*, 956-971 (1951)

4.107 G. Liebmann: Proc. Phys. Soc. London B*64*, 972-977 (1951)

4.108 G. Liebmann: Proc. Phys. Soc. London B*65*, 94-107 (1952)

4.109 G. Liebmann: Proc. Phys. Soc. London B*66*, 448-458 (1953)

4.110 G. Liebmann: Proc. Phys. Soc. London B*68*, 679-681 (1955)

4.111 G. Liebmann: Proc. Phys. Soc. London B*68*, 737-745 (1955)

4.112 B. Lischke, W. Münchmeyer: Optik *50*, 315-328 (1978)

4.113 J.J. Lopez: In *Proc. 4ième Cong. Int. Microscopie Electronique à Haute Tension*, Toulouse, 1975, ed. by B. Jouffrey, P. Favard (Société Française de Microscopie Electronique, Paris 1976) pp.43-46

4.114 N.C. Lund: In *Proc. 25th Annual EMSA Meeting*, Chicago, 1967, ed. by C.J. Arceneaux (Claitor, Baton Rouge, Louisiana 1967) pp.228-229

4.115 A. Mager: Z. Angew. Phys. *23*, 381-386 (1967)

4.116 A. Mager: Frequenz *22*, 25-27 (1968)

4.117 L.A. Marzetta: Rev. Sci. Instrum. *32*, 1192-1195 (1961)

4.118 M. McCaig: *Permanent Magnets in Theory and Practice* (Pentech, London, Plymouth 1977)

4.119 G. Möllenstedt: J. Microsc. (Paris) *4*, 413-428 (1965)

4.120 K. Müller, E. Ruska: Z. Wiss. Mikrosk. *62*, 205-219 (1955)

4.121 K. Müller: Z. Wiss. Mikrosk. *63*, 303-328 (1957)

4.122 K. Müller, E. Ruska: In *Vierter Int. Kong. Elektronenmikroskopie*, Berlin, 1958, ed. by W. Bargmann, G. Möllenstedt, H. Niehrs, D. Peters, E. Ruska, C. Wolpers (Springer, Berlin, Heidelberg, New York 1960) Vol.1, pp.184-187

4.123 K. Müller, E. Ruska: Mikroskopie *23*, 197-219 (1968)

4.124 K. Müller, E. Ruska, H. Neff: In *Proc. 27th Annual EMSA Meeting*, St. Paul, Minn., 1969, ed. by C.J. Arceneaux (Claitor, Baton Rouge, Louisiana 1969) pp.74-75

4.125 K.H. Müller, R. Schliepe: Siemens Z. *47*, 471-475 (1973)

4.126 T. Mulvey: Proc. Phys. Soc. London B*66*, 441-447 (1953)

4.127 T. Mulvey, M.J. Wallington: J. Sci. Instrum. (J. Phys. E) Ser.2, *2*, 466-472 (1969)

4.128 T. Mulvey: In *Proc. 25th Anniversary Meeting EMAG*, Cambridge, 1971, ed. by W.C. Nixon (Institute of Physics, London 1971) pp.77-83

4.129 T. Mulvey, C.D. Newman: In *Proc. 5th European Cong. Electron Microscopy*, Manchester, 1972 (Institute of Physics, London 1972) pp.116-117

4.130 T. Mulvey, C.D. Newman: In *Scanning Electron Microscopy; Systems and Applications 1973*, ed. by W.C. Nixon (Institute of Physics, London 1973) pp.16-21

4.131 T. Mulvey, C.D. Newman: In *Proc. 3rd Int. Conf. High Voltage Electron Microscopy*, Oxford 1973, ed. by P.R. Swann, C.J. Humphreys, M.J. Goringe (Academic, London, New York 1974) pp.98-102

4.132 T. Mulvey: In *Proc. 8th Int. Cong. Electron Microscopy*, Canberra, 1974, ed. by J.V. Sanders, D.J. Goodchild (Australian Academy of Science, Canberra 1974) Vol.1, pp.16-17

4.133 T. Mulvey: This volume, pp.359-412

4.134 E. Munro: "Computer-Aided Design Methods in Electron Optics"; Ph. D. Thesis, Cambridge (1971)

4.135 E. Munro: In *Image Processing and Computer-Aided Design in Electron Optics*, ed. by P.W. Hawkes (Academic, London, New York, 1973) pp.284-323
4.135a S. Ozasa, S. Katagiri: J. Electron Microsc. (Japan) *12*, 110 (1963)
4.136 R.S. Page: In *Electron Microscopy 1964. Proc. 3rd European Regional Conf.*, Prague, 1964, ed. by M. Titlbach (Czechoslovak Academy of Sciences, Prague 1964) Vol.1, pp.31-32
4.137 K. Pape: "Die Abbildungseigenschaften des Vorfeldes magnetischer Elektronenlinsen"; Diploma Thesis, Berlin (1966)
4.138 R.J. Parker, R.J. Studders: *Permanent Magnets and Their Applications* (Wiley, New York 1962)
4.139 W. Pejas: Optik *50*, 61-72 (1978)
4.140 K.G. Petzinger, J.J. Hanak: RCA Rev. *25*, 542-550 (1964)
4.141 C.J. Rakels, J. van Helden, W. Kuypers, J.C. Tiemeyer: In *Proc. 7ième Cong. Microscopie Electronique*, Grenoble, 1970, ed. by P. Favard (Société Française de Microscopie Electronique, Paris 1970) Vol.II, pp.53-54
4.142 E.G. Ramberg: US Patent No. 2 369 796 (1943)
4.143 O. Rang: Optik *5*, 518-530 (1949)
4.143a A. Recknagel. G. Haufe: Wiss. Z. Techn. Hochsch. Dresden *2*, 1-10 (1952/1953)
4.144 J.H. Reisner, E.G. Dornfeld: J. Appl. Phys. *21*, 1131-1139 (1950)
J.H. Reisner: J. Appl. Phys. *22*, 561-565 (1951)
4.145 J.H. Reisner, S.M. Zollers: Electronics *24*, No.1, 86-91 (1951)
4.146 J.H. Reisner, J.J. Schuler: In *Proc. 25th Annual EMSA Meeting*, Chicago, 1967, ed. by C.J. Arceneaux (Claitor, Baton Rouge, Louisiana 1967) pp.226-227
4.147 B.L. Rhodes, S. Legvold, F.H. Spedding: Phys. Rev. *109*, 1547-1550 (1958)
4.148 W.D. Riecke, E. Ruska: German Patent DBP 1 033 818 (1954)
4.149 W.D. Riecke, E. Ruska: Z. Wiss. Mikrosk. *63*, 288-302 (1957)
4.150 W.D. Riecke: In *Vierter Int. Kong. Elektronenmikroskopie*, Berlin 1958, ed. by W. Bargmann, G. Möllenstedt, H. Niehrs, D. Peters, E. Ruska, C. Wolpers (Springer, Berlin, Heidelberg, New York 1960) Vol.I, pp.189-194
4.151 W.D. Riecke: In *Proc. European Regional Conf. Electron Microscopy*, Delft, 1960, ed. by A.L. Houwink, B.J. Spit (Nederlandse Vereiniging voor Electronenmicroscopie, Delft, no date) Vol.1, pp.82-88
4.152 W.D. Riecke: Optik *18*, 379-401 (1961)
4.153 W.D. Riecke: Optik *19*, 81-116 (1962)
4.154 W.D. Riecke: Optik *19*, 169-183, 193-207 (1962)
4.155 W.D. Riecke: Optik *19*, 273-286 (1962)
4.156 W.D. Riecke: In *Proc. 5th Int. Cong. Electron Microscopy*, Philadelphia, 1962, ed. by S.S. Breese (Academic, New York, London 1962) Vol.1, pp.KK-5
4.157 W.D. Riecke, E. Ruska: In *Proc. 6th Int. Cong. Electron Microscopy*, Kyoto, 1966, ed. by R. Uyeda (Maruzen, Tokyo 1966) Vol.1, pp.19-20
4.158 W.D. Riecke: Optik *24*, 397-426 (1966/1967)
4.159 W.D. Riecke: US Patent 3 560 781 (filed 27.6.1967, patented 2.2.1972) German Patent DBP 1 614 123 (24.2.1967)
4.160 W.D. Riecke: US-Patent 3 526 766 (filed 27.7.1967, patented 1.7.1970)
4.160a W.D. Riecke: German Patent DBP 1614163 (1967)
4.161 W.D. Riecke, F. Stöcklein: German Patent DBP 1 614 165 (1967)
4.162 W.D. Riecke, F. Stöcklein: German Patent DBP 1 300 992 (1967)
4.163 W.D. Riecke: In *Proc. 4th European Regional Conf. Electron Microscopy*, Rome, 1968, ed. by D.S. Bocciarelli (Tipographia Poliglotta Vaticana, Rome 1968), Vol.1, pp.207-208
4.164 W.D. Riecke: Z. Angew. Physik *27*, 155-165 (1969)
4.165 W.D. Riecke: Optik *36*, 66-84, 288-308, 375-398 (1972)
4.166 W.D. Riecke: "Die Physik der magnetischen Elektronenlinse I. Der Aufbau des Polschuhsystems der konventionellen magnetischen Elektronenlinse", Lectures held at the University of Tübingen (1973/1974); lecture notes available from Institut für Angewandte Physik der Universität Tübingen
4.167 W.D. Riecke: "Quadrupol-Elektronenlinsen". Lectures at the University of Tübingen (1975)
4.168 W.D. Riecke: 17. Tagung der Deutschen Gesellschaft für Elektronenmikroskopie, Berlin 1975; abstracts: contribution J 12 (Programme of the Meeting published by Deutsche Gesellschaft für Elektronenmikroskopie, Berlin 1975) pp.41-42

4.169 W.D. Riecke: In *Proc. 4ième Cong. Int. Microscopie Electronique à Haute Tension*, Toulouse, 1975, ed. by B. Jouffrey, P. Favard (Société Française de Microscopie Electronique, Paris 1976) pp.39-42
4.170 W.D. Riecke: In *Electron Microscopy in Materials Science*, ed. by U. Valdrè, E. Ruedl (Commission of the European Communities, Luxemburg 1975) Vol.1, pp.21-50
4.171 W.D. Riecke: In *Electron Microscopy in Materials Science*, ed. by U. Valdrè, E. Ruedl (Commission of the European Communities, Luxemburg, 1975) Vol.1, pp.51-79
4.172 W.D. Riecke: In *Electron Microscopy in Materials Science*, ed. by U. Valdrè, E. Ruedl (Commission of the European Communities, Luxemburg 1975) Vol.1, pp.81-111
4.173 W.D. Riecke: In *Proc. 5th Int. Conf. High Voltage Electron Microscopy*, Kyoto, 1977, ed. by T. Imura, H. Hashimoto (Japanese Soc. of Electron Microscopy, Tokyo 1977) pp.73-78
4.174 J.W. Rommerts: German Patent DBP 1 078 703 (1958), US-Patent 2 939 955 (filed 4.3.1958; patented 7.6.1960)
4.175 E. Ruska: Z. Phys. *87*, 580-602 (1934)
4.175a E. Ruska: "Grundlagen der Elektronenoptik und ihre Anwendungen bei modernen Elektronenmikroskopen", in Convegno di Elettronica e Televisione, Milan 1954; session 4, general lecture to the third section. La Ricerca Scientifica Suppl. 3-30 (1954)
4.176 E. Ruska: Arch. Elektrotech. *38*, 102-130 (1944)
4.177 E. Ruska: Kolloid Z. *107*, 2-16 (1944)
4.178 E. Ruska: Kolloid Z. *115*, 102-120 (1950)
4.179 E. Ruska: Z. Wiss. Mikrosk. *61*, 152-171 (1952)
4.180 E. Ruska: In *Electron Physics* (National Bureau of Standards Circular 527, Washington 1954) pp.389-410
4.181 E. Ruska: In *Les Techniques Récentes en Microscopie Electronique et Corpusculaire* (C.N.R.S., Paris 1956) pp.253-260
4.182 E. Ruska, O. Wolff: Z. Wiss. Mikrosk. *62*, 465-509 (1956)
4.183 E. Ruska: J. Roy. Microsc. Soc. *84*, 77-103 (1965)
4.184 E. Ruska: In *Proc. 8th Int. Cong. Electron Microscopy*, Canberra, 1974, ed. by J.V. Sanders, D.J. Goodchild (Australian Academy of Sciences, Canberra 1974) Vol.1, pp.24-25
4.185 Y. Sakitani, S. Ozasa, H. Fujita, K. Ohji: In *Proc. 7ième Cong. Int. Microscopie Electronique*, Grenoble, 1970, ed. by P. Favard (Société Française de Microscopie Electronqiue, Paris 1970) Vol.1, pp.127-128
4.186 O. Scherzer: J. Appl. Phys. *20*, 20-28 (1949)
4.187 K. Schüler, K. Brinkmann: *Dauermagnete, Werkstoffe und Anwendungen* (Springer, Berlin, Heidelberg, New York 1970)
4.188 A. Septier, P. Bonjour: Phys. Bull. *24*, 433-434 (1973)
4.189 C. Sêverin, D. Génotel, M. Girard, A. Laberrigue: Rev. Phys. Appl. *6*, 459-465 (1971)
4.190 B.M. Siegel, N. Kitamura, R.A. Kropfli, M.P. Schulhof: In *Proc. 6th Int. Cong. Electron Microscopy*, Kyoto, 1966, ed. by R. Uyeda (Maruzen, Tokyo 1966) Vol.1, pp.151-152
4.191 B.M. Siegel, D.L. Musinski, Huei Pei Kuo: In *Proc. 8th Int. Cong. Electron Microscopy*, Canberra, 1974, ed. by J.V. Sanders, D.J. Goodchild (Australian Academy of Science, Canberra 1974) Vol.1, pp.26-27
4.192 B.M. Siegel: In *Proc. 6th Eur. Cong. Electron Microscopy*, Jerusalem, 1976, ed. by D.G. Brandon (Tal International, Jerusalem 1976) Vol.1, pp.105-108
4.192a B.M. Siegel, M. Shimoyama, E. van der Leeden: In *Proc. 9th Int. Cong. Electron Microscopy*, Toronto, 1978, ed. by J.M. Sturgess (Microscopical Society of Canada, Toronto 1978) Vol.1, pp.8-9
4.193 F. Stöcklein: German Patent DBP 1286239 (1967) and US Patent 3474247 (1967)
4.194 D.L. Strandburg, S. Legvold, F.H. Spedding: Phys. Rev. *127*, 2046-2051 (1962)
4.195 A. Strojnik: In *Proc. 27th Annual EMSA Meeting*, St. Paul, Minn., 1969, ed. by C.J. Arceneaux (Claitor, Baton Rouge, Louisiana 1969) pp.104-105
4.196 S. Suzuki: US Patent No. 3 173 005 (1965)
4.197 S. Suzuki, K. Akashi, H. Tochigi: In *Proc. 26th Annual EMSA Meeting*, New Orleans, La., 1968, ed. by C.J. Arceneaux (Claitor, Baton Rouge, Louisiana 1968) pp.320-321

4.198 B. Tadano, Y. Onuma: Hitachi Rev. 1954 (July) pp.89-102
4.199 B. Tadano, I. Makino: *Hitachi Electron Microscope,* Type HU-10 (Hitachi Ltd. Tech. Manual 1958) pp.1-24
4.200 B. Tadano, H. Kimura, Y. Onuma: Hitachi Rev. *15*, 340-344 (1966)
4.201 R.S. Tebble, D.J. Craik: *Magnetic Materials* (Wiley Interscience, New York 1969)
4.202 H. Teuchert: Elektrotechnik *2*, 353-355 (1948)
4.203 Sir William Thomson (Lord Kelvin): *Reprints of Papers on Electrostatics and Magnetism* (MacMillan, London 1872) esp. p.564
4.204 F. Thon: "Zur Deutung der Bildstrukturen in hochaufgelösten elektronenmikroskopischen Aufnahmen dünner amorpher Objekte"; Dissertation Dr. Naturwiss., Tübingen (1968)
4.205 U. Valdrè: In *Electron Microscopy in Materials Science*, ed. by U. Valdrè, E. Ruedl (Commission of the European Communities, Luxemburg 1975) Vol.1, pp.115-134
4.206 M. Watanabe, T. Someya: Optik *20*, 99-108 (1963)
4.207 M. Watanabe, T. Yanaka, M. Yamamoto, S. Suzuki, Y. Nagahama, U. Fujisawa: In *Proc. 4th European Regional Conf. Electron Microscopy,* Rome 1968, ed. by D.S. Bocciarelli (Tipografia Poliglotta Vaticana, Rome 1968) Vol.1, pp.203-204
4.208 R.C. Weast, M.J. Astle (eds.): *CRC Handbook of Chemistry and Physics* (CRC Press Inc., Palm Beach, Florida 1978/1979)
4.209 L. Wegmann: Optik *11*, 153-170 (1954)
4.210 P. Weiss: J. Phys. 4, *6*, 353-368 (1907)
4.211 J.J. Went, G.W. Rathenau, E.W. Gorter, G.W. van Oosterhout: Philips' Techn. Rundschau *13*, 361-376 (1952)
4.212 D. Willasch: Siemens Analytical Application Note No. 212 (1976)
4.213 D. Willasch: In *Proc. Vol.7 Electron Microscopy Society of Southern Africa,* Kaapstad, 1977 (Electron Microscopy Soc. of Southern Africa, Kaapstad 1977) pp.5-6
4.214 R.E. Worsham, J.E. Mann: In *Proc. 6th Eur. Cong. Electron Microscopy,* Jerusalem, 1976, ed. by D.G. Brandon (Tal International, Jerusalem 1976) Vol.1, pp.358-359
4.215 R.E. Worsham, J.E. Mann, E.G. Richardson, N.F. Ziegler: In *Proc. 28th Annual EMSA Meeting,* Houston, 1970, ed. by C.J. Arceneaux (Claitor, Baton Rouge, Louisiana 1970) pp.354-355
4.216 K. Yada, H. Kawakatsu: J. Electron Microsc. (Japan) *25*, 1-9 (1976)
4.217 F. Zemlin: "Magnetischer Fluß and magnetische Feldstärke in magnetisch übersättigten Polschuhsystemen"; Diploma Thesis, Free University of Berlin (1967)

5. Unconventional Lens Design

T. Mulvey

With 32 Figures

Conventional magnetic electron lenses have evolved to their present highly developed state under the pressing requirements of high resolution transmission electron microscopy. However, recent changes in the emphasis of electron microscopy towards quantitative analysis at high resolution have made it desirable to undertake a reappraisal of the basic design of magnetic electron lenses. This is timely and relevant now that improved numerical methods are available for calculating the electron optical performance of unconventional lens designs. Moreover, technological developments in electron field-emission sources and superconducting lens elements make it desirable to reconsider the whole basis of the subject. The present chapter is, therefore, chiefly concerned with opening up new ways of thinking about electron lenses and realising them in practice in the light of the information presented in the preceding chapters.

5.1 Introduction

In the previous chapters the theory and practical realisation of magnetic electron lenses have been discussed in some detail. The position has now been reached where the electron optical properties of all forms of magnetic electron lenses now in common use can be predicted with an accuracy entirely adequate for practical purposes. Moreover, it is possible to estimate the mechanical and electrical tolerances required in the construction of high-resolution lenses, for example, the objective lens of a high-resolution electron microscope. Similarly, it is now possible to calculate the overall electron optical behaviour of a multi-lens system, for example, the projection system of an electron microscope. Some of the practical problems of operating these more complicated microscopes can now be largely overcome by the use of microprocessors, which can be programmed to select the optimum lens settings for a given operational requirement. These factors have led to new approaches and possibilities for magnetic lenses.

In the past, the development of new forms of magnetic lenses, and certainly that of complete magnetic lens systems, was handicapped by the absence of such methods of calculation. Consequently, it was necessary to rely heavily on experimental methods, which are often time consuming and expensive.

Until recently, electron microscopes were comparatively straightforward instruments inasmuch as they were restricted to the examination of thin specimens in transmission; the image was generally recorded on a photographic plate for subsequent visual inspection. Today an electron microscope may be expected to operate not only in the transmission (TEM) mode but also in the scanning transmission (STEM) mode with transmission specimens and even in the scanning electron microscope (SEM) mode with solid specimens. In addition, a high-grade instrument is expected to be able to record characteristic X rays or Auger electrons emitted from the specimen as well as the energy-loss spectrum of the electrons transmitted by the specimen. For the examination of semiconductors it is also desirable to be able to record the photo-luminescence spectra. Furthermore, a requirement to be able to obtain convergent-beam electron diffraction patterns from micro-areas in the sample or electron channelling patterns from a solid specimen can place heavy demands on lens performance. It is not surprising therefore that it is becoming increasingly difficult to meet all these requirements simultaneously in one instrument with conventional magnetic electron lenses. To add to these difficulties, the steady demand for higher resolution and the extension of electron microscopy to thicker and thicker specimens can only be met by an appreciable increase in the accelerating voltage above the conventional 100 kV.

5.1.1 High Voltage Electron Microscopy

Electron microscopes operating at an accelerating voltage of up to 1000 kV have been available commercially for some years. These have been used with success for the examination of thick specimens but up to now have failed to achieve a comparable resolving power to that of the best 100 kV electron microscopes. These high voltage instruments have been largely scaled-up versions of standard 100 kV designs. This has led to electron optical columns weighing several tons and requiring special buildings to house them. Such installations are therefore extremely expensive and unsuitable for use in ordinary laboratories. This situation makes it desirable to reconsider the design of electron lenses and electron optical systems for high-resolution microscopy at high accelerating voltages taking into account what is now technologically feasible.

It is therefore perhaps convenient to begin by asking what improvements are possible and desirable in the design and construction of *conventional* magnetic electron lenses and then go on to enquire whether there are any advantages to be gained by a radical departure from conventional forms.

5.1.2 Conventional Lenses

It has been pointed out in previous chapters that the essential electron optical part of a conventional magnetic electron lens resides in the magnetic field concentrated in the immediate neighbourhood of the pole pieces. This volume often amounts to only a few cubic millimetres. Theory and experiment both suggest that improvements in electron optical performance follow as the maximum axial magnetic flux density in the pole-piece region is increased and as the "half-width" of the field distribution decreases. This consideration inevitably leads to lens designs with pole pieces of small bores and gaps operating at or beyond the saturation flux density of the ferromagnetic material of the pole pieces. The specimen, which must be placed between the pole pieces in order to reduce aberrations, inevitably becomes less accessible.

In order to produce a high magnetic flux density in the pole pieces, a large coil is employed (Fig.2.12a in Chapter 2) placed well away from the axis and surrounded by a massive iron circuit. The advantage of this construction is that the magnetic flux density on the lens axis due to the coil itself is negligible, the magnetic flux density along the axis being almost entirely provided by the magnetization of the tips of the pole pieces. Consequently, a crude wire winding may be used since winding asymmetries will not adversely affect lens performance and the electrical power needed to produce the necessary ampere-turns in the lens gap may be reduced below any prescribed limit by increasing the size of the coil. In the early stages of development of electron lenses, when only limited power was available from valve stabilisers, this was an important consideration. However, the electron optical disadvantages, many of which are revealed in Figs.2.12a,b of Chapter 2, greatly exceed these two advantages. The first disadvantage is that the volume of the iron circuit greatly exceeds that of the pole pieces themselves. This inevitably leads to unwanted hysteresis effects when the lens current is varied. In addition, unwanted magnetic fields are created along the axis as shown in Fig.2.12b: these arise from the magnetization of adjacent parts of the iron circuit. Since these parts cannot be machined as accurately as can the pole pieces and because of inevitable imperfections in the iron, such fields are unlikely to be aligned accurately with the axial field distribution from the pole pieces. This often introduces severe alignment problems. What is, however, more serious from the point of view of instrument design is that the coil and associated iron circuit occupy a considerable space along the axis, making it impossible to introduce additional lenses in this region. This fact alone makes conventional electron optical columns unnecessarily long and unwieldy. As a result, many electron optical columns are unduly sensitive to the effects of stray AC magnetic fields and mechanical vibration. For all these reasons it therefore seems essential to investigate ways of considerably reducing the size of the exciting coil and of eliminating non-essential parts of the iron circuit. Notice that this can only be achieved at the cost of increasing the power supplied to the coil. This is perfectly feasible with present day solid-state technology, but the extent to which it is applied to practical lens design will be largely dictated by economic considerations.

5.1.3 Electrical Power Requirements

In a magnetic electron lens, the excitation needed to produce a certain focal length at a given accelerating voltage can be specified in advance; for example, for a projector lens operating at an accelerating voltage of 1000 kV, an excitation of some 18000 ampere-turns would be required at minimum focal length in a conventional pole-piece lens irrespective of pole-piece geometry. Consider therefore a coil of arbitrary shape with a mean diameter D_m, an electrical resistivity ρ, operating at a current density σ in its windings. The electrical power W required to energise the coil is given by

$$W = \rho\sigma^2 V_w \quad , \tag{5.1a}$$

where V_w is the volume of the windings. This may be written more conveniently as

$$W = \pi\rho D_m \sigma (NI) \tag{5.1b}$$

where NI is the excitation in ampere-turns. It should be noted that the power required in a coil is not affected, other things being equal, by the particular combination of N and I. The choice of lens current and number of turns to make up a prescribed excitation NI is a matter of practical convenience. Conventional lenses operate at a current density of some 2 A mm^{-2}; a typical mean diameter for the coil would vary from some 100 mm for a 100 kV instrument to perhaps 1000 mm for the largest lens so far built (for the 3000 kV electron microscope at Toulouse). Equation (5.1b) shows that as the mean diameter D_m of the coil is reduced, the current density σ in the winding can be increased in the same ratio without increasing the power required. This leads to a substantial reduction in the size of the coil. Miniature windings of this type were first devised and realised by LE POOLE [5.22] in the form of a "mini-lens" shown schematically in Fig.5.1. This consisted of a multi-layer wire-wound solenoid of only a few millimetres in mean diameter but of some 50 mm in length; the latter was necessary in order to provide sufficient surface area to enable the heat to be extracted from the windings, which were operated at a current density of about 80 A mm^{-2}. There was no iron circuit, the magnetic flux density on the axis being produced directly by the windings. Such lenses have since been successfully applied in electron probe analysers and analytical transmission electron microscopes [5.5,10], as condenser lenses in high-voltage electron microscopes, and in other situations where it is necessary to introduce an auxiliary lens at a region on the optical axis already occupied, and therefore obstructed, by a conventional magnetic electron lens.

The electron optical properties of iron-free lenses of a more generalized form, which are also of interest in connection with superconducting lenses, are described below. It should perhaps be mentioned here, however, that it is difficult to construct iron-free lenses of small diameter with the necessary axial symmetry, especially if wire rather than tape conductors are used. Moreover, high-current density coils of this type cannot be applied to advantage in conventional magnetic

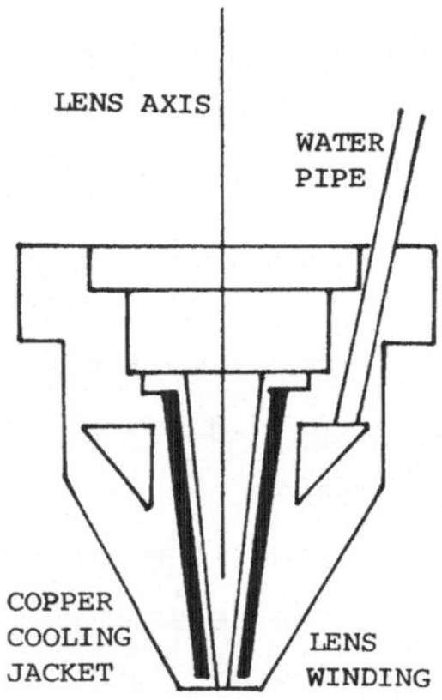

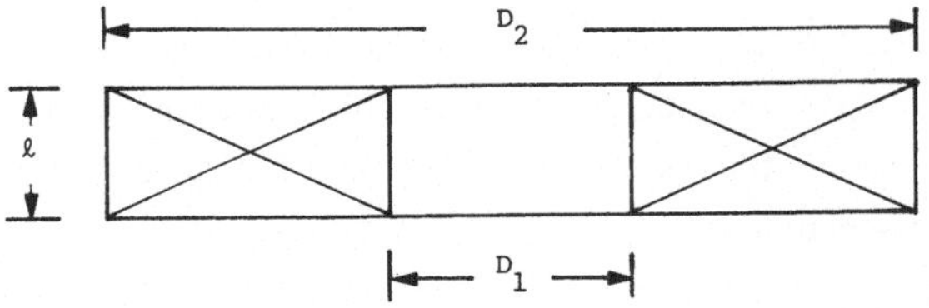

↑ Fig.5.2. Lens coil of rectangular cross-section

← Fig.5.1. Schematic cross-section of iron-free mini-lens of LE POOLE [5.22] showing long solenoid in a water-cooled jacket. Length of coil 28 mm

electron lenses because of difficulties in designing the iron circuit. In such lenses it is of greater importance to reduce the axial length of the coil rather than its diameter, since in general one is interested in shortening the height of an electron optical column or in inserting additional lenses, the column diameter being generally of secondary importance. For such lenses a narrow coil of rectangular cross-section is more appropriate. In order to understand this, consider a coil of rectangular cross-section and of axial length l as shown in Fig.5.2. Suppose the fraction of the total cross-section occupied by conductor, rather than by insulation, is λ. For a typical wire winding $\lambda \simeq 0.65$. From (5.1b) we have

$$W = \pi\rho(D_2 + D_1)(NI)^2/(D_2 - D_1)l\lambda \quad . \tag{5.2a}$$

If $D_2 >> D_1$,

$$W \simeq \pi\rho(NI)^2/l\lambda \quad . \tag{5.2b}$$

Under these conditions the power required by a coil is inversely proportional to the axial length l of the coil. It should perhaps be pointed out that, although increasing the outer diameter of the coil makes little difference to the power required for a given number of ampere-turns, it will increase the surface area available for cooling.

As an illustrative example, consider a conventional coil of axial length of 200 mm for a projector lens of a million-volt electron microscope, taking the maximum excitation as 18000 A-t, as mentioned above, and assuming a copper winding operating at 100°C ($\rho \simeq 2 \times 10^{-4}$ Ω m). According to (5.2), W = 102 W if $D_2 \gg D_1$ and 136 W if $D_2 = 3D_1$. If the axial length of the coil is now reduced by a factor of 20, namely from 200 to 10 mm, the corresponding values of W become 2040 watts and 2720 watts, respectively. Stabilized current supplies using solid-state devices are now commercially available in this power range. Provided that the heat generated within the coil can be satisfactorily removed, it is clear that the axial height of magnetic electron lenses and indeed their total volume can be substantially reduced by the use of high-current-density coils, at the expense, of course, of providing more powerful lens current supplies. In some cases it may be necessary to do this in order to meet a more exacting specification of the instrument; in other cases the saving of the cost of a special building may well justify the increased expenditure.

It might be thought that as the windings of conventional electron optical lenses are usually water-cooled (and even so, frequently operate at temperatures in the region of 100°C), an increase by an order of magnitude in the power supplied to the lens, together with a reduction of two orders of magnitude in its volume, would cause a dangerous rise in temperature of the windings. On the other hand, the basic design of the lens coils and the lens cooling arrangements in commercial electron microscopes have not changed significantly since the publication by RUSKA [5.39] in 1934 of his original design in which the comparatively small amount of heat generated in the lens windings was removed by water cooling the external iron casing of the lens. Ideally, in a magnetic electron lens, the exciting coils should be isolated both thermally and mechanically from the surrounding magnetic structure and hence from the electron optical column. The coolant should be in intimate contact with the windings. Present-day high-resolution magnetic electron lenses still leave much to be desired in these respects. In fact, current densities approaching those achieved in superconductors can be obtained in ordinary lens windings if due attention is paid to the heat transfer mechanism.

5.1.4 Heat Transfer in Lens Windings

High-current-density coils have been used for a long time in powerful electromagnets, but the operational requirements for these are significantly less stringent than those for magnetic electron lenses, especially with respect to the mechanical and electrical stability. Several interesting suggestions for improving the cooling of magnetic lenses were made by FARRANT [5.9]; these included the use of liquid nitrogen rather than water as a cooling agent. MACLACHLAN [5.23] designed and constructed a coil for a lens shown schematically in Fig.5.3 in which bare copper wire of square cross-section was embedded in Araldite. Heat was removed from the windings by the closely spaced, electrically insulated, copper discs as shown in Fig. 5.3. These discs were cooled by water flowing directly over their exposed surfaces. Current densities of 10-30 A mm^{-2} were reported by MACLACHLAN with this type of lens, a notable improvement over conventional designs.

Even higher current densities of 50 A mm^{-2} had previously been reported in 1972 by MULVEY and NEWMAN [5.31]. This was achieved by allowing the cooling water to flow directly over the windings, which consisted of a comparatively small number of turns of enamelled copper wire held in place by Araldite; the axial thickness of the winding was restricted to about 5 mm in order to reduce the temperature difference between the inside and outside parts of the winding. Figure 5.4 gives some idea of the reduction in coil size that is possible in this way. The figure shows

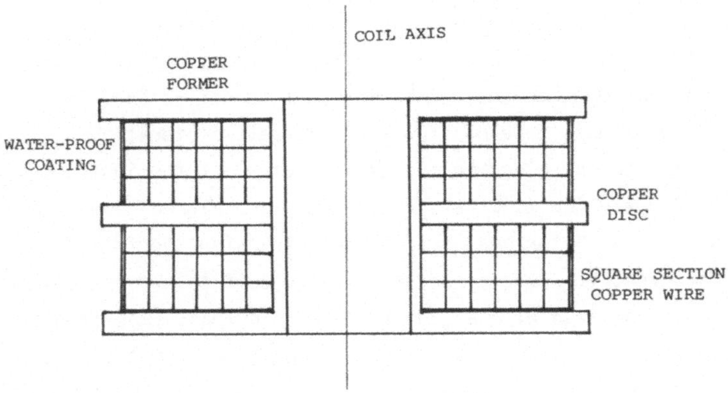

Fig.5.3. Schematic arrangement of MACLACHLAN's coil construction [5.23] with square section bare copper wire embedded in Araldite. Cooling water flows over exposed copper cooling surfaces

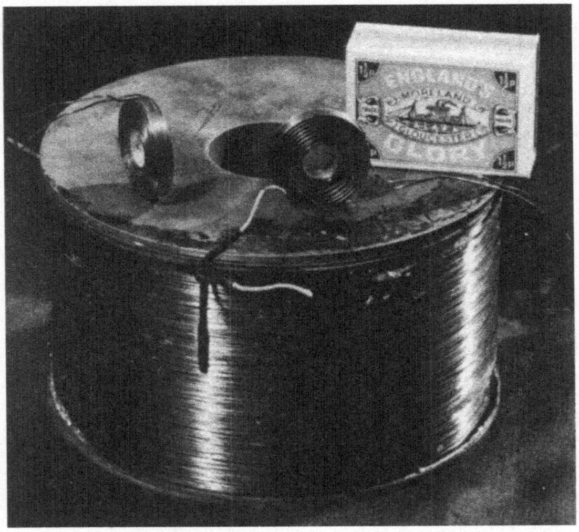

Fig.5.4. Two miniature water-cooled coils [5.31] placed on top of a conventional water-cooled lens winding of the same capacity (4000 A-t)

two water-cooled wire coils, providing between them 4000 A-t, placed on top of a conventional lens winding also designed to provide 4000 A-t. A matchbox of length 50 mm has been added for comparison.

The heat transfer in a conventional lens winding and cooling system is usually both complicated and ill-defined. In a compact winding, such as the one shown in Fig.5.4, it becomes possible to define an average heat transfer coefficient h given by

$$W = hS_A\Delta T \quad , \tag{5.3}$$

where W is the power dissipated in the coil, S_A is the surface area of the coil in contact with the coolant, and ΔT is the difference in temperature between the surface and the cooling liquid. The heat transfer coefficient h increases with increasing flow velocity of the cooling water. In high-current-density coils for electromagnets, *turbulent flow* is employed to ensure maximum heat transfer. Experiment with magnetic lenses reveals that this type of flow tends to set up vibrations, which can easily be transmitted to the electron optical structure of the instrument, leading to poor resolution. At very low flow rates and suitably designed geometries *laminar* flow is also a possibility. Unfortunately, the heat transfer coefficient for laminar flow is low. Moreover, it would be extremely difficult to ensure laminar flow in the type of geometry associated with a magnetic electron lens. Experiments by the author and his colleagues have shown that the flow region between these two extremes is the most suitable for electron microscopy and leads to convenient values for the coefficient h. It is not generally possible to ensure that the velocity of water over each part of a lens coil will be the same, and so it is not possible to define experimental conditions with great exactitude. However, as a rough guide and for flow rates in common use in present day electron lenses (about 2 l/min), experiment indicates that it is possible to extract about 5 kW $m^{-2}\,{}^{\circ}C^{-1}$ from such water-cooled lenses without running into operational difficulties such as vibration or overheating of the windings. In this flow region, the coefficient h varies approximately with the square root of the water velocity [5.14], so that the flow rate is not a critical factor. For further details of the calculation of heat transfer in water-cooled coils, the reader should consult KROON [5.20] and NEWMAN [5.36].

5.1.5 Tape Windings

Windings consisting of many turns of insulated round wire, although traditional, are thermally unsuitable for lens coils, since the packing factor is low and the thermal conductivity is also poor. Windings made from wire of square cross-section are superior, but windings made from thin copper tape are greatly to be preferred since these allow a high packing factor ($\lambda > 0.9$) which reduces the power required to produce a given excitation. More importantly, heat flow in the axial direction takes place entirely through the copper tape without traversing the electrical insulation, which has a relatively poor thermal conductivity. This is illustrated in Fig.5.5, which shows schematically the arrangement of a typical tape winding. Heat generated in the windings can readily be removed by a coolant applied to the end-faces. Since the temperature gradient across the width of the tape is small for materials such as copper or aluminium, it is only necessary to apply the coolant to one end-face.

Experiment with copper tape windings has shown that current densities of some 200 A-t mm^{-2} [5.14,30] can readily be obtained for a moderate water flow of 3 l/min. This current density is comparable with that obtained in superconducting windings. Such tape windings, made from copper tape insulated by a thin Mylar tape, can be constructed with a considerably higher axial symmetry than that of a wire winding.

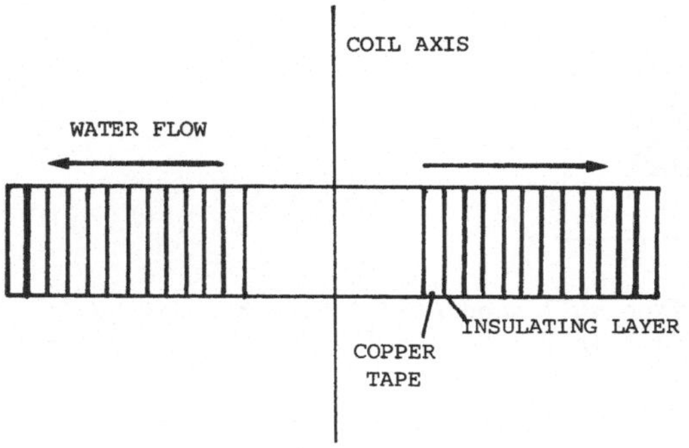

Fig.5.5. Schematic view of the cross-section of a copper tape winding with interleaved insulating tape. Cooling water flows over the exposed edges of the copper

This is particularly important in the construction of single pole-piece lenses as described below.

The manufacture of tape coils must be carried out with care if a high heat transfer coefficient is required. HERMANN et al. [5.12] have shown that if the insulation protrudes beyond the edge of the copper tape, the velocity of the cooling water relative to the tape, and hence the heat transfer coefficient, can be seriously reduced. By embedding the end surfaces of the coil in an epoxy resin and then machining the surface, a heat transfer coefficient $h = 23$ kW $m^{-2}\,{}^{o}C^{-1}$ can be obtained, i.e., some 4.5 times higher than that for a typical wire coil. Unfortunately, the authors give no indication of the water flow rate or surface velocity needed to achieve this performance. Nevertheless, this investigation and others have shown clearly that the miniaturization of tape windings is not in general limited by considerations of temperature rise, but rather by the amount of power needed to energize the coil.

Although conventional flowing-water cooling systems for electron lenses are perfectly adequate for most applications, attractive alternative methods are now available. For example, there are many advantages to be gained by allowing the cooling water to boil and removing the heat in the form of steam.

5.2 Boiling-Water Cooling Systems

If the supply of cooling water to a conventional lens is interrupted, the cooling water remaining in the lens will eventually boil, releasing latent heat at the rate of 2256 kJ per kilogram of water boiled. In order for heat transfer to take place, the windings must, of course, be at a higher temperature than that of the boiling point of water, namely 100°C at an atmospheric pressure of 101 kN/m^2 (760 Torr). However, this boiling point can be reduced if necessary to room temperature (20°C) by reducing the pressure in the cooling system to about 2.9 kN/m^2 (22 Torr). Lowering the boiling point in this way has little effect on the value of the latent heat of vaporization. The maintenance of vacuum pressures of this order of magnitude is perfectly straightforward, a simple water-jet pump being entirely adequate

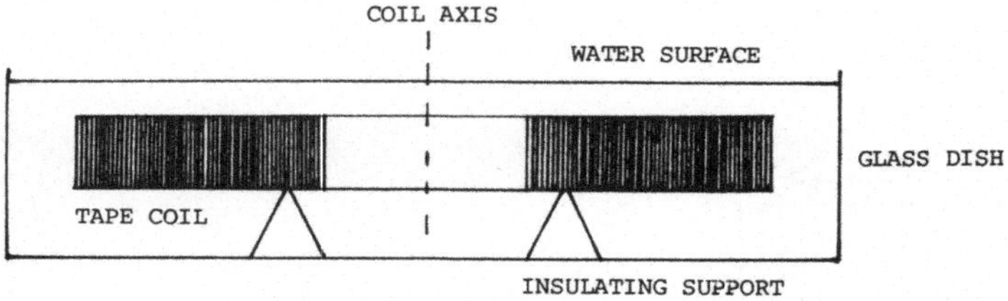

Fig.5.6. Experiment with current-carrying tape coil supported inside a glass vessel to determine the heat transfer mechanism. Coil initially immersed in distilled water

for the purpose. Although little attention has been paid in the past to this method of lens cooling, it has many advantages, and the prospect of removing large quantities of heat from a small coil maintained essentially at room temperature is very attractive from an instrumental point of view. A successful design will aim at avoiding the presence of large quantities of boiling water with its associated mechanical vibration.

The first successful boiling-water lens has been briefly described by CREWE [5.6], in connection with the final probe-forming lens of an experimental 1MeV scanning transmission electron microsope. The lens winding consists essentially of an insulated copper tape coil placed in a vacuum chamber maintained at a pressure of some 2.9 kN/m^2 (22 Torr). Water is fed into the bottom surface of the coil through a wick placed in contact with the surface of the coil; the other end of the wick dips into a reservoir of water. Steam generated by the coil passes through an exhaust tube into a water-cooled heat exchanger where it is condensed and returned to the reservoir. The surface power density is quoted by CREWE [5.6] as being about 10 W cm^{-2} with a temperature rise of only 1.5°C. This is an unusually high value of heat transfer and could possibly be in error since it is extremely difficult to evaluate the total surface area contributing to steam formation in this type of lens winding. This has been shown experimentally by the author and his colleagues in the simple experiment illustrated in Fig.5.6. A tape winding was made from a length of copper foil, 10 mm in width and 125 μm in thickness interleaved with a strip of Melinex insulation 10 μm thickness and of the same width. The coil was completely immersed in water so that boiling could take place freely from the exposed edges of the copper. A constant current was supplied to the coil so that its resistance and hence its temperature could be determined by measuring the voltage across the coil with a digital voltmeter of rapid response. A current density of some 200 A mm^{-2} was maintained in the coil and boiling soon commenced, clearly visible at both exposed surfaces of the copper. As boiling continued (at atmospheric pressure), the level of the water dropped, eventually leaving the top surface of the copper uncovered, thereby reducing the effective area of the end faces in contact with the water by a factor of two. In spite of this loss of contact area there was no measurable change in coil temperature. However, it was noticed that a mixture of steam and water droplets was issuing from the exposed surface through minute crevices between the Melinex and the copper. The temperature of the coil remained steady until the level of the water fell just below the level of the lower surface of the coil. Even at this point steam continued to issue from the coil for a short time; a few seconds later this subsided and immediately the temperature of the coil rose sharply. This experiment suggested that the main heat transfer was taking place *within* the coil and not at the end faces. This is quite plausible since the area of the end faces amounts to only 2.5% of the total area of the tape. Further confirmation of this was obtained by repeating the experiment but placing the tape coil on a pad of cotton wool immersed in water in order to impede the release of steam at the lower face. This made no appreciable difference to the heat transfer process. It therefore appears that this type of lens works by capillary action between the insulation and the conductor and the ability of steam to penetrate minute gaps which, in this case, could not amount to more than about 1 μm across. The wick does not appear to be essential to the cooling action but seems to serve a useful practical function in automatically regulating the flow of water to the coil and in inhibiting boiling at its lower surface. If this explanation is correct, one might expect that the temperature of the coil will not greatly exceed that of the steam produced, and that the temperature rise above the boiling point of water will, for a given size of coil, decrease as the tape is made thinner, since this will increase the surface area available for steam generation and hence for cooling.

This form of cooling can be explained as follows. Consider a unit length of tape of width w and thickness t, carrying a current density σ. The power W_g generated in the tape is given by [see (5.1)]

$$W_g = \rho\sigma^2 wt \quad , \tag{5.4}$$

where ρ is the resistivity of the tape. The temperature rise ΔT is given by [see (5.3)]

$$\Delta T = W_g/h\, S_A \quad , \tag{5.5}$$

$$\simeq \rho\sigma^2 t/2h \quad \text{if} \quad t \ll w \quad . \tag{5.6}$$

Experiments by the author and his colleagues [5.24] show that where boiling takes place freely at the surface, the heat transfer coefficient has, to a first approximation, the form

$$h = a\sigma^2 \quad ,$$

where a is a constant, typically of the order of $10^{-9}\ \Omega\ cm^2\,{}^\circ C^{-1}$ for current densities up to about 500 A mm^{-2}. Under these conditions (5.6) simplifies to

$$\Delta T = \rho t/2a \quad . \tag{5.7}$$

Equation (5.7) shows that the temperature rise ΔT does not depend on the current density and hence the power supplied to the coil. Similarly, the width of the tape is unimportant. However ΔT is directly proportional to the thickness of the tape, and so it is advantageous to use as thin a tape as is practicable.

As an example consider a copper tape of thickness 50 μm operating in boiling water at atmospheric pressure ($\rho \simeq 2 \times 10^{-6}$ Ω cm at 100°C). Equation (5.7) shows that $\Delta T \approx 5^{\circ}C$ irrespective of the power supplied to the coil.

The same analysis can be applied to uninsulated round wires. The result is

$$\Delta T = \rho d/2a \quad , \tag{5.8}$$

where d is the diameter of the wire.

This result suggests that, unlike the case with surface cooling by flowing water, a tape has no advantage over a wire winding provided that the winding is not steam-tight. In fact such a wire winding could be slightly better since the total surface area of a given diameter will be approximately twice as great as that of a tape winding whose thickness is equal to the diameter of the wire. In order to test this idea, a simple experiment was performed. Here, a loosely wound coil of insulated copper wire was placed on a pad of cotton wool and immersed in water. As before, a constant current was supplied, and the temperature of the coil was determined by measuring the voltage across it. It was found that current densities up to 470 A mm^{-} could be readily maintained in such wire coils; such current densities are comparable with those expected in tape coils. Even when the water level fell well below the upper surface of the coil, no change of temperature was observed and a mixture of steam and water could be seen emerging from the upper surface of the coil. Repeating these experiments at pressures below atmospheric pressure down to a pressure of some 40 Torr showed no essential differences in heat transfer, although the temperature of the steam and therefore of the coil was considerably reduced. Further work is clearly needed to establish the most convenient way of realising such high-current-density windings in practice, but it now becomes quite feasible to consider magnetic lens systems that were quite impracticable in the immediate past. Moreover, the ability to concentrate the lens excitation in a small volume opens up the possibility of designing magnetic electron lenses with a single pole piece as described below. However, before considering this type of lens, it may be useful to consider how high-current-density coils can be advantageously applied to twin pole-piece lenses. As an example consider a rotation-free miniature projector lens for a high voltage microscope.

5.3 Miniature Rotation-Free Magnetic Electron Lenses

A distinguishing feature of magnetic electron lenses is the rotation of the image with respect to the object. This angle, denoted by θ, is given by

$$\theta = \left(\frac{e}{8mU}\right)^{\frac{1}{2}} \int_{z_0}^{z_i} B \, dz \quad , \tag{5.9}$$

where e/m is the ratio of charge to mass of the electron, U is the relativistically corrected accelerating voltage, B(z) is the axial flux density distribution and z_o, z_i are the coordinates of the object and image, respectively. In a projector lens these coordinates usually lie outside the axial magnetic field and so the integral has a constant value equal to μ_o NI, where NI is the lens excitation and $\mu_o = 4\pi 10^{-7}$ H m^{-1}. Hence for a projector lens (5.9) may be written

$$\theta = \left(\frac{e}{8m}\right)^{\frac{1}{2}} \frac{NI}{U^{\frac{1}{2}}} = 0.1863 \frac{NI}{U^{\frac{1}{2}}} \quad . \tag{5.10}$$

Equation (5.10) shows that the image rotation produced by a projector lens does not depend on the form of the magnetic field distribution and hence on lens geometry. This rotation of the image with respect to the object is inconvenient in electron microscopy since the orientation of the image changes with lens current making it difficult to correlate image and object detail at different magnifications. Furthermore, fluctuations in lens current or accelerating voltage give rise to chromatic changes of rotation.

In principle, image rotation can be eliminated by arranging the exciting windings of two or more magnetic lenses in such a way that the combined excitation, and therefore the net rotation, is zero. With conventional forms of construction it is difficult to find a practical form of such a rotation-free lens. Early attempts such as those of STABENOW [5.41] in 1935 and of BECKER and WALLRAFF [5.4] in 1940 did not lead to successful designs. Interest in rotation-free lenses has, however, never ceased entirely, and KANAYA [5.18] in 1958 showed theoretically and experimentally that astigmatic and misalignment defects in an electron microscope can be reduced by the employment of a rotation-free objective-projector system. Later DER-SHVARTS and RACHKOV [5.7] in 1965 calculated the focal properties of several "twin-lenses" operating in the rotation-free mode. Such lenses were, however, never incorporated into commercial electron microscopes. In such instruments, especially those incorporating several projector lenses, it is possible to avoid gross changes in image rotation during the operation of the microscope by employing a microprocessor to optimize the choice of lens excitation at a given magnification so as to minimize overall image rotation. Unfortunately, image rotation cannot be removed entirely by this means as the inconvenient size of conventional electron microscope lenses places severe restrictions on rotation-free operation. These restrictions can be removed by the use of miniature magnetic lenses [5.15]. Such rotation-free lens units consist essentially of two adjacent magnetic lenses whose coils have equal numbers of turns. The coils are connected in series opposition so that the excitations are equal and opposite whatever the strength of the lens current. The total rotation of this "double lens" is therefore always zero.

A practical form of such a lens is shown schematically in Fig.5.7a. The main body of the lens is machined from a single piece of iron thereby eliminating unnecessary

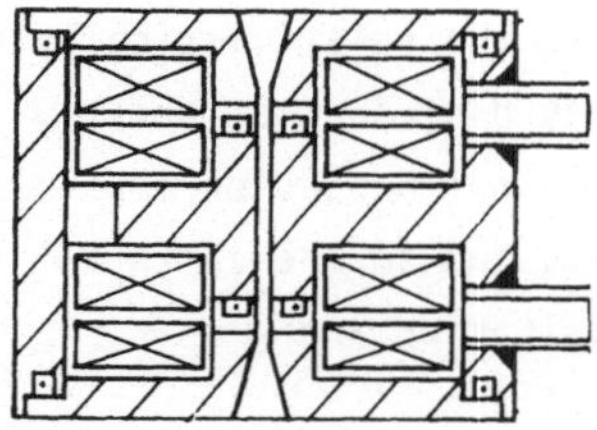

Fig.5.7a. Rotation-free 100 kV magnetic projector lens with miniature windings [5.17]. Lens bores 2 mm, lens gaps 3 mm, intergap spacing 20 mm

magnetic joints and ensuring good alignment. Miniature coils are inserted as shown in the figure and the lens connections (not shown) are led out through the water pipes. An upper and lower lid, each incorporating a pole piece of identical shape, are fitted thereby completing the magnetic circuit. A non-ferromagnetic pole-piece spacer is permanently fixed (by means of solder or in some other way) to one of the pole pieces as shown in the figure and an 'O' ring groove, machined in the other face of this spacer, completes the vacuum seal. Cooling water circulates freely past the windings at a rate of about 1 l/min as explained previously. The pole-piece design is conventional, both pole pieces having the same dimensions with a gap to bore ratio S/D = 1.5. This is, of course, not essential to the maintenance of rotation-free operation; the pole-piece design of each gap may be chosen arbitrarily since the total excitation of the "double lens" is always zero. However, it is advantageous for each gap to have the same refractive power, as explained below. The absolute size of lens will, of course, depend on the accelerating voltage of the electrons.

It should be noted that if the dimensions of a magnetic lens *and* its excitation are scaled by a factor n, the magnetic field produced at corresponding points in the original and scaled model will be the same. If the B/H properties of the iron are the same, then the magnetic flux at corresponding points in the original and scale model will also be the same, even if parts of the iron circuit are saturated. If the quantity $NI/U^{\frac{1}{2}}$ is the same in both lenses, the corresponding electron trajectories will be scaled by a factor n, so that focal lengths will also be scaled up by the same factor. Thus if the relativistically corrected accelerating voltage to be applied to the scaled up lens is U_2, and that of the original lens is U_1, then it is convenient to choose a value of n given by $n = (U_2/U_1)^{\frac{1}{2}}$. The scaling laws then enable one to predict the electron optical performance of a lens whose properties have been measured at a given voltage when it is scaled up for use at a different accelerating voltage. Image distortion on a given size of viewing screen is not affected by this scaling operation since the appropriate distortion coefficients are dimensionless.

The miniature lens shown in Fig.5.7b has been extensively tested at an accelerating voltage of 100 kV (U = 110 kV) [5.17]; it can also serve as a model for a high-voltage projector lens. The dimensions of this lens are shown in Table 5.1; its height is 40 mm with an outside diameter of 50 mm, resulting in a remarkably small weight of 0.58 kg for a 100 kV double-projector lens. The lens gap and bore

Table 5.1. Rotation-Free Projector Lens

Accelerating voltage [kV]	Lens height H [nm]	Lens outer diameter D_o [mm]	Gap S [mm]	Bore D [mm]	Lens spacing l' [mm]
100 (n=1)	40	50	3	2	20
500 (n=2.61)	104.4	130.6	7.8	5.2	52.2
1000 (n=4.26)	170.4	213	12.8	8.5	85.2

Excitation of each gap at minimum focal length [A-t]	Combined minimum focal length f_p [mm]	Top magnification at projection distance l_s = 380 mm	Distortion-free low magnification	Power needed [kW] (tape winding)	Weight [kg]
4000	0.18	2000	20	0.16	0.58
10444	0.47	766	7.7	1.09	10.25
17040	0.77	470	4.7	2.95	44.50

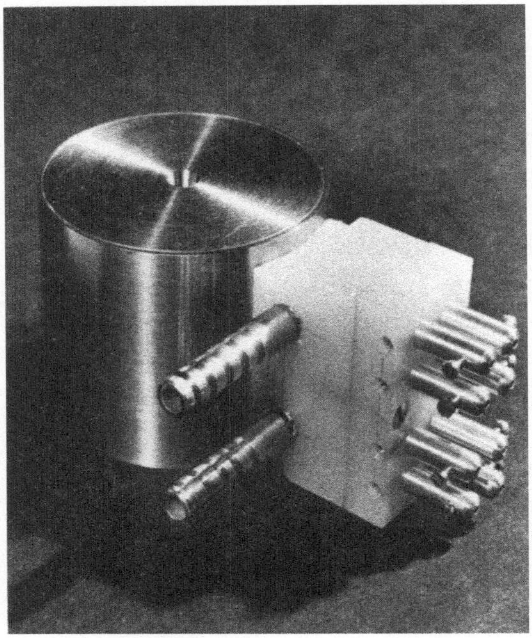

Fig.5.7b Practical realisation of the 100 kV miniature rotation-free projector lens (outside diameter 50 mm). Note the integral cooling water and individual coil connections

are of conventional design being 3 mm and 2 mm, respectively. The mean spacing l between the gaps is 20 mm. An excitation of some 4000 A-t is provided by two small coils in each gap. Each coil is 5 mm in axial thickness and has 191 turns of 24 SWG (0.56 mm diameter) insulated copper wire arranged in eight layers. The lens current at maximum excitation is thus 10.5 A, all coils being connected in series. Thus an excitation of some 8000 A-t is provided for the complete "double lens". The water flow rate is not critical but a conventional flow rate of about 1 l/min was usually maintained. The minimum focal length f_p of the combination is 0.18 mm, giving an unusually high magnification of 2000× on a photographic plate placed at a distance l_s = 380 mm away. A notable feature of a lens of this type is that it provides an image free from radial distortion at a low magnification, here about 20×. The image is, of course, free from rotation at all settings of the lens current. A careful experimental study of this lens has shown that its properties are in excellent agreement with those calculated theoretically. Operational experience with the lens over a number of years in a 100 kV electron microscope has shown the design to be very reliable in operation. Table 5.1 shows the corresponding scaled-up dimensions for operation at accelerating voltages of 500 kV (n = 2.61) and 1000 kV (n = 4.26), respectively, together with the corresponding electron optical properties. The maximum power required scales as n^2 and the weight as n^3; these figures have also been included for comparison purposes. Thus at 500 kV the top magnification is 766×; the lens dimensions are noticeably smaller than those of a standard 100 kV double-projector lens of the same magnification, the weight 10.25 kg being about half that of a standard 100 kV projector lens. The maximum power required has increased from 0.16 kW to 1.1 kW, assuming the use of a tape winding. At an accelerating voltage of 1000 kV the lens dimensions (outer diameter = 130 mm) are still comparable with those of a 100 kV double-projector lens, the weight (44.5 kg) being about double that of a standard 100 kV lens. The top magnification (470×) is still acceptably high. The power required is now just below 3 kW, within the range of presently available solid-state current supplies.

Double lenses of this type can also be used as intermediate projector lenses and are especially useful in providing rotation-free selected-area diffraction patterns [5.16]. It therefore appears perfectly feasible to make electron optical columns for high-voltage electron microscopes comparable in size and weight to those of existing 100 kV electron microscopes without introducing unduly radical changes in design and technology.

However, before turning to the subject of new types of lenses, such as single pole-piece lenses, it may be useful to refer briefly to the electron optical properties of rotation-free lenses in the low magnification, distortion-free mode, which is difficult to achieve with conventional lenses.

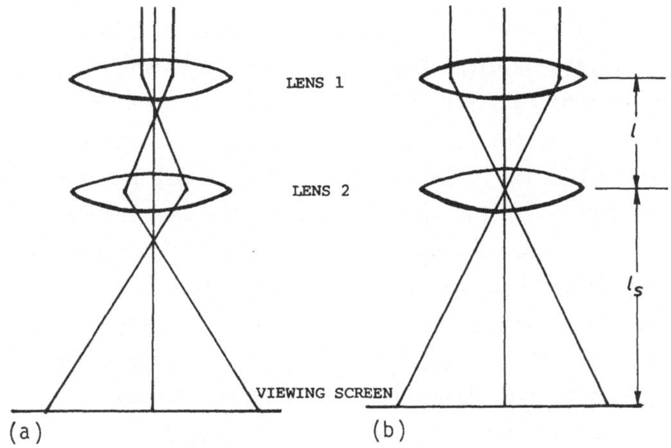

Fig.5.8a,b. Ray paths (schematic) in a rotation-free twin-lens system. (a) High-magnification mode. (b) Distortion-free low-magnification mode

5.3.1 The Distortion-Free Mode

Figure 5.8a shows schematically the double lens arranged as the final projector lens of an electron microscope. The magnification M at the fluorescent screen is given by

$$M = l_s l/f^2 - (2l_s + l)/f + 1 \tag{5.11a}$$

$$= l_s(l - 2f)/f^2 \text{ at high magnification} \quad . \tag{5.11b}$$

The projector to screen distance l_s, typically about 380 mm, is generally determined by the maximum distortion that is tolerable at the fluorescent screen. With a miniature lens construction there is a wide freedom of choice of the intergap spacing l. Thus a high magnification can be produced from the double-lens unit without the necessity for striving after very small values of focal length for each lens gap with the attendant mechanical alignment difficulties. An additional benefit of the doublet is that a distortion-free image will be produced when the focal length of the first lens is equal to the intergap spacing l as shown in Fig.5.8b. Here the rays from the upper lens pass through the centre of the lower lens; its refractive power is thus zero, and so it does not contribute to the total magnification. The magnification at the fluorescent screen is therefore given by

$$M = l_s/l \quad . \tag{5.12}$$

This magnification can be conveniently set to a suitably low value simply by choosing the appropriate value of the interlens spacing l. Although the lower lens does not contribute to the total magnification, it nevertheless serves a useful function in correcting radial distortion, since marginal rays from the upper lens will not, in general, pass through the paraxial focal point. Such rays will be bent back towards the lens axis in such a way that the radial distortion at the final screen is zero.

This method of correcting radial distortion was originally proposed and realized in an electron microscope projector lens by HILLIER [5.13] in 1946. However, because of the restrictions at that time in lens construction techniques, a serious difficulty arose. In HILLIER's arrangement the two lens-gaps were energized in series by a single coil so that a rotation-free mode of operation was not possible. This meant that although the radial distortion of the upper projector lens was corrected, the *spiral* distortion was noticeably increased since the spiral distortion produced by the correcting lens was added to that produced by the lens under correction. It may not be generally realized that spiral distortion can be quite noticeable at weak lens excitations in spite of the fact that the spiral distortion coefficient D_{sp} (denoted by e in Chaps.1,3) tends to zero under these conditions. The reason for this lies in the fact that the spiral distortion at the image depends on the *product* $D_{sp} \cdot f_p^2$, where f_p is the projector focal length [5.27]. It can be shown that under the conditions set out above, the rotation-free mode of operation can help considerably in reducing spiral distortion, but a complete correction is not possible. The correction improves as the two lenses are moved closer together, reaching a maximum of 50% correction when the two lens gaps are immediately adjacent to each other. This cannot readily be achieved in practice as partial cancellation of the lens fields would occur. It would therefore seem that even in the rotation-free mode, HILLIER's method of correction does not offer any appreciable advantages over a single projector lens working at maximum magnification, i.e., at minimum focal length, where the radial distortion is, in fact, very small. However, it can be employed very usefully at *low* magnification to correct completely what would otherwise be severe radial distortion. The partial correction of spiral distortion in the rotation-free mode can yield acceptably low values (around 3%) for low-magnification viewing.

It should perhaps be mentioned that if the low-magnification image is to fill the screen of radius ρ, the lens bore D should be chosen (Fig.5.8) so that

$$D = 2\rho \ell / \ell_s \quad . \tag{5.13}$$

In general the bore of such a double lens will be larger than that of a conventional projector lens. This is not a disadvantage since the magnification of a double lens does not depend critically on the bore diameter, and a wide lens bore means that a vacuum liner can be conveniently inserted into the bore, leading to an improvement in the vacuum in the electron optical column [5.16].

5.4 Iron-Free or Partly Shrouded Magnetic Lenses

The high-current densities that are now possible in lens windings make it possible to consider entirely different types of lenses. The best starting point for considering these is probably the iron-free solenoid shown schematically in Fig.5.9. Here

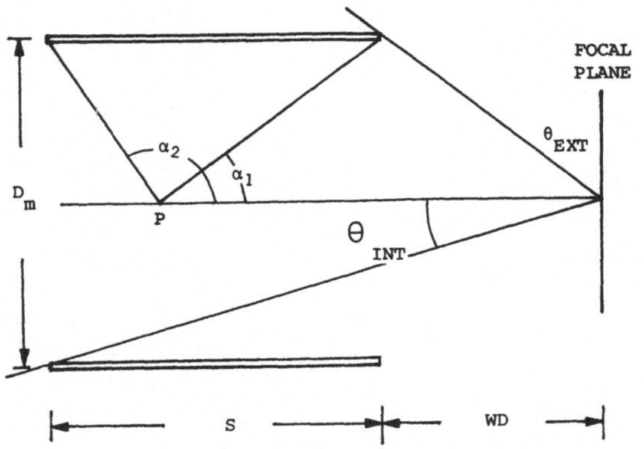

Fig.5.9. Important geometrical parameters in a thin iron-free solenoid

a thin solenoid of length S and mean diameter D_m produces a magnetic field along the z axis whose flux density B at a point P on the z axis is given, according to the Biot-Savart law, by

$$B = \mu_0 \frac{NI}{2S} (\cos\alpha_1 - \cos\alpha_2) \quad , \tag{5.14}$$

where α_1 and α_2 are the angles defined in Fig.5.9.

A parallel beam of electrons incident on the solenoid will be brought to a focus in the focal plane located at a "working distance" WD from the coil. Because of the small size of the solenoid compared with a conventional electron lens, X rays and other radiation emerging from a specimen located in the vicinity of the focal plane can pass outside the lens for exit angles greater that θ_{ext} or can pass through the lens for angles less than θ_{int}. The electron optical properties of a lens of this type can be calculated from the field distribution given by (5.14). However, according to the work of DURANDEAU [5.8], the axial field distribution of such a solenoid is approximately the same as that of an iron pole-piece lens of symmetrical construction having the same gap length S but of a diameter that is greater than that of the solenoid by a factor 3/2. Since it is known for such iron pole-piece lenses that the spherical aberration coefficient increases rapidly as the ratio S/D of gap to bore increases, one can infer that for a solenoid of given mean diameter D_m, the performance could be improved by making the length S as short as possible.

This is demonstrated in Fig.5.10, which shows the calculated values of the ratio of spherical aberration to working distance (C_s/WD) as a function of the ratio (D_m/S) of mean diameter to length of the solenoid, for two values of the take-off angle θ_{ext}. Figure 5.10 shows clearly that as the coil gets longer ($D_m/S < 0.5$) a rapid increase takes place in the spherical aberration coefficient. On the other hand, there is no great advantage to be gained by making the coil shorter than the mean diameter ($D_m/S > 1$), especially if one remembers that for a constant current density in the windings the power needed to produce a given number of ampere-turns increases inversely as the length of the coil. The minimum permissible length,

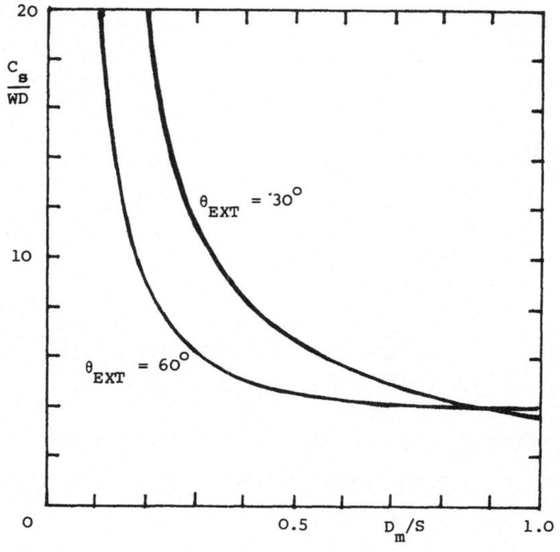

Fig.5.10. Variation of spherical aberration coefficient to working distance (C_S/WD) in an iron-free solenoid as a function of the ratio of mean diameter to coil length (D_m/S) for two take-off angles θ_{EXT}

and hence the performance of such a lens, is thus determined almost entirely by technological considerations. Perhaps the most serious of these is the difficulty of achieving adequate axial symmetry in a winding of small diameter, through which electrons must pass at relatively large off-axis distances.

5.4.1 Conical Lenses

This consideration suggests that perhaps a conical coil might have an advantage since the mean diameter of the turns would be increased without a substantial increase in the obstruction to radiation from the specimen presented by the coil. Figure 5.11 shows the geometrical arrangement and terminology. The axial flux density B at a point P on the axis [5.3] is given by

$$B = \mu_0 \frac{NI}{2S} \cos\alpha_c \left[\cos(\alpha_c - \alpha_1) - \cos(\alpha_c - \alpha_2) + \sin^2\alpha_c \ln\left\{ \frac{[1 - \cos(\alpha_c + \alpha_2)]}{[1 - \cos(\alpha_c + \alpha_1)]} \frac{D_2 \sin\alpha_1}{D_1 \sin\alpha_2} \right\} \right] , \tag{5.15a}$$

where α_c is the cone semi-angle and the mean diameter $D_m = (D_1 + D_2)/2$. Here the thickness of the winding is assumed to be small compared with the mean diameter of the coil. In practice this is an adequate representation for coils in which the depth of winding is less than one tenth of the mean diameter of the coil. Figure 5.11 shows the calculated field distribution for $\alpha_c = 41^\circ$ and $D_2/D_1 = 8$. The field distribution is now asymmetrical, the flux density rising slowly in the neighbourhood of the turns of large diameter and falling off rapidly outside the coil in the vicinity of the turns of small diameter. In this figure the axial flux density is given in terms of the quantity B/B_M, where $B_M = \mu_0\, NI/S$, the flux density that oc-

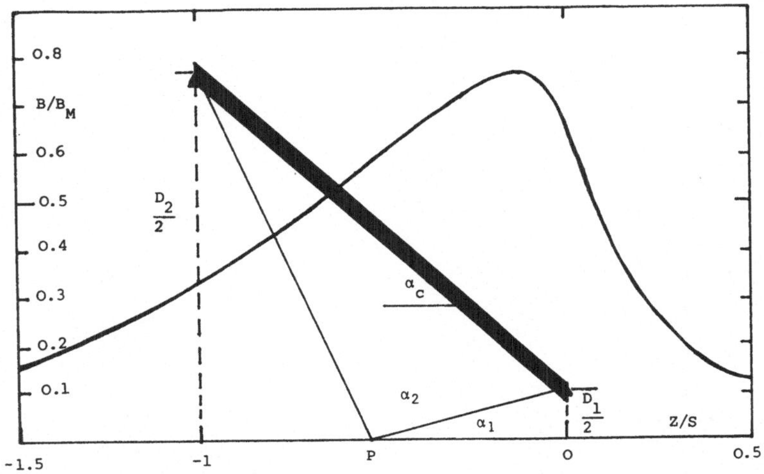

Fig.5.11. Important geometrical parameters of a conical iron-free coil and a typical asymmetrical axial flux density distribution ($\alpha_c = 41^\circ$ and $D_2/D_1 = 8$)

curs in an infinitely long solenoid with a field strength of NI/S [A-t m^{-1}]. Because of the asymmetry of the field distribution, conical lenses will have different electron optical properties if the lens is reversed in relation to the electron beam. In general terms the behaviour of such lenses is analogous to that of asymmetrical magnetic pole-piece lenses, but up to now no mathematical expression has been found enabling one to relate the properties of the two types of lenses in a simple manner. Consequently, it is necessary to calculate the properties of conical lenses directly from the axial field distribution. The construction of a conical lens is comparable in difficulty to that of a solenoid lens of the same mean diameter. As with parallel-sided solenoids, calculations indicate that the spherical aberration coefficient of the lens improves as the distance S is made smaller. Under these conditions a high internal take-off angle θ_{int} can be obtained. It therefore appears that the long solenoid and the conical solenoid do not admit of an optimum design and of necessity involve a compromise between electron optical performance and ease of construction. They may nevertheless have useful applications in particular circumstances, especially in the electron probe X-ray microanalyser. In searching for the best lens we are therefore led to consider in more detail a lens structure for which $S \to 0$ and $\alpha_c \to \pi/2$. Both conditions are met in a plane helix of vanishingly small thickness.

5.4.2 Thin Helical (Pancake) Lenses

Figure 5.12 shows a thin helix (S = 0) of internal diameter D_1 and external diameter D_2. Considering this as a conical coil with $\alpha_c = \pi/2$, the axial flux density distribution B at a point P on the axis could be found from (5.15a). However, it is preferable to redefine α_1 and α_2 as being the respective angles that the lines joining the ends of the coil to the point P make with a line drawn *parallel* to the

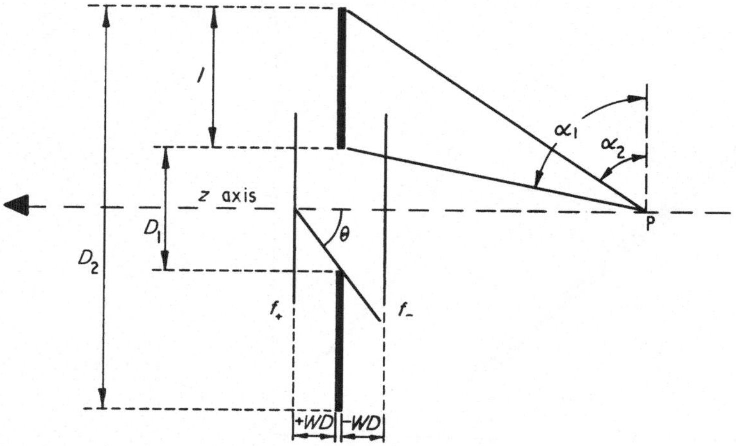

Fig.5.12. Important geometrical parameters of a thin helical (pancake) coil ($\alpha_c = \pi/2$). (Courtesy of the Institute of Physics, London)

plane of the winding. The axial flux density B can then be written as [5.33]

$$B = \mu_0 \frac{NI}{2l}\left[\cos\alpha_1 - \cos\alpha_2 + \ln\left(\frac{1+\cos\alpha_2}{1+\cos\alpha_1}\right)\frac{D_2}{D_1}\frac{\cos\alpha_1}{\cos\alpha_2}\right] , \tag{5.15b}$$

where $l = (D_2 - D_1)/2$. This form of the equation brings out more clearly the fact that in a thin helical lens the length l is closely analogous to the length S of a thin parallel-sided solenoid. However, for a given value of l, it is possible in a helical lens to increase the maximum axial flux density B_m by suitably increasing the ratio D_2/D_1 since B_m is given by

$$B_m = \mu_0 \frac{NI}{l}\left[\ln\left(\frac{D_2}{D_1}\right)^{\frac{1}{2}}\right] . \tag{5.16}$$

The term outside the brackets is the flux density in a uniform solenoid of length l but negligible diameter; the term in the brackets may be regarded as a peak-flux magnification factor and is tabulated in Table 5.2. Thus if $D_2/D_1 > 7.4$, for a given excitation, B_m will be greater than that produced by a long solenoid of length l. The peak flux density on the axis may be increased without limit by increasing the ratio D_2/D_1. Because of the logarithmic term, the magnification factor increases rapidly at first, then more slowly, reaching a value of 12.9 for $D_2/D_1 = 10^6$.

The production of a coil with a large D_2/D_1 ratio will in practice be limited mainly by technological considerations; nevertheless, this type of lens has considerable theoretical interest since it appears to be the optimum form for an iron-free lens of least spherical aberration. This arises from the fact that as the peak value of the axial flux density is increased the "half-width" 2h (defined here as the width of the axial field distribution between the points at which $B = B_m/2$) must decrease since the total area ($\int B\,dz$) under the field distribution curve is constant for a given excitation. Table 5.3 shows the relative half-width ($2h/D_m$)

Table 5.2. Peak-flux magnification factor $\ln(D_2/D_1)^{\frac{1}{2}}$ as a function of D_2/D_1 for flat helical lenses

D_2/D_1	5	10	25	100	200	10^3	10^6
$\ln(D_2/D_1)^{\frac{1}{2}}$	0.80	1.15	1.60	2.30	2.65	3.45	12.9

Table 5.3. Relative half-width ($2h/D_m$) values as a function of D_2/D_1 for flat helical lenses

D_2/D_1	5	10	25	100	200
2h/L	0.56	0.47	0.29	0.15	0.09

values as a function of the ratio D_2/D_1. It should be noted that for $D_2/D_1 > 10$ the half-widths of these "pancake coils" are considerably smaller than those of all other forms of iron-free magnetic lenses. Thus for $D_2/D_1 = 200$ the relative half-width $2h/D_m$ is less than 0.1. The ratio $2h/D_m$ is a good guide to the relative value of the spherical aberration coefficient. This means that for a given value of 1, one would expect the aberration coefficient (C_s) of a helical lens of high D_2/D_1 ratio to be an order of magnitude smaller than that of a long solenoid. It is therefore of considerable theoretical interest to determine the value of C_s as $D_2/D_1 \to \infty$.

5.4.3 The Ultimate Performance of Pancake Lenses

Pancake lenses of vanishingly small thickness would require an infinite amount of power to excite them. From a practical point of view it is therefore desirable to make the ratio S/D_m as large as possible without seriously impairing lens performance. Calculations show that the relative half-width $2h/D_m$ and the peak axial field do not change appreciably over the range $0 < S/D_m < 0.1$, so that one may suppose that the favourable electron optical properties associated with a helical coil of vanishing thickness would be retained in coils for which $S/D_m < 0.1$.

This question has been investigated by MARAI [5.25], who computed the field distribution from the Biot-Savart law for coils of different widths S and ratios D_2/D_1. His results show that the spherical aberration coefficient of such a lens actually goes through a shallow minimum at a value of S/D_m of 0.1. However, for larger values of S/D_m the spherical aberration rises rapidly, and so it is important not to exceed this value. At the optimum value of $S/D_m = 0.1$, as the ratio D_2/D_1 is increased, the universal spherical aberration parameter $C_s\ B_{max}/U^{\frac{1}{2}}$ falls rapidly to a value of 3.05×10^{-6} m T $V^{-\frac{1}{2}}$ at a value of $D_2/D_1 = 20$. For values of $D_2/D_1 > 100$ the curve levels off at a value close to 2.9×10^{-6}, the calculated value of $C_s\ B_{max}/U^{\frac{1}{2}}$ at $D_2/D_1 = 1000$ being 2.87×10^{-6}. This should be compared with the theoretical limit of 2.23×10^{-6} m T $V^{-\frac{1}{2}}$ calculated by TRETNER [5.42] and MOSES [5.29]. To put the matter into perspective, consider a lens with $D_2/D_1 = 1000$ for which the minimum value of C_s occurs at an excitation parameter $NI/U^{\frac{1}{2}}$ value of about 30. The corresponding value of the peak axial flux density B_{max} is given by (5.16) as

$$B_{max} = 3.45 \mu_0 NI/D_m$$

from which it follows that $C_s/D_m = 0.03$, an extremely low value of spherical aberration coefficient with respect to the mean diameter of the coil. Thus, for example, a coil of 30 mm outside diameter and 1.5 mm thickness would have a spherical aberration coefficient of only 0.5 mm at an excitation parameter $NI/U^{\frac{1}{2}} = 30$.

At an accelerating voltage of one million, such a coil would require an excitation of about 42000 A-t and hence a current density of some 1000 A-t mm^{-2}, which represents the present technical limit of performance of boiling-water-cooled coils. The resolution δ_0 of this lens as an objective of a million-volt electron microscope can be calculated from the relationship

$$\delta_0 = 0.7 C_s^{\frac{1}{4}}\lambda^{\frac{1}{2}} \quad , \tag{5.17}$$

where λ is the electron wavelength and δ_0 represents the spatial frequency in the object for which the transfer function of the lens attains its first zero. Equation (5.17) yields a value of $\delta_0 = 0.09$ nm. If the lens dimensions are scaled up by a factor of two in order to ease constructional problems, δ_0 becomes 0.11 nm, still an extremely favourable value compared with what can be achieved with conventional iron-pole-piece lenses.

It should also be clear from the above that when considering iron-free lenses, there is no inherent advantage in *size* from the use of superconducting windings rather than water-cooled copper windings. The chief advantages of superconducting lenses appear to lie in the possibility of employing superconducting magnetic screens or special pole pieces made of materials such as holmium, which require a a low temperature for developing high permeability.

5.4.4 Minimization of Chromatic Aberration

It should perhaps be pointed out that, as with conventional magnetic lenses, it is not possible with any non-conventional lens so far investigated to minimize, simultaneously, both chromatic *and* spherical aberration. The difficulty appears to be fundamental. For example, a long solenoid of vanishingly small diameter corresponds closely to the lens of minimum chromatic aberration [5.33], whereas a helix of large ratio of outer to inner diameter and of vanishingly small width corresponds closely to the lens of least spherical aberration. Unfortunately, the long solenoid has a C_s value approaching infinity, whereas the C_c value of the helix is of the same order as its focal length.

For this reason flat helical lenses are usually to be preferred where low aberrations are important. What is perhaps of more importance, a helical lens structure occupies a relatively short axial distance. This means, among other things,

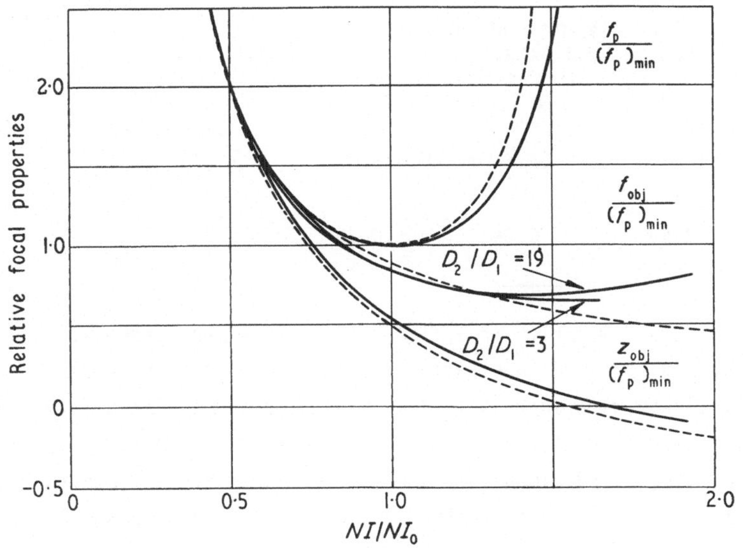

Fig.5.13. Generalized focal properties of flat helical lenses (solid lines) for various ratios D_2/D_1 of outer to inner diameter. Dashed lines indicate the corresponding properties of a long solenoid. (Courtesy of the Institute of Physics, London)

that the electrons to be imaged are relatively far removed from the windings themselves and hence from local perturbations of the magnetic field caused by mechanical imperfections. The mechanical tolerances in such lenses are thus greatly relaxed compared with those of a long solenoid. Furthermore, there is more free space available round the specimen since this need not be immersed in the lens structure. The pancake lens can therefore form the basis of new types of electron lenses for both electron probe-forming instruments and for transmission electron microscopes. It has already been pointed out that the electron optical behaviour of long solenoids is closely analogous to that of twin pole-piece lenses. This is not, in general, true of flat helical lenses; this is illustrated in Fig.5.13, which shows their chief focal properties, such as objective focal length f_{obj}, projector focal length f_p, and focal distance z_{obj}. Dashed lines indicate the corresponding focal properties of long solenoids. For convenience of comparison, the curves are normalized in terms of minimum focal length $(f_p)_{min}$ and excitation NI_0; the relative quantities $(f_p/D_m)_{min}$ and $NI_0/U^{\frac{1}{2}}$ are tabulated in Table 5.4. It may be seen that the curves are approximately "universal" for relative excitations $NI/NI_0 < 1.5$ and apply equally well to long solenoids and flat helical lenses. However, in the excitation region $NI/NI_0 > 1.5$, which is of greatest practical importance, the curves for flat helical lenses cease to be universal and begin to diverge in character from those of long solenoids. In particular, for values of $D_2/D_1 > 4$, at these excitations the objective focal length remains almost constant or increases only slowly with increasing excitation. This behaviour is indeed quite different from that of conventional pole-piece lenses where the objective focal length falls continuously

Table 5.4. Minimum relative focal length and corresponding excitation for a range of flat helical lenses

D_2/D_1	4	5.67	9	19
$NI_0/U^{1/2}$	16.2	16.4	16.6	16.8
$\left(\frac{f_p}{D_m}\right)_{min}$	0.44	0.43	0.39	0.35

as the excitation is increased. Similarly, in a flat helical lens the spherical aberration coefficient and the chromatic aberration coefficient will also remain constant or increase only slowly with increasing lens excitation. The distinctive features of such lenses are therefore fully realized only for excitation parameters $NI/NI_0 \geqq 1.5$. Under these conditions only one half of the field is used in forming the image itself, whether it be an electron probe or an image of a specimen in a transmission electron microscope. It therefore seems worthwhile to suppress the unwanted half of the field thereby saving half of the lens ampere-turns. This can be done, in principle, by halving the axial extent of the coil and placing it in contact with a ferromagnetic plate of infinite extent and high permeability; this leads to the idea of an iron-shrouded pancake lens.

5.5 Iron-Shrouded Pancake Lenses

This type of lens is shown schematically in Fig.5.14a. If the iron plate has high permeability, lines of force enter the plate at right angles and the field distribution to the right of the plate shown in Fig.5.14c (solid line) is identical with that of an iron-free coil of twice the axial extent. In practice, the diameter of the plate need not greatly exceed that of the maximum diameter of the coil. In a practical lens, the iron plate also serves a second useful purpose; in the absence of the iron plate, a small axial movement of the coil will cause the entire field distribution to be displaced by the same amount along the axis. However, with the iron plate in position a small axial movement of the coil will have a negligible effect on the field distribution since the magnetic "image" of the coil in the iron plate will move axially in the opposite direction, thereby exerting a compensating effect on the movement of the coil. In a practical magnetic lens some form of magnetic screening from transverse AC and DC fields is essential. The iron backing plate cannot provide this; for this purpose however, an iron cylinder placed round the coil can be very effective. The lines of magnetic flux produced by the coil enter such a cylinder approximately at right angles. It therefore appears plausible that the presence of an iron cylinder, of internal diameter equal to that of the external diameter of the coil, would not seriously disturb the existing

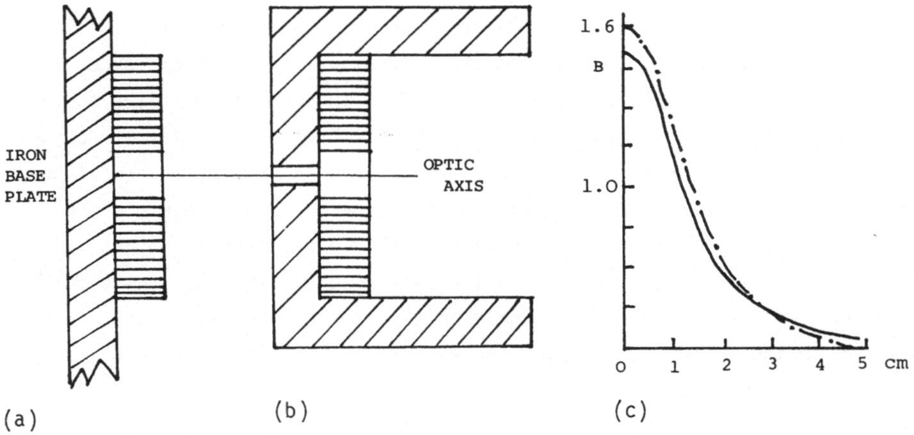

Fig.5.14a,b. The iron-shrouded pancake lens: (a) schematic principle; (b) practical realization for a high-voltage objective; (c) axial field distribution due to the coil (solid line) and of the complete lens (dashed line)

magnetic field distribution, especially along the axis. Experiment and calculation both show that this is indeed the case. The peak axial flux density appears to increase slightly, possibly by two or three percent depending on the geometry of the coil, while the half-width of the field distribution hardly changes at all. It further seems likely that the presence of such a cylinder will reduce the effects on the axial field distribution of asymmetries in the coil windings. Combining the two types of screening leads to the design shown in Fig.5.14b in which the iron circuit has little or no effect on the essential shape of the field distribution from the coil but provides magnetic screening against external fields and also helps to reduce the effect of lens imperfections caused by axial movement of the coil and asymmetries in the windings.

The design shown in Fig.5.14b is a possible form of a lens for a million-volt electron microscope proposed by MULVEY and NEWMAN [5.32]. This lens, used as an objective, has a focal length of about 10 mm, but can also serve in unmodified form as a projector lens of the same projector focal length if required. This arises from the unusual feature of pancake lenses that the minimum projector focal length coincides with the minimum objective focal length.

The design shown here is not intended to show the limit of performance of such lenses but to indicate that good electron optical properties, comparable with those of the best conventional lenses, may be achieved even at modest flux densities and with simple construction. The heart of the design is a tape coil of width 10 mm, outside diameter 50 mm, and inside diameter 10 mm, capable of providing the 20000 A-t needed at an accelerating voltage of 1000 kV. The thickness of the iron backing plate and screening cylinder circuit is 10 mm giving an outside diameter of 70 mm, an inner diameter of 50 mm, and a total axial length of 60 mm, unusually small dimensions for an objective operating at 1000 kV. For minimum objective focal length and zero pre-field, the specimen S must be placed in contact with the iron backing plate. This means that the entire flux distribution is employed in forming the image. There is therefore no pre-field in this position of the specimen ($z = 0$), the corresponding excitation being 17300 A-t. The specimen may of course be moved, if required, to an axial position further into the field; this will not result in any appreciable change of focal length or aberration. However, a pre-field will then arise, and hence more excitation will be required.

A small hole is, of course, required in the backing plate to admit the illuminating electron beam. The effect of this hole on the axial field distribution has been neglected in the following calculations since it will mainly affect the illuminating conditions rather than the image formation. It should perhaps be remarked that the determination of the axial field distribution in the neighbourhood of such a small hole in a region of high flux density is difficult both experimentally and theoretically, even for such powerful techniques as the finite-element method.

Figure 5.14c (broken line) shows the axial field distribution, kindly calculated by MUNRO [5.34] using the finite-element method. This should be compared with the other curve (solid line), which shows the axial flux density distribution calculated for a coil assumed to be in contact with an iron sheet of infinite extent and infinite permeability. It should perhaps be mentioned that for simplicity the axial field distribution for the configuration of Fig.5.14b was carried out ignoring the presence of the axial hole in the backing plate and for a constant permeability ($\mu_r = 1000$) in the iron. Both curves show essentially the same form of field distribution and focal properties. The finite-element method indicates that the peak field occurs in the region of the iron circuit in contact with the outer turns of the tape windings. The maximum field on the axis is of the order of 1.5 T at an excitation of 20000 A-t. This confirms that saturation effects are absent in the iron circuit, justifying the assumption of constant permeability in the calculation. However, the value of 1.59 T given by the finite-element calculation is likely to be slightly on the high side since a fairly coarse mesh was used to represent the iron circuit and the full iteration facilities of the programme, taking into account the local variations of permeability, were not employed on this occasion. The electron optical properties calculated independently from the curves in Fig.5.14c were in fact in good agreement. For example, for a specimen placed in the plane of maximum flux density ($z = 0$), the following properties were obtained: objective focal length $f_{obj} = 8$ mm, chromatic aberration coefficient $C_c = 5$ mm, and spherical aberration constant $C_s = 2.5$ mm for an excitation of 17300 A-t. This corresponds to a value for the resolution parameter $\delta_0 = 0.7 \times (C_s\lambda^3)^{1/4} = 1.4$ Å (0.14 nm). These excellent aberration figures are accompanied by a considerable reduction in lens size compared with conventional lenses, the calculated weight of the lens in Fig.5.14 being in the region of 2 kg.

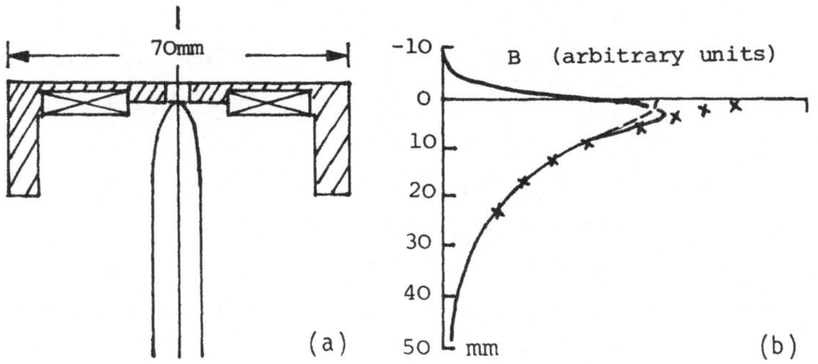

Fig.5.15. (a) Improved iron-shrouded pancake lens, 70 mm outer diameter, as a 1000 kV objective. Focal length 10 mm, C_s = 3.2 mm, δ_0 = 0.15 nm. (b) Measured axial field distribution (solid line). Axial field due to the coil (dashed line). Exponential field distribution (crosses)

A possible objection to this design is that inevitable asymmetries in the tape winding could lead to undue astigmatism. Direct measurements on tape windings with the aid of the Rank, Taylor, Hobson "Talyrond" measuring instrument have indicated that the two-fold asymmetry associated with this type of winding would be expected to limit the resolution of an electron microscope to no worse than 2 nm, an amount that can readily be corrected by the normal astigmatism control. Nevertheless, attention has been given to these points in the experimental lens shown in Fig. 5.15a. This has the same external diameter (70 mm), but the coil is shorter in axial extent (5 mm) and larger in internal diameter (20 mm). Both these measures help considerably in reducing possible axial astigmatism. In addition, the base-plate has been machined to form a small "snout" extending 1 mm into the internal bore of the coil. This snout partly screens the region around the specimen from the effect of asymmetries in the inner turns of the coil and raises the peak field in order to compensate for the reduction in field strength caused by the 5 mm bore in the iron snout. These points are illustrated in the axial field distribution shown in Fig.5.15b. The solid curve represents the axial field distribution obtained by means of Hall probe measurements in an experimental lens that had been scaled up by a factor of two. This scaling up was necessary in order to obtain an accurate field distribution in the region of the hole in the backing plate. The dashed line shows the calculated axial flux density distribution from the coil itself, assuming it to be in contact with an infinite backing plate placed at z = 0. Figure 5.15b shows that the combined effect of the 1 mm snout and the 5 mm bore is to push the peak axial flux density away from the face of the snout, a useful feature in an objective lens as it allows more freedom in the manipulation of the specimen.

Calculation of the electron optical properties of this lens shows that the minimum projector focal length and minimum objective focal length are both equal to 10 mm. The chromatic aberration coefficient C_c = 6 mm and the spherical aberration coefficient C_s = 3.2 mm. Thus at an accelerating voltage of 1000 kV, $\delta_0 = 0.7\ (C_s\lambda^3)^{1/4} = 1.5$ Å (0.15 nm). This is a remarkably low value when one considers that at the 20000 A-t excitation required, the peak axial flux density is only 1.4 T. At 100 kV the corresponding excitation would be 4300 A-t giving a peak axial flux density of 0.4 T. Of course at this lower voltage it would be feasible and desirable to scale down this lens by a factor of two with a corresponding reduction in focal length and aberration coefficients. Such lenses clearly offer the possibility of a wide solid angle around the specimen whether in the TEM or STEM mode for the insertion of detectors of various types.

5.5.1 The Exponential Field Model

In order to understand the general behaviour of lenses of this type it would clearly be convenient if there were an analytical field distribution which closely resembled that of the actual field of the lens. In Fig.5.15b, the crosses indicate the exponential field distribution according to the law

$$B(z) = B_{max} \exp(-az) \quad , \tag{5.18}$$

where a is a constant. This is the little-known exponential field model of GLASER [5.11]. It is convenient to express this axial field distribution in the form

$$B(z) = B_{max} \exp[-(\ln 2/d)z] \quad , \tag{5.19}$$

where d is the half-width defined as the axial distance from the position of maximum field B_{max} to the point where the field has fallen to half this value. The resulting focal properties [5.25,26] are shown in Fig.5.16. The exponential field distribution has the interesting property that the minimum projector focal length (f_p = 1.157d) is equal to the minimum objective focal length. The minimum projector focal length occurs at an excitation parameter $NI/U^{1/2} = 13$. However, on increasing the lens excitation, the objective focal length remains constant as the specimen plane z_{obj} moves to larger values of z. The same applies to the aberration coefficients C_c and C_s. Thus for $NI/U^{1/2} \geq 13$, $C_c = 0.72d$ ($C_c/f_{obj} = 0.63$) and $C_s = 0.36d$ ($C_s/f_{obj} = 0.315$).

In an experimental lens of this type, if the peak axial flux density and the half-width d can be measured experimentally, the exponential field model can be used to provide a useful first approximation to the focal properties. A lens of this type can also be used as a condenser-objective lens. This is illustrated in Fig.5.17, which shows the telescopic ray path which occurs when $NI/U^{1/2} = 20.6$ for the exponential field. By placing a specimen at a distance z = 0.7d along the axis, as indicated in the figure, a strong pre-field is created followed by an imaging field of low aberration in a similar manner to that achieved by the condenser-objective of Riecke and Ruska. There are two possible directions for the illuminating beam, the preferred direction being determined by the operational requirements of a particular instrument. For example, if the illuminating beam proceeds

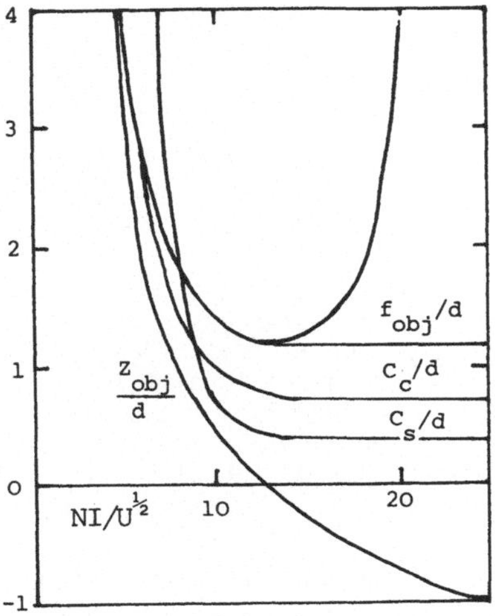

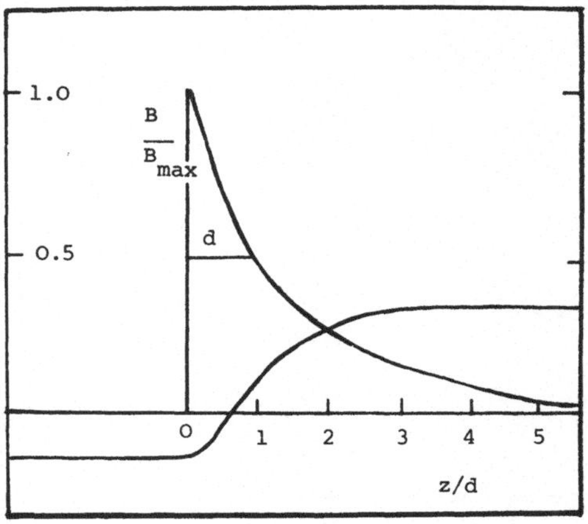

← Fig.5.16 ↑ Fig.5.17

Fig.5.16. Focal properties of the exponential field distribution $B = B_{max} \exp[-(\ln 2/d)z]$

Fig.5.17. Telescopic ray path at $NI/U^{\frac{1}{2}} = 20.6$ in the exponential field distribution. Focal point at $z = 0.7d$

in the positive direction of z towards the specimen as shown in Fig.5.17, low *spherical* aberration will be achieved in the image, whereas low *chromatic* aberration will be achieved in the probe-forming part of the pre-field which arises between $z = 0$ and $z = 0.7d$. Low spherical aberration will be produced in the probe if the direction of the rays is reversed. It is, unfortunately, not possible to use this model to calculate the spherical aberration coefficient for a ray passing through the lens in this direction since the model does not take account of the presence of the small axial hole that is needed to allow the passage of the electron beam. In this case the spherical aberration coefficient must be calculated from the actual field distribution in a particular lens. The optimum direction of the electron beam in a condenser-objective lens of this type must therefore be determined in the light of operational requirements, especially if the lens has to be capable of working both in the TEM and STEM modes. Fortunately, this consideration does not apply to the spiral distortion coefficient D_{sp} which is important in the design of projector lenses.

Figure 5.18 shows the dimensionless coefficient d^2D_{sp} as a function of the excitation parameter $NI/U^{\frac{1}{2}}$ for the two possible directions of the exponential field distribution with respect to the incoming beam. The projector focal length is not affected by reversing the direction of the lens in this way, but the spiral distortion coefficient is greatly increased, as shown in Fig.5.18, if the incoming electron beam is incident on the steeply rising part of the field distribution, compared with the values obtained when it is incident on the slowly rising part. In this more favourable orientation, the spiral distortion coefficient is appreci-

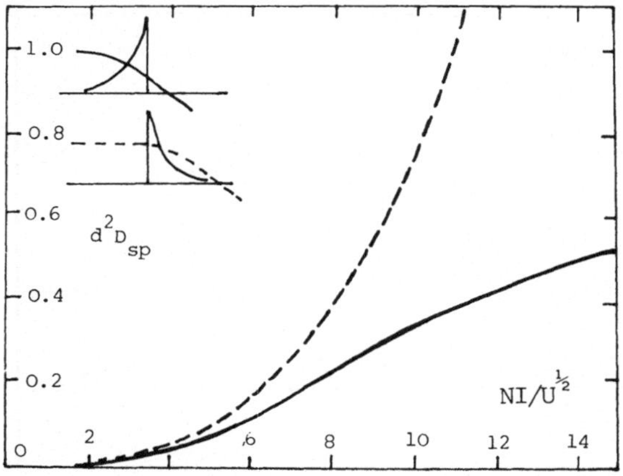

Fig.5.18. Dimensionless spiral distortion coefficient d^2D_{sp} for the exponential field distribution as a function of excitation parameter $NI/U^{\frac{1}{2}}$ and the two directions of incidence of a parallel ray

ably lower than that of conventional electron lenses. This result can be understood from a consideration of the electron trajectories through the lens field. For example, in the favourable direction, when the radial height of the electron is high, the field strength B(z) is low, and when the axial field strength is high, the radial height of the electron is low. Both of these effects lead to low values of the aberration coefficient. The reverse is true for rays proceeding in the other direction. It therefore appears that the performance of projector lenses, limited essentially by spiral distortion [5.27], could be considerably improved by the employment of iron-shrouded pancake lenses set up in the correct orientation with respect to the imaging beam. It should perhaps be pointed out that the distortion $\Delta\rho/\rho$ on a fluorescent screen of radius ρ is determined, not so much by the distortion coefficient D_{sp}, but by the quality factor Q, where $Q = (D_{sp})^{\frac{1}{2}} f_p$ and f_p is the projector focal length. Single pole-piece lenses can have significantly lower Q values than those of conventional lenses.

5.5.2 The Single Pole-Piece Lens

The advantages of concentrating the magnetic field in a lens by means of a pair of iron pole pieces have been appreciated since the beginnings of electron optics. However, it is not generally realized that a single pole piece is equally effective in producing both a high axial field strength and a small half-width and may have additional advantages. A suitable arrangement is shown schematically in Fig.5.19. Here, a single conical pole piece is placed at the centre of a flat helical winding in contact with a semi-infinite plane of ferromagnetic material of infinite permeability. The use of the term "single pole piece" is considered to be justified in this case since the magnetic flux emerging from the pole piece returns to a semi-infinite plane which would not normally be described as a "pole piece". The insertion of such a pole piece has several interesting effects on the axial flux density distribution. At the tip of the pole piece, the axial flux density will be increased and the half-width of the field distribution will be reduced. Both these effects

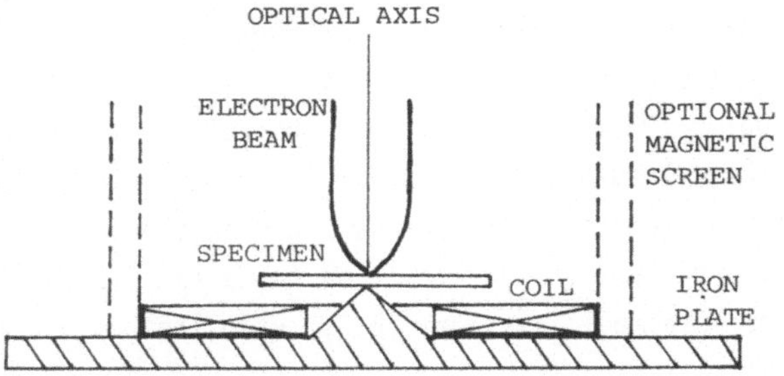

Fig.5.19. Schematic arrangement of a single pole-piece lens with conical pole piece on a semi-infinite base plate

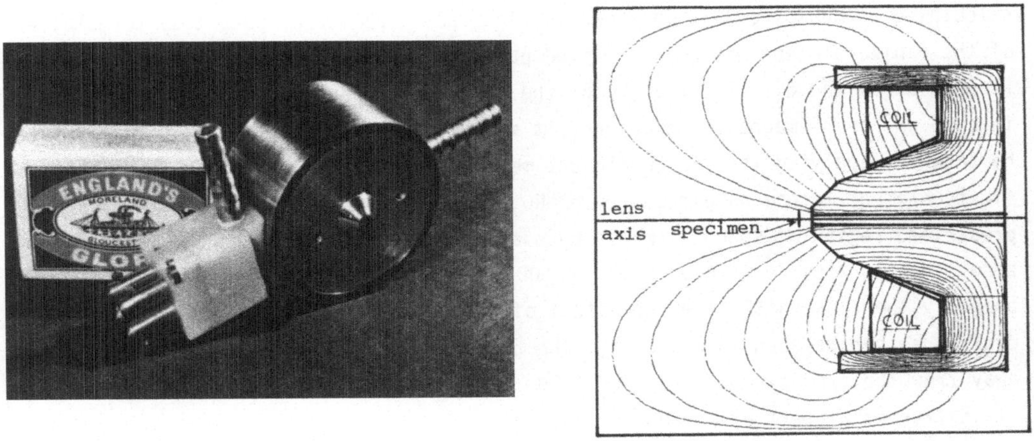

(a) (b)

Fig.5.20. (a) Practical form of a 100 kV single pole-piece lens with 2 mm bore. Focal length 4 mm, C_s = 2 mm. Excitation 4000 A-t. (b) Flux density distribution calculated by the finite-element method of MUNRO [5.35]

would be expected to lead to lower aberrations. In addition, the axial flux density distribution will be pushed away from the lens structure itself making the lens properties less dependent on imperfections in both the iron circuit and the coil, as well as providing more space to manipulate the specimen. This is of particular interest in applications such as microlithography, where microfabrication may have to be carried out on a comparatively large slice of material, as shown in Fig.5.19. It should be noted that for this kind of image formation, it is not necessary to provide a hole in the pole piece. This fact eliminates one of the chief causes of difficulty encountered in the manufacture of conventional pole-piece lenses which rely on the presence of the lens bores to produce the focusing fields. A coaxial magnetic screen, shown in dashed lines in Fig.5.19, may be employed for magnetic screening purposes without essentially changing the character of the axial distribution. For linguistic convenience, a single pole-piece lens provided with such a screen will still be referred to as a single pole-piece lens although there can be no sharp dividing line between single pole-piece lenses and conventional pole-piece lenses of extreme asymmetry. (For field measurements, see [5.17a].)

A practical and versatile form of a single pole-piece lens is shown in Fig.5.20a. It is of 50 mm outside diameter; a water-cooled wire coil provides an excitation of some 4000 A-t, enabling it to be used up to 100 kV. The snout or "snorkel" protrudes some 5 mm beyond the main magnetic structure, pushing the magnetic imaging field away from the windings and the main body of the lens. This causes imperfections in the coil winding and the magnetic circuit to have little effect on the axial flux density distribution, which in this lens is almost identical [5.31] with that of a pancake coil with $D_2/D_1 = 20$. The magnetic flux distribution in the lens, kindly calculated by MUNRO [5.35], is shown in Fig.5.20b. The vector potential formulation of the finite-element method has proved particularly useful in analysing snorkel lenses, since the direct influence of the coil must clearly be taken into account in such an "open" magnetic structure. It should be noted, however, that care should be taken in applying the finite-element method to this type of structure. In the finite-element method as devised by MUNRO, the flux density is calculated within a prescribed region of space at whose boundary the vector potential is assumed to be zero. This is true in general only for an infinitely distant boundary. Placing the boundary too close will have the effect of compressing the flux lines in its vicinity, as can be seen in Fig.5.20b. The boundary should therefore be placed as far away from the lens structure as possible. However, the finite-element method in its differential form also requires that not only the coil and the magnetic circuit should be divided into finite elements of volume but also the entire space within the boundary. The accuracy of the calculation improves considerably as the number of these elements is increased, but this results in a sharp increase in core store required and in computational time. A decision on the choice of boundary position and the number of finite elements to be used within that boundary therefore calls for nice judgment.

A possible way round this difficulty has recently been suggested [5.30a]. The method is analogous to that previously employed in the solution of scalar potential problems in electrolytic tanks and resistance networks, in which the potential at a convenient boundary near the lens structure is first determined using a small model of the lens with a distant boundary. A larger model of the lens is then set up with the near boundary set to the appropriate potential so determined. This method of calculation is most easily explained by reference to an actual calculation of the axial flux density distribution of a single pole-piece lens as shown in Fig. 5.21 at an excitation of 9000 A-t. There are 29 mesh points in the axial (Z) direction and 19 mesh points in the radial (R) direction. The boundary is first chosen in the usual way so as to be at a distance large compared with the lens dimensions. The vector potential A is arbitrarily set to zero at this boundary. The magnetic field is then determined by the standard finite-element program (RUN 1).

The resulting axial field distribution is shown in Fig.5.21. The discontinuities in the axial field are characteristic of any finite-element method and would nor-

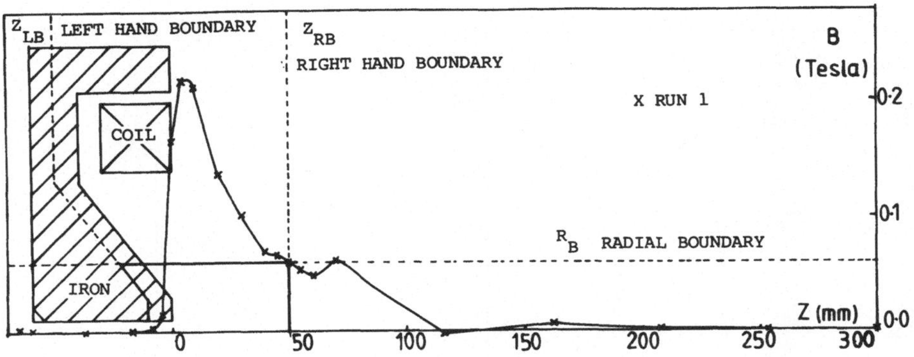

Fig.5.21. Axial magnetic flux density distribution of a single pole-piece lens (RUN 1) by the standard Munro program. Lens excitation 9000 A-t. Note the intermediate Z boundaries Z_{LB}, Z_{RB} and the R boundary R_B defining the selected area (heavy line). (Upper half of lens only)

mally be reduced by increasing the number of mesh points and hence the core memory requirements of the computer. In the new method, intermediate boundaries, Z_{LB} and Z_{RB} (left-hand and right-hand Z boundaries, respectively) and R_B (radial boundary) are introduced as shown in Fig.5.21. The intersection of these boundaries defines a "selected area" in the region of the pole face as indicated by the heavy black line in Fig.5.21. Next, the 29 axial mesh points are transferred to the axial region between the new left-hand and right-hand boundaries Z_{LB} and Z_{RB}, respectively. At this stage the radial boundary is not altered, so the vector potentials at the 19 radial points to the new boundaries are known directly from RUN 1. The axial distribution is then recalculated between the new boundaries, the results being shown in Fig.5.22 (RUN 2) by open circles. A considerable improvement in smoothness and accuracy is apparent. If necessary, a further improvement can be obtained by repeating the above procedure, restoring the original 29 axial points but transferring the 19 radial points to the region between the axis and the intermediate radial boundary R_B and re-running the calculation (RUN 3) to improve the vector potential distribution on the radial boundary. The results of RUN 3 are not shown on Fig.5.22 because the improvement is not significant in the critical region near the pole piece. However, it is now possible to set the vector potential accurately at both the axial and radial boundary of the selected area. The finally calculated axial flux density distribution (RUN 4) is shown by the solid circles in Fig.5.22. Careful inspection shows that a small, but significant, improvement in smoothness of the peak flux density has occurred.

It is clearly possible to place the selected area anywhere in the lens field. If, for example, it is transferred to the axial region between Z_{LB} = 50 mm and Z_{RB} = 300 mm (RUN 5), a considerably improved flux density distribution in this region can be obtained. The triangles in Fig.5.22 show the values obtained in the axial region between Z = 50 and Z = 120 mm. These computing procedures for

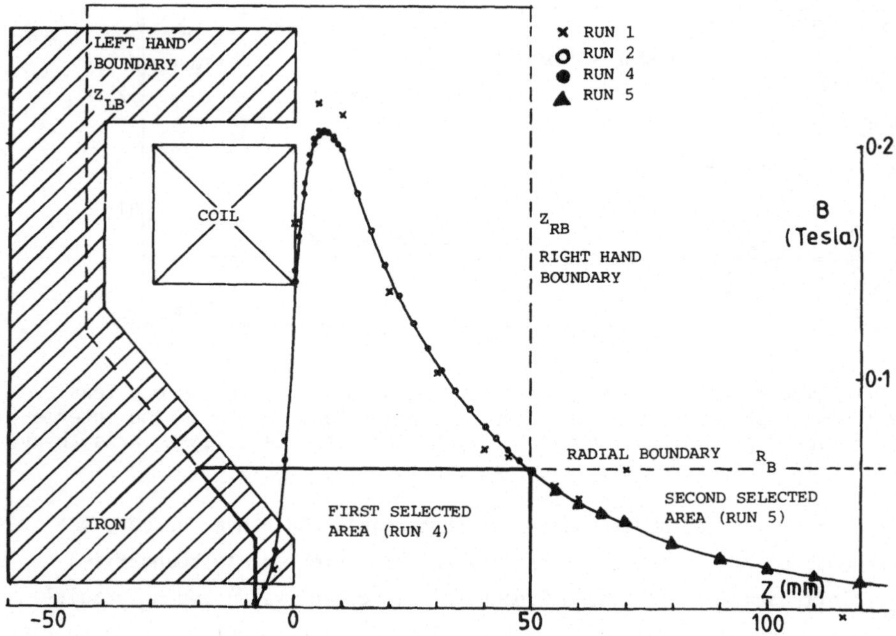

Fig.5.22. Improved axial magnetic flux density distribution within the intermediate boundaries by the selected area method (RUNS 2, 4, and 5). Calculated points of RUN 1 added for comparison purposes. (Upper half of lens only)

changing the boundary condition, devised by the author's colleague H. Nasr, may at first seem unnecessarily cumbersome, but they avoid the need for difficult and often unsatisfactory interpolation schemes at the critical boundary. The main advantage of the above method is that for a given accuracy in field distribution, the number of mesh points can be reduced fivefold compared with that of previous methods of calculation. It should perhaps be mentioned that the above calculations were carried out on a Commodore 2001 PET mini-computer with an active store of 32 kilobytes and two floppy discs providing an additional capacity of 340 kilobytes. The method lends itself, therefore, to interactive computer-aided design.

Some of the above-mentioned difficulties can also be avoided by making use of the "integral form" of the finite-element method of NEWMAN et al. [5.37]. In this method the magnetic circuit and the current-carrying conductors are divided into finite elements as before. The magnetic field in the iron caused by the current-carrying elements is then calculated by the Biot-Savart relation. It is then assumed as in the finite-element method, that the resultant intensity of magnetization for each magnetic element is constant in magnitude and direction. The magnetization of a particular element will, of course, depend on the magnetization of all the other elements, and this can be allowed for. The resulting set of linear equations can be expressed in matrix form and solved in the usual way in a digital computer. The magnetic flux density at any given point is then simply the sum of the magnetic flux due to the current-carrying conductors and that due to the magnetization of the iron circuit. With this method, therefore, it is not necessary to make special assumptions about the magnetic medium outside the magnetic circuit or to impose an artificial boundary. It appears that the computational time involved is comparable with that of the differential form of the finite-element method. The two methods are in fact complementary, and it has recently become possible to combine some of the useful features of both methods [5.30b]. Thus when using the Munro method, it is a simple matter to calculate the axial magnetic field due to the coil along the axis by the Biot-Savart law. This field can then be subtracted from the total B(z) field already calculated. The result will yield the field due to the iron, which in general will exhibit fluctuations due to the presence of the finite elements. Since the field due to the iron decays along the axis more rapidly than that due to the coil, it is easier to smooth out any fluctuations. Finally, the smoothed field due to the iron can be added to the true field produced by the coil to yield a smooth total field extending as far along the axis as required.

Returning to the lens shown in Fig.5.20, one can surmise from the form of the flux density distribution that the presence of an external magnetic screen will have very little effect on the axial field distribution, although it may well affect the flux density in the immediate vicinity of the screen. This is in line with practical experience. The calculated focal properties of this lens are f_{obj} = 4 mm, C_s = 2mm. The peak flux density at the pole face of 4 mm diameter is approximately 1.2 T at an excitation of some 4000 A-t. This is sufficient to bring a 100 keV electron beam to a focus at the pole face. For electron probe applications, where the electron beam is focused in front of the pole face, no hole is necessary in the pole piece. For use in TEM or STEM a small bore will be necessary. Such a hole will not greatly affect the lens properties but may increase the flux density near the pole face. Thus the effect of a bore of 2 mm in the pole face of this lens is to *decrease* the focal length to about 3.6 mm and to reduce the spherical aberration coefficient C_s to 1.8 mm.

Figure 5.23 illustrates what can be achieved in practice in transmission electron microscopy by fitting non-conventional lenses [5.30,36]. This figure shows an experimental miniature 100 kV transmission electron microscope column. The beam from an electron gun proceeds upwards through the snorkel objective lens of 50 mm outside diameter described previously. By moving the specimen holder in the z direction different conditions of illumination can be obtained without appreciably changing focal length or aberrations. Thus illuminating conditions ranging from no pre-field to the probe-forming conditions of the condenser-objective mode can be obtained within an axial movement of less than 3 mm. Immediately above the specimen stage is mounted the rotation-free projector lens previously described; this projects a rotation-free image onto the transmission fluorescent screen at the top of the instrument. The total height of the column is some 170 mm, compar-

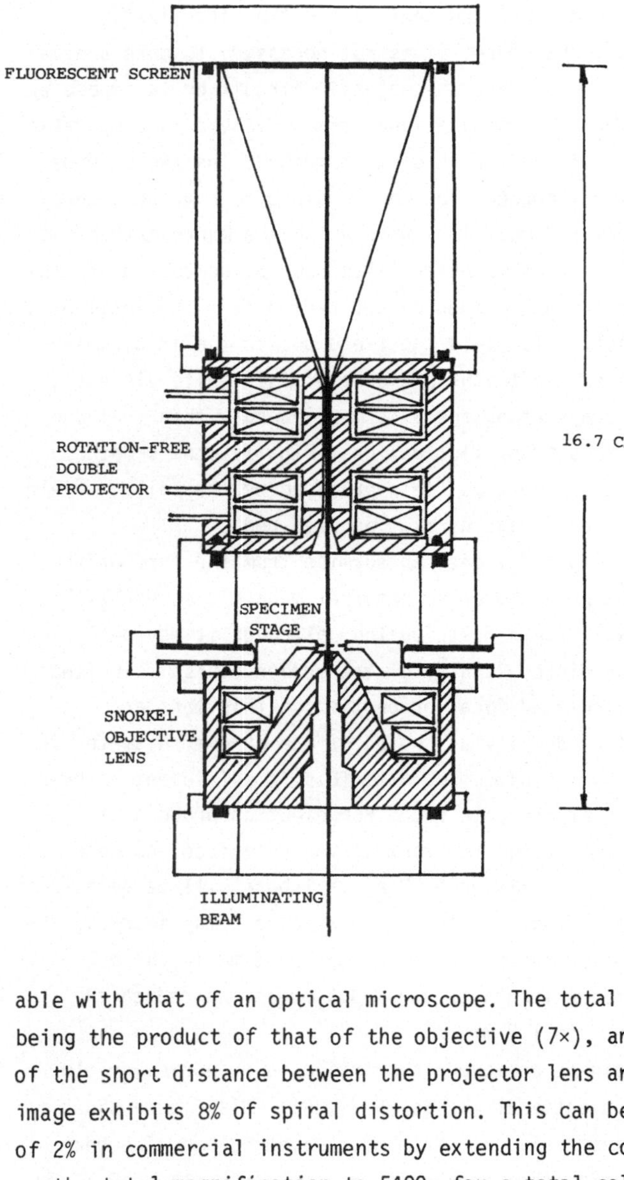

Fig.5.23. Experimental miniature 100 kV TEM with snorkel objective and rotation-free projector lens unit. Length of column 170 mm, magnification 2500×

able with that of an optical microscope. The total magnification is 2500×, this being the product of that of the objective (7×), and the projector (350×). Because of the short distance between the projector lens and the fluorescent screen the image exhibits 8% of spiral distortion. This can be reduced to the standard value of 2% in commercial instruments by extending the column length by 80 mm, bringing up the total magnification to 5400× for a total column length of 250 mm, still an unusually short column for an electron microscope. The distortion-free low-magnification image, discussed previously, occurs at a magnification of 50×, a useful value for search purposes. The measured resolution of this instrument (5 nm) is limited essentially by the resolution of the fluorescent screen. On the other hand, the resolving power of the objective lens $\delta_0 = 0.7\ (C_s\lambda^3)^{\frac{1}{4}} = 0.38$ nm. To achieve this resolution, an extra magnification of some 20× would be required. This could readily be obtained with a further rotation-free lens of total height 30 mm bringing the total column length of the instrument to 280 mm.

If the electron source and the fluorescent screen are interchanged [5.36], the instrument becomes a versatile scanning electron microscope or scanning transmission electron microscope. For the latter, a suitable detector must be provided below the snorkel objective lens. This construction allows additional lenses to be placed below the objective lens to assist in the formation of diffraction patterns and to act as an interface to an energy-loss spectrometer. In electron probe-forming applications, the advantage of the rotation-free projector lens is that it is less sensitive than a pair of independent lenses to misalignment errors. In addition, operation is simplified since there is only one lens control. It should perhaps be emphasized that these two non-conventional instruments have been built without resorting to any essentially new technology. It should also be remembered that for comparison purposes, a design of this type can be scaled up exactly, i.e., so that the magnetic flux density is the same at corresponding points in the original and scaled lens. At a voltage of 1000 kV, for example, the scaling factor is 4.264, so that the scaled up column diameter would be 213.2 mm only, slightly larger than that of conventional 100 kV magnetic lenses. It is sometimes asserted that the large size of conventional magnetic lenses is an advantage in high-voltage microscopes since it provides the necessary X-ray screening. Although this statement is true, it should be remembered that the substantial reduction in column length brought about by miniature lenses results in a corresponding reduction in the volume of screening material required. Moreover, an electron optical column does not usually provide uniform X-ray screening along its length, especially in the regions between lenses. It is therefore advantageous to separate the design of the X-ray screening from that of the electron optical system. Ideally, the biological shield should make use of high-density material, in a form adapted to the use to which the instrument is to be put. In general, the X-ray screening should not be rigidly attached to the column as this tends to make the instrument unnecessarily cumbersome and unwieldy.

5.5.3 The Magnetized Iron Sphere Model

The exponential field model, although extremely useful in predicting the general form of the properties of single pole-piece magnetic lenses, nevertheless suffers from certain defects. The most serious is that the field distribution cannot be related directly to pole-piece structure nor can it explain certain curious features of single pole-piece lenses, such as the slight, but measurable, increase in objective focal length at high lens excitations. For this reason it seems desirable to look for a more realistic model, preferably in the form of a simple geometrical shape whose magnetic properties could be determined analytically. The uniformly magnetized sphere seems capable of fulfilling these conditions [5.1]. Figure 5.24 shows schematically the magnetic field distribution from an iron sphere, uniformly magnetized to a flux density B_0 in the direction of the z axis [5.19]. From con-

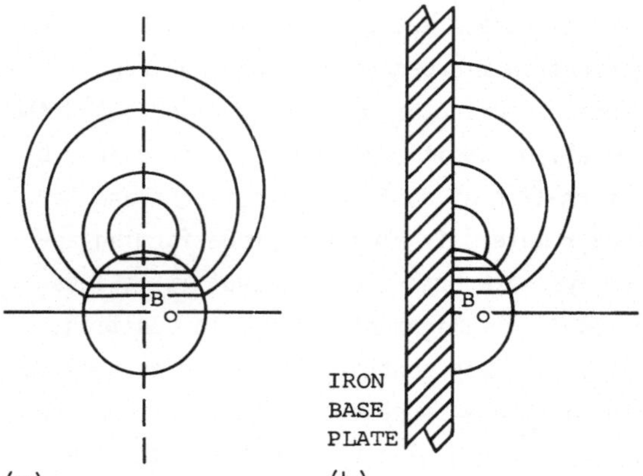

(a) (b)

Fig.5.24. (a) Magnetic field distribution (schematic) from a uniformly magnetized sphere of radius a. (b) Model of a single pole-piece lens

siderations of symmetry, the left-hand half of the sphere may be replaced by an infinite sheet of infinite permeability as shown in the figure without disturbing the remaining field, thereby forming a physical model of a single pole-piece magnetic lens. The axial flux density distribution of this lens is given by

$$B(z) = B_0(a/z)^3 \quad \text{or} \quad z \geqq a \quad , \tag{5.20}$$

where a is the radius of the sphere and z is the axial distance measured from the centre of the sphere. Equation (5.20) shows that the maximum flux density occurs at the tip of the sphere and is equal to the flux density B_0 within the sphere. The flux density falls off inversely as z^3, the half-width d being equal to 0.26a. Thus an iron hemisphere of 5 mm radius would produce a peak field of some 2 T and a half-width d = 1.3 mm; this is a considerably smaller half-value than could be produced in a conventional lens with the same excitation.

Inserting the axial flux density according to (5.20) into the paraxial ray equation

$$r'' + (e\,B^2/8mU)r = 0 \quad , \tag{5.21}$$

one obtains the trajectory [5.1] of a paraxial ray entering the field parallel to the axis:

$$\frac{r}{a} = \left(\frac{z}{a}\right)^{\frac{1}{2}} \frac{1.6406}{(NI/U^{\frac{1}{2}})^{\frac{1}{4}}} J_{\frac{1}{4}}\left[0.186\,\frac{a^2}{z^2}\left(\frac{NI}{U^{\frac{1}{2}}}\right)\right] \quad , \tag{5.22}$$

where $J_{\frac{1}{4}}$ is the Bessel function of order $\frac{1}{4}$, U is the relativistically corrected accelerating voltage, and NI is the excitation required to maintain the axial field distribution, the iron being assumed to have infinite permeability. Equation (5.22) allows the paraxial focal properties to be determined conveniently by reference to tables of Bessel functions. The results are similar in form to those predicted by the exponential field model, but there are important differences. Thus, although the projector and objective have the same minimum focal length at an excitation parameter $NI/U^{\frac{1}{2}} = 14$, for larger excitations the objective focal length increases slowly according to the relation

$$f_{obj}/a = 0.116\ (NI/U^{\frac{1}{2}})^{\frac{1}{2}} \quad . \tag{5.23}$$

This is in excellent agreement with experimentally observed values. The minimum focal length is given by $(f/a)_{min} = 0.43$.

5.5.4 Aberrations of the Uniformly Magnetized Sphere Model

The corresponding spherical and chromatic aberration coefficients are most conveniently calculated numerically by Scherzer's formula. In particular, for $NI/U^{\frac{1}{2}} > 14$, $C_c = 0.6\ f_{obj}$, a ratio normally obtained in conventional magnetic lenses. However, the spherical aberration coefficient is significantly lower; the minimum value of C_s/a is 0.1 ($C_s/d = 0.4$). Thus a sphere of 5 mm radius would have a C_s value of 0.5 mm, an encouragingly small value. This corresponds to a ratio $C_s/f_{obj} = 0.23$, a low value compared with that of conventional lenses but quite usual in pancake lenses. There is thus a close relationship between pancake lenses and single pole-piece lenses. This is confirmed when one compares the corresponding absolute parameter $C_s\ B_0/U^{\frac{1}{2}}$ for each lens. For a single pole-piece lens the minimum value of $C_s\ B_0/U^{\frac{1}{2}} = 3.5 \times 10^{-6}$, compared with 3×10^{-6} for the optimum iron-free coil; both are considerably better than the value of 5×10^{-6} of symmetrical iron pole-piece lenses. However, since the accurate manufacture of small iron hemispheres is easier than the winding of the inner turns of flat helical solenoids, the single pole-piece lens seems to have a distinct practical advantage for the achievement of high resolution.

The focal properties of such a single pole-piece lens are shown in Fig.5.25, plotted in terms of the half-width d, rather than the radius a, in order to facilitate comparison with other field distributions. The corresponding values relative to the radius a can readily be obtained from the relation a = 3.846 d. It should be emphasized that these properties relate exclusively to the magnetic field produced by the pole piece itself. As in all other magnetic lenses the pole piece must be magnetized by means of a coil which will, in general, produce an additional field on the axis, which must be taken into consideration. In order to magnetize an iron sphere uniformly to a flux density B_0, it must be placed in a uniform field of flux density $B_0/3$ extending over the whole sphere. Ideally the stray field from the coil should not extend appreciably along the optical axis. These conditions can be achieved to a good approximation by a short high-current-density coil and a suitably arranged external magnetic circuit.

Practical experience with single pole-piece lenses indeed shows that the focusing action in the neighbourhood of the iron pole piece is predominantly controlled by the pole piece rather than by the stray field of the coil. However, once a choice of pole piece and coil design has been arrived at on the lines set out above, a detailed calculation of the field distribution is always advisable, preferably by the finite-element method.

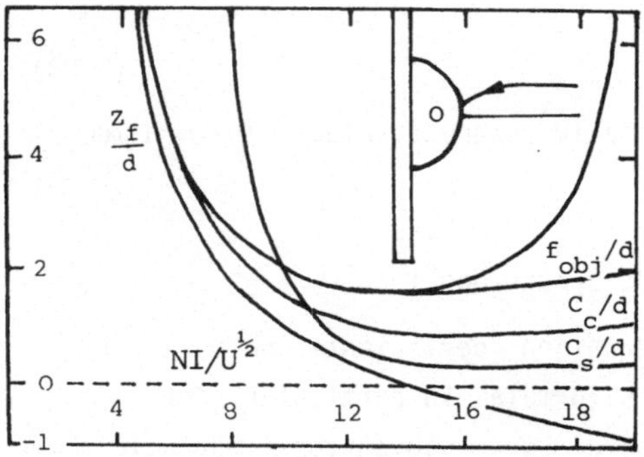

Fig.5.25. Relative focal properties of the uniformly magnetized sphere model in terms of the half-width d (d = a/3.846)

5.5.5 The Preferred Form of Single Pole-Piece Lenses

The main advantage of the spherical pole-piece model is that it provides a basis for the preferred form of such lenses at different accelerating voltages. Thus, for example, by integrating the axial field distribution over the optic axis, one obtains the relation

$$B_0 = 2\mu_0 NI/a \quad . \tag{5.24}$$

If the hemisphere is magnetized to a constant flux density, limited ultimately by iron saturation, the radius of the hemisphere must be increased as the accelerating voltage, and hence the excitation NI, is raised. Table 5.5 shows the electron optical characteristics of the spherical pole-piece lens for accelerating voltages between 100 and 1000 kV. For B_0 = 2 T, the radius of the pole piece must increase from 6 mm at 100 kV to 25 mm at 1000 kV, the corresponding resolution δ_0 going from 0.29 nm at 100 kV to 0.14 nm at 1000 kV. Since the absolute minimum spherical aberration parameter $C_sB_0/U^{1/2}$ for single pole-piece lenses is some 40% better than those of symmetrical iron pole-piece lenses, the resolving power δ_0 will be some 10% better at a given electron wavelength. Such an improvement, although small, can be significant at the 0.3 nm level in imaging the structures of oxides and at the 0.2 nm level in imaging metallic structures. In order to obtain the ultimate performance, the pole piece must be magnetized to the limit of saturation. It would also be advantageous to make use of ferromagnetic materials such as cobalt iron alloys with higher saturation levels. Figure 5.26 shows a practical form of such a lens intended for use at 1000 kV as either an objective or a projector lens, since for single pole-piece lenses the design is essentially the same. The radius of the pole piece is 25 mm with an excitation of 20000 A-t, sufficient to produce a peak axial flux density B_0 of some 2 T. The outside diameter of the lens is 120 mm and the axial length 50 mm. The lens windings are directly water-cooled and may be in the form either of wires or of tapes. The estimated weight of this lens is of the order of 4 kg. This is smaller, by nearly two orders of magnitude, than that of a comparable conventional lens, so that doubts could reasonably arise as to its practicability.

Table 5.5. Electron optical characteristics of the spherical pole-piece lens for B_0 = 2 T

Accelerating voltage [kV]	NI[A-t]	a[mm]	f_{obj}[mm]	C_s[mm]	$\delta_0 = 0.7(C_s\lambda^3)^{1/4}$ [Å]
100	4640	6	1.1	0.6	2.9
250	7825	10	4.2	1.0	2.2
500	12120	15	6.5	1.5	1.8
1000	19800	25	10.7	2.5	1.4

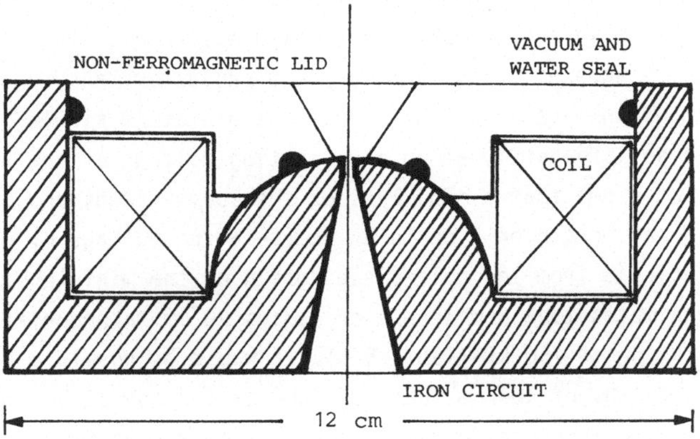

Fig.5.26. Practical form of the magnetized iron hemisphere lens

It is, of course, easy to measure the electron optical properties at a lower accelerating voltage and also to confirm by measurement that the flux density required at 1000 kV can, in fact, be achieved. However, there is no substitute for a direct test; unfortunately, it is not easy to incorporate such a lens in a conventional 1000 kV electron microscope without extensive modifications to the instrument.

However, in the Hitachi HU1000 million-volt microscope there is a cylindrical port of 100 mm diameter housing a Faraday cage mounted immediately above the projector lens. It was therefore decided to design, in collaboration with the C.E.G.B. Research Laboratories at Berkeley, Gloucestershire, an experimental single-pole mini-projector lens of only 98 mm outside diameter and 12000 A-t for insertion into the HU1000 microscope [5.32] located at the above laboratory. The design of the lens was essentially a scaled-down version of the projector lens shown in Fig. 5.26. The main differences were that the exit cone had a smaller semiangle (9°) and the outside diameter of the lens was rounded off to allow it to enter the cylindrical side port of 100 mm diameter in the Hitachi HU1000. Construction was essentially that of the projector lens shown in Fig.5.26 the magnetic circuit being turned from a single piece of iron. Four wire coils were provided, driven by a constant current supply of 3 kW capacity. The brass top plate of the lens was provided with two 'O' rings which acted both as vacuum and water seals. The weight of the lens was just over 2 kg. The minimum projector focal length calculated from the axial field distribution was approximately 10 mm. However, with the severe restrictions on the size of this lens, it was only possible to reach an excitation parameter $NI/U^{\frac{1}{2}} = 8.84$ compared with 15 needed to attain minimum projector focal length at 1000 kV. The observed projector focal length at 1000 kV at reduced excitation was 17 mm, in good agreement with the calculated value. The images obtained on the fluorescent screen and on the photographic plate with the mini-projector lens were entirely comparable with those of the normal projector at the same $NI/U^{\frac{1}{2}}$ value. A full analysis of the experimental performance of this lens has been given by NEWMAN [5.36]. It should perhaps be mentioned that in these experiments the projection distance L was approximately 600 mm. With such a large projection distance it is impossible to realise the full advantages of a single pole-piece projector lens. However, the experiment was an important one in establishing that miniature single-pole projector lenses can operate satisfactorily at an accelerating voltage of one million. Subsequent calculation has shown that a mini-lens of the above dimensions can readily achieve its minimum focal length only up to an accelerating voltage of 750 kV. This may readily be appreciated from the pole-piece design of the mini-lens as shown in Fig.5.26, which approximates to a hemisphere of radius 20 mm, appropriat for a single-pole-piece lens operating at 750 kV (Table 5.5). It would be possible as a tour de force to design a mini-lens of 100 mm outside diameter for operation at 1000 kV but with a disproportionate increase in the electrical power consumed in the windings.

The above experiments suggested that apart from the appreciable reduction in lens size, considerable improvements could also be made to the image viewing system of conventional electron microscopes by the optimum use of single pole-piece lenses.

5.6 Improved Image Viewing Systems

Ideally, the viewing system of a transmission electron microscope should present a correctly magnified image of the object on a fluorescent screen, from which a suitable field of view can be selected for photographic recording. In the high-resolution electron microscopy of specimens in which radiation damage is a limiting factor, the minimum acceptable magnification and hence the minimum number of electrons needed to establish the required detail should be used. In many electron microscopes stray alternating magnetic fields in the vicinity of the viewing chamber may preclude this mode of operation since the distance between the final projector lens and the photographic plate may be large (typically of the order of 500 mm) and virtually unscreened magnetically because of the presence of large viewing windows. This limits the minimum size of image detail that can be recorded and hence the minimum permissible magnification. Thus in a typical laboratory with a horizontal AC magnetic field of the order of 0.5 μT (peak to peak), the unscreene beam path should not exceed some 200 mm at an accelerating voltage of 100 kV if the aberration of the image due to stray magnetic fields is not to exceed 10 μm, i.e., about one third of the resolution of the photographic plate. The sensitivity to stray magnetic fields decreases as the square of the projection distance L. A reduction in the projection distance L would not only reduce the sensitivity to external AC magnetic fields but, by reducing the total height of the column, would also reduce the effects of mechanical vibrations. Unfortunately, the final projector lens produces spiral distortion which, unlike radial distortion, is surprisingly difficult to correct [5.27].

The spiral distortion $\Delta\rho/\rho$ at an image point of radius ρ is given by

$$\Delta\rho/\rho = (f_p^2 D_{sp})(\rho/L)^2 = Q_{sp}^2 \tan^2\alpha_p \quad , \tag{5.25}$$

where f_p is the projector focal length, D_{sp} is the spiral distortion coefficient, and L is the distance between the focal point of the projector lens and the fluorescent screen. Equation (5.25) shows that image distortion depends solely on a quality factor $Q_{sp} = f_p(D_{sp})^{\frac{1}{2}}$ and the projection semi-angle α_p that the outermost ray to the image makes with the optic axis. An analogous expression for radial distortion results simply by substituting the corresponding quality factor Q_{rad} for Q_{sp} in (5.25). Figure 5.27 shows the quality factors Q_{sp} and Q_{rad} for the hemispherical pole-piece lens as a function of NI/NI_0, the relative excitation parameter. The relative magnification M/M_0 is also shown, where M_0 is the magnification at minimum projector focal length which occurs at a relative excitation $NI/NI_0 = 1$. This is the preferred operating point for a final projector lens, since high magnification is accompanied by low radial distortion ($Q_{rad} \approx 0$). For comparison purposes, the quality factor Q_{sp} for spiral distortion in symmetrical pole-piece lenses is shown as a dashed line. (N.B. the quality factor for spiral distortion of the hemispherical pole-piece lens refers, of course, to the favourable direction of the electron beam into the lens, namely facing the pole piece.) A comparison of these curves reveals that in the region of maximum magnification the single pole-piece lens has a Q_{sp} value of 0.75 compared with 1.0 for a symmetrical iron pole-piece lens. This would correspond to a reduction of spiral distortion by a factor of 1.5 for a given size of screen and a fixed value of L. Furthermore, the variation of spiral distortion with lens excitation is much less pronounced with a single pole-piece lens than for a twin iron pole-piece lens. In practice when photographing an image of standard size (100 mm in diameter), it is more useful to reduce the projection distance L from 500 mm ($\alpha_p = 5.7^\circ$) required by conventional lenses to 380 mm ($\alpha_p = 7.6^\circ$) for the spherical pole-piece lens for a maximum spiral distortion of 1% or from 350 mm ($\alpha_p = 8.1^\circ$) to 270 mm ($\alpha_p = 10.8^\circ$) for 2% spiral distortion, the maximum amount tolerable in critical microscopy.

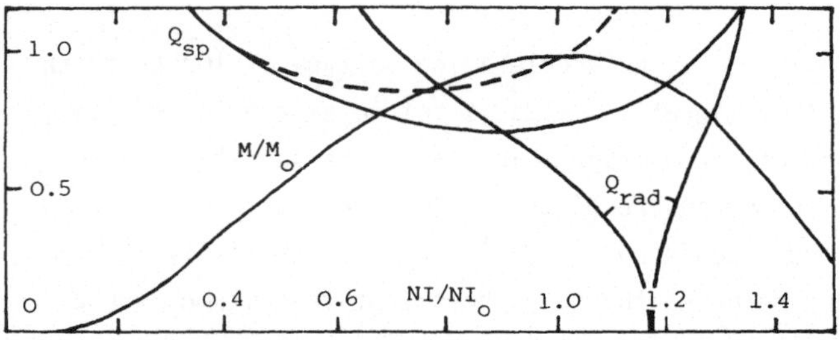

Fig.5.27. Single-pole-piece projector lens quality factors Q_{sp} and Q_{rad} and relative magnification M/M_0 as a function of the excitation parameter NI/NI_0 (for the preferred lens direction). Dashed line: Q_{sp} for conventional lenses

5.7 Correction of Spiral Distortion

If the spiral distortion of the projector lens could be corrected, new possibilities would be opened for the design of the viewing system in electron microscopes. It is possible, in principle, to correct spiral distortion by introducing a compensating distortion from a preceding lens. Unfortunately, the spiral distortion introduced into the final image by the correcting lens is diminished by the square of the magnification M_c introduced by the correcting lens. The correcting lens should therefore have a quality factor Q_{sp} some M_c^2 times greater than that of the lens to be corrected. For a symmetrical pole-piece lens this can only be achieved by greatly increasing the excitation, perhaps by operating in the second or third focal zone. However, if a single pole-piece lens is operated in the unfavourable direction of the electron beam, the Q_{sp} value is increased by about a factor of three. Such a lens should therefore be more effective as a corrector than would a symmetrical iron pole-piece lens.

Figure 5.28 shows an experimental arrangement by LAMBRAKIS et al. [5.21] for investigating this possibility. The final projector lens was a 750 kV mini-projector lens of diameter 98 mm and axial extent 36 mm. The exit cone from the pole piece was increased from the normal maximum value of α_p of 8° to 22°, corresponding to a spiral distortion of 12% at the edge of the viewing screen in the absence of any correction. For experimental convenience the viewing screen (ρ = 22.2 mm) was placed at a relatively short distance L = 55 mm from the tip of the pole piece. This is equivalent to a projection distance L = 124 mm for a standard screen, i.e., about a quarter of the normal projection distance of commercial instruments. Above this lens was mounted a single pole-piece rotation-free doublet of the type previously described by JUMA and MULVEY [5.16] mounted outside the vacuum so that it could readily slide up and down a tube through which the incoming electron beam passed down the column. This arrangement allowed the distance between correcting lens and projector to be varied continuously whilst observing the final image. The

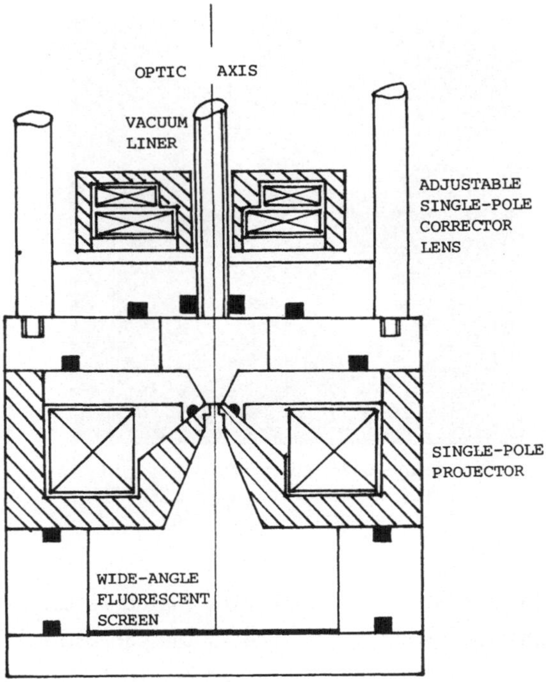

Fig.5.28. Experimental arrangement [5.21] of two single pole-piece lenses for the correction of spiral distortion

correcting lens was originally designed as a rotation-free doublet, but in these experiments only the lower half of the unit was energized. This produced a magnetic field distribution in the unfavourable direction with respect to the incoming electron beam and hence large spiral distortion. For experimental convenience a 30 kV electron gun was used as a source and the projected image of an electron microscope grid on the transmission fluorescent screen was photographed externally. The projected image on the fluorescent screen was thus magnified by both lenses. The spiral distortion produced at the edge of the final screen (cone semi-angle 22°) by the final projector acting alone was 12%, compared with 15% for a double pole-piece lens. When the corrector lens was energized so that its image rotation was in the opposite sense to that of the projector, the distortion was reduced. In the first focal zone, the spiral distortion coefficient of the corrector is some three times greater than that of the projector; correction would therefore be expected to take place only if the magnification of the corrector is greater than about $\sqrt{3}$, i.e., 1.7×. However, to fulfil this condition, the corrector lens must be brought close to the projector, and so the opposing magnetic field distributions will tend to cancel each other. It was confirmed experimentally [5.21] that even with the single-pole-piece corrector lens, correction could not be achieved in the first focal zone. However, by increasing the separation between the lenses and operating the corrector lens at a magnification of just below 4× and an increased excitation parameter ($NI/U^{\frac{1}{2}} = 34$), a substantial reduction of spiral distortion took place as shown in Fig.5.29.

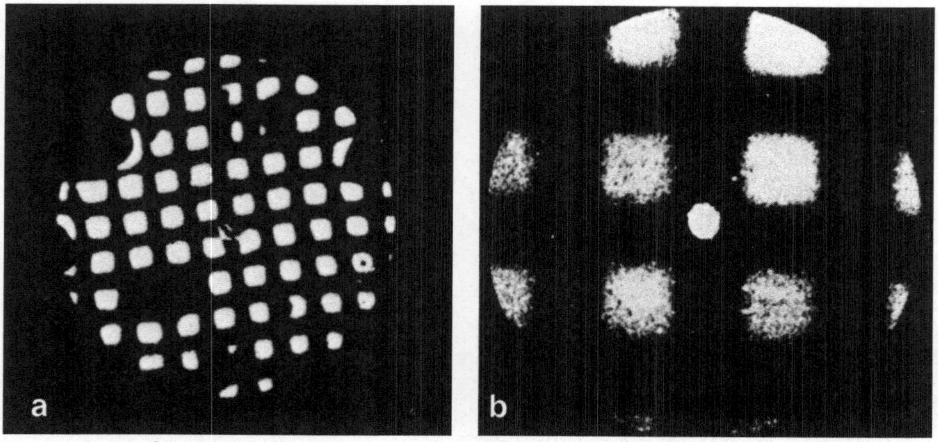

Fig.5.29. (a) Extended angle (α_p = 17.4°) TEM image with corrector lens off, showing 7% spiral distortion at edge of field. (b) Wide-angle (α_p = 22°) TEM image with corrector lens energized, showing further magnification (4×) and low distortion

Figure 5.29a shows the image of a standard microscope grid with the corrector lens de-energized. Under these conditions the vacuum liner tube in the corrector lens (Fig.5.26) reduces the effective projector semi-angle α_p from 22° to 17.4°, thereby reducing the spiral distortion at the edge of the field from 12% to 7%. On switching on the corrector lens as described above, the full semi-angle of 22° is attained as shown in Fig.5.29b. Because of the relatively poor contrast and resolution of the transmission screen, which was photographed externally, it is not possible to make an accurate measurement of the residual distortion. However, some idea of the magnitude of the correction may be gained by noting that at the standard projection distance (L = 50 cm), the image in Fig.5.29b would be over 50 cm in diameter. Further experiments under conditions of practical electron microscopy [5.8a] have shown that operating the corrector lens at its minimum focal length in order to avoid radial distortion is unduly restrictive and wasteful of ampere-turns. Figure 5.30 shows the general layout of the revised corrector system designed for distortion-free operation at a projection semi-angle α_p = 30°. Typical calculated electron trajectories through the system are also shown in Fig.5.30. The projector lens now has a carefully shaped pole-piece snout which will allow a cone of electrons of 35° semi-angle to emerge from the lens. The lens excitation at 100 kV is about 5000 A-t. In the absence of a corrector lens the spiral distortion at the edge of the image for a projection semi-angle of 30° would be about 20%.

The corrector lens, mounted at a fixed distance above the projector lens, is also a single pole-piece lens with its pole piece oriented so as to produce the maximum possible spiral distortion. Between the pole pieces there is an iron screening plate with an axial hole whose function is to minimize magnetic field cancellation effects whilst retaining the desired asymmetrical field distributions.

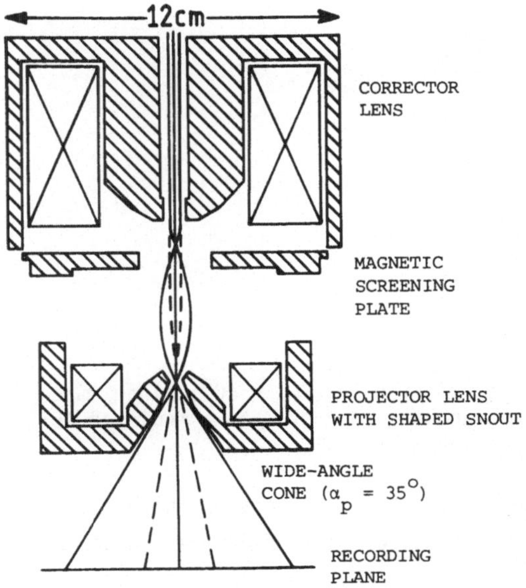

Fig.5.30. Wide-angle ($\alpha_p = 30^\circ$) projection system showing calculated electron trajectories. Full line, corrector lens on. Dashed line, corrector lens off

The corrector lens operates at an excitation parameter $NI/U^{\frac{1}{2}} \simeq 24$, i.e., fairly high but still just within the first focal zone, corresponding to an excitation of 8000 A-t at 100 kV. This produces a virtual intermediate image of 3× magnification with spiral distortion of 180%, sufficient to cancel 20% of the spiral distortion of the projector lens. This high corrector lens excitation inevitably produces considerable barrel distortion. This, however, is readily removed from the final image by operating the final projector at an excitation slightly below that needed for minimum focal length. This in turn reduces the spiral distortion to be corrected. The correction of both aberrations is therefore not critically dependent on lens excitation. Similarly, both calculation and experiment indicate that this method is economical in lens excitation and not critically dependent on the lens separation or the position of the iron screening plate. The axial adjustment of the corrector lens, employed previously, can therefore be dispensed with. The combined focal length and hence the magnification of the unit can be varied over wide limits. For the unit shown in Fig.5.30, the combined focal length is 3.5 mm. For a standard photographic image of 100 mm in diameter, the distance between the upper surface of the corrector lens and the photographic recording plane is only 21 cm, corresponding to a magnification of 25×. Figure 5.31 shows a wide-angle ($\alpha_p = 30^\circ$) image of a rectangular grid obtained in a modified EM6 electron microscope [5.8b]. It should be pointed out that at the standard projection distance ($L = 400$ mm) the corrected image shown in Fig.5.31 would be about 460 mm in diameter, an increase of distortion-free viewing area by a factor of seventeen over a conventional lens system. Alternatively, the viewing area could be kept the same and the projection distance reduced by a factor of four. It therefore seems likely that in the future, with the aid of single pole-piece lenses, the design of the

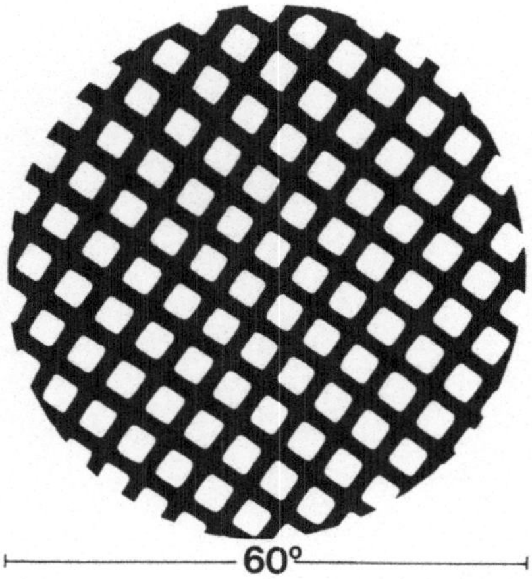

Fig.5.31. Wide-angle ($2\alpha_D = 60^\circ$) final image of a rectangular mesh grid in a modified AEI EM6 electron microscope

viewing system of the TEM need not be dictated by the aberrations of the projector lens, but solely by the operational requirements of viewing and recording the image.

It should be added that in the design of wide-angle projection systems, third-order aberration theory, which is based on paraxial rays, gives only a first approximation to the image distortion. A more satisfactory approach, but difficult to implement in practice, is to use the general ray equation to determine the actual trajectories and thereby simulate the actual appearance of the image on the viewing screen or photographic plate [5.0].

The employment of miniature single pole-piece corrector systems is not, of course, restricted to transmission electron microscopes. In many scanning transmission microscopes it is now becoming common practice not only to place a velocity spectrometer immediately after the final probe-forming lens, but also to insert photographic film for the parallel recording of diffraction patterns. In both cases it is desirable to provide an "interfacing" projector lens to ensure that the angular cone of rays leaving the specimen is matched to the optimum angle for the spectrometer and to ensure that a suitable range of camera lengths is available in the taking of diffraction patterns.

5.8 Concluding Remarks

Quite recently, the need for electron microscopes to be capable of operating efficiently and conveniently in both the TEM and STEM mode has led to a comparatively rapid introduction of unconventional and miniaturized lenses in commercial instruments. These may take the form of an important miniature auxiliary lens [5.28a], which enables an objective lens to function optimally in both TEM and STEM modes or even forms the basis of a completely miniaturized analytical TEM column [5.38], making use of tape windings cooled by a temperature-stabilized flow of oil.

However, significant improvements in electron optical performance are unlikely to occur until the electron optical systems of field emission sources are considerably improved. It was pointed out by VENEKLASEN [5.46] in 1971 that it would be advantageous to replace the traditional electrostatic focusing lens in a field emission gun by a suitable magnetic lens, thereby reducing considerably the spherical and chromatic aberration of the system. The technical difficulties of realising this in practice are considerable since a conventional lens would have to be placed in the immediate vicinity of the field-emitting tip and hence would have to be capable of withstanding vacuum bake-out procedures. Although TROYON and LABERRIGUE [5.44] succeeded in constructing an experimental gun operating at 4kV and TROYON [5.43] later raised this to 90 kV, operational difficulties still remain.

In an alternative approach SMITH and SWANN [5.40] and CLEAVER [5.4a] proposed to achieve the same result more conveniently by the use of a single-pole-piece lens mounted outside the high-vacuum system; the magnetic field of the lens would be concentrated in the region of the emitting tip. VENABLES and ARCHER [5.45] attack the same problem by suggesting that a single pole-piece condenser lens be placed outside the vacuum system in the vicinity of the first anode.

Such developments, if successful, would enable the advantages of field emission guns to be incorporated into a wide variety of electron optical instruments.

It should also be clear that the unconventional lenses described in this chapter are complementary in design to lenses making use of superconducting windings. It might be expected therefore that the future development of both types of lenses would follow similar paths. This can perhaps be illustrated by reference to the recently developed unconventional laminated lens of BALLADORE and MURILLO [5.2] shown in Fig.5.32. Here the pole pieces are conventional, but the traditional magnetic yoke is replaced by six sets of laminated silicon iron transformer stampings. This ferromagnetic material is crystallographically oriented so that maximum permeability occurs in a direction parallel to the face of the sheet, resulting in marked reduction in the amount of iron needed. In this version of the lens a superconducting winding is employed. Tests carried out by the above authors have shown that up to an excitation of 25000 A-t the performance of this lens is indistinguishable from that of a classical lens.

It is perhaps of interest to note that this excitation could also be provided, for example, by a copper wire or tape winding of the same size operating around room temperature at reduced pressure. The power consumption is estimated at 3.8 kW, well within the range of available current stabilizers.

We can therefore expect to see, in the not too distant future, radical departures from the traditional design of electron optical instruments, in which "non-conventional" lens designs play a leading role.

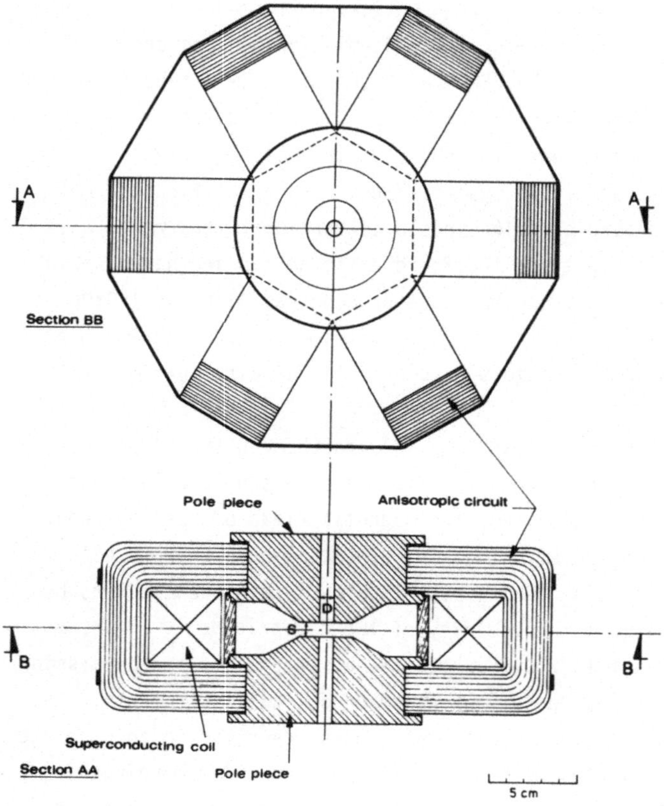

Fig.5.32. Laminated miniature lens of BALLADORE and MURILLO [5.2]. (Courtesy of Drs. J.L. Balladore and R. Murillo and the SFME)

References

5.0 S.M. Al-Hilly, T. Mulvey: In *Electron Microscopy and Analysis, 1981*, ed. by M.J. Goringe (Institute of Physics, Bristol, Conference series No. 61, 1982) pp.103-106

5.1 A. Alshwaikh, T. Mulvey: In *Developments in Electron Microscopy and Analysis 1977*, ed. by D.L. Misell (Institute of Physics, Bristol, Conference series No. 36, 1977) pp.25-28

5.2 J.L. Balladore, R. Murillo: J. Microsc. Spectrosc. Electron. *2*, 211-222 (1977)

5.3 R. Bassett, T. Mulvey: *Proc. 5th Int. Cong. X-Ray Optics and Microanalysis, Tübingen 1968*, ed. by G. Möllenstedt, K.H. Gaukler (Springer, Berlin, Heidelberg, New York 1969) pp.224-230

5.4 H. Becker, A. Wallraff: Arch. Elektrotech. *34*, 115-120 (1940)

5.4a J.R.A. Cleaver: Optik *52*, 293-303 (1978/1979)

5.5 C.J. Cooke, P. Duncumb: *Proc. 5th Int. Cong. X-Ray Optics and Microanalysis, Tübingen 1968*, ed. by G. Möllenstedt, K.H. Gaukler (Springer, Berlin, Heidelberg, New York 1969) pp.245-247

5.6 A.V. Crewe: *Proc. 6th Eur. Cong. Electron Microscopy, Jerusalem, 1976*, ed. by D.G. Brandon (Tal International, Jerusalem 1976) Vol.1, pp.65-66

5.7 G.V. Der-Shvarts, V.P. Rachov: Radio Eng. Electron. Phys. (USSR) *10*, 783-789 (1965)

5.8 P. Durandeau: Ph. D. Thesis, University of Toulouse: Ann. Fac. Sci. Univ. Toulouse Math. Sci. Phys. *21*, 1-88 (1957)

5.8a H.H. Elkamali, T. Mulvey: *Proc. 7th Eur. Cong. Electron Microscopy, The Hague, 1980*, ed. by P. Brederoo, G. Boom (7th Eur. Cong. Electron Microscopy Foundation, Leiden 1980) Vol.1, p.74-75
5.8b H.H. Elkamali, T. Mulvey: In *Developments in Electron Microscopy and Analysis, 1979*, ed. by T. Mulvey (Institute of Physics, Bristol Conference series No.52, 1980) pp.63-64
5.9 J.L. Farrant: *Proc. 7th Int. Cong. Electron Microscopy, Microscope Electronique 1970, résumés des communications présentées au septième congrès international*, ed. by P. Favard (Société Francaise de Microscopie Eléctronique, Paris 1970) Vol.1, pp.141-142
5.10 L.A. Fontijn, A.B. Bok, J.G. Kornet: *Proc. 5th Int. Cong. X-Ray Optics and Microanalysis, Tübingen 1968*, ed. by G. Möllenstedt, K.H. Gaukler (Springer, Berlin, Heidelberg, New York 1969) pp.261-268
5.11 W. Glaser: *Grundlagen der Elektronenoptik* (Springer, Wien 1952) pp.306-307
5.12 K.H. Herrmann, K. Ihmann, D. Krahl: *Proc. 8th Int. Cong. Electron Microscopy, Canberra, 1974*, ed. by J.V. Sanders, D.J. Goodchild (Australian Academy of Science, Canberra 1974) Vol.1, pp.132-133
5.13 J. Hillier: J. Appl. Phys. *17*, 411-419 (1946)
5.14 S.M. Juma: Ph. D. Thesis, University of Aston in Birmingham (1975)
5.15 S.M. Juma, T. Mulvey: *Proc. 8th Int. Cong. Electron Microscopy, Canberra, 1974*, ed. by J.V. Sanders, D.J. Goodchild (Australian Academy of Science, Canberra 1974) Vol.1, pp.134-135
5.16 S.M. Juma, T. Mulvey: In *Developments in Electron Microscopy and Analysis*, ed. by A. Venables (Academic, London, New York 1976) pp.45-48
5.17 S.M. Juma, T. Mulvey: J. Phys. E *11*, 759-764 (1978)
5.17a S.M. Juma, A.D. Faisal: J. Phys. E *14*, 1389-1393 (1981)
5.18 K. Kanaya: Bull. Electrotech. Lab. *22*, 55-62 (1958)
5.19 Lord Kelvin (Thomson, W) : In *Reprints of Papers on Electricity and Magnetism*, 2nd ed. (Macmillan, London 1884)
5.20 D.J. Kroon: *Laboratory Magnets* (Philips, Eindhoven 1968)
5.21 E. Lambrakis, F.Z. Marai, T. Mulvey: In *Developments in Electron Microscopy and Analysis 1977*, ed. by D.L. Misell (Institute of Physics, Bristol, Conference series No. 36, 1977) pp.35-36
5.22 J.B. Le Poole: *Proc. 3rd Eur. Conf. Electron Microscopy, Prague 1964*, ed. by M. Titlbach (Publishing House of the Czechoslovak Academy of Sciences, Prague 1964) Appendix p.6
5.23 M.E.C. Maclachlan: In *Scanning Electron Microscopy – Systems and Applications*, ed. by W.C. Nixon (Institute of Physics, London, Conference series No. 18, 1973) pp.46-49
5.24 S. Mahmoud, M. Muhammad, T. Mulvey: Unpublished results
5.25 F.Z. Marai: Ph. D. Thesis, University of Aston in Birmingham (1977)
5.26 F.Z. Marai, T. Mulvey: *Proc. 8th Int. Cong. Electron Microscopy, Canberra 1974*, ed. by J.V. Sanders, D.J. Goodchild (Australian Academy of Science, Canberra 1974), Vol.1, pp.130-131
5.27 F.Z. Marai, T. Mulvey: Ultramicroscopy *2*, 187-192 (1977)
5.28 F.Z. Marai, T. Mulvey, C.D. Newmann, G.K. Rickards: To be published
5.28a K.D. van der Mast, C.J. Rackels, J.B. Le Poole: *Proc. 7th Eur. Cong. Electron Microscopy, The Hague, 1980*, ed. by P. Brederoo, G. Boom (7th Eur. Cong. Electron Microscopy Foundation, Leiden 1980) Vol.1, pp.72-73
5.29 R.W. Moses: *Proc. 5th Eur. Cong. Electron Microscopy, Manchester 1972* (Institute of Physics, London 1972) pp.86-87
5.30 T. Mulvey: In *Scanning Electron Microscopy 1974*, ed. by O. Johari, I. Corvin (ITTRI, Chicago 1974) Part 1, pp.43-49
5.30a T. Mulvey, H. Nasr: *Proc. 7th Eur. Cong. Electron Microscopy, The Hague, 1980*, ed. by P. Brederoo, G. Boom (7th Eur. Cong. Electron Microscopy Foundation, Leiden 1980) Vol.1, pp.64-65
5.30b T. Mulvey, H. Nasr: Nucl. Instrum. Meth. *187*, 201-208 (1981)
5.31 T. Mulvey, C.D. Newman: In *Scanning Electron Microscopy – Systems and Applications*, ed. by W.C. Nixon (Institute of Physics, London, Conference series No. 18, 1973) pp.16-21
5.32 T. Mulvey, C.D. Newman: In *High-Voltage Electron Microscopy*, ed. by P. Swann, C. Humphreys, M. Goringe (Academic, London, New York 1974) pp.98-102
5.33 T. Mulvey, M.J. Wallington: Repts. Prog. Phys. *36*, 347-421 (1973)

5.34 E. Munro: In *Scanning Electron Microscopy 1974*, ed. by O. Johari, I. Corvin (ITTRI, Chicago 1974) Part 1, pp.35-42
E. Munro, O.C. Wells: In *Scanning Electron Microscopy 1976*, ed. by O. Johari, I. Corvin (ITTRI, Chicago 1976) Part 1, pp.27-36
5.35 E. Munro: *Proc. COMPUMAG Conference on the Computation of Magnetic Fields*, Oxford, 31 March - 2 April 1976 (The Rutherford Laboratory, Science Research Council, Oxon OX11 0QX, 1976) pp.35-46
5.36 C.D. Newman: Ph. D. Thesis, University of Aston in Birmingham (1976)
5.37 M.J. Newman, C.W. Trowbridge, L.R. Turner: *Proc. 4th Int. Conf. Magnet Technology Brookhaven 1972*
M.J. Newman: *Proc. COMPUMAG Conference on the Computation of Magnet Fields*, Oxford, 31 March - 2 April 1976 (The Rutherford Laboratory, Science Research Council, Oxon OXII 0QX, 1976) pp.144-148
5.38 J. Podbrdský: *Proc. 7th Eur. Cong. Electron Microscopy, The Hague, 1980*, ed. by P. Brederoo, G. Boom (7th Eur. Cong. Electron Microscopy Foundation, Leiden 1980) Vol.1, pp.66-67
5.39 E. Ruska: Z. Phys. *87*, 580-602 (1934)
5.40 K.C.A. Smith, D.J. Swann: UK Patent 1 291 221 (1969)
5.41 G. Stabenow: Z. Phys. *96*, 634-642 (1935)
5.42 W. Tretner: Optik *11*, 312-326 (1954) (Erratum, Optik *12*, 293 (1955)); Optik *13*, 516-519 (1956); Optik *16*, 155-184 (1959)
5.43 M. Troyon: *Proc. 7th Eur. Cong. Electron Microscopy, The Hague, 1980*, ed. by P. Brederoo, G. Boom (7th Eur. Cong. Electron Microscopy Foundation, Leiden 1980) Vol.1, pp.56-57
5.44 M. Troyon, A. Laberrigue: J. Microsc. Spectrosc. Electron *2*, 7-11 (1977)
5.45 J.A. Venables, G.D. Archer: *Proc. 7th Eur. Cong. Electron Microscopy, The Hague, 1980*, ed. by P. Brederoo, G. Boom (7th Eur. Cong. Electron Microscopy Foundation, Leiden 1980) Vol.1, pp.67-72
5.46 L.H. Veneklasen: Ph. D. Thesis, Cornell University, Ithaca NY (1971)

Appendix A
Some Earlier Sets of Curves Representing Lens Properties

Chosen by P. W. Hawkes

With 5 Figures

Chapter 3 by Lenz contains a comprehensive set of curves from which the paraxial properties and aberration coefficients of all normal lens geometries may be read directly. Various sets of such curves have been published in the past, a few of which are reproduced here as it was felt that readers might like to examine these results, so heavily used in the past. It is to be stressed that this is merely an anthology, very far from comprehensive, of earlier findings.

Figure A.1a shows the curves published in 1947 by VAN MENTS and LE POOLE [A.9] for focal length and position of the focus for objective lenses, for various values of S/D, as a function of the excitation $K = (NI)^2/U$, where U is the relativistic accelerating voltage. Figure A.1b shows the curves for projector focal length obtained by the same authors. These results were obtained by field measurement followed by numerical calculation, and it was assumed that there was no saturation.

In 1950, LIEBMANN [A.6] described a resistance network for solving Laplace's equation and by combining this with a numerical trajectory tracing technique [A.5] calculated the paraxial properties and main aberration coefficients of a wide range of magnetic lenses. The results were published in a series of papers, culminating in "A unified representation of magnetic electron lens properties" [A.8]; this, together with the earlier paper of LIEBMANN and GRAD [A.7], was for many years a standard source of information in this domain. The first of these, [A.7], is concerned with symmetrical lenses and contains graphs showing the dependence of objective and projector focal length and focal position for a range of values of S/D between 0.2 and 2 on excitation k^2, defined as $k^2 = 0.022\, H_0^2 R^2/U$, where H_0 denotes the maximum axial field (in gauss) and R = D/2 (in cm). The aberration coefficients C_s/R and C_c/R are plotted against k^2 for the same values of S/D, as are the isotropic and anisotropic coma coefficients. Some of these curves are reproduced in Fig.A.2. The second paper, [A.8], contains curves applicable to asymmetric as well as symmetric lenses; unified representations are given for focal length, focal distance, and spherical aberration coefficient as functions of $U/(NI)^2$ and its reciprocal (Fig.A.3).

A series of papers by DURANDEAU and FERT culminated in two long detailed studies of lens properties by DURANDEAU and FERT [A.3] and DUGAS et al. [A.2].

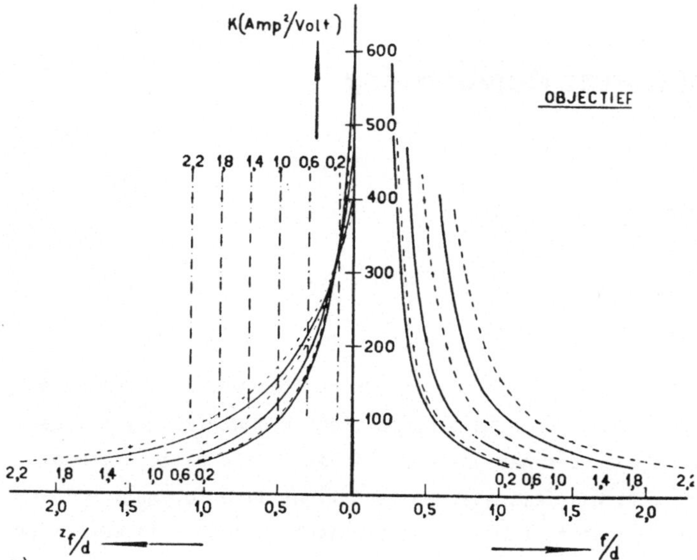

a)

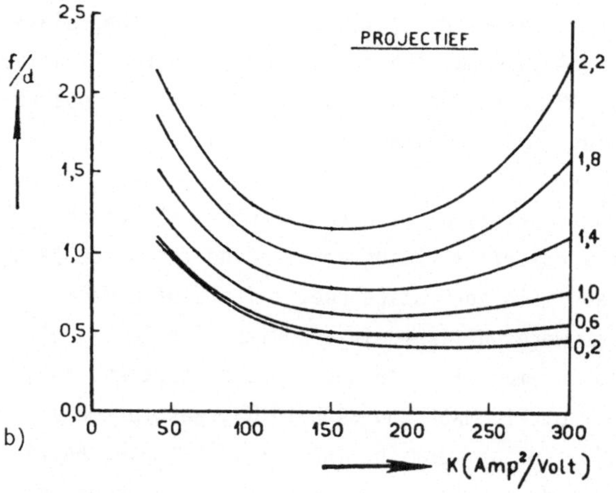

b)

Fig.A.1. (a) Objective focal length and focal distance (measured from the mid-plane) as a function of $K = (NI)^2/U$; d is the bore diameter (now denoted by D). The curves correspond to different values of S/D. (b) Projector focal length as a function of K for various values of S/D. ([A.9], courtesy of the authors and Martinus Nijhoff Publishers)

The first of these is restricted to symmetric lenses while the second extends the earlier findings to the asymmetric case. As discussed in some detail by Lenz (Chap.3), these papers introduced a more satisfactory scaling unit than D; extensive sets of curves showing the paraxial properties, C_s and C_c, as functions of excitation are given. The unit of length is the minimum projector focal length and the excitation is scaled with respect to the corresponding excitation $(NI)_0$. Figure A.4 contains a small selection from these papers.

A new way of presenting design curves was introduced almost simultaneously by KAMMINGA et al. [A.4] and BROOKES et al. [A.1] in 1968/1969. They perceived that the constraint imposed by the fact that the magnetic field cannot exceed a certain

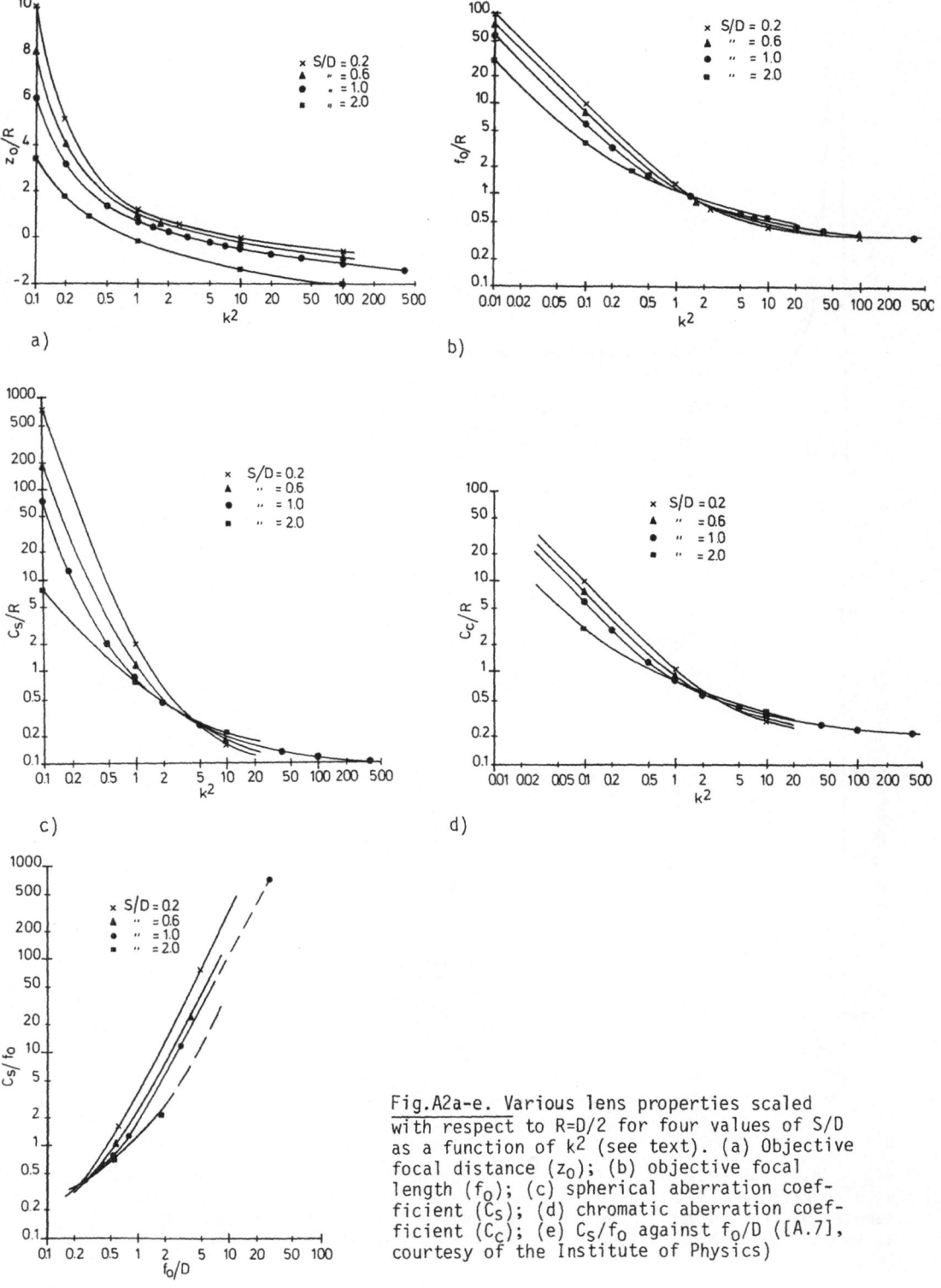

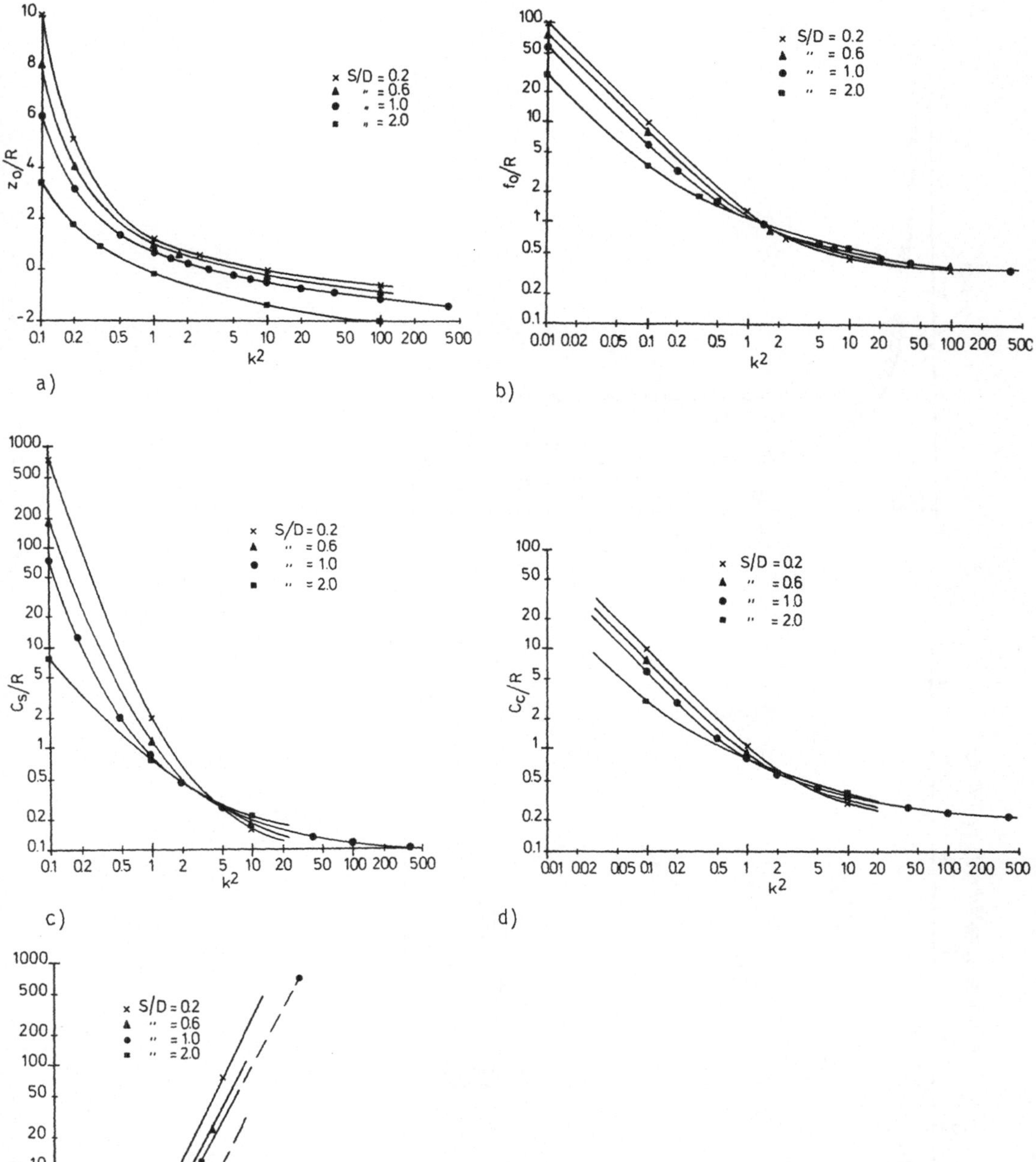

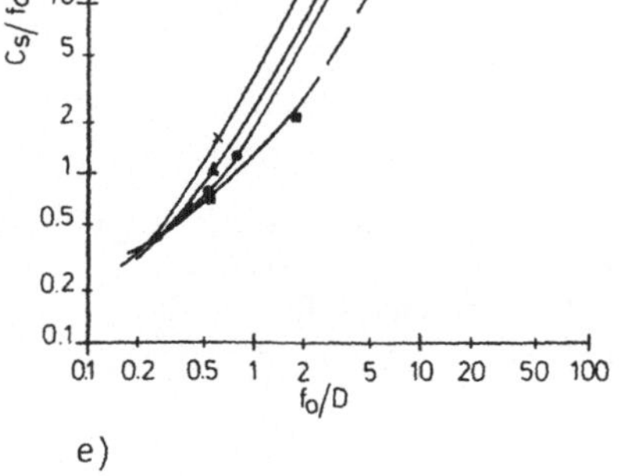

Fig.A2a-e. Various lens properties scaled with respect to R=D/2 for four values of S/D as a function of k^2 (see text). (a) Objective focal distance (z_0); (b) objective focal length (f_0); (c) spherical aberration coefficient (C_s); (d) chromatic aberration coefficient (C_c); (e) C_s/f_0 against f_0/D ([A.7], courtesy of the Institute of Physics)

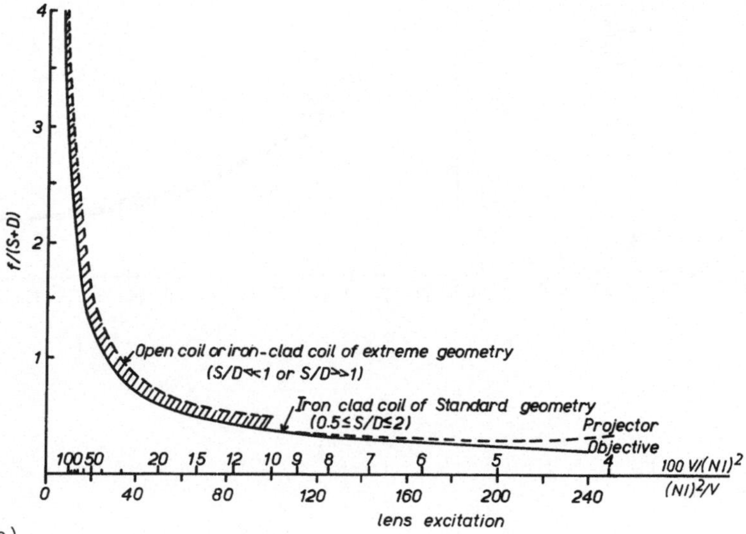

a)

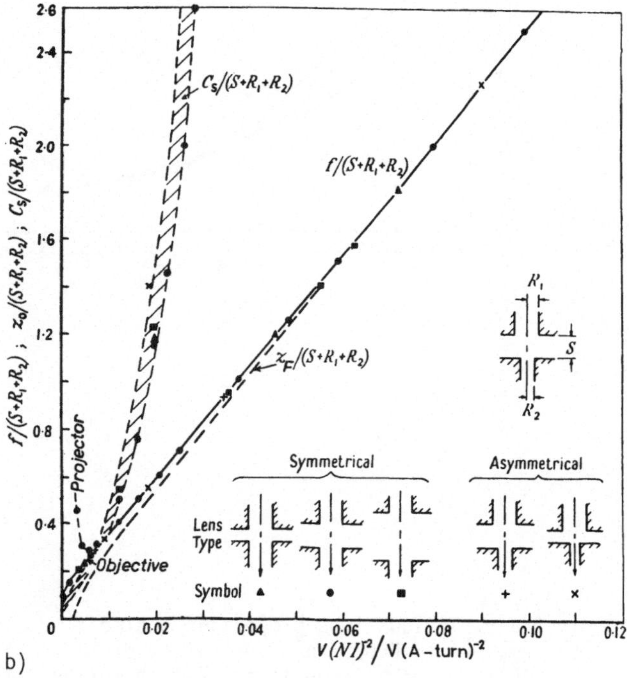

b)

Fig.A.3a,b. Two sets of "unified" curves for magnetic lens properties. (a) f/(S+D) as a function of $(NI)^2/V$ (V is the relativistic accelerating voltage, elsewhere denoted U); (b) Curves suitable for asymmetric lenses with bore radii R_1 and R_2, as shown inset ([A.8], courtesy of the Institute of Physics)

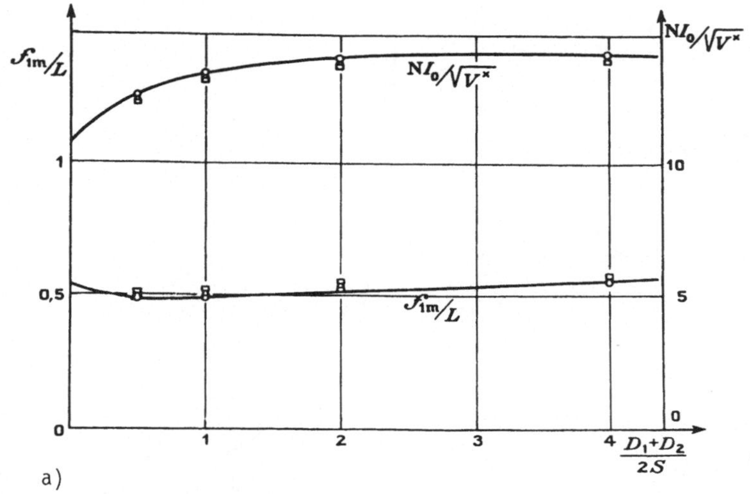

a)

(Fig.A4b-i: see pages 418-419)

Fig.A.4a-i. A selection of the results of DUGAS et al. [A.2]. (a) Minimum projector-focal length (f_{1m}) and corresponding excitation $NI_0/V^{1/2}$; the length L (cf. Chap.3) is defined by $L^2 = S^2 + 0.45\ [(D_1 + D_2)/2]^2$; (b) *Symmetric lenses.* C_c/f_{1m} and C_c/f_0 as functions of NI/NI_0. o: D/S=0.5; x: D/S=1; Δ: D/S=2; □:D/S=0; (c) *Symmetric lenses.* C_s/f_{1m} as a function of NI/NI_0. I:D/S=0.5; II:D/S=1; III:D/S=2; (d) *Asymmetric lenses.* $(D_1 + D_2)/2S=0.5$. Paraxial properties (suffix 0, objective; suffix 1, projector). Δ:$D_1/D_2=0.5$; □=$D_1/D_2=2$; (e) *Asymmetric lenses.* $(D_1 + D_2)/2S=0.5$. C_s/f_0 against NI/NI_0 for several values of D_1/D_2. o:$D_1/D_2=3$; □:$D_1/D_2=2$; •:$D_1/D_2=1$; Δ:$D_1/D_2=0.5$; x:$D_1/D_2=0.33$; (f) *Asymmetric lenses.* $(D_1+D_2)/2S=1$. Otherwise, as (d); (g) *Asymmetric lenses.* $(D_1+D_2)/2S=1$. Otherwise as (e); (h) *Asymmetric lenses.* $(D_1+D_2)/2S=2$. Otherwise as (d); (i) *Asymmetric lenses.* $(D_1+D_2)/2S=2$. Otherwise as (e) ([A.2], courtesy of the authors and Revue d'Optique)

maximum value (B_m) if saturation is to be avoided can be incorporated into these curves. In Fig.A.5, a number of curves from [A.4] are reproduced. Figure A.5a shows $C_s\ B_m/\mu_0\ U^{1/2}$ as a function of β = S/D for various values of lens excitation, $K = (NI)^2/U$. Figure A.5b shows $C_c\ B_m/\mu_0\ U^{1/2}$, again as a function of β for several values of K. Other curves, not reproduced here, enable the designer to meet the conflicting requirements of the illumination system, high-resolution operation, and absence of saturation.

Further discussion of the use of such design curves is to be found in the article by MULVEY and WALLINGTON [A.10] and in particular in their long survey [A.11], where many such curves are reproduced and replotted and some errors indicated. The few curves reproduced here are intended merely to give some flavour of the earlier work, since the curves in Chap.3 have all been newly prepared for this volume.

Fig.A.4b-e.

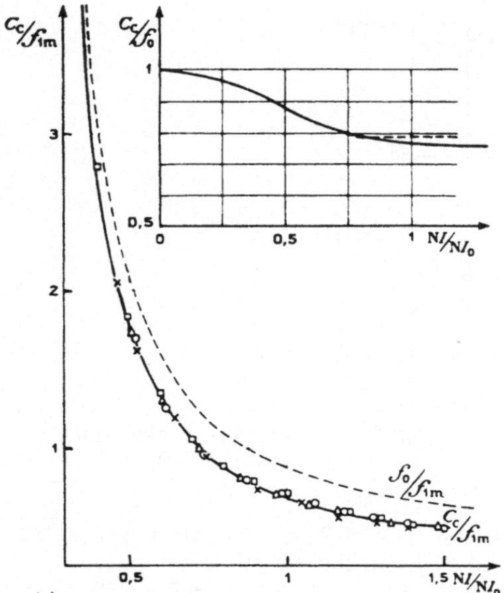

b)

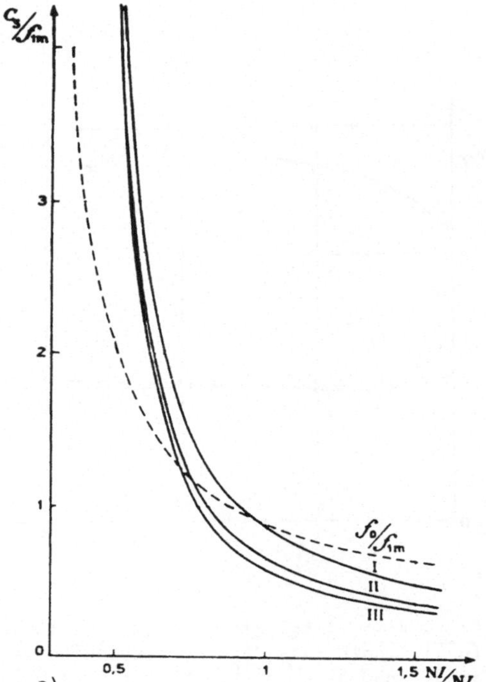

c)

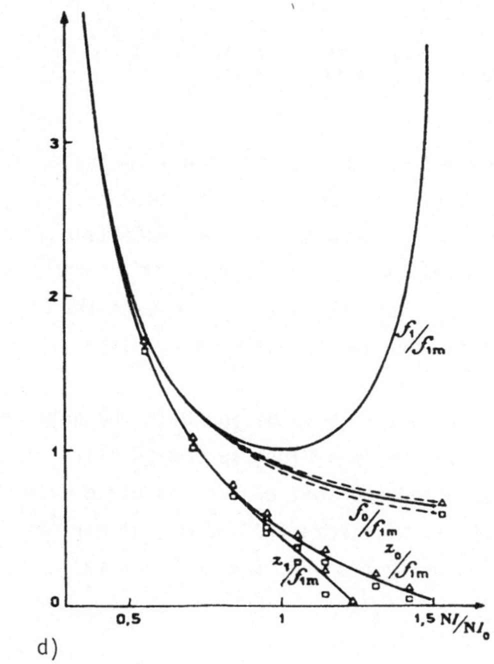

d)

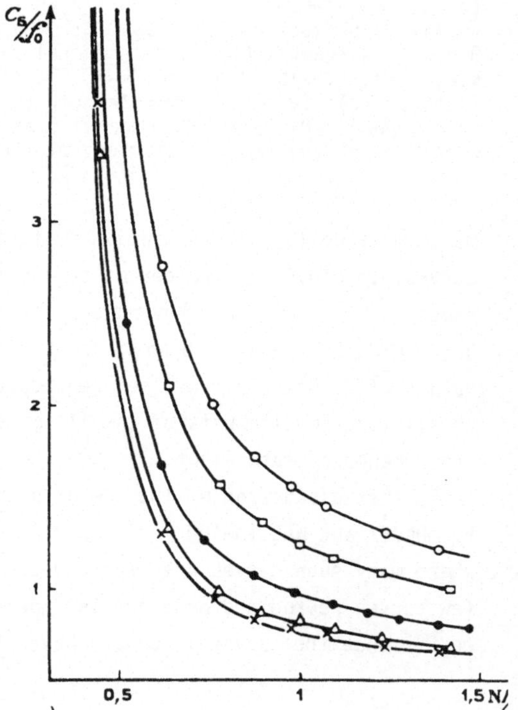

e)

Fig.A.4f-i.

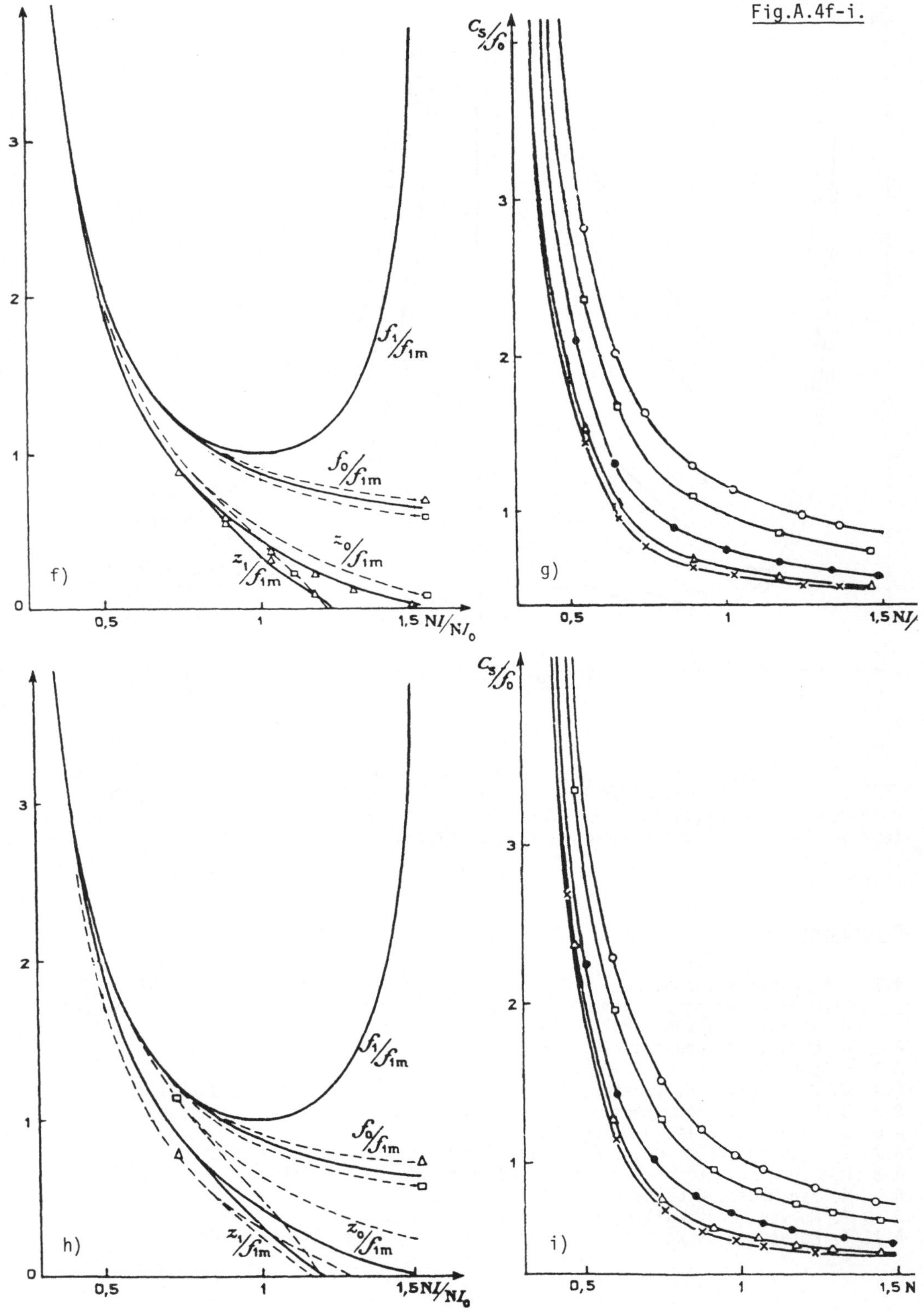

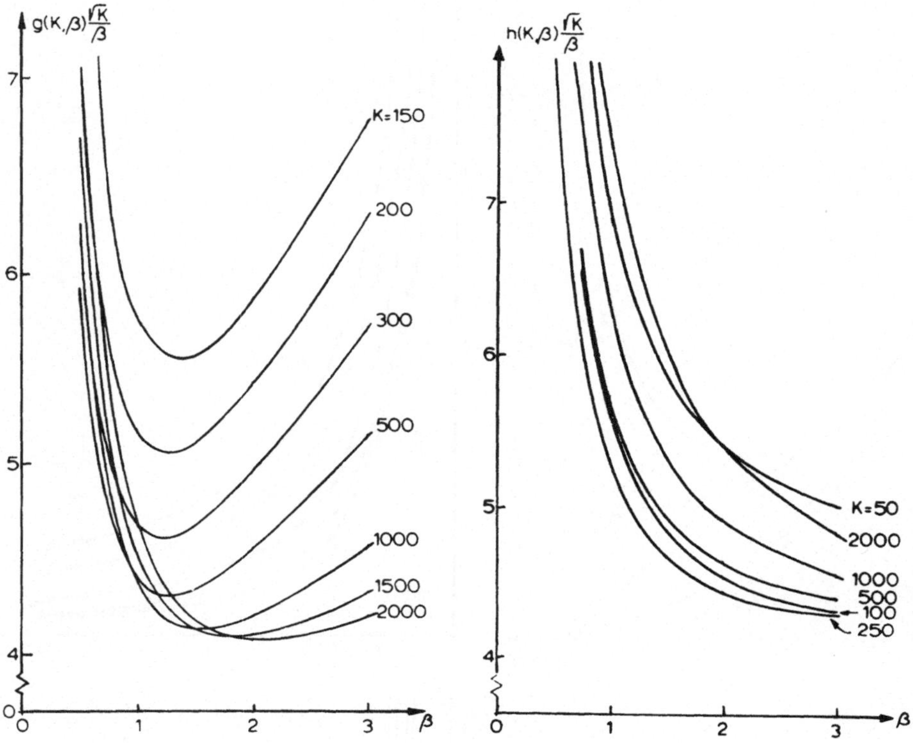

a) b)

Fig.A.5. (a) Graphs of $C_s\ B_m/\mu_0\ U^{\frac{1}{2}}$, here denoted by $g(K,\beta)\sqrt{K}/\beta$, as a function of $\beta = S/D$ for various values of the excitation $K = (NI)^2/U$; (b) Graphs of $C_c/B_m/\mu_0\ U^{\frac{1}{2}}$, here denoted by $h(K,\beta)\sqrt{K}/\beta$, as a function of β for various values of K ([A.4], courtesy of the authors and Wissenschaftliche Verlagsgesellschaft)

References

A.1 K.A. Brookes, T. Mulvey, M.J. Wallington: *Proc. 4th European Regional Conf. Electron Microscopy*, Rome, 1968, ed. by D.S. Bocciarelli (Tipografia Poliglotta Vaticana, Rome 1968) Vol.I, pp.165-166

A.2 J. Dugas, P. Durandeau, C. Fert: Rev. Opt. *40*, 277-305 (1961)

A.3 P. Durandeau, C. Fert: Rev. Opt. *36*, 205-234 (1957)

A.4 W. Kamminga, J.L. Verster, J.C. Francken: Optik *28*, 442-461 (1968/1969)

A.5 G. Liebmann: Proc. Phys. Soc. Lond. B*62*, 753-772 (1949)

A.6 G. Liebmann: Brit. J. Appl. Phys. *1*, 92-103 (1950)

A.7 G. Liebmann, E.M. Grad: Proc. Phys. Soc. Lond. B*64*, 956-971 (1951)

A.8 G. Liebmann: Proc. Phys. Soc. Lond. B*68*, 737-745 (1955)

A.9 M. Van Ments, J.B. Le Poole: Appl. Sci. Res. B*1*, 3-17 (1947)

A.10 T. Mulvey, M.J. Wallington: J. Phys. E *2*, 466-472 (1969)

A.11 T. Mulvey, M.J. Wallington: Repts. Prog. Phys. *36*, 347-421 (1973)

Appendix B
Bibliography of Publications on Magnetic Electron Lens Properties

Compiled by P. W. Hawkes

As explained in the preface, this bibliography is restricted to papers dealing, however briefly, with the relation between lens geometry, lens excitation, paraxial properties, and aberration coefficients. It includes, for example, theoretical papers giving formulae for the latter but excludes papers on field calculation, trajectory tracing (Chap.2), and many topics dealt with in Chap.4. Nevertheless, these criteria have not been applied too severely: borderline cases have usually been left in and a number of papers are included as much because they have frequently been referred to in this connection as because their actual contents justify their presence.

The list includes publications in journals, however recondite, and in published conference proceedings but largely excludes unpublished material, such as internal company memoranda and theses. It also excludes all but a very few books, for it would otherwise have been necessary – and not very helpful – to include every book on electron optics and many on electron microscopy. It is hoped that the list is reasonably complete so far as Western Europe and the United States are concerned. The coverage for Eastern Europe, Japan and, to a lesser extent, the Soviet Union is probably much more patchy, and the editor will be most grateful to have errors and omissions drawn to his attention.

References are listed alphabetically by first author and chronologically thereafter. Two series of conference proceedings are referred to merely by location and date; these are the European and International Congresses on Electron Microscopy, which have alternated biennially since 1954, prior to which they were not quite so regular. Bibliographic details of these conferences are given at the end of the list of references. The proceedings of the annual meetings of the Electron Microscope Society of America, which have been published since 1967 (with the exception of 1978 when a separate conference was not held since the International Congress met in Toronto), are referred to as EMSA followed by the number of the meeting and the date. These EMSA volumes are all published by Claitor, Baton Rouge (LA) and edited first by C.J. Arceneaux (25th - 32nd meeting) and subsequently by G.W. Bailey (33rd meeting onwards). Cyrillic characters have been transliterated according to the convention adopted by the journals containing English translations of Russian material; note that the French and German conventions are not quite identical with each other or with the English one.

Many people have contributed to the compilation of this list by sending photocopies or the missing details of elusive papers. I take this opportunity of thanking them all for their help; by far my largest single debt of gratitude is due to the Librarian and Staff of the Scientific Periodicals Library in Cambridge, who went of out of their way to be helpful and to whom I particularly wish to express my thanks.

Bibliography

Akashi, K., Mashimo, M., Tochigi, H., Uchida, H., Shirai, S.: " A New Akashi High Resolution Transmission Electron Microscope 'Model S-500', Provided with a New Second-Zone Objective Lens", Grenoble 1970, Vol.1, pp.143-144

Al-Hilly, S.M., Mulvey, T.: "Wide-Angle Projector Systems for the Transmission Electron Microscope", in *Electron Microscopy and Analysis 1981*, ed. by M.J. Goringe (Institute of Physics, Bristol 1982) pp.103-106

Alshwaikh, A., Mulvey, T.: "The Magnetised Iron Sphere, a Realistic Theoretical Model for Single-Polepiece Lenses", in *Developments in Electron Microscopy and Analysis, 1977*, ed. by D.L. Misell (Institute of Physics, Bristol 1977) pp.25-28

Andersen, W.H.J.: "New Aspects in the Design of Electron Optical Magnification Systems", EMSA 30, 1972, pp.606-607

Andersen, W.H.J.: "New Aspects in the Design of Electron Optical Magnification Systems", Manchester 1972, pp.128-129

Anderson, K., Brookes, K.A., Fibow, D.C.: "The Double-Field Condenser Objective Lens", EMSA 33, 1975, pp.136-137

Anger, K., Frosien, J., Lischke, B.: Design and testing of a corrected projection lens system for electron-beam lithography. Siemens Forsch. Entwicklungsber. *9*, 174-178 (1980)

Archard, G.D.: Magnetic lens aberrations due to mechanical defects. J. Sci. Instrum. *30*, 352-358 (1953)

Archard, G.D.: On the spherical aberration constant. Rev. Sci. Instrum. *29*, 1049-1050 (1958)

Archard, G.D.: Fifth order spherical aberration of magnetic lenses. Brit. J. Appl. Phys. *11*, 521-522 (1960)

Ardenne, von, M.: Zur Größe des Öffnungsfehlers beim Elektronenmikroskop. Z. Tech. Phys. *20*, 289-290 (1939)

Ardenne, von, M.: Über ein Universal-Elektronenmikroskop für Hellfeld-, Dunkelfeld- und Stereobild-Betrieb. Z. Phys. *115*, 339-368 (1940)

Ardenne, von, M.: Zur Prüfung von kurzbrennweitigen Elektronenlinsen. Eine einfache Methode und ihre Ergebnisse. Z. Phys. *117*, 602-611 (1941)

Ardenne, von, M.: Über ein 200 kV-Universal-Elektronenmikroskop mit Objektabschattungsvorrichtung. Z. Phys. *117*, 657-688 (1941)

Ardenne, von, M.: Ergänzung zu den Arbeiten 'Zur Prüfung von kurzbrennweitigen Elektronenlinsen' und 'Über ein 200-kV-Universal-Elektronenmikroskop mit Objektabschattungsvorrichtung'. Z. Phys. *118*, 384-388 (1941)

Ardenne, von, M.: Über ein neues Universal-Elektronenmikroskop mit Hochleistungsmagnet-Objektiv und herabgesetzter thermischer Objektbelastung. Kolloid. Z. *108*, 195-208 (1944)

Ardenne, von, M.: *Tabellen der Elektronenphysik, Ionenphysik und Übermikroskopie*, Vols.1,2 (Deutscher Verlag der Wissenschaften, Berlin 1956) [Volume 1 contains information about lens properties (pp.51-63) and about aberrations and their effect on image quality (pp.287-326)]

Ardenne, von, M.: *Tabellen zur angewandten Physik. I. Elektronenphysik, Übermikroskopie, Ionenphysik* (Deutscher Verlag der Wissenschaften, Berlin 1962; 2nd ed. 1973) [Magnetic lens properties are described on pp.53-64]

Asmus, A., Herrmann, K.-H., Wolff, O.: Elmiskop 101 – a new high-power electron microscope. Siemens Rev. *36*, No.2, 57-67 (1969)

Baba, N., Oikawa, T., Kanaya, K.: "Properties of Magnetic Lenses Whose Axial Field Distribution Given by $B(z/a)=B_{max}(1-\mu^2)^{(2m-1)/2}$ for HVEM", in *High Voltage Electron Microscopy, 1977*, ed. by T. Imura, H. Hashimoto (Japanese Society of Electron Microscopy, Kyoto 1977) pp.45-48
Baba, N., Nagashima, K., Kanaya, K.: "Magnetic Einzel Lens Satisfying Anastigmatic and Achromatic Conditions", Toronto 1978, Vol.1, pp.26-27
Baba, N., Kanaya, K.: General theory of the magnetic einzel lens based on the axial field distribution model $H(z)=H_0\, C|z/a|^{m-1}/(1+|z/a|^{2m})$. J. Phys. E *12*, 525-537 (1979)
Balladore, J.L., Murillo, R.: Nouveau type de lentille magnétique pour microscope électronique. J. Microsc. Spectrosc. Electron. *2*, 211-222 (1977)
Balladore, J.L., Murillo, R., Trinquier, J., Jouffrey, B.: "New Type of Magnetic Lens for Electron Microscopy", in *High Voltage Electron Microscopy, 1977*, ed. by T. Imura, H. Hashimoto (Japanese Society of Electron Microscopy, Tokyo 1977) pp.41-44
Balladore, J.L., Murillo, R., Hawkes, P.W., Trinquier, J., Jouffrey, B.: Magnetic electron lenses for high voltage microscopes. Nucl. Instrum. Meth. *187*, 209-215 (1981)
Balossier, G., Laberrigue, A., Génotel, D., Homo, J.C.: "Développement et amélioration d'un microscope 400 kV à lentilles supraconductrices", Jerusalem 1976, Vol.1, pp.338-339
Barbier, M.: Eine Näherungsmethode zur Bildfehlerberechnung der Elektronenoptik. Mitt. Inst. Tech. Phys. ETH Zürich *3*, 28 (1951)
Barbier, M.: Calcul pratique des aberrations du troisième ordre dans les systèmes centrés de l'optique électronique. Ann. Radioélectr. *8*, 111-133 (1953)
Barnes, R.L., Openshaw, I.K.: A comparison of experimental and theoretical C_s values for some probe-forming lenses. J. Phys. E *1*, 628-630 (1968)
Barthère, J., Dugas, J., Durandeau, P.: Coefficient d'aberration de sphéricité des lentilles électroniques magnétiques dissymétriques. C.R. Acad. Sci. Paris *250*, 3461-3463 (1960)
Basin, L.A., Bobykin, B.V., Davydovskii, V.Ya., Kel'man, V.M.: Zh. Tekh. Fiz. *40*, 2106-2111 (1970) [English transl.: Magnetic lenses with parabolic, triangular and trapezoidal distributions of linear ampere-turn density. Sov. Phys.-Tech. Phys. *15*, 1640-1645 (1970)]
Basin, L.A., Bobykin, B.V., Davydovskii, V.Ya., Ermekov, N.T., Kel'man, V.M.: Zh. Tekh. Fiz. *41*, 1485-1488 (1971) [English transl.: Effect of the winding thickness on the electron optical parameters of air-core magnetic lenses. Sov. Phys.-Tech. Phys. *16*, 1168-1170 (1971)]
Bassett, R., Mulvey, T.: "Iron-Free Lenses for Electron Probe-Forming Systems", in *Fifth International Congress on X-Ray Optics and Microanalysis, Tübingen, 1968*, ed. by G. Möllenstedt, K.-H. Gaukler (Springer, Berlin, Heidelberg, New York 1969) pp.224-230
Bassett, R., Mulvey, T.: Flat helical electron-optical lenses. Z. Angew. Phys. *27*, 142-145 (1969)
Bauer, M.: Magnetické pole v objektivu elektronového mikroskopu při nasycení pólových nástavců. Elektrotech. Čas. *18*, 10-26 (1967)
Bauer, M., Kamenický, J.: Zařízení pro měření magnetického pole na ose rotačně souměrných magnetických čoček. Elektrotech. Čas. *19*, 221-227 (1968)
Bauer, M.: Některé poznatky z řešení polí v magnetické čočce elektronového mikroskopu. Slaboproudý Obzor *29*, 132-137 (1968)
Bauer, M.: Pole piece saturation in magnetic electron lenses. J. Phys. E *1*, 1081-1089 (1968)
Beaumont, S.P., Ahmed, H.: "Combined CTEM and STEM using a Condenser-objective Lens and 100 kV LaB_6 Gun", Toronto 1978, Vol.1, pp.2-3
Becker, H., Wallraff, A.: Über die sphärische Aberration magnetischer Linsen. Arch. Elektrotech. *32*, 664-675 (1938)
Becker, H., Wallraff, A.: Die Bildwölbung bei magnetischen Linsen. Arch. Elektrotech. *33*, 491-505 (1939)
Becker, H., Wallraff, A.: Der Astigmatismus magnetischer Linsen. Arch. Elektrotech. *34*, 43-48 (1940)
Becker, H., Wallraff, A.: Bildfehleruntersuchungen an einer bilddrehungsfreien magnetischen Linse. Arch. Elektrotech. *34*, 115-120 (1940)

Becker, H., Wallraff, A.: Über Bildfehlermessungen an einer eisengekapselten Linse mit veränderlichem Luftspalt. Arch. Elektrotech. *34*, 230-236 (1940)
Berjot, G., Bonhomme, P., Payen, F., Beorchia, A., Mouchet, J, Laberrigue, A.: "Résultats en microscopie électronique avec objectifs supraconducteurs", Grenoble 1970, Vol.2, pp.103-104
Bernhard, W., Koops, H.: Kompensation der Farbabhängigkeit der Vergrösserung und der Farbabhängigkeit der Bilddrehung eines Elektronenmikroskops. Optik *47*, 55-64 (1977)
Boersch, H., Bostanjoglo, O., Grohmann, K.: Supraleitender Hohlzylinder als magnetische Linse. Z. Angew. Phys. *20*, 193-194 (1966)
Boersch, H., Bostanjoglo, O., Lischke, B.: Helium-Kühlbauteil für hochauflösende Elektronenmikroskopie und supraleitender Hohlzylinder als magnetische Linse. Optik *24*, 460-465 (1966/1967)
Bonhomme, P., Laberrigue, A.: Hautes stabilités électriques et mécaniques d'un objectif supraconducteur. J. Microsc. (Paris) *8*, 795-798 (1969)
Bonhomme, P., Beorchia, A., Laberrigue, A.: Expériences sur les fonctions de transfert du contraste de phase d'un objectif supraconducteur. C.R. Acad. Sci. Paris B*274*, 1217-1220 (1972)
Bonjour, P., Septier, A.: Lentilles magnétiques supraconductrices à pièces polaires en métaux des terres rares. C.R. Acad. Sci. Paris B*264*, 747-750 (1967)
Bonjour, P., Septier, A.: Caractéristiques optiques d'une lentille supraconductrice à pôles de dysprosium. C.R. Acad. Sci. Paris B*265*, 1392-1395 (1967)
Bonjour, P., Septier, A.: "Eléments cardinaux et coefficients d'aberration des lentilles supraconductrices à pôles en métal de terres rares", Rome 1968, Vol.1, pp.189-190
Bonjour, P.: Eléments cardinaux et coefficients d'aberration de lentilles supraconductrices à pôles d'holmium. C.R. Acad. Sci. Paris B*268*, 23-26 (1969)
Bonjour, P.: "Premiers résultats obtenus avec une lentille supraconductrice à pôles d'holmium", Grenoble 1970, Vol.2, pp.105-106
Bonjour, P.: Une lentille magnétique supraconductrice à pôles d'holmium pour microscopie à très haute tension. J. Microsc. (Paris) *20*, 219-240 (1974)
Bonjour, P.: "Experimental Study of a Superconductive Objective Lens", Canberra 1974, Vol.1, pp.142-143
Bonjour, P.: "The Design of Highly Saturated Magnetic Lenses", Canberra 1974, Vol.1, 148-149
Bonjour, P.: New shapes for magnetic superconducting lenses for high voltage microscopy. IEEE Trans. MAG-*11*, 1470-1471 (1975)
Bonjour, P.: Proposed designs for superconducting lenses for high voltage electron microscopes. J. Phys. E *8*, 761-763 (1975)
Bonjour, P.: High field superconducting electron lenses with holmium polepieces. J. Phys. E *8*, 764-768 (1975)
Bonjour, P.: "Superconducting Lenses: Present Trends and Design", Jerusalem 1976, Vol.1, 73-78
Bonshtedt, B.E.: Method of determining a broad class of electrostatic and magnetic fields, for which the solutions of the fundamental equations of motion of electron optics are expressible in terms of known functions. Zh. Tekh. Fiz. *25*, 541-543 (1955) [in Russian]
Borries, von, B., Ruska, E.: Aufbau und Leistung des Siemens-Übermikroskops. Z. Wiss. Mikrosk. *56*, 317-333 (1939)
Borries, von, B., Ruska, E.: Die Technik des Siemens-Übermikroskops. Siemens Z. *20*, 217-227 (1940)
Borries, von, B., Ruska, E.: Mikroskopie hoher Auflösung mit schnellen Elektronen. Ergeb. Exakten Naturwiss. *19*, 237-322 (1940)
Borries, von, B., Ruska, E., Krumm, J., Müller, H.O.: Übermikroskopische Abbildung mittels magnetostatischer Linsen. Naturwissenschaften *28*, 350-351 (1940)
Borries, von, B.: Ein magnetostatisches Objektiv-Projektiv-System für das Elektronenmikroskop. Kolloid Z. *114*, 164-167 (1949)
Borries, von, B.: Ein magnetostatisches Gebrauchs-Elektronenmikroskop für 60 kV Strahlspannung. Z. Wiss. Mikrosk. *60*, 329-358 (1952)
Borries, von, B., Lenz, F., Opfer, G.: Über die Remanenz in Weicheisenkreisen magnetischer Elektronenlinsen. Optik *10*, 132-136 (1953)

Borries, von, B.: "The Physical Situation and the Performance of High Resolving Microscopy Using Fast Corpuscles", London 1954, pp.4-25
Borries, von, B., Johann, I., Huppertz, J., Langner, G., Lenz, F., Scheffels, W.: Die Entwicklung regelbarer permanentmagnetischer Elektronenlinsen hoher Brechkraft und eines mit ihnen ausgerüsteten Elektronenmikroskopes neuer Bauart. Forschungsber. Wirtsch. Verkehrsminist. Nordrhein-Westfalen 156, 88pp. (1956)
Borries, von, B., Lenz, F.: Vorausberechnung magnetostatischer elektronenoptischer Abbildungssysteme. Optik *13*, 264-276 (1956)
Borries, von, B., Langner, G.: "Sur les lentilles électroniques à aimant permanent", in *Les techniques récentes en microscopie électronique et corpusculaire* (C.N.R.S., Paris 1956) pp.285-295
Borries, von, B., Langner, G., Scheffels, W.: "Imaging Elements Operating with Permanent Magnets", Stockholm 1956, pp.14-17
Bremmer, H.: The derivation of paraxial constants of electron lenses from an integral equation. Appl. Sci. Res. B*2*, 416-428 (1952)
Bremmer, H.: Eine einfache Näherungsformel für die Feldverteilung längs der Achse magnetischer Elektronenlinsen mit ungesättigten Polschuhen. Optik *10*, 1-4 (1953)
Bremmer, H.: "Numerical Analysis of Magnetic Lens Parameters on a Theoretical Basis", London 1954, pp.89-92
Brookes, K.A., Mulvey, T., Wallington, M.J.: "Magnetic Objective Lenses of Minimum Spherical and Chromatic Aberration", Rome 1968, Vol.1, pp.165-166
Busch, H.: Berechnung der Bahn von Kathodenstrahlen im axialsymmetrischen elektromagnetischen Felde. Ann. Phys. (Leipzig) *81*, 974-993 (1926)
Busch, H.: Über die Wirkungsweise der Konzentrierungsspule bei der Braunschen Röhre. Arch. Elektrotech. *18*, 583-594 (1927)
Chapman, P.F., Stobbs, W.M.: A microprobe in a high voltage electron microscope. J. Phys. E *6*, 373-376 (1973)
Chiang Man-Ying: On the Petzval coefficient in electron optics. Acta Phys. Sinica *12*, 439-446 (1956) [in Chinese]
Christofides, S., Mulvey, T.: "A High Flux Density Single Polepiece Objective Lens", The Hague 1980, Vol.1, pp.70-71
Cleaver, J.R.A., Smith, K.C.A.: "Two-Lens Probe Forming Systems Employing Field Emission Guns", in *Scanning Electron Microscopy 1973*, ed. by O. Johari (IIT Research Institute, Chicago 1973) pp.49-56
Cleaver, J.R.A.: "The Design and Construction of the Condenser-Objective Lens for the Cambridge High Resolution Electron Microscope", in *Developments in Electron Microscopy and Analysis, 1977*, ed. by D.L. Misell (Institute of Physics, Bristol 1977) pp.17-20
Cleaver, J.R.A.: Some optical characteristics of the saturated symmetrical condenser-objective lens. Optik *49*, 413-431 (1978)
Cleaver, J.R.A.: Field emission electron gun systems incorporating single-pole magnetic lenses. Optik *52*, 293-303 (1978/1979)
Cleaver, J.R.A.: "The Single-Pole Magnetic Lens as a Condenser for Use with Field-Emission Electron Guns", in *Developments in Electron Microscopy and Analysis, 1979*, ed. by T. Mulvey (Institute of Physics, Bristol 1980) pp.55-58
Cleaver, J.R.A.: The choice of polepiece shape and lens operating mode for magnetic objective lenses with saturated polepieces. Optik *57*, 9-34 (1980)
Cleaver, J.R.A.: "Polepiece Shapes and Operating Modes for Saturated Magnetic Objective Lenses", The Hague 1980, Vol.1, 68-69
Cleaver, J.R.A.: The off-axis aberrations of magnetic objective lenses with saturated polepieces. Optik *58*, 409-432 (1981)
Coleman, J.W.: "Use of the Magnetostatic Analogue in Electromagnetic Lens Engineering", EMSA 28, 1970, pp.360-361
Cosslett, V.E.: A magnet lens for β-rays of high energy. J. Sci. Instrum. *17* 259-264 (1940)
Cosslett, V.E.: The resolving power of the magnetic electron lens used as a beta-ray spectrometer. Proc. Phys. Soc. London *52*, 511-517 (1940)
Cosslett, V.E.: The resolving power of the electron microscope. J. Sci. Instrum. *22* 170-174 (1945)
Cosslett, V.E.: The variation of resolution with voltage in the magnetic electron microscope. Proc. Phys. Soc. London *58*, 443-455 (1946)

Cosslett, V.E.: Conditions for extending the resolution limit of the electron microscope. J. Sci. Instrum. *24*, 40-43 (1947)
Cosslett, V.E.: "Present trends in electron microscopy", in Electron Physics, Nat. Bur. Stand. Cir. *527*, 291-303 (1954)
Cosslett, V.E.: "Introduction and State of the Art" [The Instrument Today and Tomorrow], Toronto 1978, Vol.3, pp.163-172
Cowley, J.M., Strojnik, A.: "A 600-kV Transmission Scanning Electron Microscope with Energy Analysis", Rome 1968, Vol.1, pp.71-72
Crewe, A.V., Parker, N.W.: Correction of third-order aberrations in the scanning electron microscope. Optik *46*, 183-194 (1976)
Crewe, A.V.: "Design Trends in the STEM", Jerusalem 1976, Vol.1, pp.65-66
Crewe, A.V.: Ideal lenses and the Scherzer theorem. Ultramicroscopy *2*, 281-284 (1977
De, M.L., Saha, D.K.: Distortion in electron lens. Ind. J. Phys. *28*, 263-268 (1954)
De, M.L.: On an analytical expression for the axial field of electromagnetic lenses. Philos. Mag. *7*, 1065-1067 (1962)
Dekker, H.: A new method of measuring the axial field distribution in a superconducting electron lens by means of the Faraday effect. J. Phys. E *5*, 368-372 (1972)
Der-Shvarts, G.V.: The effect of departures from rotational symmetry of the focusing field on the resolution of magnetic objective lenses for electron microscopes. Zh. Tekh. Fiz. *24*, 859-870 (1954) [in Russian]
Der-Shvartz, G.V., Rachkov, V.P.: Izv. Akad. Nauk SSSR Ser. Fiz. *25*, 676-679 (1961) [English transl.: Design of magnets for the lenses of magnetostatic electron microscopes. Bull. Acad. Sci. USSR Phys. Ser. *25*, 691-694 (1961)]
Der-Shvartz, G.V., Kushnir, Yu.M., Rozenfel'd, L.B., Zaitsev, P.V., Bezlepkin, S.V.: Izv. Akad. Nauk SSSR Ser. Fiz. *25*, 721-724 (1961) [English transl.: Modernization of the UEM-100 electron microscope. Bull. Acad. Sci. USSR Phys. Ser. *25*, 733-736 (1961)]
Der-Shvarts, G.V., Rachkov, V.P.: Radiotekh. Elektron. *10*, 922-928 (1965) [English transl.: Some electron optical properties of double-gap lenses. Radio Eng. Electron. Phys. (USSR) *10*, 783-789 (1965)]
Der-Shvarts, G.V., Makarova, I.S.: Radiotekh. Elektron. *11*, 89-93 (1966) [English transl.: Concerning the calculation of the spherical aberration of axisymmetric magnetic lenses. Radio Eng. Electron. Phys. (USSR) *11*, 72-75 (1966)]
Der-Shvarts, G.V., Makarova, I.S.: Radiotekh. Elektron. *12*, 168-171 (1967) [English transl.: Calculation of the spherical aberration of axisymmetrical magnetic lenses. Radio Eng. Electron. Phys. (USSR) *12*, 161-163 (1967)]
Der-Shvarts, G.V., Makarova, I.S.: Radiotekh. Elektron. *13*, 1268-1272 (1968) [English transl.: Electron optical properties of cryogenic objectives for high resolution electron microscopes. Radio Eng. Electron. Phys. (USSR) *13*, 1100-1103 (1968)]
Der-Shvarts, G.V., Makarova, I.S.: Izv. Akad. Nauk SSSR Ser. Fiz. *32*, 932-936 (1968) [English transl.: Evaluation of the electron optical characteristics of cryogenic electron microscope objective solenoids. Bull. Acad. Sci. USSR Phys. Ser. *32*, 866-870 (1968)]
Der-Shvarts, G.V.: Izv. Akad. Nauk SSSR Ser. Fiz. *32*, 937-941 (1968)[English transl.: Simulation of flux functions and design of magnetic circuits of axially symmetric electron microscope lenses. Bull. Acad. Sci. USSR Phys. Ser. *32*, 871-875 (1968)]
Der-Shvarts, G.V., Arkhipova, N.V., Krypnova, E.A.: Radiotekh. Elektron. *14*, 738-740 (1969) [English transl.: Fifth order spherical aberration of axially symmetric magnetic lenses. Radio Eng. Electron Phys. *14*, 639-641 (1969)]
Der-Shvarts, G.V.: "Approximate Analytic Expression for Calculating the Spherical Aberration Constant of Axially Symmetric Magnetic Lens with Yokes", in *Elektrofizicheskie i Elektrokhimicheskie Metody Obrabotki* [Electrophysical and Electrochemical Processing Methods] (Moscow 1970) Vol.4, pp.7-10 [in Russian]
Der-Shvarts, G.V.: Radiotekh. Elektron. *16*, 1305-1306 (1971) [English transl.: Calculation of the spherical aberration of axially symmetric magnetic lenses. Radio Eng. Electron. Phys. (USSR) *16*, 1240-1241 (1971)]
Der-Shvarts, G.V., Makarova, I.S.: Izv. Akad. Nauk SSSR Ser. Fiz. *36*, 1304-1311 (1972) [English transl.: Third-order field aberrations of axially symmetrical magnetic lenses. Bull. Acad. Sci. USSR Phys. Ser. *36*, 1164-1170 (1972)]

Diels, K., Knoll, M.: Nachweis der Bildfehler von Elektronenlinsen bei Abbildung eines Punktes. Z. Tech. Phys. *16*, 617-621 (1935)
Diels, K., Wendt, G.: Die 8 Bildfehler dritter Ordnung magnetischer Elektronenlinsen. Z. Tech. Phys. *18*, 65-69 (1937); also published in *Beiträge zur Elektronenoptik*, ed. by H. Busch, E. Brüche (Barth, Leipzig 1936) pp.19-24
Dietrich, I., Weyl, R., Zerbst, H.: High magnetic field gradient for electron microscopy. Cryogenics *7*, 178-179 (1967)
Dietrich, I., Weyl, R.: "A Superconducting Electron Microscope Lens", in *Proc. 2nd Int. Cryogenic Engineering Conf.*, Brighton 1968 (Iliffe, Guildford 1968) pp.210-212
Dietrich, I., Maier, R.G., Weyl, R., Zerbst, H.: "Investigations on a Superconducting Shielding Container with an Enclosed Coil", in *Proc. 3rd Int. Cryogenic Engineering Conf.*, Berlin 1970 (Iliffe, Guildford 1970) pp.422-425
Dietrich, I., Weyl, R., Zerbst, H.: "Electronoptical Testing on the Rotational Symmetry of a Superconducting Lens", Grenoble 1970, Vol.2, pp.101-102
Dietrich, I., Koller, A., Lefranc, G.: Feldverteilung in eisenfreien supraleitenden Linsen. Optik *35*, 468-478 (1972)
Dietrich, I., Weyl, R., Zerbst, H.: "An Electron Optical 400-kV System for Testing a Superconducting Lens", Manchester 1972, pp.120-121
Dietrich, I., Lefranc, G., Weyl, R., Zerbst, H.: Supraleitendes Abbildungssystem hoher Auflösung für die Elektronenmikroskopie. Optik *38*, 449-453 (1973)
Dietrich, I., Fox, F., Weyl, R., Zerbst, H.: "A High-Voltage Superconducting Lens", in *High Voltage Electron Microscopy, Proc. 3rd Int. Conf.*, ed. by P.R. Swann, C.J. Humphreys, M.J. Goringe (Academic, London 1974) pp.103-107
Dietrich, I.: "Superconducting Lenses for Electron Microscopy", in *Proc. 5th Int. Cryogenic Engineering Conf.*, Kyoto 1974 (IPC, Guildford 1974) pp.319-321
Dietrich, I., Herrmann, K.-H., Passow, C.: A proposal for a high voltage electron microscope with superconducting microwave linear accelerator and superconducting lenses. Optik *42*, 439-462 (1975)
Dietrich, I., Knapek, E., Weyl, R., Zerbst, H.: Superconducting lenses in electron microscopy. Cryogenics *15*, 691-699 (1975)
Dietrich, I.: *Superconducting Electron-optic Devices* (Plenum, New York 1976)
Dietrich, I., Fox, F., Knapek, E., Lefranc, G., Nachtrieb, K., Weyl, R., Zerbst, H.: "High Resolution Imaging with a Superconducting Electron Microscope Lens System with Helium Cooled Specimen Stage", Jerusalem 1976, Vol.1, pp.405-407
Dietrich, I., Fox, F., Knapek, E., Lefranc, G., Nachtrieb, K., Weyl, R., Zerbst, H.: Improvements in electron microscopy by application of superconductivity. Ultramicroscopy *2*, 241-249 (1977)
Dietrich, I.: "Superconducting Lenses", Toronto 1978, Vol.3, pp.173-184
Dietrich, I., Lefranc, G., Müller, K.-H., Stemmer, A.: "Testing of a Superconducting Lens System for Commercial Microscopes", The Hague 1980, Vol.1, pp.84-85
Dorsten van, A.C., Oosterkamp, W.J., Le Poole, J.B.: An experimental electron microscope for 400 kilovolts. Philips Tech. Rev. *9*, 193-201 (1947)
Dorsten van, A.C., Nieuwdorp, H., Verhoeff, A.: The Philips 100 kV electron microscope. Philips Tech. Rev. *12*, 33-51 (1950/1951)
Dorsten van, A.C., Le Poole, J.B.: The EM 75 kV, an electron microscope of simplified construction. Philips Tech. Rev. *17*, 47-59 (1955/1956)
Dosse, J.: Strenge Berechnung magnetischer Linsen mit unsymmetrischer Feldform nach $H = H_0/[1 + (z/a)^2]$. Z. Phys. *117*, 316-321 (1941)
Dosse, J.: Über optische Kenngrößen starker Elektronenlinsen. Z. Phys. *117*, 722-753 (1941)
Dosse, J.: Ergänzung zur Arbeit "Über optische Kenngrößen starker Elektronenlinsen". Z. Phys. *118*, 375-383 (1941)
Drahoš, V.: "Electron Optical Imaging Systems", Manchester 1972, pp.34-39
Dugas, J., Durandeau, P., Fagot, B.: "Caractéristiques électrooptiques paraxiales des lentilles électroniques magnétiques dissymétriques", Delft 1960, Vol.1, pp.35-40
Dugas, J., Durandeau, P., Fert, C.: Lentilles électroniques magnétiques symétriques et dissymétriques. Rev. Opt. *40*, 277-305 (1961)
DuMond, J.W.M.: Conditions for optimum luminosity and energy resolution in an axial β-ray spectrometer with homogeneous magnetic field. Rev. Sci. Instrum. *20*, 160-169 (1949)

Dupouy, G.: "L'optique électronique magnétique, application au microscope électronique magnétique", in *L'optique électronique, Réunions Louis de Broglie* (Editions Revue d'Optique, Paris 1946) pp.161-208
Dupouy, G.: Microscope électronique magnétique à grand pouvoir de résolution. J. Phys. Radium *7*, 320-329 (1946)
Dupouy, G., Perrier, F., Trinquier, J.: Contribution à la correction de l'aberration chromatique du microscope électronique magnétique. C.R. Acad. Sci. Paris *257*, 4099-4104 (1963)
Dupouy, G., Perrier, F.: Microscope électronique à très haute tension. Ann. Phys. (Paris) *8*, 251-260 (1963)
Dupouy, G., Perrier, F., Tringuier, J.: Contribution à la correction de l'aberration chromatique du microscope électronique magnétique. J. Microsc. (Paris) *3*, 115-132 (1964)
Dupouy, G., Perrier, F.: Microscopie électronique - variation de l'aberration chromatique en fonction de l'énergie des électrons. J. Microsc. (Paris) *5*, 369-376 (1966)
Dupouy, G., Perrier, F., Trinquier, J.: Fayet, Y.: Condenseur-objectif pour microscope à très haute tension. C.R. Acad. Sci. Paris B*265*, 676-680 (1967)
Dupouy, G., Perrier, F., Trinquier, J., Murillo, R.: Aberration sphérique et pouvoir de résolution d'un "condenseur-objectif" pour microscope à très haute tension. C.R. Acad. Sci. Paris B*265*, 1221-1225 (1967)
Dupouy, G., Perrier, F.: Le microscope électronique à 1,5 million de volts du Laboratoire d'Optique Electronique du C.N.R.S. Toulouse. Onde Electr. *483*, 3-11 (1967)
Dupouy, G., Perrier, F., Trinquier, J.: Nouvelle méthode de mesure du coefficient d'aberration chromatique d'une lentille électronique magnétique. C.R. Acad. Sci. Paris B*266*,1102-1106 (1968)
Dupouy, G.: Electron microscopy at very high voltages. Adv. Opt. Electron Microsc. *2*, 167-250 (1968)
Dupouy, G., Perrier, F.: "Caractéristiques de l'objectif d'un microscope électronique de 3 millions de volts", Rome 1968, Vol.1, pp.9-10
Dupouy, G., Perrier, F., Fabre, R., Durrieu, L., Cathelinaud, R.: Microscope électronique 3 millions de volts. C.R. Acad. Sci. Paris B*269*, 867-874 (1969)
Dupouy, G., Perrier, F.: Microscope électronique 3 MV. Z. Angew. Phys. *27*, 224-227 (1969)
Dupouy, G.: Le microscope électronique 1.5 MV. J. Phys. D *2*, 769-774 (1969)
Dupouy, G., Perrier, F., Trinquier, J.: "Caractéristiques optiques de la lentille objectif du microscope 1000 kV de Toulouse", Grenoble 1970, Vol.1, pp.119-120
Dupouy, G., Perrier, F., Durrieu, L.: Microscope électronique 3 millions de volts. J. Microsc. (Paris) *9*, 575-592 (1970)
Durand, E.: Une présentation simple de la théorie générale des systèmes de révolution en optique électronique (relativité et aberrations comprises). Rev. Opt. *33*, 617-629 (1954)
Durand, E.: "Calcul numérique des lentilles magnétiques", in *Les techniques récentes en microscopie électronique et corpusculaire* (C.N.R.S., Paris 1956) pp.235-251
Durandeau, P.: "Lentilles électroniques magnétiques - règles de leur construction et expressions universelles de leurs caractéristiques électro-optiques", Stockholm 1956, pp.55-58
Durandeau, P.: Expressions générales et courbes représentatives des caractéristiques électro-optiques des lentilles électroniques magnétiques. C.R. Acad. Sci. Paris *242*, 1710-1712 (1956)
Durandeau, P.: Construction des lentilles électroniques magnétiques. J. Phys. Radium *17*, 18A-25A (1956)
Durandeau, P.: Lentilles électroniques magnétiques: expressions simples et générales du champ sur l'axe et des caractéristiques électrooptiques. J. Phys. Radium *17*, 33S-35S (1956)
Durandeau, P.: Etude sur les lentilles électroniques magnétiques. Ann. Fac. Sci. Univ. Toulouse. Sci. Math. Sci. Phys. *21*, 1-88 (1957)
Durandeau, P., Fert, C.: Lentilles électroniques magnétiques. Rev. Opt. *36*, 205-234 (1957)

Durandeau, P., Fert, C., Tardieu, P.: Les lentilles électroniques magnétiques dissymétriques. C.R. Acad. Sci. Paris *246*, 79-81 (1958)
Durandeau, P., Fagot, B., Fert, C.: Aberration de sphéricité des lentilles magnétiques de faible convergence. C.R. Acad. Sci. Paris *248*, 946-949 (1959)
Dutov, G.G.: Izv. Akad. Nauk SSSR Ser. Fiz. *25*, 668-671 (1961) [English transl.: Electron optics of probe systems. Bull. Acad. Sci. USSR Phys. Ser. *25*, 683-686 (1961)]
Elkamali, H.H., Mulvey, T.: "Improved Viewing Arrangements in the TEM", in *Developments in Electron Microscopy and Analysis, 1977*, ed. by D.L. Misell (Institute of Physics, Bristol 1977) pp.33-34
Elkamali, H.H., Mulvey, T.: "A Double Lens System for the Correction of Spiral Distortion", in *Developments in Electron Microscopy and Analysis, 1979*, ed. by T. Mulvey (Institute of Physics, Bristol 1980) pp.63-64
Elkamali, H.H., Mulvey, T.: "A Wide-Angle TEM Projection System", The Hague 1980, Vol.1, pp.74-75
El-Kareh, A.B., Parks, H.G.: "Computer Analysis of Symmetrical Magnetic Lenses with Unsaturated Pole Pieces", in *Rec. 10th Symp. Electron, Ion and Laser Beam Technology*, ed. by R.F.W. Pease (San Francisco Press, San Francisco 1967) pp.407-428
El-Kareh, A.B., El-Kareh, J.C.J.: *Electron Beams, Lenses and Optics*, Vols.1,2 (Academic, New York 1970)
Engler, P.E., Parsons, D.F.: The accurate determination of spherical aberration and focal length of an objective polepiece using a miniature magnetic field probe. Optik *41*, 309-318 (1974)
Fairbanks, J.R.: "Modular Concepts in the Design of Electron Optical Systems", EMSA 28, 1970, pp.362-363
Farrant, J.L.: "The Design of Magnetic Electron Lenses", Grenoble 1970, Vol.1, pp.141-142
Fernández-Morán, H.: Electron microscopy with high-field superconducting solenoid lenses. Proc. Natl. Acad. Sci. USA *53*, 445-451 (1965)
Fernández-Morán, H.: "Electron Microscopy with Superconducting Lenses", in *Proc. AMU-ANL Workshop on High-Voltage Electron Microscopy*, ed. by J. Gilroy, R.K. Hart, M. Weber, J. Kopta, Argonne National Laboratory Report ANL-7275 (1966) pp.51-58
Fernández-Morán, H.: High-resolution electron microscopy with superconducting lenses at liquid helium temperatures. Proc. Natl. Acad. Sci. USA *56*, 801-808 (1966)
Fert, C., Durandeau, P.: "Magnetic Electron Lenses", in *Focusing of Charged Particles*, Vol.1, ed. by A. Septier (Academic, New York 1967) pp.309-352
Fisher, R.M., Lally, J.S.: "Characteristics of 1 MeV Electron Microscope", Rome 1968, Vol.1, pp.15-18
Fontijn, L.A.: Imaging conditions for electron-beam micromachining. J. Vac. Sci. Technol. *15*, 1053-1055 (1978)
Foss, M.H.: "High-Field Superconducting Lenses", in *Proc. AMU-ANL Workshop on High-Voltage Electron Microscopy*, ed. by J. Gilroy, R.K. Hart, M. Weber, J. Kopta, Argonne National Laboratory Report ANL-7275 (1966) pp.160-162
Francken, J.C.: "On the Design of a Condensor-Objective Lens for a Low-Voltage Electron Microscope", Manchester 1972, pp.124-125
Francken, J.C., Heeres, A.: Telescopic condensor-objective lenses for electrons. Optik *37*, 483-500 (1973)
Funk, P.: Über die Seidelsche Fehlertheorie in der Elektronenoptik. Monatsh. Math. Phys. *43*, 305-316 (1936); *45*, 314-319 (1937)
Gábor, D.: "Oszillographieren von Wanderwellen mit dem Kathoden-Oszillographen", in *Forschungshefte der Studiengesellschaft für Höchstspannungsanlagen, Sonderheft: Kathodenoszillograph*, ed. by A. Matthias (Verlag der Vereinigung der Elektrizitätswerke, Berlin 1927) Heft 1, pp.7-46 [Part II, Sect.5, "Die Hilfsmittel zur Konzentrierung des Elektronenstrahles" contains a description and diagram of an "eisengekapselte Konzentrationsspule"]
Gabor, D.: Die Entwicklungsgeschichte des Elektronenmikroskops. Elektrotech. Z. A*78*, 522-530 (1957)
Gautier, P.: Contribution à l'étude des champs magnétiques de l'optique électronique. Ann. Fac. Sci. Univ. Toulouse Sci. Math. Sci. Phys. *21*, 89-184 (1957)

Génotel, D., Séverin, C., Laberrigue, A.: Lentilles magnétiques à enroulements supraconducteurs en microscopie électronique. J. Microsc. (Paris) *6*, 933-944 (1967)
Génotel, D., Laberrigue, A., Levinson, P., Séverin, C.: Caractéristiques optiques de lentilles magnétiques supraconductrices à très grand champ. Application en microscopie électronique à très haute tension. C.R. Acad. Sci. Paris B*265*, 226-229 (1967)
Génotel, D., Laberrigue, A., Payen, F., Séverin, C.: "Lentilles magnétiques supraconductrices pour microscopes à très haute tension", Rome 1968, Vol.1, pp.187-188
Gianola, U.F.: Investigation of magnetic lenses having the axial field $H(0,z) = \gamma/z^n$. Proc. Phys. Soc. London B*65*, 597-603 (1952)
Glaser, W.: Zur geometrischen Elektronenoptik des axialsymmetrischen elektromagnetischen Feldes. Z. Phys. *81*, 647-686 (1933)
Glaser, W.: Theorie des Elektronenmikroskops. Z. Phys. *83*, 104-122 (1933)
Glaser, W.: Über optische Abbildung durch mechanische Systeme und die Optik allgemeiner Medien. Ann. Phys. (Leipzig) *18*, 557-585 (1933)
Glaser, W.: Zur Bildfehlertheorie des Elektronenmikroskops. Z. Phys. *97*, 177-201 (1935)
Glaser, W.: Die kurze Magnetlinse von kleinstem Öffnungsfehler. Z. Phys. *109*, 700-721 (1938)
Glaser, W.: Über ein von sphärischer Aberration freies Magnetfeld. Z. Phys. *116*, 19-33 (1940)
Glaser, W.: Die Farbabweichung bei Elektronenlinsen. Z. Phys. *116*, 56-67 (1940)
Glaser, W.: Über den Öffnungsfehler der Elektronenlinsen, Bemerkungen zu vorstehender Erwiderung. Z. Phys. *116*, 734-735 (1940)
Glaser, W.: Strenge Berechnung magnetischer Linsen der Feldform $H=H_0/[1 + (z/a)^2]$. Z. Phys. *117*, 285-315 (1941)
Glaser, W.: Über die zu einem vorgegebenen Magnetfeld gehörende Windungsdichte einer Kreisspule. Z. Phys. *118*, 264-268 (1941)
Glaser, W.: Über elektronenoptische Abbildung mit gestörter Rotationssymmetrie. Z. Phys. *120*, 1-15 (1942)
Glaser, W., Lammel, W.: Strenge Berechnung der elektronenoptischen Aberrationskurven eines typischen Magnetfeldes. Arch. Elektrotech. *37*, 347-356 (1943)
Glaser, W.: Maxwell's "fish-eye" as an ideal electron lens. Nature *162*, 455-456 (1948)
Glaser, W.: Berechnung der optischen Konstanten starker magnetischer Elektronenlinsen. Ann. Phys. (Leipzig) *7*, 213-227 (1950)
Glaser, W.: "Signification de la distance focale en optique électronique", Paris 1950, pp.158-164
Glaser, W.: "Calcul de valeurs propres dans la figuration en optique électronique", Paris 1950, pp.165-170
Glaser, W., Lenz, F.: Berechnung der elektronenoptischen Abbildung durch drei typische, starke Magnetlinsen und ihr Zusammenhang mit der gewöhnlichen Linsengleichung. Ann. Phys. (Leipzig) *9*, 19-28 (1951)
Glaser, W., Robl, H.: Apertur und Blenden magnetischer Übermikroskope. Österr. Ing.-Arch. *5*, 36-48 (1951)
Glaser, W., Grümm, H.: Die Aberrationskonstanten des elektronenoptischen Abbildungssystems ohne Blende. Österr. Ing.-Arch. *6*, 360-372 (1952)
Glaser, W., Schiske, P.: Bildstörungen durch Polschuhasymmetrien bei Elektronenlinsen. Z. Angew. Phys. *5*, 329-339 (1953)
Glaser, W.: Zum Öffnungsfehler magnetischer Elektronenlinsen. Optik *13*, 7-12, 478 (1956)
Glaser, W.: "Elektronen- und Ionenoptik", in *Handbuch der Physik*, Vol.XXXIII Korpuskularoptik, ed. by S. Flügge (Springer, Berlin, Heidelberg, New York 1956) pp.123-395
Goddard, L.S.: A note on the Petzval field curvature in electron-optical systems. Proc. Cambridge Philos. Soc. *42*, 127-131 (1946)
Goncharenko, I.I., Sidorenko, I.S., Skorumnyi, G.M.: An iron-yoked superconducting magnetic lens. Cryogenics *11*, 315-316 (1971)
Green, B.A., Parsons, D.F., Ratkowski, A.J.: "Toward a 1-MV Microprobe", EMSA 37, 1979, pp.584-585

Grinberg, G.A.: On some classes of static axially symmetric electrostatic and magnetic fields, for which the basic equation of electron optics is soluble in terms of known functions. Zh. Tekh. Fiz. *23*, 1904-1914 (1953) [in Russian]
Grivet, P.: Un nouveau modèle mathématique de lentille électronique. C.R. Acad. Sci. Paris *233*, 921-923 (1951)
Grivet, P.: Eléments cardinaux d'un nouveau modèle de lentille électronique. C.R. Acad. Sci. Paris *234*, 73-75 (1952)
Grivet, P.: Un nouveau modèle mathématique de lentille électronique. J. Phys. Radium *13*, 1A-9A (1952)
Grivet, P.: "Lentilles électroniques", in Electron Physics, Nat. Bur. Stand. Cir. *527*, 167-196 (1954)
Grümm, H.: Zur Frage der unteren Grenze des Öffnungsfehlers bei magnetischen Elektronenlinsen. Optik *13*, 92-93 (1956)
Hahn, E.: Feldfunktionen und Kardinalelemente der Magnetlinsen vom Typus des Glaserschen Glockenfeldes und des Stufenfeldes. Exp. Tech. Phys. *13*, 375-392 (1965)
Hahn, E.: Theorie der elektrisch-magnetischen Linsen bei Zugrundelegung einer natürlichen Massbestimmung der Achsenabszisse. Jenaer Jahrb. 107-145 (1965)
Hahn, E.: Theorie der elektrisch-magnetischen Linsen mit gestörter Feldsymmetrie bei Zugrundelegung einer natürlichen Massbestimmung der Achse. Jenaer Jahrb. 145-172 (1966)
Hahn, E.: Über eine einheitliche Theorie der elektrisch-magnetischen Linsen vom rotationssymmetrischen Typ. Wiss. Z. Tech. Univ. Dresden *20*, 361-363 (1971)
Hahn, E.: "Lösungsverfahren der paraxialen Bahngleichung, wenn die Achsfeldstärke nur quantisiert definiert ist", Abstracts VIII. Arbeitstagung Elektronenmikroskopie, Berlin, 1975, pp.249-250
Haine, M.E.: The electron optical system of the electron microscope. J. Sci. Instrum. *24*, 61-66 (1947)
Haine, M.E.: "Some Simplified Magnetic Lens Design Features", London 1954, pp.92-97
Haine, M.E.: The electron microscope - a review. Adv. Electron. Electron Phys. *6*, 295-370 (1954)
Haine, M.E., Agar, A.W., Mulvey, T.: An electrostatic-magnetic alinement section for the electron microscope. J. Sci. Instrum. *35*, 357-358 (1958)
Hänsel, H.: Über eine elektronenoptische Feldverteilung mit vorgegebenen Abbildungseigenschaften. Optik *19*, 67-75 (1962)
Hänsel, H.: Über eine magnetische Feldverteilung mit exakt lösbarer Paraxialgleichung [$B(z) = B_0/z^n$]. Optik *21*, 273-280 (1964)
Hanszen, K.-J., Lauer, R.: Der Einfluß der Pupillenlage auf den chromatischen Fehler der Vergrößerung von elektrostatischen und magnetischen Elektronenlinsen. Optik *23*, 478-494 (1965/1966)
Hanszen, K.-J., Ade, G., Lauer, R.: Genauere Angaben über sphärische Längsaberration, Verzeichnung in der Pupillenebene und über die Wellenaberration von Elektronenlinsen. Optik *35*, 567-590 (1972)
Hanszen, K.-J., Lauer, R., Ade, G.: Eine Beziehung zwischen der Abhängigkeit des Öffnungsfehlers von der Objektlage und der Koma, ausgedrückt in der Terminologie der Elektronenoptik. Optik *36*, 156-159 (1972)
Harada, Y., Tsuno, K., Arai, Y.: "Theoretical and Practical Design Considerations for Ultra-High Resolution Electron Lenses", EMSA 38, 1980, pp.266-269
Hardy, D.F.: Superconducting electron lenses. Adv. Opt. Electron Microsc. *5*, 201-237 (1973)
Harris, P., Mulvey, T.: "An Experimental Single-Pole Magnetic Lens Probe-Forming System", in *Developments in Electron Microscopy and Analysis, 1975*, ed. by J.A. Venables (Academic, London 1976) pp.49-50
Hawkes, P.W.: "Lens Aberrations", in *The Focusing of Charged Particles*, ed. by A. Septier (Academic, New York 1967) Vol.1, pp.411-468
Hawkes, P.W.: General expressions for aperture aberration coefficients. J. Microsc. (Paris) *6*, 917-932 (1967)
Hawkes, P.W.: Asymptotic aberration coefficients, magnification and object position in electron optics. Optik *27*, 287-304 (1968)

Hawkes, P.W.: The dependence of the spherical aberration coefficient of an electron-optical objective lens on object position and magnification. J. Phys. D *1*, 131-133 (1968)
Hawkes, P.W.: The relation between the spherical aberration and distortion coefficients of electron probe-forming and projector lenses. J. Phys. D *1*, 1549-1558 (1968)
Hawkes, P.W.: Asymptotic aberration integrals for round lenses. Optik *31*, 213-219 (1970)
Hawkes, P.W.: The addition of round electron lens aberrations. Optik *31*, 592-599 (1970)
Hawkes, P.W.: Can objective spherical aberration be expressed as a polynomial of fourth degree in reciprocal magnification? J. Microsc. (Paris) *9*, 435-454 (1970)
Hawkes, P.W.: "The Dependence of Objective Spherical Aberration upon Magnification", Grenoble 1970, Vol.2, pp.17-18
Hawkes, P.W.: "Computer-Aided Design of Electron Lens Combinations", in *Image Processing and Computer-Aided Design in Electron Optics,* ed. by P.W. Hawkes (Academic, London 1973) pp.230-248
Hawkes, P.W.: "Note on the Structure of the Aberration Coefficients of Bell-Shaped Fields", Canberra 1974, Vol.1, pp.156-157
Hawkes, P.W., Valdrè, U.: Superconductivity and electron microscopy. J. Phys. E *10*, 309-328 (1977)
Hawkes, P.W.: "Developments in Techniques for Focusing Charged Particles", EMSA 38, 1980, pp.262-265
Hawkes, P.W.: "Methods of Computing Optical Properties and Combating Aberrations for Low-Intensity Beams", in *Applied Charged Particle Optics,* ed. by A. Septier (Academic, New York 1980) Vol.A, pp.45-157 [Supplement 13A to Adv. Electron. Electron Phys.]
Hawkes, P.W.: Some approximate magnetic lens aberration formulae: a commentary. Optik *56*, 293-320 (1980)
Heritage, M.B.: "Asymptotic Aberration Coefficients of Magnetic Lenses", Manchester 1972, pp.88-89
Heritage, M.B.: "The Computation of the Third-Order Aberration Coefficients of Electron Lenses", in *Image Processing and Computer-Aided Design in Electron Optics,* ed. by P.W. Hawkes (Academic, London 1973) pp.324-338
Herrmann, K.-H., Ihmann, K., Krahl, D.: "Ribbon Coils for Electromagnetic Electron Lenses of Small Dimensions", Canberra 1974, Vol.1, pp.132-133
Herrmann, K.-H., Kowalsky, U., Kunath, W., Weiss, K., Zemlin, F.: "Advantages and Limitations of a Condenser Objective for Minimum Exposure", Toronto 1978, Vol.1, pp.34-35
Hesse, M.B.: The calculation of magnetic lens fields by relaxation methods. Proc. Phys. Soc. London B*63*, 386-401 (1950)
Hildebrandt, H.-J., Riecke, W.D.: Zur Dimensionierung des Eisenkreises magnetischer Elektronenlinsen. I. Der magnetische Fluß in nicht gesättigten Polschuhsystemen. Z. Angew. Phys. *20*, 336-342 (1966)
Hill, R., Smith, K.C.A.: "Analysis of the Single-Pole Lens by Finite-Element Computation", in *Developments in Electron Microscopy and Analysis, 1979,* ed. by T. Mulvey (Institute of Physics, Bristol 1980) pp.49-52
Hill, R., Smith, K.C.A.: "An Interactive Computer System for Magnetic Lens Analysis", The Hague 1980, Vol.1, pp.60-61
Hill, R., Smith, K.C.A.: "Applications of the Cambridge Interactive Electron Lens Analysis System: CIELAS", in *Electron Microscopy and Analysis 1981*, ed. by M.J. Goringe (Institute of Physics, Bristol 1982) pp.71-74
Hillier, J.: A study of distortion in electron microscope projection lenses. J. Appl Phys. *17*, 411-419 (1946)
Hillier, J.: A removable intermediate lens for extending the magnification range of an electron microscope. J. Appl. Phys. *21*, 785-790 (1950)
Hillier, J.: An objective for use in the electron microscopy of ultra-thin sections. J. Appl. Phys. *22*, 135-137 (1951)
Honda, T., Watanabe, H., Etoh, T., Hiraga, K.: "A High Resolution Ultra-High Voltage Electron Microscope", in *High Voltage Electron Microscopy* (*Electron Microscopy 1980,* Vol.IV), ed. by P. Brederoo, J. van Landuyt (Seventh European Congress of Electron Microscopy Foundation, Leiden 1980) pp.18-21

Honda, T., Tsuno, K., Watanabe, H.: "Improvement of Resolution of the High Voltage Electron Microscope by Means of an Asymmetrical Electron Lens", in *Electron Microscopy and Analysis 1981*, ed. by M.J. Goringe (Institute of Physics, Bristol 1982) pp.99-102
Honjo, G., Yagi, K., Takayanagi, K., Nagakura, S., Katagiri, S., Kubozoe, M., Matsui, I.: "Design Features of a New Ultra-High-Vacuum and High-Resolution 1-MV Electron Microscope", in *High-Voltage Electron Microscopy* (*Electron Microscopy 1980*, Vol.IV), ed. by P. Brederoo, J. van Landuyt (Seventh European Congress of Electron Microscopy Foundation, Leiden 1980) pp.22-25
Howie, A.: "Wave Mechanics of Strong Magnetic Lenses", Rome 1968, Vol.1, pp.151-152
Hutter, R.G.E.: The class of electron lenses which satisfy Newton's image relation. J. Appl. Phys. *16*, 670-699 (1945)
Inoue, T.: Denki Gakkai Zashi [J. Inst. Electr. Eng. Jpn.] *59*, 120-125 (1939) [English transl.: On the aberration of geometrical electron optics of fifth order. Electrotech. J. Jpn. *3*, 178-180 (1939)]
Ito, K., Ito, T., Kanaya, K., Morito, N.: "Calculations on the Magnetic Lens", Paris 1950, pp.207-212
Ito, K., Ito,T.: On the design of electron lens for electron microscope use of magnetic type. Denshikenbikyo [Electron Microscopy] *1*, No.1, 47-51 (1950) [in Japanese]
Ito, K., Ito, T.: Studies on the design of magnetic electron lenses for electron microscope. Denshikenbikyo [Electron Microscopy] *1*, No.2, 45-46 (1950) [in Japanese]
Ito, K., Ito, T.: Correction of lens defects in three-stage electron microscope. Denshikenbikyo [Electron Microscopy] *3*, No.1, 12-16 (1953) [in Japanese]
Jandeleit, O., Lenz, F.: Berechnung der Bildfehlerkoeffizienten magnetischer Elektronenlinsen in Abhängigkeit von der Polschuhgeometrie und den Betriebsdaten. Optik *16*, 87-107 (1959)
Johannson, H., Knecht, W.: Beitrag zur kombinierten Benutzung elektrischer und magnetischer Elektronenlinsen. Z. Phys. *86*, 367-372 (1933)
Jouffrey, B., Trinquier, J., Balladore, J.L.: Double focus electron probe forming lens for STEM. J. Microsc. Spectrosc. Electron. *4*, 89-93 (1979)
Juma, S.M., Mulvey, T.: "New Rotation-Free Magnetic Electron Lenses", Canberra 1974, Vol.1, pp.134-135
Juma, S.M., Mulvey, T.: "A New Experimental Electron Microscope with a Rotation-Free Projector System", in *Developments in Electron Microscopy and Analysis 1975*, ed. by J.A. Venables (Academic, London 1976) pp.45-48
Juma, S.M., Mulvey, T.: Miniature rotation-free magnetic electron lenses for the electron microscope. J. Phys. E *11*, 759-764 (1978)
Juma, S.M., Mulvey, T.: "The Axial Field Distribution of Single-Polepiece Lenses", in *Developments in Electron Microscopy and Analysis, 1979*, ed. by T. Mulvey (Institute of Physics, Bristol 1980) pp.59-60
Juma, S.M., al-Shwaikh, A.A.: "Single-Polepiece Objective and Projector Lenses for the Electron Microscope", in *Developments in Electron Microscopy and Analysis, 1979*, ed. by T. Mulvey (Institute of Physics, Bristol 1980) pp.61-62
Juma, S.M., Mulvey, T.: "Some Electron Optical Properties of Single-Polepiece Projector Lenses", The Hague 1980, Vol.1, pp.78-79
Juma, S.M., Faisal, A.D.: Properties of the axial flux density distribution of magnetic electron lenses with a single cylindrical polepiece. J. Phys. E *14*, 1389-1393 (1981)
Kamminga, W., Verster, J.L., Francken, J.C.: Design consideration for magnetic objective lenses with unsaturated pole pieces. Optik *28*, 442-461 (1968/1969)
Kamminga, W.: Finite-element solutions for devices with permanent magnets. J. Phys. D *8*, 841-855 (1975)
Kamminga, W.: Properties of magnetic objective lenses with highly saturated pole pieces. Optik *45*, 39-54 (1976)
Kanaya, K.: Aberrations of compound magnetic lenses with pole pieces of double gaps. Bull. Electrotech. Lab. *13*, 453-458 (1949) [in Japanese]
Kanaya, K.: Theory of aberrations for an axisymmetrical magnetic type electron microscope. Res. Electrotech. Lab. *495*, 37pp. (1949) [in Japanese]
Kanaya, K.: Abérrations of magnetic lens having bell-shaped field with various degrees of asymmetry. Denshikenbikyo [Electron Microscopy] *1*, No.2, 53-55 (1950) [in Japanese]

Kanaya, K.: Spherical and chromatic aberrations in magnetic lenses for electron microscopes. Bull. Electrotech. Lab. *15*, 86-91 (1951) [in Japanese]
Kanaya, K.: Distortion in magnetic lenses for electron microscopes. Bull. Electrotech. Lab. *15*, 91-94 (1951) [in Japanese]
Kanaya, K.: Curvature of field and astigmatism in magnetic lenses for electron microscopes. Bull. Electrotech. Lab. *15*, 193-198 (1951) [in Japanese]
Kanaya, K.: Coma in magnetic lenses for electron microscopes. Bull. Electrotech. Lab. *15*, 199-202 (1951) [in Japanese]
Kanaya, K., Kato, A.: Measurement of distortion in magnetic lenses for electron microscopes. Bull. Electrotech. Lab. *15*, 827-833 (1951) [in Japanese]
Kanaya, K.: Three-stage electron microscopes. I. Spherical and Chromatic aberrations of magnetic lenses. Bull. Electrotech. Lab. *16*, 25-30 (1952) [in Japanese]
Kanaya, K.: Three-stage electron microscopes. II. Distortion of magnetic lenses. Bull. Electrotech. Lab. *16*, 135-142 (1952) [in Japanese]
Kanaya, K.: Three-stage electron microscopes. III. Curvature of field and astigmatism of magnetic lenses. Bull. Electrotech. Lab. *16*, 184-192 (1952) [in Japanese]
Kanaya, K.: On the manufacture accuracy of magnetic lenses in electron microscopes. J. Electronmicrosc. *1*, 7-12 (1953)
Kanaya, K.: Further improvement in resolution of electron microscopes by using compound condenser lens system. I. Compound condenser lens system. Bull. Elec. trotech. Lab. *17*, 529-536 (1953) [in Japanese]
Kanaya, K.: Further improvement in resolution of electron microscopes by using compound condenser lens system. II. Theoretical treatment on aberrations of magnetic lens systems, considering the effect of front lens system. Bull. Electrotech. Lab. *17*, 582-590 (1953) [in Japanese]
Kanaya, K.: Further improvement of resolution in electron microscopes using compound condenser lens system. III. Chromatic field-aberrations of compound magnetic lens system. Bull. Electrotech. Lab. *18*, 13-22 (1954) [in Japanese]
Kanaya, K.: On the field aberrations in the electron microscope. Denshikenbikyo [Electron Microscopy] *3*, No.3, 1-4 (1954) [in Japanese]
Kanaya, K.: Electron optic theory on the design of magnetic lenses for electron microscopes. Res. Electrotech. Lab. *548*, 70 pp. (1955)
Kanaya, K., Kawakatsu, H., Ishikawa, Y.: Electron-optical design of the demagnifying system to provide micro- and angström probes. Bull. Electrotech. Lab. *33*, 1233-1255 (1969)
Kanaya, K., Ishikawa, Y.: Demagnifying properties to provide electron micro- and ångström probes. J. Electron Microsc. *19*, 133-140 (1970)
Kanaya, K., Baba, N., Ono, S.: Rigorous treatment of magnetic lens whose axial field distribution is given by $H(z/a) = H_0(1-\mu^2)^{(2m-1)/2}$. Optik *46*, 125-148 (1976)
Kasper, E., Lenz, F.: "Numerical Methods in Geometrical Electron Optics", The Hague 1980, Vol.1, pp.10-15
Kas'yankov, P.P., Dutova, K.P.: Izv. Akad. Nauk SSSR Ser. Fiz. *25*, 665-667 (1961) [English transl.: Concerning aberrations of electron optical systems. Bull. Acad. Sci. USSR Phys. Ser. *25*, 680-682 (1961)]
Katagiri, S.: Experimental study of axial chromatic aberration. J. Electron Microsc. *1*, 13-18 (1953)
Kato, N., Inoue, T.: Denki Gakkai Zashi [J. Inst. Electr. Eng. Jpn.] *60*, 121-123 (1940) [English transl.: On the aberration of the electron microscope. Electrotech. J. Jpn. *4*, 219-222 (1940)]
Kato, N., Inoue, T.: Denki Gakkai Zashi [J. Inst. Electr. Eng. Jpn] *61*, 33-36 (1941) [English transl.: Discussion about the aberration formulae of geometrical electron optics. Electrotech. J. Jpn. *5*, No.4, 68-71 (1941)]
Kawakatsu, H., Plomp, F.H., Siegel, B.M.: "Imaging Properties of a High Resolution Superconducting Electron Lens", Rome 1968, Vol.1, pp.193-194
Kawakatsu, H.: Quadrupole and superconductor lenses. J. Electron Microsc. *19*, 121-132 (1970) [in Japanese]
Kel'man, V.M., Peregud, B.P., Skopina, V.I.: Zh. Tekh. Fiz. *29*, 1219-1224 (1959) [English transl.: Short magnetic lens with distributed winding. Sov. Phys.: Tech. Phys. *4*, 1118-1122 (1959)]

Kempin, H.-J., Müller, K.-H., Rauch, M.v., Schäfer, N., Vollborn, K.: Ein Hochauflösungsobjektiv mit Goniometer. Beitr. elektronenmikroskop. Direktabb. Oberfl. *8*, 339-346 (1975)
Kil-hong Kim: Investigations on electron lenses by the method of successive approximations. Sukhvak gva Mulli [J. Math. Phys. Korea] *6*, No.3,28-41 (1962) [in Korean]
Kimura, H.: "A New Electron Microscope Excited with Permanent Magnets", *Electron-Microscopy*, Proc. 1st Reg. Conf. Asia Oceania, Tokyo, 1956 (Electrotechnical Laboratory, Tokyo 1957) pp.108-113
Kimura, H., Idei, Y.: Permanent magnetic lens systems for electron microscope. I. Various lens systems and their characteristics. Denshikenbikyo [Electron Microsc.] *5*, No 2, 7-14 (1957) [in Japanese]
Kimura, H., Kikuchi, Y.: "Permanent Lens Systems and Their Characteristics", Berlin 1958, Vol.1, pp.53-57
Kimura, H.: Permanent magnet lens systems of electron microscope. II. Various methods of changing the focal length and performances. Denshikenbikyo [Electron Microsc.] *8*, No 1, 36-43 (1959) [in Japanese]
Kimura, H.: Double-gap lens and its characteristics. Denshikenbikyo [Electron Microsc.] *8*, No 2-3, 2-3 (1959) [in Japanese]
Kimura, H., Kikuchi, Y.: Permanent magnet lens systems and their characteristics. J. Electron Microsc. *7*, 1-4 (1959)
Kimura, H., Katagiri, S.: An electron lens system excited by permanent magnets with a new astigmatism compensator. Optik *16*, 50-55 (1959)
Kimura, H.: Electron lens system excited by permanent magnet. J. Electron Microsc. *11*, 10-17 (1962) [in Japanese]
Kinder, E., Pendzich, A.: Eine neue magnetische Linse kleiner Brennweite. Jahrbuch AEG-Forsch. *7*, 23-26 (1940)
Kinder, E.: Über das magnetische Jochlinsen-Übermikroskop und einige Anwendungen in der Kolloidchemie. Kolloid Z. *95*, 326-336 (1941)
Kinder, E.: Das Jochlinsen-Kombinations-Übermikroskop für Durchstrahlungs- und Selbststrahlungsaufnahmen. Z. Phys. *122*, 192-208 (1944)
Kinder, E.: Über die Dimensionierung von Kleinmikroskopen. Optik *10*, 171-191 (1953)
Kitamura, N., Schulhof, M.P., Siegel, B.M.: Superconducting lens for electron microscopy. Appl. Phys. Lett. *9*, 377-380 (1966)
Knapek, E.: Modellversuche zur elektromagnetischen Justierung der supraleitenden Abschirmlinse. Optik *41*, 506-514 (1975)
Knapek, E.: Objektfreie Prüfung supraleitender Abschirmlinsen mit zwei verschiedenen teleskopischen Strahlengängen. Optik *46*, 97-106 (1976)
Knoll, M., Ruska, E.: Das Elektronenmikroskop. Z. Phys. *78*, 318-339 (1932)
Kohonen, T., Puolakka, H., Väyrynen, H.: Investigation of a Siegbahn-Slätis beta-ray spectrometer. Ann. Acad. Sci. Fenn. Ser. A6 *89*, 23 pp (1962)
Konopáč, J., Petr. J., Rychnovský, M.: "Optical System of the Scanning Electron Microscope BS 300", in *Proc. XVth Czechoslovak Conf. Electron Microscopy* Prague, 1977, ed. by V. Viklický and J. Ludvîk (Czechoslovak Academy of Sciences, Prague 1977) Vol.B, pp.547-548
Koops, H.: Elektronenoptische Verkleinerung durch Kombination einer langbrennweitigen Feldlinse mit einer starken kurzbrennweitigen Magnetlinse. Optik *36*, 93-110 (1972)
Koops, H.: "Zur verkleinernden elektronenoptischen Übertragung großer Bildpunktzahlen mit Magnetlinsen", Manchester 1972, pp.126-127
Koops, H.: Zur elektronenoptischen Herstellung von 100-nm-Transmissionsgittern für weiche Röntgenstrahlen. Optik *38*, 246-260 (1973)
Koops, H.: On electron projection systems. J. Vac. Sci. Technol. *10*, 909-912 (1973)
Koops, H., Bernhard, W.: Electron-beam projection systems with compensated chromatic field aberrations. J. Vac. Sci. Technol. *12*, 1141-1145 (1975)
Koops, H., Bernhard, W.: "An Objective Lens for an Electron Microscope with Compensated Axial Chromatic Aberration", Toronto 1978, Vol.1, pp.36-37
Koops, H.: "Aberration Correction in Electron Microscopy", Toronto 1978, Vol.3, pp.185-196
Koops, H.: Erprobung eines chromatisch korrigierten elektronenmikroskopischen Objektives. Optik *52*, 1-18 (1978)

Koops, H., Walter, G.: "Automated Compensation of Lens Aberrations, a Simulation", The Hague 1980, Vol.1, pp.40-41
Koops, H.: "Electron Beam Projection Techniques", in *Fine Line Lithography*, ed. by R. Newman (North Holland, Amsterdam 1980) pp.232-335
Kostenko, V.P., Konovalenko, V.A., Zelev, S.F.: Opt.-Mekh. Prom. *37*, No.1, 76-77 (1970) [English transl.: The maximum magnetic flux density of electron microscope lenses with polepieces of different shapes. Sov. J. Opt. Technol. *37*, 68-69 (1970)]
Kubozoe, M., Kimura, T., Ozasa, S., Kasai, S.: "Aberrations of Four Stage Image Forming System", Grenoble, 1970, Vol.2, pp.47-48
Kubozoe, M., Sato, H., Kasai, S., Ozasa, S.: An electron microscope for low magnification with a wide field. J. Electron. Microsc. *27*, 83-87 (1978)
Kucherov, G.V., Tsyganenko, V.V.: Radiotekh. Elektron. *22*, 1699-1705 (1977) [English transl.: Problems in the design and optimization of electron-optical systems with specified image parameters. Radio Eng. Electron Phys. (USSR) *22*, No.8, 117-122 (1977)]
Kunath, W., Riecke, W.D.: Zur Bestimmung der Öffnungsfehlerkoeffizienten magnetische Objektivlinsen. Optik *23*, 322-342 (1965/1966)
Kunath, W., Riecke, W.D., Ruska, E.: "Spherical Aberration of Saturated Strong Objective Lenses", Kyoto 1966, Vol.1, pp.139-140
Kushnir, Yu.M.: Physico-technical foundations of electron microscopy. Elektrichestvo No.5, 3-16 (1947) [in Russian]
Kushnir, Yu.M.: Construction and use of electron microscopes. Elektrichestvo No.7, 17-31 (1947) [in Russian]
Kuypers, W., Thompson, M.N., Andersen, W.H.J.: "A Scanning Transmission Electron Microscope", in *Scanning Electron Microscopy 1973*, ed. by O. Johari (IIT Research Institute, Chicago 1973) pp.9-16
Kynaston, D., Mulvey, T.: "Distortion-Free Operation of the Electron Microscope", Philadelphia 1962, Vol.1, D-2
Kynaston, D., Mulvey, T.: The correction of distortion in the electron microscope. Brit. J. Appl. Phys. *14*, 199-206 (1963)
Laberrigue, A., Levinson, P.: Utilisation des propriétés des fils supraconducteurs en microscopie électronique. C.R. Acad. Sci. Paris *259*, 530-532 (1964)
Laberrigue, A., Levinson, P., Bergeot, G., Bonhomme, P., Payen, F., Séverin, C.: "Lentilles magnétiques en fil supraconducteur. Performances en microscopie électronique", Kyoto 1966, Vol.1, pp.153-154
Laberrigue, A., Séverin, C.: Caractéristiques optiques de lentilles magnétiques sans fer à enroulements supraconducteurs en microscopie électronique. J. Microsc. (Paris) *6*, 123-134 (1967)
Laberrigue, A., Levinson, P., Berjot, G., Bonhomme, P.: Objectifs supraconducteurs utilisés en microscopie électronique; premiers résultats. Ann. Univ. ARERS *5*, 50-54 (1967)
Laberrigue, A., Génotel, D., Homo, J.C., Levinson, P., Séverin, C., Testard, O.: "Microscope à haute tension à lentilles magnétiques supraconductrices", Rome 1968, Vol.1, p.29
Laberrigue, A., Génotel, D., Girard, M., Séverin, C., Balossier, G., Homo, J.C.: "A 400-kV Superconducting Electron Microscope: Performance, Viability and Possibilities in the 3-MV Range", in *High Voltage Electron Microscopy*, Proc. 3rd Int. Conf., ed. by P.R. Swann, C.J. Humphreys, M.J. Goringe (Academic, London 1974) pp.108-113
Lambrakis, E., Marai, F.Z., Mulvey, T.: "Correction of Spiral Distortion in the Transmission Electron Microscope", in *Developments in Electron Microscopy and Analysis, 1977*, ed. by D.L. Misell (Institute of Physics, Bristol 1977) pp.35-38
Langner, G., Lenz, F.: Die Ausmessung magnetischer Felder an maßstäblichen Modellen unter Berücksichtigung der Sättigungserscheinungen. Optik *11*, 171-180 (1954)
Langner, G.: Die Erprobung einiger regelbarer permanentmagnetischer Elektronenlinsen. Optik *12*, 554-562 (1955)
Laplume, J.: Les lentilles électroniques en mécanique relativiste. Cah. Phys. *29-30*, 55-66 (1947)
Laplume, J.: Les lentilles électroniques en mécanique relativiste. Rev. Opt. *29*, 106-111 (1950)

Laverick, C.: "Superconducting Lens Studies", in *Proc. AMU-ANL Workshop on High-Voltage Electron Microscopy,* ed. by J. Gilroy, R.K. Hart, M. Weber, J. Kopta, Argonne National Laboratory Report ANL-7275 (1966) pp.38-48
Lefranc, G., Nachtrieb, K.: "A Superconducting Lens System for Electron Microscopes with Beam Voltages Between 100 and 500 kV", in *Developments in Electron Microscopy and Analysis, 1979,* ed. by T. Mulvey (Institute of Physics, Bristol 1980) pp.31-34
Lefranc, G., Müller, K.-H., Dietrich, I.: A superconducting lens system operated in the fixed-beam and the scanning mode. Ultramicroscopy *6*, 81-84 (1981)
Lefranc, G., Müller, K.-H.: "Asymmetric Superconducting Shielding Lens for TEM and STEM", in *Electron Microscopy and Analysis 1981,* ed. by M.J. Goringe (Institute of Physics, Bristol 1982) pp.91-94
Leisegang, S.: Zum Astigmatismus von Elektronenlinsen. Optik *10*, 5-14 (1953)
Leisegang, S.: Zur Zentrierung magnetischer Elektronenlinsen. Optik *11*, 397-406 (1954)
Leisegang, S.: "Elektronenmikroskope", in *Handbuch der Physik,* Vol.XXXIII Korpuskularoptik, ed. by S. Flügge (Springer, Berlin, Göttingen, Heidelberg 1956) pp.396-545
Lencová, B.: Metoda konečných prvků v elektronové optice, Čs. Čas. Fyz. A*25*, 57-61 (1975)
Lencová, B., Lenc, M.: "The Probe Forming System of a Field Emission Scanning Microscope", in *Proc. XVth Czechoslovak Conf. Electron Microscopy* Prague, 1977, ed. by V. Viklický and J. Ludvík (Czechoslovak Academy of Sciences, Prague 1977) Vol.B, pp.563-564
Lencová, B.: "On the Use of Finite Element Method for the Computation of Electron Optical Elements", *Proc. 6th Int. Conf. Magnet. Technology,* Bratislava, 1977, (ALFA, Bratislava 1978) pp.813-819
Lencová, B.: Numerical computation of electron lenses by the finite element method. Comp. Phys. Commun. *20*, 127-132 (1980)
Lenz, F.: Berechnung der Feldverteilung längs der Achse magnetischer Elektronenlinsen aus Polschuhabmessungen und Durchflutung. Optik *7*, 243-253 (1950)
Lenz, F.: Annäherung von rotationssymmetrischen Potentialfeldern mit zylindrischen Äquipotentialflächen durch eine analytische Funktion. Ann. Phys. (Leipzig) *8*, 124-128 (1950)
Lenz, F.: Berechnung optischer Kenngrößen magnetischer Elektronenlinsen vom erweiterten Glockenfeldtyp. Z. Angew. Phys. *2*, 337-340 (1950)
Lenz, F.: Berechnung optischer Kenngrößen magnetischer Elektronenlinsen aus Polschuhabmessungen und Betriebsdaten. Z. Angew. Phys. *2*, 448-453 (1950)
Lenz, F.: Berechnung der elektronenoptischen Kenngrößen eines speziellen magnetischen Linsenfeldes ohne numerische Bahnintegrationen. Ann. Phys. (Leipzig) *9*, 245-258 (1951)
Lenz, F.: Ein einfaches Verfahren zur Bestimmung der Feldform magnetischer Elektronenlinsen. Optik *9*, 3-18 (1952)
Lenz, F., Hahn, M.: Die Untersuchung der Unsymmetrie magnetischer Polschuhlinsen im schwach unterteleskopischen Strahlengang. Optik *10*, 15-27 (1953)
Lenz, F.: Ein mathematisches Modellfeld für eine permanentmagnetische Einzellinse. Z. Angew. Phys. *8*, 492-496 (1956)
Lenz, F.: "Über asymptotische Bildfehler", Stockholm 1956, pp.48-51
Lenz, F.: Über asymptotische Bildfehler. Optik *14*, 74-82 (1957)
Lenz, F.: Die Bildfehler im elektronenoptischen Beugungsbild. Optik *16*, 457-460 (1959)
Lenz, F.: "Computer-Aided Design of Electron Optical Systems", in *Image Processing and Computer-Aided Design in Electron Optics,* ed. by P.W. Hawkes (Academic, London 1973) pp.274-282
Le Poole, J.B.: A new electron microscope with continuously variable magnification. Philips Tech. Rev. *9*, 33-45 (1947)
Le Poole, J.B., Salvat, M.: "Lentille achromatique pour diffratographe électronique", Paris 1950, pp.252-255
Le Poole, J.B., Van Dorsten, A.C.: "Magnetic lenses of extremely short focal length", in *Electron Physics*, Nat. Bur. Stand. Cir. *527*, 410 (1954)
Le Poole, J.B.: "Miniature Lens", Appendix, Prague 1964, p.6
Le Poole, J.B., Van de Vrie, J.B., Revallier, L.J.: "Permanent Magnetic Lenses", Manchester 1972, pp.114-155

Liebmann, G.: The limiting resolving power of the electron microscope. Philos. Mag. *37*, 677-685 (1946)
Liebmann, G., Grad, E.M.: "Imaging Properties of a Series of Magnetic Electron Lenses", Paris 1950, pp.138-147
Liebmann, G., Grad, E.M.: Imaging properties of a series of magnetic electron lenses. Proc. Phys. Soc. London B*64*, 956-971 (1951)
Liebmann, G.: The symmetrical magnetic electron microscope objective lens with lowest spherical aberration. Proc. Phys. Soc. London B*64*, 972-977 (1951)
Liebmann, G.: Magnetic electron microscope projector lenses. Proc. Phys. Soc. London B*65*, 94-108 (1952)
Liebmann, G.: The magnetic electron microscope objective lens of lowest chromatic aberration. Proc. Phys. Soc. London B*65*, 188-192 (1952)
Liebmann, G.: "Characteristics of symmetrical magnetic electron lenses", in *Electro Physics*, Nat. Bur. Stand. Cir. *527*, 283-290 (1954)
Liebmann, G.: The field distribution in asymmetrical magnetic electron lenses. Proc. Phys. Soc. London B*68*, 679-681 (1955)
Liebmann, G.: The magnetic pinhole electron lens. Proc. Phys. Soc. London B*68*, 682-685 (1955)
Liebmann, G.: A unified representation of magnetic electron lens properties. Proc. Phys. Soc. London B*68*, 737-745 (1955)
Lindgren, I.: Investigation of a beta-ray spectrometer with a triangular field. Nucl. Instrum. *3*, 104-108 (1958)
Lindgren, I., Schneider, W.: The focusing properties of long magnetic lenses. I. Nucl. Instrum. Methods *22*, 48-60 (1963)
Lindgren, I., Pettersson, G., Schneider, W.: The focusing properties of long magnetic lenses. II. Nucl. Instrum. Methods *22*, 61-72 (1963)
Lindgren, I., Olsen, B., Petterson, G. [sic], Schneider, W.: The focusing properties of long magnetic lenses. III. Nucl. Instrum. Methods *41*, 331-337 (1966)
Lippert, W.: Erfahrungen mit einer Zwischenlinse bei einem Siemens-Elektronenmikroskop älterer Bauart. Optik *12*, 274-280 (1955)
Lund, N.C.: "High Ampère-Turn Density Coils for the Reduction of Magnetic Lens Size", EMSA 25, 1967, pp.228-229
Lyubchik, Ya.G., Mokhnatkin, A.V.: Radiotekh. Elektron. *17*, 2234-2237 (1972) [English transl.: An experimental determination of the spherical aberration of a magnetic axially symmetrical lens. Radio Eng. Electron. Phys. (USSR) *17*, 1795-1798 (1972)]
Maclachlan, M.E.C., Hawkes, P.W.: "Objective-Projector System Design for t.e.m", Grenoble 1970, Vol.2, pp.23-24
Maclachlan, M.E.C.: "Objective-Intermediate-Projector System Design", in *Electron Microscopy and Analysis*, ed. by W.C. Nixon (Institute of Physics, London 1971) pp.98-99
Makarova, I.S.: Radiotekh. Elektron. *20*, 213-216 (1975) [English transl.: Relationship between the number of resolvable elements and the resolution of electron and ion optical systems. Radio Eng. Electron Phys. (USSR) *20*, No.1, 150-153 (1975
Makarova, I.S.: Radiotekh. Elektron. *22*, 2220-2223 (1977) [English transl.: Third-order field geometrical aberrations of axially symmetrical single-gap magnetic lenses, forming an electron probe in the field of the lens itself. Radio Eng. Electron Phys. (USSR) *22*, No.10, 156-158 (1977)]
Makarova, I.S.: Opt. Mekh. Prom. *45*, No.1, 15-16 (1978) [English transl.: Procedure for calculating the imaging system of transmission electron microscopes with minimum distortion of the image at low magnification. Sov. J. Opt. Technol. *45*, 13-14 (1978)]
Makarova, I.S., Myakinkova, G.G., Yarmusevich, Ya.S.: Izv. Akad. Nauk SSSR Ser. Fiz. *44*, 1298-1301 (1980) [English transl.: Calculation of the electron optical characteristics of many-lens imaging systems for TEM, operating in the pancratic mode and with minimum distortion at low magnification. Bull. Acad. Sci. USSR Phys. Ser. *44*, No.6, 144-147 (1980)]
Marai, F.Z., Mulvey, T.: "Electron Optical Characteristics of Single-Pole Magnetic Lenses", Canberra 1974, Vol.1, pp.130-131
Marai, F.Z., Mulvey, T.: "Electron-Optical Characteristics of Single-Pole and Related Magnetic Electron Lenses", in *Electron Microscopy and Analysis, 1975*, ed. by J.A. Venables (Academic, London 1976) pp.43-44

Marai, F.Z., Mulvey, T.: Scherzer's formula and the correction of spiral distortion in the electron microscope. Ultramicroscopy *2*, 187-192 (1977)
Marschall, H.: Über die sphärische Aberration magnetischer Konzentrierspulen. Telefunken-Röhre *16*, 190-197 (1939)
Marton, L., Banca, M.C., Bender, J.F.: A new electron microscope. RCA Rev. *5*, 232-243 (1940)
Marton, L., Hutter, R.G.E.: The transmission type of electron microscope and its optics. Proc. Inst. Radio Eng. *32*, 3-12 (1944)
Marton, L., Hutter, R.G.E.: Optical constants of a magnetic-type electron microscope. Proc. Inst. Radio Eng. *32*, 546-552 (1944)
Marton, L., Hutter, R.G.E.: On apertures of transmission-type electron microscopes using magnetic lenses. Phys. Rev. *65*, 161-167 (1944)
Marton, L.: Electron microscopy. Repts. Prog. Phys. *10*, 204-252 (1946)
Marton, L., Bol, K.: Spherical aberration of compound magnetic lenses. J. Appl. Phys. *18*, 522-529 (1947)
Marton, L.: *Early History of the Electron Microscope* (San Francisco Press, San Francisco and Heffer, Cambridge 1968)
Mast, van der, K.D., Rakels, C.J., Le Poole, J.B.: "A High Quality Multipurpose Objective Lens", The Hague 1980, Vol.1, pp.72-73
Matsuda, T., Komoda, T.: Measurement of aberration coefficients of magnetic electron lenses. J. Electron Microsc. *21*, 163-168 (1972) [in Japanese]
Ments, van, M., Le Poole, J.B.: Numerical computation of the constants of magnetic electron lenses. Appl. Sci. Res. B*1*, 3-17 (1947)
Merli, P.G., Valdrè, U.: "New Possibilities Offered by Superconducting Lenses to the Design of Specimen Stages", Rome 1968, Vol.1, pp.197-198
Merli, P.G.: "Electron Optical Properties of Systems of Superconducting Coils", Grenoble 1970, Vol.2, pp.99-100
Morito, N.: On the chromatic field aberration of the magnetic electron lens in the electron microscope. J. Appl. Phys. *25*, 986-993 (1954)
Morito, N.: On the chromatic field aberration of the magnetic electron lens in the electron microscope. Denshikenbikyo [Electron Microsc.] *3*, No 2, 1-5 (1954) [in Japanese]
Morito, N.: On the chromatic aberration of magnetic projector lenses. J. Electron Microsc. *5*, 1-2 (1957)
Morito, N., Tadano, B., Katagiri, S.: "The Development of the Lens System in the Hitachi Electron Microscope", Berlin 1958, Vol.1, pp.51-53
Moseev, V.V., Stoyanov, P.A., Makarova, I.S., Yarmusevich, Ya.S., Der-Shvarts, G.V.: Izv. Akad. Nauk SSSR Ser. Fiz. *36*, 1275-1280 (1972) [English transl.: Investigation and calculation of optical characteristic of condenser-objective for a 300-kV electron microscope. Bull. Acad. Sci. USSR Phys. Ser. *36*, 1139-1143 (1972)]
Moses, R.W.: "Minimum Spherical Aberration of Coma-Free Magnetic Round Lenses", Manchester 1972, pp.86-87
Moses, R.W.: "Lens Optimization by Direct Application of the Calculus of Variations", in *Image Processing and Computer-Aided Design in Electron Optics*, ed. by P.W. Hawkes (Academic, London 1973) pp.250-272
Mulak, A.: Pola elektryczne i magnetyczne realizujace zadane rodziny torów elektronów [Electric and magnetic fields realizing the required families of electron trajectories]. Pr. Nauk. Inst. Technol. Elektron. Politech. Wroclaw. *16*, No.4, 39 pp. (1975)
Müller, K.: "Über magnetostatische Linsenanordnungen mit mechanischen Regelgliedern", Stockholm 1956, pp.17-20
Müller, K.: Regelbare magnetostatische Linsensysteme für Elektronenmikroskope. Z. Wiss. Mikrosk. *63*, 303-328 (1957)
Müller, K., Ruska, E.: Über ein einfaches und leistungsfähiges permanentmagnetisches Durchstrahlungs-Elektronenmikroskop. Mikroskopie *23*, 197-219 (1968)
Mulvey, T.: The magnetic circuit in electron microscope lenses. Proc. Phys. Soc. London B*66*, 441-447 (1953)
Mulvey, T.: "Elektronenoptisches System zur Herstellung feiner Elektronensonden", Berlin 1958, Vol.1, pp.68-71
Mulvey, T.: Electron-optical design of an X-ray microanalyser. J. Sci. Instrum. *36*, 350-355 (1959)

Mulvey, T.: "Electron Microprobes", in *Focusing of Charged Particles,* ed. by A. Septier (Academic, New York 1967) Vol.I, pp.469-494
Mulvey, T., Wallington, M.J.: The focal properties and aberrations of magnetic electron lenses. J. Phys. E *2*, 466-472 (1969)
Mulvey, T.: "Developments in Magnetic Electron Lenses for Electron Microscopes and Probe-Forming Lenses", in *Electron Microscopy and Analysis,* ed. by W.C. Nixon (Institute of Physics, London 1971) pp.78-83
Mulvey, T.: "Progress in Electron Optics", Manchester 1972, pp.64-69
Mulvey, T., Newman, C.D.: "Versatile Miniature Electron Lenses", Manchester 1972, pp.116-117
Mulvey, T., Wallington, M.J.: Electron lenses. Repts. Prog. Phys. *36*, 347-421 (1973)
Mulvey, T., Newman, C.D.: "New Electron-Optical Systems for SEM and STEM", in *Scanning Electron Microscopy, Systems and Applications 1973,* ed. by W.C. Nixon (Institute of Physics, London 1973) pp.16-21
Mulvey, T., Newman, C.D.: "New Experimental Lens Designs for High Voltage Electron Microscopes", in *High Voltage Electron Microscopy,* Proc. 3rd Int. Conf., ed. by P.R. Swann, C.J. Humphreys, M.J. Goringe (Academic, London 1974) pp.98-102
Mulvey, T.: "Mini-Lenses and the SEM", in *Scanning Electron Microscopy 1974,* ed. by O. Johari (IIT Research Institute, Chicago 1974) pp.43-49
Mulvey, T.: "Design Trends in TEM, STEM and SEM", Jerusalem 1976, Vol.1, pp.59-64
Mulvey, T.: "Instrumental Trends in TEM, STEM and SEM", in *Proc. XVth Czechoslovak Conf. Electron Microscopy* Prague, 1977, ed. by V. Viklický and J. Ludvík (Czechoslovak Academy of Sciences, Prague 1977) Vol.B, pp.577-580
Mulvey, T., Nasr, H.: "Limitation of the Finite-Element Method", in *Developments in Electron Microscopy and Analysis, 1979,* ed. by T. Mulvey (Institute of Physics, Bristol 1980) pp.53-54
Mulvey, T.: "Electron Guns and Instrumentation", The Hague 1980, Vol.1, pp.46-53
Mulvey, T., Nasr, H.: "An Improved Finite Element Program for Calculating the Field Distribution in Magnetic Lenses", The Hague 1980, Vol.1, pp.64-65
Munro, E.: "Computer-Aided Design of Magnetic Electron Lenses Using the Finite Element Method", Grenoble 1970, Vol.2, pp.55-56
Munro, E.: "A Low-Aberration Probe-Forming Lens for a Scanning Electron Microscope", in *Electron Microscopy and Analysis,* ed. by W.C. Nixon (Institute of Physics, London 1971) pp.84-87
Munro, E.: "Computer-Aided Design of Electron Lenses by the Finite Element Method", in *Image Processing and Computer-Aided Design in Electron Optics,* ed. by P.W. Hawkes (Academic, London 1973) pp.284-323
Munro, E.: "Scanning Electron Microscope Lens Design", in *Scanning Electron Microscopy 1974*, ed. by O. Johari (IIT Research Institute, Chicago 1974) pp.35-42
Munro, E.: Design and optimization of magnetic lenses and deflection systems for electron beams. J. Vac. Sci. Technol. *12*, 1146-1150 (1975)
Munro, E., Wells, O.C.: "Some Comments on the Design of Magnetic Lenses for the Scanning Electron Microscope", in *Scanning Electron Microscopy 1976,* Vol.1, ed. by O. Johari (IIT Research Institute, Chicago 1976) pp.27-36
Munro, E.: "Electron Beam Lithography", in *Applied Charged Particle Optics,* ed. by A. Septier (Academic, New York 1980), Vol.B, pp.73-131 [Supplement 13B to Adv. Electron. Electron Phys.]
Munro, E.: "Electron Optics for Microcircuit Engineering", in *Microcircuit Engineering",* ed. by H. Ahmed, W.C. Nixon (Cambridge University Press, Cambridge 1980) pp.513-534
Murillo, R., Balladore, J.L.: Etude des éléments cardinaux d'un objectif à pièces polaires saturées. J. Microsc. (Paris) *20*, 1-13 (1974)
Nakagawa, S.: "A Method for Measuring the Spherical and Chromatic Aberration Coefficients of an Objective Lens", in *Scanning Electron Microscopy 1977,* ed. by O. Johari (IIT Research Institute, Chicago 1977) Vol.I, pp.33-39
Nakagawa, S., Miyokawa, T., Noguchi, Y.: "Axial Magnetic Corrected Field Lens (C-F lens). Principle and Characteristics, Minimizing C_s and C_c", The Hague 1980, Vol.1, pp.80-81
Naruse, M., Watanabe, E., Harada, Y., Sakurai, S., Etoh, T.: Development of a 200 kV high-resolution electron microscope. J. Electron Microsc. *29*, 54-58 (1980)

Nasr, H., Chen, W., Mulvey, T.: "Improved Programs for Calculating Optical Properties of Electron Lenses", in *Electron Microscopy and Analysis, 1981*, ed. by M.J. Goringe (Institute of Physics, Bristol 1982) pp.75-78
Nishigaki, M., Katagiri, S., Kimura, H., Tadano, B.: "A New 1,000-kV Electron Microscope", EMSA 25, 1967, pp.260-261
Oostrum, van, K.J.: Design parameters for non-immersion lenses. Jernkontorets Ann. *155*, 491-492 (1971)
Ozasa, S., Katagiri, S., Kimura, H., Tadano, B.: "Superconducting Electron Lens", Kyoto 1966, Vol.1, pp.149-150
Ozasa, S., Kitamura, N., Katagiri, S., Kimura, H.: "The Superconducting Electron Lens for the Ultra High Voltage Electron Microscope", Rome 1968, Vol.1, pp.185-186
Ozasa, S., Sakitani, Y., Katagiri, S., Kimura, H., Sugata, E., Fukai, K., Fujita, E., Fujita, H., Ura, K.: "Development of 3-MeV Electron Microscope Column", Grenoble 1970, Vol.1, pp.123-124
Parker, N.W., Golladay, S.D., Crewe, A.V.: "A Theoretical Analysis of Third-Order Aberration Correction in the SEM and STEM", in *Scanning Electron Microscopy 1976*, Vol.1, ed. by O. Johari (IIT Research Institute, Chicago 1976) pp.37-44
Parker, N.W., Utlaut, M., Crewe, A.V.: "Evaporative Cooling of High Current Density Mini-Lenses", EMSA 34, 1976, pp.536-537
Parker, N.W., Crewe, A.V., Isaacson, M.S., Mankawich, W.: "A New Analytical Electron Microscope", Toronto 1978, Vol.1, pp.18-19
Paszkowski, B.: "Object-image" characteristics of short magnetic lenses. Bull. Acad. Pol. Sci. (Ser. Sci. Tech.) *13*, 721-724 (1965)
Pavlík, K.: Supravodivé magnetické čočky elektronového mikroskopu. Slaboproudý Obz. *34*, 405-416 (1973)
Pavlík, K., Studeník, J.: Objektiv elektronového mikroskopu se supravodivým budicím vinutím. Knižnice Odb. Věd. Spisů Vys. Učeni Tech. Brně B*59*, 129-136 (1975)
Payen, F., Laberrigue, A.: Courants induits dans des anneaux supraconducteurs, application en microscopie électronique. C.R. Acad. Sci. Paris B*272*, 405-408 (1971)
Payen, F., Laberrigue, A.: Propriétés optiques optimales de lentilles à anneaux supraconducteurs. C.R. Acad. Sci. Paris B*273*, 116-119 (1971)
Petrie, D.P.R.: "The Dependence of Spherical Aberration of Electron Lenses on the Conjugate Positions", Philadelphia 1962, Vol.1, pp.KK-2
Picht, J.: Über eine Methode zur systematischen Errechnung einer elektronenoptisch abbildenden Feldverteilung mit bestimmten geforderten Abbildungseigenschaften. Optik *12*, 433-440 (1955)
Podbrdský, J.: "The Optical System of the Analytical Electron Microscope with Mini-Lenses", The Hague 1980, Vol.1, pp.66-67
Prebus, A.F.: Improved pole piece construction of the objective lens of a magnetic electron microscope. Can. J. Res.*18* A, 175-177 (1940)
Preisberg, W.: "Über die Möglichkeit der Simulation elektronenoptischer Systeme mit Hilfe digitaler Rechenanlagen", Rome 1968, Vol.1, pp.171-172
Preisberg, W.: "Untersuchung über den Einfluß des Unsymmetriegrades einer rotationssymmetrischen magnetischen Linse auf ihren Öffnungs- und Farbfehler durch Simulation mit Hilfe einer digitalen Rechenanlage", Rome 1968, Vol.1, pp.173-174
Rakels, C.J., Tiemeijer, J.C., Witteveen, K.W.: The Philips electron microscope EM 300. Philips Tech. Rev. *29*, 370-386 (1968)
Rakhimov, Sh.M., Sushkin, N.G.: Selection of the polepiece material for electron microscopes. Zh. Tekh. Fiz. *18*, 1166-1172 (1948) [in Russian]
Ramberg, E.G.: Variation of the axial aberrations of electron lenses with lens strength. J. Appl. Phys. *13*, 582-594 (1942)
Rebsch, R.: Das theoretische Auflösungsvermögen des Elektronenmikroskops. Ann. Phys. (Leipzig) *31*, 551-560 (1937)
Rebsch, R., Schneider, W.: Der Öffnungsfehler schwacher Elektronenlinsen. Z. Phys. *107*, 138-143 (1937)
Rebsch, R.: Über den Öffnungsfehler der Elektronenlinsen. Z. Phys. *116*, 729-733 (1940)
Recknagel, A.: Über Fehler von Elektronenlinsen. Jahrb. AEG Forsch. *7*, 15-22 (1940)
Recknagel, A.: Über die sphärische Aberration bei elektronenoptischer Abbildung. Z. Phys. *117*, 67-73 (1941)

Reisner, J.H., Dornfeld, E.G.: A small electron microscope. J. Appl. Phys. *21*, 1131-1139 (1950)
Reisner, J.H.: Permanent magnet lenses. J. Appl. Phys. *22*, 561-565 (1951)
Reisner, J.H., Zollers, S.M.: Permanent-magnet electron microscope. Electronics *24*, 86-91 (1951)
Reisner, J.H., Coleman, J.W.: "The Optical System for the One Million Volt U.S. Steel Microscope Column", EMSA 25, 1967, pp.266-267
Reisner, J.H.: "Measurement of Radial Distortion", EMSA 28, 1970, pp.350-351
Riecke, W.D.: Die Abbildungseigenschaften des Vorfeldes einer magnetischen Objektivlinse vom Glaserschen Glockenfeldtyp. Optik *19*, 169-183, 193-207 (1962)
Riecke, W.D.: "Ein Kondensorsystem für eine starke Objektivlinse", Philadelphia 1962, Vol.1, pp.KK-5
Riecke, W.D.: "Einige Bemerkungen zur Gaußschen Dioptrik von magnetischen Objektivlinsen mit dezentrierten Polschuhen", Prague 1964, Vol.A, pp.7- 8
Riecke, W.D., Ruska, E.: "A 100-kV Transmission Electron Microscope with Single-Field Condenser Objective", Kyoto 1966, Vol.1, pp.19-20
Riecke, W.D.: Zur Zentrierung des magnetischen Elektronenmikroskops. Optik *24*, 397-426 (1966/1967)
Riecke, W.D.: "On the Alignment of an Electron Microscope with Condenser-Objective Lens", Rome 1968, Vol.1, pp.207-208
Riecke, W.D.: Zur Zentrierung des magnetischen Elektronenmikroskops. Optik *36*, 66-84, 288-308, 375-398 (1972)
Riecke, W.D.: "Procedures for the Alignment of Pole Piece Systems of Magnetic Electron Lenses", Manchester 1972, pp.76-77
Riecke, W.D.: "Objective Lens Design for Transmission Electron Microscopes - a Review of the Present State of the Art", Manchester 1972, pp.98-103
Riecke, W.D.: "Instrument Operation for Microscopy and Microdiffraction", in *Electron Microscopy in Materials Science*, ed. by U. Valdrè, E. Ruedl (Commission of the European Communities, Luxemburg 1976) pp.19-111
Riecke, W.D.: "Instrumentation in HVEM", in *High Voltage Electron Microscopy, 1977*, ed. by T. Imura, H. Hashimoto (Japanese Society of Electron Microscopy, Kyoto 1977) pp.73-78
Rogowski, W.: Über Fehler von Elektronenbildern. Arch Elektrotech. *31*, 555-593 (1937)
Rose, H.: Der Zusammenhang der Bildfehler-Koeffizienten mit den Entwicklungs-Koeffizienten des Eikonals. Optik *28*, 462-474 (1968/1969)
Rose, H., Petri, U.: Zur systematischen Berechnung elektronenoptischer Bildfehler. Optik *33*, 151-165 (1971)
Rose, H., Moses, R.W.: Minimaler Öffnungsfehler magnetischer Rund- und Zylinderlinsen bei feldfreiem Objektraum. Optik *37*, 316-336 (1973)
Rozenfel'd, A.M., Zaitsev, P.V.: Izv. Akad. Nauk SSSR Ser. Fiz. *25*, 713-716 (1961) [English transl.: Magnetic objective lens for an emission electron microscope. Bull. Acad. Sci. USSR Phys. Ser. *25*, 726-729 (1961)]
Ruska, E., Knoll, M.: Die magnetische Sammelspule für schnelle Elektronenstrahlen. Z. Tech. Phys. *12*, 389-400, 448 (1931)
Ruska, E.: Über Fortschritte im Bau und in der Leistung des magnetischen Elektronenmikroskops. Z. Phys. *87*, 580-602 (1934)
Ruska, E.: Über ein magnetisches Objektiv für das Elektronenmikroskop. Z. Phys. *89*, 90-128 (1934)
Ruska, E.: Über die Linsen hochauflösender Elektronenmikroskope. Arch. Elektrotech. *36*, 431-454 (1942)
Ruska, E.: Über den Bau und die Bemessung von Polschuhlinsen für hochauflösende Elektronenmikroskope. Arch. Elektrotech. *38*, 102-130 (1944)
Ruska, E.: Zur Entwicklung der Übermikroskopie und über ihre Beziehungen zur Kolloidforschung. Kolloid Z. *107*, 2-16 (1944)
Ruska, E.: Über neue magnetische Durchstrahlungs-Elektronenmikroskope im Strahlspannungsbereich von 40 zu 220 kV, Teil I. Kolloid Z. *116*, 102-120 (1950)
Ruska, E.: Untersuchungen über regelbare magnetostatische Elektronenlinsen. Z. Wiss. Mikrosk. *61*, 152-171 (1952)
Ruska, E.: "Experiments with adjustable magnetostatic electron lenses", in *Electron Physics*, Nat. Bur. Stand. Cir. *527*, 389-409 (1954)
Ruska, E.: Grundlagen der Elektronenoptik and ihre Anwendung bei modernen Elektronenmikroskopen. La Ricerca Scientifica Suppl., 3-30 (1954) [4th Session of the Giornate della Scienza, Conf. on Electronics and Television, Milan 1954]

Ruska, E., Wolff, O.: Ein hochauflösendes 100-kV-Elektronenmikroskop mit Kleinfelddurchstrahlung. Z. Wiss. Mikrosk. *62*, 465-509 (1956)
Ruska, E.: "Microscope électronique 100 kV à haute résolution et avec limitation étroite du champ éclairé, in *Les techniques récentes en microscopie électronique et corpusculaire* (C.N.R.S., Paris 1956) pp.45-57
Ruska, E.: 25 Jahre Elektronenmikroskopie. Elektrotech. Z. A*78*, 531-543 (1957)
Ruska, E.: Travaux récents destinés à améliorer le pouvoir séparateur expérimental du microscope électronique. J. Microsc. (Paris) *3*, 357-372 (1964)
Ruska, E.: Current efforts to attain the resolution limit of the transmission electron microscope. J. R. Microsc. Soc. *84*, 77-103 (1964)
Ruska, E.: Über die Auflösungsgrenzen des Durchstrahlungs-Elektronenmikroskops. Optik *22*, 319-348 (1965)
Ruska, E.: Past and present attempts to attain the resolution limit of the transmission electron microscope. Adv. Opt. Electron Microsc. *1*, 115-179 (1966)
Ruska, E.: Die frühe Entwicklung der Elektronenlinsen und der Elektronenmikroskopie. Acta Hist. Leopold. *12*, 136 pp. (1979) [English transl. by T. Mulvey: The early development of electron lenses and electron microscopy. Microsc. Acta Suppl. *5*, 140 pp. (1980) also published separately by Hirzel, Stuttgart]
Sandor, A.: Messvorrichtungen zur Bestimmung der elektronenoptischen Hauptdaten von rotationssymmetrischen Elektronenlinsen mit dem Ziel der geometrischen Bildkonstruktion. Arch. Elektrotech. *35*, 401-423 (1941)
Scherzer, O.: Die Aufgaben der theoretischen Elektronenoptik. Z. Tech. Phys. *17*, 593-596 (1936); also published in *Beiträge zur Elektronenoptik*, ed. by H. Busch, E. Brüche (Barth, Leipzig 1936) pp.14-18
Scherzer, O.: "Berechnung der Bildfehler dritter Ordnung nach der Bahnmethode", in *Beiträge zur Elektronenoptik*, ed. by H. Busch, E. Brüche (Barth, Leipzig 1936) pp.33-41
Scherzer, O.: Über einige Fehler von Elektronenlinsen. Z. Phys. *101*, 593-603 (1936)
Scherzer, O.: Die unteren Grenzen der Brennweite und des chromatischen Fehlers von magnetischen Elektronenlinsen. Z. Phys. *118*, 461-466 (1941)
Schiske, P.: Die untere Grenze des Farbfehlers magnetischer Linsen bei vorgeschriebener Maximalfeldstärke. Optik *13*, 502-505 (1956)
Schiske, P.: Partial fraction expansion of electron optical cardinal elements. Ultramicroscopy *2*, 193-198 (1977)
Schlögl, J.: Zur Elektronenoptik magnetischer Linsen von Glaserschem Typus. Sitzungsber. Österr. Akad. Wiss. Math.-Naturwiss. Kl., Abt. IIa *157*, 237-262 (1949)
Seman, O.I.: Optical power of short electron lenses. Zh. Tekh. Fiz. *20*, 1180-1193 (1950) [in Russian]
Seman, O.I.: Transformation of the form of the fourth-order eikonal and aberration coefficients in electron optics. Dokl. Akad. Nauk. SSSR *81*, 775-778 (1951) [in Russian]
Seman, O.I.: On the question of the existence of an extremum of the spherical aberration coefficient for axially symmetric systems in electron optics. Zh. Eksp. Teoret. Fiz. *24*, 581-588 (1953) [in Russian]
Seman, O.I.: Limiting values of electron optical aberration coefficients. Dokl. Akad. Nauk SSSR *93*, 443-445 (1953) [in Russian]
Seman, O.I.: Relativistic aberration functions and normal forms of electron-optical aberration coefficients. Dokl. Akad. Nauk. SSSR *96*, 1151-1154 (1954) [in Russian]
Seman, O.I.: Relativistic theory of rotationally symmetric electron optics on the basis of eikonal theory. Trudy Inst. Fiz. Astron. Akad. Nauk. Est. SSR [Eesti NSV Teaduste Akad. Füüsika Astron. Inst. Uurimused] *2*, 3-29 (1955) [in Russian]
Seman, O.I.: Expansion of the fourth-order eikonal and the relativistic theory of aberrations. Trudy Inst. Fiz. Astron. Akad. Nauk. Est. SSR [Eesti NSV Teaduste Akad. Füüsika Astron. Inst. Uurimused] *2*, 30-49 (1955) [in Russian]
Seman, O.I.: Expansion of the fourth-order eikonal and the relativistic theory of aberrations. Trudy Inst. Fiz. Astron. Akad. Nauk. Est. SSR *2*, 30-49 (1955) [in Russian]
Seman, O.I.: On the theory of virtual and transverse aberrations in electron optics. Uch. Zap. Rostov. na Donu Gos. Univ. *68*, No 8, 63-75 (1958) [in Russian]
Seman, O.I.: On the theory of the fourth-order point eikonal in electron optics. Uch. Zap. Rostov. na Donu Gos. Univ. *68*, No 8, 77-90 (1958) [in Russian]

Seman, O.I.: Radiotekh. Elektron. *3*, 283-287 (1958) [English transl.: On normal coefficients of electron-optical aberrations of magnetic immersion systems. Radio Eng. Electron. Phys. (USSR) *3*, 402-409 (1958)]
Seman, O.I.: Radiotekh. Elektron. *4*, 1213-1214 (1959) [English transl.: Dependence of anisotropic distortion of the image on the curvature of the object field in electron optics. Radio Eng. Electron. Phys. (USSR) *4*, 227-230 (1959)]
Seman, O.I.: Radiotekh. Elektron. *4*, 1702-1707 (1959) [English transl.: On the theoretical limits of the chromatic aberration coefficient in limited electron-optical magnetic fields. Radio Eng. Electron. Phys. (USSR) *4*, 235-244 (1959)]
Seman, O.I.: Opt. Spektrosk. *7*, 113-115 (1959) [English transl.: On some properties of the image curvature in short electron optical lenses. Opt. Spectrosc. *7*, 68-69 (1959)]
Seman, O.I.: Some properties of the curvature of the electron optical image in magnetic fields. Trudy Inst. Fiz. Astron. Akad. Nauk. Eston. SSR [Eesti NSV Teaduste Akad. Füüsika Astron. Inst. Uurimused] *13*, 20-37 (1961) [in Russian]
Seman, O.I.: Radiotekh. Elektron. *13*, 907-912 (1968) [English transl.: Properties of Petzval curvature in electron optics. Radio Eng. Electron. Phys. (USSR) *13*, 784-788 (1968)]
Septier, A.: "Superconducting Lenses", Manchester 1972, pp.104-109
Séverin, C., Gênotel, D., Girard, M., Laberrigue, A.: Microscope électronique 400 kV à lentilles supraconductrices. 2. Caractéristiques électrooptiques et fonctionnement. Rev. Phys. Appl. *6*, 459-465 (1971)
Shchetnev, Yu.F., Gravel', V.M., Chentsova, N.V.: Izv. Akad. Nauk SSSR Ser. Fiz. *44*, 1155-1158 (1980) [English transl.: Cooled winding for an electron lens. Bull. Acad. Sci. USSR Phys. Ser. *44*, No.6, 27-29 (1980)]
Shiraishi, K., Katsuta, T., Ozasa, S., Todokoro, H.: "A Newly Designed 650-kV Electron Microscope", EMSA 26, 1967, pp.304-305
Shirota, K., Yonezawa, A., Shibatomi, K., Yanaka, T.: Ferro-magnetic material observation lens system for CTEM with a eucentric goniometer. J. Electron. Microsc. *25*, 303-304 (1976)
Siegbahn, K.: Untersuchungen über die Verwendung der magnetischen Linse für β-Spektroskopie. Ark. Mat. Astron. Fys. *28*A, No.17, 27 pp. (1942)
Siegbahn, K.: Formation of image in a strong magnetic lens. Ark. Mat. Astron. Fys. *30*A, No.1, 12 pp. (1944)
Siegbahn, K.: A magnetic lens of special field form for β- and γ-ray investigations; designs and applications. Philos. Mag. *37*, 162-184 (1946)
Siegel, B.M., Kitamura, N., Kropfli, R.A., Schulhof, M.P.: "High Resolution Objective Lenses Using Superconducting Materials", Kyoto 1966, Vol.1, pp.151-152
Siegel, B.: "Electron Optics for Scanning Electron Microscopy", in *Scanning Electron Microscopy 1975*, ed. by O. Johari (IIT Research Institute, Chicago 1975) pp.647-660
Siemens AG: "High Performance Electron Microscope Elmiskop 102", Focus Information No 1 (n.d.)
Stabenow, G.: Eine magnetische Elektronenlinse ohne Bilddrehung. Z. Phys. *96*, 634-642 (1935)
Stokes, A.T.: "Computational Analysis of a Field Emission Gun System Incorporating a Single-Pole Lens", The Hague 1980, Vol.1, pp.58-59
Störmer, C.: Über die Bahnen von Elektronen im axialsymmetrischen elektrischen und magnetischen Felde. Ann. Phys. (Leipzig) *16*, 685-696 (1933)
Stoyanov, P.A.: Izv. Akad. Nauk SSSR Ser. Fiz. *25*, 672-675 (1961) [English transl.: Achromatization of lenses in multiple-lens magnetic electron microscopes. Bull. Acad. Sci. USSR Phys. Ser. *25*, 687-690 (1961)]
Stoyanov, P.A., Vol'fson, L.Yu.: Izv. Akad. Nauk SSSR Ser. Fiz. *25*, 717-720 (1961) [English transl.: Investigation of the magnetic circuits of electron microscope lenses. Bull. Acad. Sci. USSR Phys. Ser. *25*, 730-732 (1961)]
Stoyanov, P.A., Artemova, T.V., Susov, E.V., Talashev, A.A.: Izv. Akad. Nauk SSSR Ser. Fiz. *32*, 928-931 (1968) [English transl.: An electron microscope objective of superconducting material. Bull. Acad. Sci. USSR Phys. Ser. *32*, 863-866 (1968)]
Stoyanov, P.A., Moseev, V.V., Tikhonov, D.D.: Izv. Akad. Nauk. SSSR Ser. Fiz. *32*, 1120-1123 (1968) [English transl.: An electron microscope objective lens with a large magnetomotive force. Bull. Acad. Sci. USSR Phys. Ser. *32*, 1043-1046 (1968)]

Strojnik, A.: "The Arizona 1-MeV Transmission Scanning Electron Microscope - Some Design Features", in *Scanning Electron Microscopy 1972*, ed. by O. Johari (IIT Research Institute, Chicago 1972) pp.215-223
Strojnik, A.: "The Theoretical Resolving Power of the Transmission SEM at High Voltages", in *Scanning Electron Microscopy 1973*, ed. by O. Johari (IIT Research Institute, Chicago 1973) pp.17-24
Strojnik, A., Strojnik, M.: "On the Design of Electron Lenses Operating at Partial Magnetic Saturation", in *Electron Microscopy and Analysis, 1975*, ed. by J.A. Venables (Academic, London 1976) pp.51-52
Strojnik, A.: "Optimization of the Objective Lens in a HV Microscope Operating in the Scanning Mode", in *Microscopie électronique à haute tension, 1975*, ed. by B. Jouffrey, P. Favard (Société Française de Microscopie Electronique, Paris 1976) pp.27-30
Strojnik, A.: "Compact Magnetic Lenses for HV Electron Microscopy", EMSA 27, 1978, pp.104-105
Strojnik, A., Passow, C.: "Scanning System for a 5-MeV Electron Microscope", in *Scanning Electron Microscopy 1978*, ed. by O. Johari (SEM Inc, AMF O'Hare 1978) Vol.I, pp.319-324
Strojnik, M.: "Field Distribution of Strongly Excited Magnetic Lenses", EMSA 33, 1975, pp.140-141
Studeník, J.: Objektiv elektronového mikroskopu se supravodivým vinutím. Slaboproudý Obz. *34*, 417-420 (1973)
Studeník, J., Pavlík, K.: Nová supravodičová magnetická čočka. Čs. Čas. Fyz. A*25*, 154-155 (1975)
Sturrock, P.A.: "The Aberrations of Magnetic Electron Lenses due to Asymmetries", Delft 1949, pp.89-93
Sturrock, P.: Formules nouvelles pour les aberrations du troisième ordre des lentilles magnétiques. C.R. Acad. Sci. Paris *233*, 146-147 (1951)
Sturrock, P.: Propriétés optiques des champs magnétiques de révolution de la forme $H = H_0/[1-(z/a^2)]$ et $H = H_0/[(z/a)^2 - 1]$[sic]. C.R.Acad. Sci. Paris *233*, 401-403 (1951)
Sturrock, P.A.: Perturbation characteristic functions and their application to electron optics. Proc. R. Soc. London A*210*, 269-289 (1951)
Sturrock, P.A.: The aberrations of magnetic electron lenses due to asymmetries. Phil. Trans. R. Soc. London A*243*, 387-429 (1951)
Sugata, E., Nishitani, Y., Hamada, H.: The effects of pole piece saturation on the constants of magnetic lenses. J. Electron. Microsc. *3*, 9-14 (1955)
Sugata, E., Nishitani, Y., Hirosue, S.: Leakage magnetic field by the saturation of pole-pieces of magnetic lenses. J. Electron Microsc. *4*, 49-50 (1956)
Sugiura, Y., Suzuki, S.: On the magnetic lens of minimum spherical aberration. Proc. Imp. Acad. Jpn. (Tokyo) *19*, 293-302 (1943)
Šuhel, P., Avčin, F.: "Coreless Electromagnetic Lenses for High Acceleration Voltage Electron Microscopes", Grenoble 1970, Vol.1, pp.133-134
Šuhel, P., Avčin, F.: Nouvelle conception d'une puissante lentille électronique magnétique sans noyau de fer destinée aux microscopes électroniques haute tension. Elektrotehn. Vest. *41*, No.5-8, 139-147 (1974)
Šuhel, P.: Détermination de la trajectoire paraxiale des électrons de la nouvelle lentille électronique magnétique de puissance élevée (II). Elektrotehn. Vest. *41*, No.9-10, 207-215 (1974)
Šuhel, P.: Optimisation des caractéristiques électro-optiques de la nouvelle lentille (III). Elektrotehn. Vest. *41*, No.11-12, 267-273 (1974). *Note.* The page numbers in the last three references correspond to the text in Slovene
Sushkin, N.G., Sushkin, V.N.: Radiotekh. Electron *15*, 1764-1765 (1970) [English transl.: Diverging magnetic lenses. Radio Eng. Electron. Phys. (USSR) *15*, 1523-1524 (1970)]
Sushkin, V.N., Alferova, Ye.V., Sushkin, N.G.: Radiotekh. Elektron. *16*, 1095-1096 (1971) [English transl.: A calculation of the aberrations of a system with a diverging magnetic lens. Radio Eng. Electron Phys. (USSR) *16*, 1072-1074 (1971)]
Suzuki, S., Akashi, K., Tochigi, H.: "Objective Lens Properties of Very High Excitation", EMSA 26, 1968, pp.320-321
Suzuki, S. Ishikawa, A.: "On the Magnetic Electron Lens of Minimum Spherical Aberration", Toronto 1978, Vol.1, pp.24-25

Svartholm, N.: Berechnung magnetischer Linsen mit gegebener Feldform. Ark. Mat. Astron. Fys. *28*B, No.16, 8 pp. (1942)
Svartholm, N.: The focal length of a long magnetic lens. Ark. Mat. Astron. Fys. *35*A, No.6, 9 pp. (1948)
Szilágyi, M.: Electron motion in a stationary axially symmetric magnetic field varying linearly along the axis. Period. Polytech. Electr. Eng. *13*, 221-233 (1969)
Szilágyi, M.: "A New Magnetic Lens Model" [$B(z) = B_0 + B_1 z$], Grenoble 1970, Vol.2, pp.61-62
Szilágyi, M.: A new approach to electron optical optimization. Optik *48*, 215-224 (1977)
Szilágyi, M.: A dynamic programming search for magnetic field distributions with minimum spherical aberration. Optik *49*, 223-246 (1977)
Szilágyi, M.: "Reduction of Aberrations by New Optimization Techniques", Toronto 1978, Vol.1, pp.30-31
Szymánski, H.: Aberracje elektronooptyczne w urzadzeniach z wiązka elektronową [Electron optical aberration in electron beam equipments], Pr. Nauk. Inst. Technol. Elektron. Politech. Wroclaw *15*, No.3, 77 pp. (1975)
Tadano, B.: Experimental investigations on electromagnetic lens aberrations. Denshikenbikyo [Electron Microsc.] *5*, No 3, 1-13 (1957) [in Japanese]
Tadano, B., Laverick, C.: "Current Superconducting Lens Studies for Electron Microscopes", in *Proc. AMU-ANL Workshop on High-Voltage Electron Microscopy*, ed. by J. Gilroy, R.K. Hart, M. Weber, J. Kopta, Argonne National Laboratory Report ANL-7275 (1966) pp.154-159
Tamura, N., Saito, H., Ohyama, J. Aihara, R., Kabaya, A.: "Field Emission SEM Using Strongly Excited Objective Lens", EMSA 38, 1980, pp.68-69
Tochigi, H., Uchida, H., Shirai, S., Akashi, K., Evins, D.J., Bonnici, J.: "A New High Resolution Transmission Electron Microscope with Second-Zone Objective Lens", EMSA 27, 1968, pp.176-177
Tochigi, H., Nakajima, H.: "An Experiment Obtaining Very Strong Magnetic Field on a Standard Electron Microscope", Grenoble 1970, Vol.1, pp.7-8
Tretner, W.: Die untere Grenze des Öffnungsfehlers magnetischer Elektronenobjektive. Optik *7*, 242 (1950)
Tretner, W.: Die untere Grenze des Öffnungsfehlers bei magnetischen Elektronenlinsen. Optik *11*, 312-326 (1954);
Berichtigung. Optik *12*, 293-294 (1955)
Tretner, W.: Zur unteren Grenze des Öffnungsfehlers magnetischer Elektronenlinsen. Optik *13*, 516-519 (1956)
Tretner, W.: Existenzbereiche rotationssymmetrischer Elektronenlinsen. Optik *16*, 155-184 (1959)
Trinquier, J., Balladore, J.L.: "Caractéristique optiques et coefficients d'aberrations de quelques lentilles à supraconducteurs", Rome 1968, Vol.1, pp.191-192
Trinquier, J., Balladore, J.L., Murillo, R.: Influence de cylindres creux supraconducteurs sur les caractéristiques optiques des lentilles électroniques magnétiques. C.R. Acad. Sci. Paris B*268*, 1707-1710 (1969)
Trinquier, J., Balladore, J.L.: "Microscope 300 kV à objectif supraconducteur", Grenoble 1970, Vol.2, pp.97-98
Trinquier, J., Balladore, J.L., Martinez, J.P.: "Propriétés et applications d'un objectif supraconducteur pour microscope 300 kV", Manchester 1972, pp.118-119
Trinquier, J., Balladore, J.L., Murillo, R., Martinez, J.P.: Influence de la saturation du fer des pièces polaires sur les caractéristiques optiques des lentilles électroniques magnétiques. C.R. Acad. Sci. Paris B*276*, 203-206 (1973)
Troyon, M., Laberrigue, A.: A field emission gun with high beam current capability. J. Microsc. Spectrosc. Electron. *2*, 7-11 (1977)
Troyon, M.: "A Magnetic Field Emission Electron Probe-Forming System", The Hague 1980, Vol.1, pp.56-57
Troyon, M.: High current efficiency field emission gun system incorporating a preaccelerator magnetic lens. Its use in CTEM. Optik *57*, 401-419 (1980)
Tsuno, K., Ishida, Y., Harada, Y.: "Spiral Distortion in Electron Microscopy During Use of Three-Pole-Piece Lens", EMSA 38, 1980, pp.280-281
Tsuno, K., Arai, Y., Harada, Y.: "Elimination of Spiral Distortion in Electron Microscope by Means of a Three-Pole-Piece Lens", The Hague 1980, Vol.1, pp.76-77

Tsuno, K., Harada, Y.: Minimisation of radial and spiral distortion in electron microscopy through the use of a triple pole-piece lens. J. Phys. E *14*, 313-319 (1981)
Tsuno, K., Harada, Y.: Elimination of spiral distortion in electron microscopy using an asymmetrical triple pole-piece lens. J. Phys. E *14*, 955-960 (1981)
Tsuno, K., Harada, Y.: "Calculation of Field Distribution and Optical Properties of Saturated Asymmetrical Objective Lenses", in *Electron Microscopy and Analysis 1981*, ed. by M.J. Goringe (Institute of Physics, Bristol 1982) pp.95-98
Tszu-I: Investigation on perturbations of Gaussian electron optics. Acta Sci. Natl. Univ. Pekin *3*, 337-349 (1957) [in Russian]
Typke, D., Burger, M., Lemke, N., Lefranc, G.: "Determination of the Shielding Geometry of a Superconducting Two-Lens-System with Off-Axial Ray Paths for 3D Imaging, The Hague 1980, Vol.1, pp.82-83
U My-Chzhen': Investigations on aberrations of fifth order in rotationally symmetric electron optical systems. Sci. Sinica *6*, 833-846 (1957) [in Russian]; Acta Phys. Sinica *13*, 181-206 (1957) [in Chinese]
Venables, J.A.: "Design Trends in Scanning Electron Microscopy", Jerusalem 1976, Vol.1, pp.67-72
Vertsner, V.N.: The electron microscope of the State Optical Institute. Izv. Akad. Nauk SSSR Ser. Fiz. *8*, 232-234 (1944) [in Russian]
Vertsner, V.N., Vorona, Yu.M., Vorob'ev, Yu.V., Bogdanovskii, G.A., Chentsov, Yu.V.: Izv. Akad. Nauk SSSR Ser. Fiz. *25*, 680-682 (1961) [English transl.: Optics of the EM-5 and EM-7 electron microscopes. Bull. Acad. Sci. USSR Phys. Ser. *25*, 695-698 (1961)]
Vlasov, A.G.: A magnetic lens with minimum spherical aberration. Izv. Akad. Nauk SSSR Ser. Fiz. *8*, 235-239 (1944) [in Russian]
Voit, H.: Über die elektronenoptischen Bildfehler dritter Ordnung. Z. Instrumentenkde. *59*, 71-82 (1939)
Vorob'ev, Yu.V.: Analysis of alignment methods for magnetic electron microscopes. Opt. Spektrosk. *10*, 257-264 (1961) [in Russian]
Wallauschek, R., Bergmann, P.: Zur Theorie des Elektronenmikroskops mit Anwendung auf rein magnetische Felder. Z. Phys. *94*, 329-334 (1935)
Watanabe, A., Morito, N.: Zur Messung der Farbfehlerkonstante von magnetischen Linsen. Optik *12*, 166-172, 564 (1955)
Watanabe, M., Yanaka, T., Nagahama, Y.: "A New Objective Pole Piece for Special Observation Method of the JEM", Philadelphia 1962, Vol.1, pp.E-1
Wegman, L.: Ein verzeichnungsarmes elektromagnetisches Doppelprojektiv für Elektronenmikroskope. Helv. Phys. Acta *26*, 448-449 (1953)
L. Wegmann: Ein verzeichnungsarmes elektromagnetisches Doppelprojektiv für Elektronenmikroskope. Optik *11*, 153-170 (1954)
Weyl, R., Dietrich, I., Zerbst, H.: Die supraleitende Abschirmlinse. Optik *35*, 280-286 (1972)
Wiggins, J.W.: "Practical Considerations for STEM Optics", in *Scanning Electron Microscopy 1977*, ed. by O. Johari (IIT Research Institute, Chicago 1977) Vol.I, pp.673-682
Worsham, R.E., Mann, J.E., Richardson, E.G., Ziegler, N.F.: "Two Developments - a Superconducting Microscope Conversion and a Precision High Voltage Divider", EMSA 28, 1970, pp.354-355
Worsham, R.E., Mann, J.E., Richardson, E.G.: "A Superconducting Microscope", EMSA 29, 1971, pp.10-11
Worster, J.: Aberrations at multiple foci in magnetic electron lenses. J. Phys. D *8*, L29-30 (1975)
Ximen Jiye [or Hsimen Chi-yeh]: Chinese J. Sci. Instrum. *2*, 1-11 (1981) [English transl.: On the linear transformations for Gaussian trajectory parameters in the combined electron optical systems. Optik *59*, 237-249 (1981)]
Yada, K., Kawakatsu, H.: Magnetic objective lens with small bores. J. Electron Microsc. *25*, 1-9 (1976)
Yada, K., Kawakatsu, H.: "Aberrations of Objective Lens for HVEM", in *High Voltage Electron Microscopy, 1977*, ed. by T. Imura, H. Hashimoto (Japanese Society of Electron Microscopy, Tokyo 1977) pp.35-40
Yanaka, T., Watanabe, M.: "Aberration Coefficients of Extremely Asymmetrical Objective Lenses", Kyoto 1966, Vol.1, pp.141-142

Yanaka, T., Shirota, K.: "High Resolution Specimen Area Imaged Under Negligible Field Aberrations", Grenoble 1970, Vol.2, pp.59-60
Yanaka, T., Shirota, K., Yonezawa, A., Arai, Y.: "Capability of High Resolution Objective Lens with Miniaturized Lower Pole Piece Top. I", Canberra 1974, Vol.1, pp.128-129
Yanaka, T., Shirota, K., Iwaki, T., Sakai, T., Watanabe, M.: "Extremely Low Magnification Imaging Method with Very Small Field Aberrations", The Hague 1980, Vol.1, pp.536-537
Zworykin, V.K., Hillier, J., Vance, A.W.: A preliminary report on the development of a 300-kilovolt magnetic electron microscope. J. Appl. Phys. *12*, 738-742 (1941)
Zworykin, V.K., Hillier, J.: A compact high resolving power electron microscope. J. Appl. Phys. *14*, 658-673 (1943)

Publication Details of International and European Congresses on Electron Microscopy

Delft 1949: *Proceedings of the Conference on Electron Microscopy*, Delft, 4-8 July 1949, ed. by A.L. Houwink, J.B. Le Poole, W.A. Le Rütte (Hoogland, Delft 1950)

Paris 1950: *Comptes Rendus du Premier Congrès International de Microscopie Electronique*, Paris, 14-22 September, 1950 (Editions de la Revue d'Optique théorique et instrumentale, Paris 1953)

London 1954: *The Proceedings of the Third International Conference on Electron Microscopy*, London, 1954, ed. by R. Ross (Royal Microscopical Society, London 1956)

Stockholm 1956: *Electron Microscopy: Proceedings of the Stockholm Conference*, September 1956, ed. by F.J. Sjöstrand, J. Rhodin (Almqvist and Wiksells, Stockholm 1957)

Berlin 1958: *Vierter Internationaler Kongress für Elektronenmikroskopie*, Berlin, 10-17 September 1958, *Verhandlungen* ed. by W. Bargmann, G. Möllenstedt, H. Niehrs, D. Peters, E. Ruska, C. Wolpers (Springer, Berlin, Göttingen, Heidelberg, 1960) 2 Vols

Delft 1960: *The Proceedings of the European Regional Conference on Electron Microscopy*, Delft, 1960, ed. by A.L. Houwink, B.J. Spit (Nederlandse Vereniging voor Electronenmicroscopie, Delft n.d.) 2 Vols.

Philadelphia 1962: *Electron Microscopy, Fifth International Congress for Electron Microscopy*, Philadelphia, Pennsylvania, 29 August to 5 September 1962, ed. by S.S. Breese (Academic, New York 1962) 2 Vols.

Prague 1964: *Electron Microscopy 1964, Proceedings of the Third European Regional Conference*, Prague, ed. by M. Titlbach (Publishing House of the Czechoslovak Academy of Sciences, Prague 1964) 2 Vols

Kyoto 1966: *Electron Microscopy 1966, Sixth International Congress for Electron Microscopy*, Kyoto, ed. by R. Uyeda (Maruzen, Tokyo 1966) 2 Vols

Rome 1968: *Electron Microscopy 1968, Pre-Congress Abstracts of Papers Presented at the Fourth Regional Conference*, Rome, ed. by D.S. Bocciarelli (Tipografia Poliglotta Vaticana, Rome 1968) 2 Vols

Grenoble 1970: *Microscopie électronique 1970, résumés des communications présentées au septième congrès international*, Grenoble, ed. by P. Favard (Société Française de Microscopie Electronique, Paris 1970) 3 Vols

Manchester 1972: *Electron Microscopy, 1972, Proceedings of the Fifth European Congress on Electron Microscopy*, Manchester (Institute of Physics, London 1972)

Canberra 1974: *Electron Microscopy 1974, Abstracts of Papers Presented to the Eighth International Congress on Electron Microscopy*, Canberra, ed. by J.V. Sanders, D.J. Goodchild (Australian Academy of Science, Canberra 1974) 2 Vols

Jerusalem 1976: *Electron Microscopy 1976, Proceedings of the Sixth European Congress on Electron Microscopy*, Jerusalem, ed. by D.G. Brandon, (Vol.I), Y. Ben-Shaul (Vol.II) (Tal International, Jerusalem 1976) 2 Vols

Toronto 1978: *Electron Microscopy, 1978, Papers Presented at the Ninth International Congress on Electron Microscopy*, Toronto, ed. by J.M. Sturgess (Microscopical Society of Canada, Toronto 1978) 3 Vols

The Hague 1980: *Electron Microscopy, 1980, Proceedings of the Seventh European Congress on Electron Microscopy*, The Hague, ed. by P. Brederoo, G. Boom (Vol.I), P. Brederoo, W. de Priester (Vol.II) (Seventh European Congress on Electron Microscopy Foundation, Leiden 1980) [Two volumes of a series of four with the general title *Electron Microscopy 1980*]

Subject Index